Grundlehren der
mathematischen Wissenschaften 300

A Series of Comprehensive Studies in Mathematics

Editors

M. Artin S.S. Chern J.L. Doob A.Grothendieck
E. Heinz F. Hirzebruch L. Hörmander
C.C. Moore J.K. Moser M. Nagata W. Schmidt
D.S. Scott Ya. G. Sinai J. Tits M. Waldschmidt
S. Watanabe

Managing Editors

M. Berger B. Eckmann S.R.S. Varadhan

Peter Orlik   Hiroaki Terao

Arrangements of Hyperplanes

With 43 Figures

Springer-Verlag

Berlin Heidelberg New York
London Paris Tokyo
Hong Kong Barcelona
Budapest

Peter Orlik   Hiroaki Terao
Department of Mathematics
University of Wisconsin
Madison, WI 53706, USA

Mathematics Subject Classification (1991):
05B35, 32S25, 57N65, 14F35, 14F40, 20F36, 20F55

ISBN 3-540-55259-6 Springer-Verlag Berlin Heidelberg New York
ISBN 0-387-55259-6 Springer-Verlag New York Berlin Heidelberg

Library of Congress Cataloging-in-Publication Data
Orlik, Peter, 1938-
Arrangements of hyperplanes / Peter Orlik, Hiroaki Terao.
p. cm. – (Grundlehren der mathematischen Wissenschaften ; 300)
Includes bibliographical references and index.
ISBN 3-540-55259-6. – ISBN 0-387-55259-6
1. Combinatorial geometry. 2. Combinatorial enumeration problems.
3. Lattice theory. I. Terao, Hiroaki, 1951- . II. Title. III. Series.
QA167.O67 1992 516'.13–dc20 92-6674

This work is subject to copyright. All rights are reserved, whether the whole or part of the material is concerned, specifically the rights of translation, reprinting, reuse of illustrations, recitation, broadcasting, reproduction on microfilm or in any other way, and storage in data banks. Duplication of this publication or parts thereof is permitted only under the provisions of the German Copyright Law of September 9, 1965, in its current version, and permission for use must always be obtained from Springer-Verlag. Violations are liable for prosecution under the German Copyright Law.

© Springer-Verlag Berlin Heidelberg 1992
Printed in the United States of America

Data conversion: EDV-Beratung K. Mattes, Heidelberg
Typesetting output: Type 2000, Mill Valley, California, USA
41/3140-543210 - Printed on acid-free paper

To our parents

Preface

An arrangement of hyperplanes is a finite collection of codimension one affine subspaces in a finite dimensional vector space. In this book we study arrangements with methods from combinatorics, algebra, algebraic geometry, topology, and group actions. These first sentences illustrate the two aspects of our subject that attract us most. Arrangements are easily defined and may be enjoyed at levels ranging from the recreational to the expert, yet these simple objects lead to deep and beautiful results. Their study combines methods from many areas of mathematics and reveals unexpected connections.

The idea to write a book on arrangements followed three semesters of lectures on the topic at the University of Wisconsin, Madison. Louis Solomon lectured on the combinatorial aspects in the fall of 1981. Peter Orlik continued the course with the topological properties of arrangements in the spring of 1982. Hiroaki Terao visited Madison for the academic year 1982–83 and gave a course on free arrangements in the fall of 1982. The original project was to enlarge the lecture notes into a book and a long but incomplete manuscript was typed up in 1984.

The book was revived in 1988 when P. Orlik was invited to give a CBMS lecture series on arrangements. The resulting lecture notes provided a core for the present work. Also in 1988, H. Terao gave a course on arrangements at Ohio State University. Parts of his lectures are used in this book. The fresh start allowed us to take advantage of the recent TEXnical innovations of word processing and computer–aided typesetting.

P. Orlik had the opportunity to lecture on arrangements at the Swiss Seminar in Bern. He wishes to thank the mathematicians in Geneva for their hospitality and the participants of the seminar for many helpful comments. H. Terao gave a graduate course on arrangements at International Christian University in Tokyo in the fall of 1989. He wishes to thank all the participants of the course.

Time constraints forced L. Solomon to leave the completion to us. Much of the material in the book is his joint work with one or both of us, and a large part of Chapter 2 was written by him. We are grateful for permission to use his work, and for his support and friendship.

We thank W. Arvola for permission to include a modified version of his presentation of the fundamental group of the complement, P. Edelman and V. Reiner for permission to include their example, and M. Falk and R. Randell for help on many technical points. In addition, we owe thanks to C. Greene, L. Paris,

M. Salvetti, S. Yuzvinsky, T. Zaslavsky, and G. Ziegler for valuable suggestions, to V. I. Arnold for references on the M-property, and to N. Spaltenstein for references on connections with representation theory.

We thank A. B. Orlik for editing and proofreading the manuscript.

University of Wisconsin   *P. Orlik and H. Terao*
February 1992

Table of Contents

1. Introduction 1
   1.1 Introduction 1
       History 1
       Recent Advances 8
   1.2 Definitions and Examples 10
       Examples 11
       Basic Constructions 13
       The Module of $\mathcal{A}$–Derivations 15
       The Complement of a Complex Arrangement 16
       Reflection Arrangements 16
   1.3 Outline 17
       Combinatorics 17
       Algebras 17
       The Module of $\mathcal{A}$–Derivations 18
       Topology 19
       Reflection Arrangements 20

2. Combinatorics 23
   2.1 The Poset $L(\mathcal{A})$ 23
       Definitions 23
       Examples 26
       Oriented Matroids 28
       Supersolvable Arrangements 30
   2.2 The Möbius Function 32
       The Möbius Function 32
       Möbius Inversion 34
       The Function $\mu(X)$ 35
       Historical Notes 38
   2.3 The Poincaré Polynomial 42
       Examples 42
       The Deletion–Restriction Theorem 46
       Supersolvable Arrangements 48
       Nice Partitions 50
       Counting Functions 51

Table of Contents

- 2.4 Graphic Arrangements 52
  - Definitions 52
  - Deletion–Contraction 54
  - Acyclic Orientations 57

3. **Algebras** 59
   - 3.1 $A(\mathcal{A})$ for Central Arrangements 60
     - Construction of $A(\mathcal{A})$ 60
     - An Acyclic Complex 62
     - The Structure of $A(\mathcal{A})$ 63
     - The Injective Map $A(\mathcal{A}_X) \to A(\mathcal{A})$ 65
     - The Broken Circuit Basis 67
   - 3.2 $A(\mathcal{A})$ for Affine Arrangements 70
     - Construction of $A(\mathcal{A})$ 70
     - The Broken Circuit Basis 72
     - Deletion and Restriction 74
     - The Structure of $A(\mathcal{A})$ 77
     - A–equivalence 78
   - 3.3 Algebra Factorizations 79
     - Supersolvable Arrangements 80
     - Nice Partitions of Central Arrangements 82
     - Nice Partitions of Affine Arrangements 85
   - 3.4 The Algebra $B(\mathcal{A})$ 86
     - The Shuffle Product 86
     - The Algebra $B(\mathcal{A})$ 88
     - The Isomorphism of B and A 89
   - 3.5 Differential Forms 92
     - The de Rham Complex 92
     - The Algebra $R(\mathcal{A})$ 93
     - Deletion and Restriction 95
     - The Isomorphism of R and A 97

4. **Free Arrangements** 99
   - 4.1 The Module $D(\mathcal{A})$ 100
     - Derivations 100
     - Basic Properties 102
   - 4.2 Free Arrangements 104
     - Saito's Criterion 104
     - Exponents 107
     - Examples 111
   - 4.3 The Addition–Deletion Theorem 113
     - Basis Extension 114
     - The Map from $D(\mathcal{A})$ to $D(\mathcal{A}'')$ 115
     - The Addition–Deletion Theorem 117
     - Inductively Free Arrangements 119
     - Supersolvable Arrangements 121

		Factorization Theorem	122
	4.4	The Modules $\Omega^p(\mathcal{A})$	123
		Definition of $\Omega^p(\mathcal{A})$	124
		Basic Properties of $\Omega^p(\mathcal{A})$	124
		The Acyclic Complex $(\Omega^{\cdot}(\mathcal{A}), \partial)$	133
		The η-Complex $(\Omega^{\cdot}(\mathcal{A}), \partial_h)$	133
	4.5	Lattice Homology	135
		The Order Complex	136
		The Folkman Complex	137
		The Homology Groups	140
		The Homotopy Type	141
		Whitney Homology	142
		Connection with the Folkman Complex	144
	4.6	The Characteristic Polynomial	145
		The Order Complex with Functors	145
		Local Functors	147
		The Homology $H_p(\mathcal{A}, F)$	148
		The Polynomial $\Psi(\mathcal{A}, x, t)$	150
		The Factorization Theorem	154
5.	Topology		157
	5.1	The Complement $M(\mathcal{A})$	158
		$K(\pi, 1)$–Arrangements	159
		Free Arrangements	163
		Generic Arrangements	164
		Deformation	166
		Arnold's Conjectures	167
	5.2	The Homotopy Type of $M(\mathcal{A})$	168
		Real Arrangements	168
		The Homotopy Type	171
		Complexified Real Arrangements	173
		Salvetti's Complex	175
		The Homotopy Equivalence	176
	5.3	The Fundamental Group	177
		Admissible Graphs	179
		Arvola's Presentation	184
	5.4	The Cohomology of $M(\mathcal{A})$	190
		The Thom Isomorphism	191
		Brieskorn's Lemma	195
	5.5	The Fibration Theorem	196
		Horizontal Subspaces	197
		Good Subspaces	198
		Good Lines	199
	5.6	Related Research	202
		Minimal Models	202

	Discriminantal Arrangements	205
	Alexander Duality	207
	The Milnor Fiber of a Generic Arrangement	209
	Arrangements of Subspaces	211

6. Reflection Arrangements 215

- 6.1 Equivariant Theory 216
  - The Action of G 216
  - Matrices 218
  - Character Formulas 219
  - Topological Interpretation 222
- 6.2 Reflection Arrangements 223
  - Basic Properties 223
  - Examples 225
  - Relative Invariants 228
  - Jacobian and Discriminant 229
  - Classification 231
- 6.3 Free Arrangements 232
  - Invariant Theory 232
  - The Hessian 234
  - $D_R(\delta)$ Is Free 235
  - $D(\mathcal{A})$ Is Free 237
  - The Discriminant Matrix 238
  - A Character Formula 241
- 6.4 The Structure of $L(\mathcal{A})$ 243
  - The Symmetric Group 243
  - The Full Monomial Group 244
  - The Monomial Group $G(r,r,\ell)$ 247
  - The Exceptional Groups 251
- 6.5 Restrictions 254
  - The Cardinality of $\mathcal{A}^H$ 254
  - $\mathcal{A}^H$ Is Free In Coxeter Arrangements 256
- 6.6 Topology 259
  - Stratification of the Discriminant 259
  - Shephard Groups 265
  - The $K(\pi,1)$ Problem 267

A. Some Commutative Algebra 271
- A.1 Free Modules 271
- A.2 Krull Dimension 272
- A.3 Graded Modules 274
- A.4 Associated Primes and Regular Sequences 276

B. Basic Derivations 279
   B.1 The Infinite Families 279
   B.2 Exceptional Groups of Rank 2 280
   B.3 Exceptional Groups of Rank ≥ 3 280
   B.4 The Coexponents 286

C. Orbit Types 289

D. Three–Dimensional Restrictions 301

References 303

Index 315

Index of Symbols 323

List of Figures

1.1	An illustration of chamber counting	4
1.2	$Q(\mathcal{A}) = xy(x+y)$ and $Q(\mathcal{A}) = xy(x+y-1)$	12
1.3	The B_3–arrangement	13
2.1	The Hasse diagram of $Q(\mathcal{A}) = xy(x+y)$	25
2.2	The Hasse diagram of $Q(\mathcal{A}) = xy(x+y-1)$	25
2.3	The Hasse diagram of the B_3–arrangement	25
2.4	The chambers of $Q(\mathcal{A}) = xy(x+y-1)$	28
2.5	The face poset of $Q(\mathcal{A}) = xy(x+y-1)$	29
2.6	Different face posets	30
2.7	The values of $\mu(X)$ for the B_3–arrangement	37
2.8	π–equivalent but not L–equivalent	48
2.9	A graph with three vertices	52
2.10	A complete graph	53
2.11	Deletion and contraction	55
2.12	An oriented graph	57
2.13	Not acyclic orientations	58
3.1	A–equivalent but not L–equivalent	79
4.1	Free but not inductively free	122
4.2	Subdivision of $\Delta^2 \times I$	137
4.3	Folkman complexes for $Q = xyz(x+y)(x+y-z)$	138
4.4	Complexes for the Boolean arrangement	139
4.5	$\pi(\mathcal{A}, t)$ factors, but $\mathcal{A}$ is not free	155
5.1	A braid on three strands	160
5.2	A pure braid on three strands	160
5.3	The generator a_i	161
5.4	$K(\pi, 1)$, but not free	164
5.5	Three lines in general position	166
5.6	The critical half-line	169
5.7	Dual cells	171
5.8	Three concurrent lines in $\mathbb{R}^2$ and in $\mathbb{C}^2$	174
5.9	A 3–graph	181

5.10	A flat graphing map	182
5.11	An admissible graph	184
5.12	Generators	186
5.13	Loop passing	187
5.14	A pencil of lines	188
5.15	Linking	189
5.16	The arrangement $\mathcal{C}^*(5)$	206
5.17	Falk's Linking	208
6.1	The lattice $L = L^{(1)}$	222
6.2	The lattices $L^{(12)}$, $L^{(123)}$, $L^{(12)(34)}$, $L^{(1234)}$	222
6.3	The Hessian configuration	227
6.4	A tetrahedron in the cube	268

List of Tables

4.1	Induction table	119
4.2	The braid arrangement	121
5.1	A subspace arrangement	212
6.1	Poincaré polynomials	223
6.2	Restrictions	249
6.3	Orbits in G_{25}	253
6.4	Induction table for G_{25}	254
6.5	The associated pairs (G, W)	267
B.1	Exponents and coexponents	287
B.2	The monomial groups	288
C.1	The rank 2 groups (I)	289
C.2	The rank 2 groups (II)	290
C.3	Orbits in G_{15}	290
C.4	Orbits in H_3	290
C.5	Orbits in G_{24}	291
C.6	Orbits in G_{25}	291
C.7	Orbits in G_{26}	291
C.8	Orbits in G_{27}	291
C.9	Orbits in F_4	292
C.10	Orbits in G_{29}	292
C.11	Orbits in H_4	292
C.12	Orbits in G_{31}	293
C.13	Orbits in G_{32}	293
C.14	Orbits in G_{33}	293
C.15	Orbits in G_{34} (I)	294
C.16	Orbits in E_6 (I)	294
C.17	Orbits in G_{34} (II)	295
C.18	Orbits in E_6 (II)	295
C.19	Orbits in E_7 (I)	296
C.20	Orbits in E_7 (II)	297
C.21	Orbits in E_8 (I)	298

C.22 Orbits in E_8 (II) 299
C.23 Orbits in E_8 (III) 300

D.1 The infinite families 302
D.2 Coxeter groups 302
D.3 The remaining groups 302

1. Introduction

1.1 Introduction

Show that n cuts can divide a cheese into as many as $(n+1)(n^2-n+6)/6$ pieces.

Problem E 554. *Amer. Math. Monthly* **50** (1943), p. 59.
Proposed by J. L. Woodbridge, Philadelphia

Solution by Free Jamison, Pittsburgh. [*ibid* pp. 564–5] Since n straight lines can divide a plane into $(n^2+n+2)/2$ areas, the $(n+1)$st plane can be divided by the first n planes into that number of areas. For each of these areas the $(n+1)$st plane divides a piece of cheese already formed into two, and increases the total number of pieces by $(n^2+n+2)/2$. Since $(n^3+5n+6)/6$ gives the number of pieces for $n = 1$ or 2, and since

$$\frac{n^3+5n+6}{6} + \frac{n^2+n+2}{2} = \frac{(n+1)^3+5(n+1)+6}{6},$$

the expression $(n^3+5n+6)/6$ holds for every n.

It is interesting to note that

(1) n points can divide a line into $1+n$ parts,
(2) n lines can divide a plane into $1+n+\binom{n}{2}$ parts,
(3) n planes can divide space into $1+n+\binom{n}{2}+\binom{n}{3}$ parts.

Editorial Note. The general formula

$$1+n+\binom{n}{2}+\binom{n}{3}+\cdots+\binom{n}{m},$$

for the case of an m-dimensional cheese, was obtained by L. Schläfli on page 39 of his great posthumous work, *Theorie der vielfachen Kontinuität* (Denkschriften der Schweizerischen naturforschenden Gesellschaft, vol. 38, 1901).

History

Such are the humble origins of our subject. In order to maximize the number of pieces, the arrangement of planes in the problem must be in "general position."

This means that any two planes have a common line, and these lines are distinct, and that any three planes have a common point, and these points are distinct. Allowing degeneracy makes the problem of counting parts much harder. In 1889, S. Roberts [191] gave a formula for the number of regions formed by an arbitrary arrangement of n lines in the plane. It is "the number of regions formed by n lines in general position" minus "the number of regions lost because of multiple points" minus "the number of regions lost because of parallels." See J. Wetzel's article [249] for a modern treatment. There is an extensive literature on partition problems in Euclidean space and projective spaces. B. Grünbaum summarized much of what was known in 1971 in [100, 101]. We quote from the introduction of his paper [100], whose title we borrowed for this book.

> ... I would like to survey the somewhat related field of *arrangements of hyperplanes*, which I expect to become increasingly popular during the next few years ... the theory of arrangements may be developed, much like topology, in rectilinear or curved versions as well as in discrete and continuous variants, and that in these developments it impinges upon many aspects of convexity, topology, and geometry which seemed to be quite unrelated.

The complement of certain hyperplanes in complex space had been studied by E. Fadell, R. Fox, and L. Neuwirth [73, 86] in connection with the braid space. The braid arrangement consists of the hyperplanes $H_{i,j} = \ker(z_i - z_j)$. Let $M = \{z \in \mathbb{C}^\ell \mid z_i \neq z_j \text{ for } i \neq j\}$ be the complement of these hyperplanes, called the pure braid space. They proved that M is a $K(\pi, 1)$ space. Let $\text{Poin}(M, t) = \sum_{k \geq 0} \text{rank} H^k(M) t^k$ be its Poincaré polynomial. In 1969, V. I. Arnold [6] proved the beautiful formula

$$(1) \qquad \text{Poin}(M, t) = (1 + t)(1 + 2t) \cdots (1 + (\ell - 1)t)$$

in connection with his work on Hilbert's 13th problem. He also constructed a graded algebra A as the quotient of an exterior algebra by a homogeneous ideal, and showed that there is an isomorphism of graded algebras

$$(2) \qquad H^*(M) \simeq A.$$

This gives a presentation of the cohomology ring of the pure braid space in terms of generators and relations. The study of the topological properties of the complement of an arbitrary arrangement over the complex numbers was launched by Arnold with the following remark at the end of his paper:

> Let M be the manifold obtained from $\mathbb{C}^n$ by discarding an arbitrary number of hyperplanes
>
> $$M = \{z \in \mathbb{C}^n \mid \alpha_k(z) \neq 0, \ k = 1, \ldots, N\}.$$
>
> Probably, the ring $H^*(M, \mathbb{Z})$ is torsion free and is generated by the one-dimensional classes $\omega_k = (1/2\pi i)(d\alpha_k/\alpha_k)$, a polynomial in ω_k being cohomologous to 0 in H^* only when it is zero.

1.1 Introduction

E. Brieskorn [41] proved these conjectures in a 1971 Bourbaki Seminar talk. One of his results captured an essential topological feature of arrangements.

> ...une famille finie quelconque d'hyperplans affines complexes V_i, $i \in I$, dans un espace affine complexe V. Pour calculer le p-iéme groupe de cohomologie, $0 \leq p \leq n$, on considére les sousensembles maximaux $I_{p,1}, \ldots, I_{p,k_p}$ de I pour lequels on ait la propriété:
> $$\operatorname{codim} \bigcap_{i \in I_{p.k}} V_i = p.$$
>
> **Lemme 3.** *Pour les complémentaires d'union d'hyperplans $Y = V - \cup_{i \in I} V_i$ et $Y_{p,k} = V - \cup_{i \in I_{p,k}} V_i$ les inclusions $i_k : Y \to Y_{p,k}$ induisent un isomorphisme:*
> $$H^p(Y, \mathbb{Z}) = \bigoplus_{k=1}^{k_p} H^p(Y_{p,k}, \mathbb{Z}).$$

Brieskorn also generalized Arnold's results in another direction. He replaced the symmetric group and the braid arrangement by a finite Coxeter group W and its reflection representation in a real vector space $V_\mathbb{R}$ of dimension ℓ. Let V be the complexification of $V_\mathbb{R}$. Then W acts as a reflection group in V. Let $M_W \subset V$ be the complement of the reflecting hyperplanes of W. He proved that the analog of (1) involves the exponents $m_1, \ldots m_\ell$ of W.

(3) $$\operatorname{Poin}(M_W, t) = (1 + m_1 t)(1 + m_2 t) \cdots (1 + m_\ell t).$$

Brieskorn conjectured that M_W is a $K(\pi, 1)$ space for all Coxeter groups W. He proved this for some of the groups by representing M_W as the total space of a sequence of fibrations.

In the 1971 paper quoted above, Grünbaum [100] reported the

> ...finding of a rock (or rather, unpolished gem) discovered thirty years ago by one of the lone wanderers in the wilderness of specialization. The "simplicial arrangements" which will be discussed below were first discovered by Melchior [154]; though they are a very natural notion and appear in the solutions of many problems about arrangements, they remained unnoticed.

Grünbaum listed all known simplicial arrangements in the affine and projective planes with ≤ 38 lines. It seems like poetic justice that these "unpolished gems" became the central objects in the 1972 solution of Brieskorn's conjecture by P. Deligne [61].

> **Théorème.** *Soit V un espace vectoriel réel de dimension finie, $\mathcal{M}$ un ensemble fini d'hyperplans homogénes de V, $V_\mathbb{C}$ le complexifié de V et $Y = V_\mathbb{C} - \cup_{M \in \mathcal{M}} M_\mathbb{C}$. On suppose que les composants connexes de $V - \cup_{M \in \mathcal{M}} M$ sont des cônes simpliciaux ouverts. Alors, Y est un $K(\pi, 1)$.*

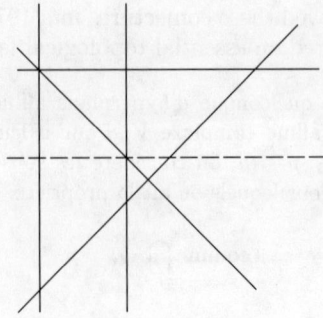

Fig. 1.1. An illustration of chamber counting

The next significant advance was made by T. Zaslavsky [256] in 1975. The title of his AMS Memoir tells it all: *"Facing up to Arrangements: Face-Count Formulas for Partitions of Space by Hyperplanes."* He introduced the method of deletion and restriction to obtain recursion formulas for counting problems. A similar result was obtained independently by M. Las Vergnas [136]. Let $\mathcal{A}$ be an arrangement and let $H \in \mathcal{A}$ be a hyperplane. Then $\mathcal{A}' = \mathcal{A} \setminus \{H\}$ is called the deleted arrangement. The arrangement in H defined by $\mathcal{A}'' = \{K \cap H \mid K \in \mathcal{A}'\}$ is called the restricted arrangement. The triple $(\mathcal{A}, \mathcal{A}', \mathcal{A}'')$ may be used to solve the problem of counting the parts of the complement of the hyperplanes of an arbitrary real arrangement. The parts are called **chambers** in modern terminology. Let $\mathcal{C}(\mathcal{A})$ be the set of chambers of $\mathcal{A}$. Let $(\mathcal{A}, \mathcal{A}', \mathcal{A}'')$ be a triple of real arrangements. Then

$$|\mathcal{C}(\mathcal{A})| = |\mathcal{C}(\mathcal{A}')| + |\mathcal{C}(\mathcal{A}'')|.$$

To prove this recursion, let P be the set of those chambers in $\mathcal{C}(\mathcal{A}')$ which intersect the distinguished hyperplane H. Let Q be the set of those chambers in $\mathcal{C}(\mathcal{A}')$ which do not intersect H. Evidently $|\mathcal{C}(\mathcal{A}')| = |P| + |Q|$. The hyperplane H divides each chamber of P into two chambers of $\mathcal{C}(\mathcal{A})$ and leaves the chambers of Q unchanged. Thus $|\mathcal{C}(\mathcal{A})| = 2|P| + |Q|$. Finally, there is a bijection between P and $\mathcal{C}(\mathcal{A}'')$ given by $C \mapsto C \cap H$. Thus $|\mathcal{C}(\mathcal{A}'')| = |P|$. Figure 1.1 illustrates this in the plane. Let H be the broken line. Then $|P| = 4$ and $|Q| = 10$, so we get $|\mathcal{C}(\mathcal{A})| = 18$.

Zaslavsky defined the set $L(\mathcal{A})$ of intersections of elements of $\mathcal{A}$ and partially ordered $L(\mathcal{A})$ by reverse inclusion. He used the Möbius function of $L(\mathcal{A})$ to define the characteristic polynomial of $L(\mathcal{A})$. There is a closely related polynomial $\pi(\mathcal{A}, t)$, defined on $L(\mathcal{A})$, which we call the Poincaré polynomial. It follows from the definition that for the empty arrangement $\pi(\mathcal{A}, t) = 1$. He proved a result about the characteristic polynomial, which amounts to the following recursion for the Poincaré polynomial: $\pi(\mathcal{A}, t) = \pi(\mathcal{A}', t) + t\pi(\mathcal{A}'', t)$. Since $|\mathcal{C}(\mathcal{A})|$ and $\pi(\mathcal{A}, 1)$ agree on the empty arrangement and satisfy the same recursion for deletion and restriction, this proves Zaslavsky's beautiful result:

(4) $$|\mathcal{C}(\mathcal{A})| = \pi(\mathcal{A}, 1).$$

1.1 Introduction

Analysis led to development in a different direction. The classical hypergeometric function $F(a, b; c; z)$ is defined by the series

$$F(a, b; c; z) = \sum_{m=0}^{\infty} \frac{(a, m)(b, m)}{(c, m)(1, m)} z^m,$$

where (a, m) denotes the factorial function

$$a(a+1) \cdots (a+m-1) = \frac{\Gamma(a+m)}{\Gamma(a)}.$$

The hypergeometric function satisfies the differential equation

$$z(1-z)F'' + (c - (a+b+1)z)F' - abF = 0,$$

and it has the Euler integral representation

$$F(a, b; c; z) = \frac{\Gamma(c)}{\Gamma(b)\Gamma(c-b)} \int_0^1 t^{b-1}(1-t)^{c-b-1}(1-zt)^{-a} dt.$$

The function is normalized to depend on the arrangement of the three points $0, 1, z^{-1}$ in the complex line. Hypergeometric functions have been defined in several complex variables, and they have analogous integral representations. These generalizations naturally depend on arrangements of hyperplanes in affine space; see [2, 109]. Much work has been done in studying these integrals over various domains. This was the motivation for A. Hattori's 1975 investigation [108] of the homotopy type of the complement of an arrangement of complex hyperplanes in general position.

We denote by **k** the set $\{1, 2, \ldots, k\}$. If I is a subset of **k**, we denote by $|I|$ the cardinal number of I. We define the subtorus T_I of T^k by

$$T_I = \{z \mid z = (z_1, \ldots, z_k) \in T^k,\ z_j = 1 \text{ for } j \notin I\}.$$

The dimension of T_I is equal to $|I|$.

THEOREM 1. *Let $L_1, \ldots, L_k$ be affine hyperplanes in $\mathbb{C}^n$ in general position, where $n + 1 \leq k$. Then the space $X = \mathbb{C}^n - L_1 \cup \cdots \cup L_k$ has the same homotopy type as the space*

$$X_0 = \bigcup_{\substack{I \subset \mathbf{k} \\ |I| = n}} T_I.$$

This is the complex analog of the cheese cutting problem. There the number of parts in the complement of a real arrangement in general position depends only on the number of hyperplanes, but not on their location. Here the homotopy type of the complement of a complex arrangement in general position depends only on the number of hyperplanes, but not on their location.

More tools were added to the study of arrangements in 1980. P. Orlik and L. Solomon [171] used combinatorial methods to study the complement $M(\mathcal{A})$ of a complex hyperplane arrangement $\mathcal{A}$. They used Brieskorn's results to compute the Poincaré polynomial of the complement of an arbitrary complex arrangement:

(5) $$\text{Poin}(M(\mathcal{A}), t) = \pi(\mathcal{A}, t).$$

Thus the Betti numbers of the complement depend only on the lattice of intersections of the hyperplanes. I. Petrowsky [184] defined the M–property for real algebraic curves.

> As early as 1876, Harnack [107] showed that the maximal number of components (maximal connected subsets) of a real algebraic curve of order n in the projective plane is precisely $\frac{1}{2}(n-1)(n-2)+1$. At the same time Harnack proposed a process for the construction of curves with this maximal number of components. Such curves we shall call in the sequel, M–curves. ...
>
> In 1891, D. Hilbert [111] proposed a new method of constructing M–curves. ...In his report to the International Mathematical Congress in 1900 on modern problems of mathematics Hilbert considers the investigation of the topology of M–curves and of the corresponding algebraic surfaces as most timely.

See [9, 103, 245] for recent progress on Hilbert's 16th problem.

Suppose $X_{\mathbb{R}}$ is a real algebraic variety and $X_{\mathbb{C}}$ is its complexification. Complex conjugation induces an involution on $X_{\mathbb{C}}$ whose fixed point set is $X_{\mathbb{R}}$. Let $b_i(X_{\mathbb{R}})$ and $b_i(X_{\mathbb{C}})$ be their respective Betti numbers with $\mathbb{Z}/2$ coefficients. An application of Smith theory, see [84, Theorem 4.4], provides the following inequality:

(6) $$\sum_{i \geq 0} b_i(X_{\mathbb{R}}) \leq \sum_{i \geq 0} b_i(X_{\mathbb{C}}).$$

The natural generalization of an M–curve is to say that the real algebraic variety $X_{\mathbb{R}}$ has the M–property if equality holds in (6).

Let $\mathcal{A}_{\mathbb{R}}$ be a real arrangement and let $\mathcal{A}_{\mathbb{C}}$ be its complexification. Let $M_{\mathbb{R}}$ and $M_{\mathbb{C}}$ be the real and complex complements. Let $Q \in \mathbb{R}[x_1, \ldots, x_\ell]$ be a product of linear polynomials whose zero set is the union of the hyperplanes in $\mathcal{A}_{\mathbb{R}}$. Note that the complement is also an algebraic variety, $M_{\mathbb{R}} \approx \{v \in \mathbb{R}^{\ell+1} \mid vQ(v) = 1\}$. It follows that $M_{\mathbb{R}}$ has the M–property

$$\sum_{i \geq 0} b_i(M_{\mathbb{R}}) = b_0(M_{\mathbb{R}}) = \pi(\mathcal{A}_{\mathbb{R}}, 1) = \pi(\mathcal{A}_{\mathbb{C}}, 1) = \sum_{i \geq 0} b_i(M_{\mathbb{C}}).$$

Here we used (4), (5), the fact that $M_{\mathbb{C}}$ has no torsion in homology, and the fact that $L(\mathcal{A}_{\mathbb{R}}) = L(\mathcal{A}_{\mathbb{C}})$, so their Poincaré polynomials are equal.

Orlik and Solomon [171] also defined a graded algebra $A(\mathcal{A})$, which is constructed using only $L(\mathcal{A})$. It is the quotient of the exterior algebra $E(\mathcal{A})$ based on $\mathcal{A}$ by a homogeneous ideal $I(\mathcal{A})$, $A(\mathcal{A}) = E(\mathcal{A})/I(\mathcal{A})$. They showed that there is an isomorphism of graded algebras

(7) $$H^*(M(\mathcal{A})) \simeq A(\mathcal{A}).$$

This generalizes (2) by giving a presentation of the cohomology ring of the complement of a complex arrangement in terms of generators and relations. In [172] they considered complex reflection arrangements. If G is a complex reflection group acting in a complex vector space V of dimension ℓ, then it has exponents $m_1, \ldots m_\ell$. However, if $M_G \subset V$ is the complement of the reflecting hyperplanes of G, then formula (3) does not hold for M_G. Orlik and Solomon [172] defined coexponents $n_1, \ldots n_\ell$ for G and showed that for real groups $m_i = n_i$. They proved the following generalization of (3) for complex reflection groups:

(8) $$\text{Poin}(M_G, t) = (1 + n_1 t)(1 + n_2 t) \cdots (1 + n_\ell t).$$

A different line of investigation was inspired by singularity theory. In the present context, the study of logarithmic vector fields and logarithmic differential forms on a hypersurface was initiated by K. Saito [199, 201]. He defined free hypersurfaces in the analytic category. Arrangements represent a special case. Here the hypersurface is the union of the hyperplanes of $\mathcal{A}$. Its singular set consists of linear subspaces. This special case was studied by H. Terao [226]. He showed that we can pass from analytic to algebraic considerations.

Let S denote the polynomial algebra of V, and let Q be a product of defining linear forms for $\mathcal{A}$. Suppose $\theta : S \to S$ is a derivation. It is called an $\mathcal{A}$–derivation if $\theta(Q) \in QS$. The set of $\mathcal{A}$–derivations is an S–module, $D(\mathcal{A})$. The arrangement $\mathcal{A}$ is called free if $D(\mathcal{A})$ is a free S–module. It is an enduring mystery of the subject just what makes an arrangement free, but it is known that the property is not generic. If $\mathcal{A}$ is free, then Terao [226] associated to $\mathcal{A}$ a collection of nonnegative integers called its exponents, $\exp \mathcal{A} = \{b_1, \ldots, b_\ell\}$. These integers are unique up to order, but they are not necessarily distinct. Terao [226] proved several results concerning a triple $(\mathcal{A}, \mathcal{A}', \mathcal{A}'')$. These results may be combined to assert that any two of the following statements imply the third:

$$\begin{aligned} \mathcal{A} \text{ is free with } \exp \mathcal{A} &= \{b_1, \ldots, b_{\ell-1}, b_\ell\}, \\ \mathcal{A}' \text{ is free with } \exp \mathcal{A}' &= \{b_1, \ldots, b_{\ell-1}, b_\ell - 1\}, \\ \mathcal{A}'' \text{ is free with } \exp \mathcal{A}'' &= \{b_1, \ldots, b_{\ell-1}\}. \end{aligned}$$

The following year Terao [228] proved that if $\mathcal{A}$ is a free ℓ-arrangement with $\exp \mathcal{A} = \{b_1, \ldots, b_\ell\}$, then

(9) $$\pi(\mathcal{A}, t) = (1 + b_1 t) \cdots (1 + b_\ell t).$$

He also proved [227] that complex reflection arrangements are free. Thus (8) is a consequence of (9). These early developments were reviewed by P. Cartier [48] in a Bourbaki Seminar talk in November 1980. Much exciting work by many authors followed. We can only describe a fraction of it here.

Recent Advances

The combinatorial vein is carried on by Zaslavsky [257, 258, 261] and others. There are several papers on various combinatorial problems by A. Björner, P. Edelman and G. Ziegler [29, 32, 265, 266]. There are close connections with matroid theory; see [30]. Note also related work by N. E. Mnëv [157] and by A. V. Zelevinsky [263]. Some of this combinatorial research is closely related to computer science; see articles in *Discrete and Combinatorial Geometry*, particularly the special issue on "*The Complexity of Arrangements*," Volume 5, Number 2, 1990. There are also connections with coding theory [52, Chapter 21]. For connections with the theory of box splines, see the work of C. de Boor [35, 36].

Much of the recent topological work is focused on the homotopy type of the complement of a complex arrangement. The rational homotopy type of the complement was studied by T. Kohno [126, 127, 129, 133], M. Falk [75] and R. Randell [80]. This work is reviewed in the survey article by Falk and Randell [81]. A presentation of the fundamental group of the complement of a complexified real arrangement was obtained by Randell [189] and M. Salvetti [203]. The problem was solved for arbitrary complex arrangements by W. Arvola [13]. In their work on stratified Morse theory, M. Goresky and R. MacPherson [91] generalized the notion of hyperplane arrangements to arrangements of affine subspaces, without restriction on their dimensions. They computed the Betti numbers of the complement and showed that the complement of a real subspace arrangement has the M-property. Salvetti [203] constructed a finite cell complex of the homotopy type of the complement of a complexified real arrangement. Orlik [169] constructed a finite cell complex of the homotopy type of the complement of an arbitrary arrangement of affine subspaces. Despite these constructions, there is still no good criterion to detect nonzero elements in the homotopy groups of the complement or to determine if the complement is a $K(\pi, 1)$ space. Even the special case of complex reflection arrangements is open. Orlik and Solomon [178] defined a subclass of complex reflection groups, called Shephard groups, and proved that the complements of their reflection arrangements are $K(\pi, 1)$ spaces. Using the classification of irreducible complex reflection groups [210], this leaves an infinite family of groups where a fibration argument may be used, some exceptional groups of rank two where the complement is always $K(\pi, 1)$, and six exceptional groups of rank ≥ 3 where the problem is still open.

The study of algebraic geometry over the complex numbers leads to different questions. The general plane cubic curve has nine inflection points, which have the property that a line through any two contains a third; see Figure 6.3. In 1893, J. J. Sylvester conjectured that it was impossible to have a nonlinear finite set of points in real space with this property. This was proved by T. Gallai.

> Let A_n be the projective n-space ($n \geq 2$) over some field $\mathbb{K}$. A finite subset X of A_n is called a Sylvester-Gallai (S-G) configuration if it verifies the following condition:
>
> (*) If $P, Q \in X$, with $P \neq Q$, the line joining P and Q contains at least one more point of X. (Equivalently: no line intersects X in exactly two points.)
>
> An S-G configuration is called linear (planar) if it is contained in a line (plane). If $\mathbb{K}$ is the field of real numbers, it is known that any S-G configuration is linear. Over the field of complex numbers there are well-known examples of nonlinear S-G configurations (e. g. the nine inflection points of a nonsingular cubic.)
>
> Is there a nonplanar Sylvester-Gallai configuration over the field of complex numbers?
>
> Problem 5359. *Amer. Math. Monthly* **73** (1966), p. 89.
> *Proposed by Jean-Pierre Serre, Paris, France*

In 1977, Y. Miyaoka and Sh. -T. Yau proved the inequality $c_1^2 \leq 3c_2$ for the Chern numbers of an algebraic surface of general type. Equality occurs if and only if the universal cover of the surface is the complex ball. The corresponding surfaces are called ball quotients. Several explicit examples of ball quotients were constructed by F. Hirzebruch [113]. Related results were obtained by A. Sommese [213] and B. Hunt [114]. We quote from the introduction of [21], where there is an excellent description of this work:

> Die hier untersuchten Flächen erhalten wir hauptsächlich als "Kummer–Überlagerungen" zu Geradenkonfigurationen in der komplex–projektiven Ebene. Diese Überlagerungen der projektiven Ebene sind entlang einer Menge von Geraden lokal mit der Ordnung $n \geq 2$ verzweigt. Die Chernschen Zahlen (der minimalen Desingularisierungen) dieser Flächen sind dann durch die kombinatorischen Invarianten der Geradenkonfiguration bestimmt. Wenn wir nun die Miyaoka–Yau–Ungleichung $c_1^2 \leq 3c_2$ anwenden, so erhalten wir wiederum Ungleichungen für diese kombinatorischen Invarianten, aus denen sich überraschenderweise Sätze über Geradenkonfigurationen in der Ebene (und durch Dualisierung auch über Punktkonfigurationen) ergeben, die sich bisher nicht auf elementarem Wege beweisen lassen.

Let $\mathcal{A}$ be an arrangement of lines in the plane. Let t_j be the number of points which lie on exactly j lines. For a real arrangement, Melchior [154] showed that if $\mathcal{A}$ has at least three noncollinear points, then $t_2 \geq 3 + t_4 + 2t_5 + 3t_6 + \cdots$. As a consequence of his work on ball quotients, Hirzebruch [112] proved the following inequality for a complex arrangement of k lines with $t_k = t_{k-1} = 0$:

$$t_2 + t_3 \geq k + t_5 + 2t_6 + 3t_7 + \cdots$$

This enabled L. M. Kelly [124] to show that every S-G configuration over the field of complex numbers is planar.

The work on hypergeometric functions in several variables has become a major new area of research. In this generalization, there are linear functions $f = \{f_i\}$, complex exponents $\alpha = \{\alpha_i\}$, and a real polytope $\Delta \subset \mathbb{R}^n$. The integral

$$I(\Delta, f, \alpha) = \int_\Delta f_1^{\alpha_1} \cdots f_N^{\alpha_N} dx_1 \cdots dx_n$$

is a function of these variables. The linear functions $\{f_i\}$ determine an arrangement $\mathcal{A}$, and the algebra $A(\mathcal{A})$ enters this work. There are unexpected connections with algebraic K-theory and conformal field theory. Among the principal practitioners are K. Aomoto [3, 4], I. M. Gelfand [87, 90], V. Schechtman, and A. Varchenko [207]. For a recent survey see [242].

In order to study the structure of the algebra $A(\mathcal{A})$, it is sometimes useful to have a standard way to choose a basis. Such a basis, called the broken circuit basis, was constructed independently by Björner [29], Gelfand and Zelevinsky [90], and Jambu and Terao [121].

It was conjectured in 1981 that the restriction of a free arrangement is free. Edelman and Reiner [70] found a counterexample in 1991. Most of the recent work on free arrangements is attempting to solve one of these conjectures: (i) the complement of a complex free arrangement is a $K(\pi, 1)$ space, (ii) the property that $\mathcal{A}$ is free depends only on $L(\mathcal{A})$. Ziegler [268] gave an example of a free arrangement and a nonfree arrangement such that their lattices are isomorphic. One is defined over a field of characteristic three, while the other is over characteristic not equal to three. Conjecture (ii) is still open for arrangements defined over the same field. S. Yuzvinsky [254, 255] gave interesting necessary conditions in terms of lattice homology for an arrangement to be free.

A general formula for $\pi(\mathcal{A}, t)$ in terms of derivations was proved by Solomon and Terao [212]. It yields the factorization theorem when applied to free arrangements. The holonomy Lie algebra associated with an arrangement was studied by Kohno [128], who also studied the Hecke algebra representations of braid groups in [131]. Other applications include the work of T. tom Dieck and T. Petrie [65, 66, 64]. Arrangements and group representations are combined in work of Orlik and Solomon [171], H. Barcelo [18, 19], G. Lehrer [139, 140], Lehrer and T. Shoji [143], and Lehrer and Solomon [144].

1.2 Definitions and Examples

The following special symbols are used throughout this book: natural numbers $\mathbb{N}$, integers $\mathbb{Z}$, rational numbers $\mathbb{Q}$, real numbers $\mathbb{R}$, complex numbers $\mathbb{C}$, the field of q elements $\mathbb{F}_q$, and arbitrary fields $\mathbb{K}, \mathbb{L}$.

Definition 1.1 *Let $\mathbb{K}$ be a field and let $V_\mathbb{K}$ be a vector space of dimension ℓ. A* **hyperplane** *H in $V_\mathbb{K}$ is an affine subspace of dimension $(\ell - 1)$. A* **hyperplane arrangement** *$\mathcal{A}_\mathbb{K} = (\mathcal{A}_\mathbb{K}, V_\mathbb{K})$ is a finite set of hyperplanes in $V_\mathbb{K}$.*

More generally, a **subspace arrangement** is a finite set of affine subspaces of V with no dimension restrictions. Since this book is mostly about hyperplane arrangements, we agree to use "arrangement" in place of "hyperplane arrangement."

The subscript $\mathbb{K}$ will be used only when we want to call attention to the field. We call $\mathcal{A}$ an ℓ-**arrangement** when we want to emphasize the dimension of V. Let Φ_ℓ denote the empty ℓ-arrangement. Let V^* be the dual space of V, the space of linear forms on V. Let $S = S(V^*)$ be the symmetric algebra of V^*. Choose a basis $\{e_1, \ldots, e_\ell\}$ in V and let $\{x_1, \ldots, x_\ell\}$ be the dual basis in V^* so $x_i(e_j) = \delta_{i,j}$. We may identify $S(V^*)$ with the polynomial algebra $S = \mathbb{K}[x_1, \ldots, x_\ell]$. Each hyperplane $H \in \mathcal{A}$ is the kernel of a polynomial α_H of degree 1 defined up to a constant. It is convenient to write $p \doteq q$ for $p, q \in S$ if $p = cq$ for some $c \in \mathbb{K}^*$. Thus if $\ker(\alpha_H) = H = \ker(\alpha'_H)$, then $\alpha_H \doteq \alpha'_H$.

Definition 1.2 *The product*

$$Q(\mathcal{A}) \doteq \prod_{H \in \mathcal{A}} \alpha_H$$

is called a **defining polynomial** *of $\mathcal{A}$. We agree that $Q(\Phi_\ell) = 1$ is the defining polynomial of the empty arrangement.*

Definition 1.3 *We call $\mathcal{A}$ centerless if $\cap_{H \in \mathcal{A}} H = \emptyset$. If $T = \cap_{H \in \mathcal{A}} H \neq \emptyset$, we call $\mathcal{A}$* **centered** *with center T. If $\mathcal{A}$ is centered, then coordinates may be chosen so that each hyperplane contains the origin. In this case we call $\mathcal{A}$* **central**. *If $\mathcal{A}$ is central, then each α_H is a linear form and $Q(\mathcal{A})$ is a homogeneous polynomial whose degree is the cardinality of $\mathcal{A}$. We agree that the empty arrangement Φ_ℓ is central. When we want to emphasize that an arrangement can be either centered or centerless we call it* **affine**.

Definition 1.4 *A* **projective arrangement** *is a finite set of projective hyperplanes in projective space.*

Since the complement of a hyperplane in projective space is affine space, the complement of a nonempty projective arrangement may be viewed as the complement of an affine arrangement. We shall not discuss projective arrangements separately.

Examples

First consider some real arrangements. The only central 1–arrangement consists of the hyperplane $\{0\}$. An affine 1–arrangement consists of a finite set of points. For $\ell = 2, 3$ we agree to use x, y, z in place of x_1, x_2, x_3. A central real 2–arrangement is a finite set of lines through the origin. An affine 2–arrangement is a finite set of lines in the plane.

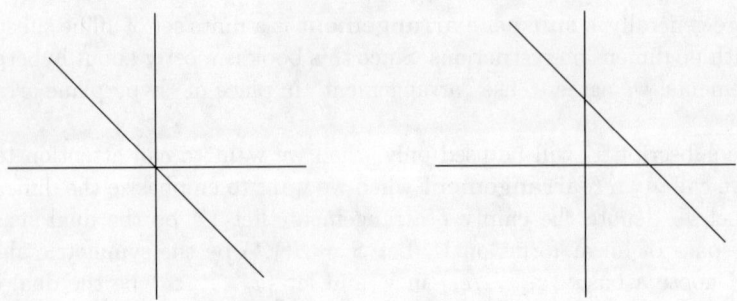

Fig. 1.2. $Q(\mathcal{A}) = xy(x+y)$ and $Q(\mathcal{A}) = xy(x+y-1)$

Example 1.5 Define $\mathcal{A}$ by $Q(\mathcal{A}) = xy(x+y)$. Then $\mathcal{A}$ is central and consists of three lines through the origin; see Figure 1.2.

Example 1.6 Define $\mathcal{A}$ by $Q(\mathcal{A}) = xy(x+y-1)$. Then $\mathcal{A}$ is centerless and consists of three affine lines; see Figure 1.2.

Real 3–arrangements are examples which display some of the intricacies of the general case.

Example 1.7 Let $\mathbb{R}^3$ have its usual basis. Consider the cube with vertices $(\pm 1, \pm 1, \pm 1)$. Its nine planes of symmetry form a central 3–arrangement defined by

$$Q(\mathcal{A}) = xyz(x-y)(x+y)(x-z)(x+z)(y-z)(y+z).$$

These nine planes intersect in lines which are axes of rotational symmetry for the cube. The group of symmetries of the cube is the Coxeter group of type B_3. We shall refer to this arrangement as the B_3–arrangement.

We can visualize central real 3–arrangements by the deconing construction of Definition 1.15. As usual, we think of the projective plane $P_\mathbb{R}^2$ as the disk with identification of diametrically opposite points on the boundary. The picture we draw is therefore the intersection of the arrangement with the upper hemisphere. The same idea may be conveyed by a slightly different picture which is easier to draw. Since we assume that the line at infinity is in $\mathcal{A}$, we may identify its complement in $P_\mathbb{R}^2$ with $\mathbb{R}^2$ and draw the corresponding affine arrangement. **Here we must remember that parallel lines meet at infinity.** In Example 1.7 we let the plane $z = 0$ go to the line at infinity. If we substitute $z = 1$ in the remaining linear forms, we get an affine 2–arrangement. In order to remember that the line at infinity is in our arrangement we draw a frame at "infinity" in Figure 1.3. It is easy to find and label the 13 lines of intersection in the 3–arrangement of Example 1.7 by finding the 13 points of intersection in Figure 1.3.

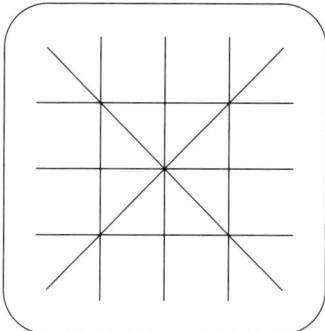

Fig. 1.3. The B_3–arrangement

Example 1.8 Let $\mathcal{A}_{\mathbb{R}}$ be the **Boolean** arrangement defined by
$$Q(\mathcal{A}) = x_1 x_2 \cdots x_\ell.$$
This is the arrangement of the coordinate hyperplanes in $\mathbb{R}^\ell$.

Example 1.9 For $1 \leq i < j \leq \ell$ let $H_{i,j} = \ker(x_i - x_j)$. Let $\mathcal{A}_{\mathbb{R}}$ be the **braid** arrangement defined by
$$Q(\mathcal{A}) = \prod_{1 \leq i < j \leq \ell} (x_i - x_j).$$

While the examples above also make sense over finite fields, the next example can only be stated in that setting.

Example 1.10 Let V be an ℓ–dimensional vector space over the finite field of q elements, $\mathbb{F}_q$. Let $\mathcal{A}$ be the central arrangement in V consisting of all hyperplanes through the origin.

Basic Constructions

Definition 1.11 *Let $|\mathcal{A}|$ denote the* **cardinality** *of $\mathcal{A}$.*

In Example 1.8 we have $|\mathcal{A}| = \ell$. In Example 1.9 we have $|\mathcal{A}| = \ell(\ell-1)/2$. In Example 1.10 we have
$$|\mathcal{A}| = 1 + q + q^2 + \cdots + q^{\ell-1}.$$
The last assertion may be seen in two ways. Using induction, we count the number of distinct linear forms $\alpha = c_1 x_1 + \cdots + c_\ell x_\ell$. If $c_\ell \neq 0$, then there are $q^{\ell-1}$ such forms. If $c_\ell = 0$, then by induction there are $1 + q + \cdots + q^{\ell-2}$ such

forms. If we introduce an inner product in V, then we may count the number of lines instead. Since V has $q^\ell - 1$ nonzero elements and each line contains $q - 1$ nonzero elements, there are $(q^\ell - 1)/(q - 1)$ lines in V.

It is clear from these examples that some of the complexity of $\mathcal{A}$ may be captured by knowledge of the intersection pattern of its hyperplanes.

Definition 1.12 *Let $L(\mathcal{A})$ be the set of all nonempty intersections of elements of $\mathcal{A}$. We agree that $L(\mathcal{A})$ includes V as the intersection of the empty collection of hyperplanes.*

We should remember that if $X \in L(\mathcal{A})$, then $X \subseteq V$. Strictly speaking, these objects should have different names, but it is always clear from the context which one is in consideration.

Definition 1.13 *Let $(\mathcal{A}, V)$ be an arrangement. If $\mathcal{B} \subseteq \mathcal{A}$ is a subset, then $(\mathcal{B}, V)$ is called a* **subarrangement***. For $X \in L(\mathcal{A})$ define a subarrangement $\mathcal{A}_X$ of $\mathcal{A}$ by*

$$\mathcal{A}_X = \{H \in \mathcal{A} \mid X \subseteq H\}.$$

Note that $\mathcal{A}_V = \Phi_\ell$ and if $X \neq V$, then $\mathcal{A}_X$ has center X in any arrangement. Define an arrangement $(\mathcal{A}^X, X)$ in X by

$$\mathcal{A}^X = \{X \cap H \mid H \in \mathcal{A} \setminus \mathcal{A}_X \text{ and } X \cap H \neq \emptyset\}.$$

We call $\mathcal{A}^X$ the **restriction** *of $\mathcal{A}$ to X. Note that $\mathcal{A}^V = \mathcal{A}$.*

The method of **deletion and restriction** is a basic construction in this book. It allows for induction on the cardinality of $\mathcal{A}$.

Definition 1.14 *Let $\mathcal{A}$ be a nonempty arrangement and let $H_0 \in \mathcal{A}$. Let $\mathcal{A}' = \mathcal{A} \setminus \{H_0\}$ and let $\mathcal{A}'' = \mathcal{A}^{H_0}$. We call $(\mathcal{A}, \mathcal{A}', \mathcal{A}'')$ a* **triple** *of arrangements and H_0 the* **distinguished** *hyperplane.*

The method of **coning** is another basic construction. It allows for comparing affine and central arrangements.

Definition 1.15 *An affine ℓ-arrangement $\mathcal{A}$ defined by $Q(\mathcal{A}) \in S$ gives rise to a central $(\ell + 1)$-arrangement $\mathbf{c}\mathcal{A}$, called the* **cone** *over $\mathcal{A}$. Let $Q' \in \mathbb{K}[x_0, x_1, \ldots, x_\ell]$ be the polynomial $Q(\mathcal{A})$ homogenized and define $Q(\mathbf{c}\mathcal{A}) = x_0 Q'$. Note that $|\mathbf{c}\mathcal{A}| = |\mathcal{A}| + 1$. We call $K_0 = \ker(x_0)$ the* **additional** *hyperplane.*

Note that in the coning construction, the arrangement $\mathcal{A}$ is embedded in $\mathbf{c}\mathcal{A}$ by identifying its total space with the affine subspace $\ker(x_0 - 1)$ in the total space of $\mathbf{c}\mathcal{A}$. There is an inverse operation. A nonempty central $(\ell + 1)$-arrangement $\mathcal{A}$ gives rise to an ℓ-arrangement $\mathbf{d}\mathcal{A}$, which is in general not centered, by the following **deconing** construction. Choose a hyperplane $K_0 \in \mathcal{A}$. Choose coordinates so that $K_0 = \ker(x_0)$. Let $Q(\mathcal{A}) \in \mathbb{K}[x_0, x_1, \ldots, x_\ell]$ be a defining polynomial for $\mathcal{A}$. The defining polynomial $Q(\mathbf{d}\mathcal{A})$ is obtained by substituting 1 for x_0 in $Q(\mathcal{A})$. The deconing construction may be viewed as first projectivizing the central arrangement $\mathcal{A}$ then removing the image of K_0, the hyperplane at infinity, and identifying its complement with affine space.

There are two sets of fundamental interest in the study of arrangements: the variety of $\mathcal{A}$ and the complement of $\mathcal{A}$.

Definition 1.16 *Define the* **variety** *of $\mathcal{A}$ by*

$$N(\mathcal{A}) = \bigcup_{H \in \mathcal{A}} H = \{v \in V \mid Q(\mathcal{A})(v) = 0\}.$$

Definition 1.17 *Define the* **complement** *of $\mathcal{A}$ by*

$$M(\mathcal{A}) = V \setminus N(\mathcal{A}).$$

It is sometimes convenient to suppress dependence on $\mathcal{A}$ and write $Q = Q(\mathcal{A})$, $L = L(\mathcal{A})$, $N = N(\mathcal{A})$, $M = M(\mathcal{A})$, etc.

The Module of $\mathcal{A}$–Derivations

The variety $N(\mathcal{A})$ is a hypersurface with a very complicated singular set. In the present context, the study of logarithmic vector fields and logarithmic differential forms on a hypersurface was initiated by Saito [199, 201]. It was shown in [232] that in the case of an arrangement, we can pass from analytic to algebraic considerations.

Definition 1.18 *A $\mathbb{K}$-linear map $\theta : S \to S$ is a* **derivation** *if for $f, g \in S$*

$$\theta(fg) = f\theta(g) + g\theta(f).$$

Let $\mathrm{Der}_{\mathbb{K}}(S)$ be the S–module of derivations of S.

Definition 1.19 *Define an S-submodule of $\mathrm{Der}_{\mathbb{K}}(S)$, called the* **module of $\mathcal{A}$–derivations**, *by*

$$D(\mathcal{A}) = \{\theta \in \mathrm{Der}_{\mathbb{K}}(S) \mid \theta(Q) \in QS\}.$$

Definition 1.20 *The arrangement $\mathcal{A}$ is called* **free** *if $D(\mathcal{A})$ is a free S-module.*

The Complement of a Complex Arrangement

Next we consider field extensions. Let $\mathbb{L}$ be an extension of $\mathbb{K}$. Every arrangement $(\mathcal{A}_{\mathbb{K}}, V_{\mathbb{K}})$ gives rise to an arrangement over $\mathbb{L}$.

Definition 1.21 Let $(\mathcal{A}_{\mathbb{K}}, V_{\mathbb{K}})$ be an arrangement with defining polynomial $Q(\mathcal{A}_{\mathbb{K}})$. The $\mathbb{L}$-extended arrangement is in $V = V_{\mathbb{K}} \otimes_{\mathbb{K}} \mathbb{L}$. It consists of the hyperplanes $\mathcal{A}_{\mathbb{L}} = \{H \otimes_{\mathbb{K}} \mathbb{L} \mid H \in \mathcal{A}_{\mathbb{K}}\}$. Thus $Q(\mathcal{A}_{\mathbb{L}}) = Q(\mathcal{A}_{\mathbb{K}})$.

One example of this is a **complexified** real arrangement. It is already quite difficult to visualize the complexification of Example 1.5. In real dimensions we have three 2–planes in 4–space which meet only at the origin. The complexification of the Boolean arrangement is the arrangement of the coordinate hyperplanes in $\mathbb{C}^{\ell}$.

The complexified braid arrangement occurs in the theory of configuration spaces and braids. Recall that a **braid** on ℓ strands may be viewed as the graph of the motion of ℓ distinct points in the complex line between times $t = 0$ and $t = 1$, subject to the condition that the points remain distinct throughout the motion and that the sets of points at $t = 0$ and $t = 1$ are equal. Thus we have a map $f : [0, 1] \to \mathbb{C}^{\ell}$ such that for each t the image point $(f_1(t), \ldots, f_\ell(t))$ satisfies the condition $f_i(t) \neq f_j(t)$. The braid is **pure** if $f(0) = f(1)$. Thus a pure braid is the image of a circle in the complement of the hyperplanes $H_{i,j}$. The variety $N(\mathcal{A})$ is called the **superdiagonal** and its complement $M(\mathcal{A}) = V \setminus N(\mathcal{A})$ is the **pure braid space**.

Reflection Arrangements

Next we define a collection of arrangements with particularly nice properties. Let $\mathbb{K} = \mathbb{R}$ or $\mathbb{C}$ and let $GL(V)$ denote the general linear group of V.

Definition 1.22 An element $s \in GL(V)$ is a **reflection** if it has finite order and its fixed point set is a hyperplane H_s. We call H_s the **reflecting hyperplane** of s. A finite subgroup $G \subset GL(V)$ is called a **reflection group** if it is generated by reflections.

Definition 1.23 Let $G \subset GL(V)$ be a finite reflection group. The set $\mathcal{A} = \mathcal{A}(G)$ of reflecting hyperplanes of G is called the **reflection arrangement** of G.

Note that the braid arrangement is the reflection arrangement of the symmetric group.

1.3 Outline

Combinatorics

The **intersection poset** $L(\mathcal{A})$ is an important combinatorial invariant of the arrangement $\mathcal{A}$. We study its properties in Chapter 2. In Section 2.1 we give $L(\mathcal{A})$ a partial order by reverse inclusion and show that it is a geometric lattice when $\mathcal{A}$ is a central arrangement. We construct the face poset of a real arrangement and show its connection with oriented matroids. We also define supersolvable arrangements here, and a generalization called arrangements with a nice partition. In Section 2.2 we define the Möbius function and study its properties. We also present notes on the interesting history of this function dating back to Euler. In Section 2.3 we define the Poincaré polynomial of $\mathcal{A}$, $\pi(\mathcal{A}, t)$, which is related to another combinatorial function called the characteristic polynomial. We show that if $\mathbf{c}\mathcal{A}$ is the cone over the affine arrangement $\mathcal{A}$, then

$$\pi(\mathbf{c}\mathcal{A}, t) = (1+t)\pi(\mathcal{A}, t). \tag{1}$$

A fundamental technical tool in this book is the method of **deletion and restriction**, which allows induction on the number of hyperplanes in the arrangement. It uses the triple $(\mathcal{A}, \mathcal{A}', \mathcal{A}'')$ of Definition 1.14. We prove a theorem of Brylawski [43] about the Poincaré polynomial under deletion and restriction:

$$\pi(\mathcal{A}, t) = \pi(\mathcal{A}', t) + t\pi(\mathcal{A}'', t). \tag{2}$$

In Section 2.3 we also prove a theorem of Stanley [218], which asserts that if $L(\mathcal{A})$ is supersolvable, then

$$\pi(\mathcal{A}, t) = (1 + b_1 t) \cdots (1 + b_\ell t), \tag{3}$$

where the b_i are nonnegative integers. We close the chapter with a section on the connections between arrangements and graph theory. This section includes a discussion of the chromatic polynomial, a precursor of the characteristic and Poincaré polynomials.

Algebras

In Chapter 3 let $\mathcal{K}$ be a commutative ring. We construct certain algebras over $\mathcal{K}$ associated with $\mathcal{A}$. We construct the graded algebra $A(\mathcal{A})$ for a central arrangement $\mathcal{A}$ in Section 3.1. This construction is generalized to affine arrangements in Section 3.2. The algebra $A(\mathcal{A})$ is the quotient of the exterior algebra $E(\mathcal{A})$ based on $\mathcal{A}$ by a homogeneous ideal $I(\mathcal{A})$, $A(\mathcal{A}) = E(\mathcal{A})/I(\mathcal{A})$. This algebra is constructed using only $L(\mathcal{A})$. In the literature $A(\mathcal{A})$ is sometimes called the Orlik–Solomon algebra. It will reappear in Chapter 5 with a topological significance. We prove that the $\mathcal{K}$–algebra $A(\mathcal{A})$ is a free graded $\mathcal{K}$–module and that its Poincaré polynomial is equal to $\pi(\mathcal{A}, t)$. This gives an interpretation of the coefficients of $\pi(\mathcal{A}, t)$. We construct a $\mathcal{K}$-basis for $A(\mathcal{A})$ as a free graded $\mathcal{K}$-module using broken circuits. We also show that given a triple $(\mathcal{A}, \mathcal{A}', \mathcal{A}'')$, there is an exact sequence of $\mathcal{K}$-modules

(4) $$0 \to A(\mathcal{A}') \to A(\mathcal{A}) \to A(\mathcal{A}'') \to 0.$$

Inspection of the maps shows that (2) is a consequence of (4). We prove some algebra factorization theorems in Section 3.3. If $\mathbf{c}\mathcal{A}$ is the cone over $\mathcal{A}$, then there exists an element $a_0 \in A(\mathbf{c}\mathcal{A})$ so that there is an isomorphism of graded $\mathcal{K}$-modules

(5) $$A(\mathbf{c}\mathcal{A}) \simeq (\mathcal{K} + \mathcal{K}a_0) \otimes A(\mathcal{A}).$$

In fact (1) is a consequence of (5). If $\mathcal{A}$ is supersolvable, then $A(\mathcal{A})$ has a tensor product decomposition

(6) $$(\mathcal{K} + B_1) \otimes \cdots \otimes (\mathcal{K} + B_\ell)$$

as graded $\mathcal{K}$-module with $b_i = \mathrm{rank} B_i$. In fact (3) is a consequence of (6). This decomposition of $A(\mathcal{A})$ is generalized to arrangements with a nice partition. In Section 3.4 we define another graded algebra, $B(\mathcal{A})$, whose multiplication is a shuffle product. We prove that $B(\mathcal{A})$ is algebra isomorphic to $A(\mathcal{A})$. In Section 3.5 we assume that $\mathcal{K}$ is a subring of $\mathbb{K}$. We associate to the arrangement $\mathcal{A}$ the $\mathcal{K}$-algebra $R(\mathcal{A})$ generated by the differential forms $\omega_H = d\alpha_H/\alpha_H$. (Note that this algebra is not a purely combinatorial object, since the defining polynomials α_H enter the definition.) The main result of Section 3.5 is that there exists an isomorphism of algebras $A(\mathcal{A}) \simeq R(\mathcal{A})$, which shows that $R(\mathcal{A})$ depends only on $L(\mathcal{A})$. The argument uses the fact that there is a short exact sequence of $\mathcal{K}$-modules

$$0 \to R(\mathcal{A}') \to R(\mathcal{A}) \to R(\mathcal{A}'') \to 0.$$

The Module of $\mathcal{A}$-Derivations

The most important algebraic geometric invariant of $\mathcal{A}$ is the module $D(\mathcal{A})$. In Chapter 4 we assume that $\mathcal{A}$ is a central arrangement and study the algebraic properties of $D(\mathcal{A})$. Section 4.1 contains basic definitions. In Section 4.2 we define free arrangements and establish their fundamental properties. If $\mathcal{A}$ is free, then we can associate with it a collection of nonnegative integers, called its exponents, $\exp \mathcal{A} = \{b_1, \ldots, b_\ell\}$. These integers are unique up to order, but they are not necessarily distinct. In Section 4.3 we prove the Addition–Deletion Theorem, which asserts that if $(\mathcal{A}, \mathcal{A}', \mathcal{A}'')$ is a triple, then any two of the following statements imply the third:

$$\begin{aligned}
\mathcal{A} \text{ is free with } \exp \mathcal{A} &= \{b_1, \ldots, b_{\ell-1}, b_\ell\}, \\
\mathcal{A}' \text{ is free with } \exp \mathcal{A}' &= \{b_1, \ldots, b_{\ell-1}, b_\ell - 1\}, \\
\mathcal{A}'' \text{ is free with } \exp \mathcal{A}'' &= \{b_1, \ldots, b_{\ell-1}\}.
\end{aligned}$$

This result leads to the definition of inductively free arrangements. We give several examples and prove that a supersolvable arrangement is inductively free. In Section 4.4 we define the module $\Omega^p(\mathcal{A})$ of logarithmic p-forms with poles on the hypersurface $N(\mathcal{A})$. We show that the complex $\Omega^{\cdot}(\mathcal{A})$ is closed under exterior product and that $\Omega^1(\mathcal{A})$ is the dual of $D(\mathcal{A})$. In Section 4.5 we construct a simplicial complex $\mathsf{F}(\mathcal{A})$ associated to $L(\mathcal{A})$ by Folkman [85]. We compute its homology groups and show that $\mathsf{F}(\mathcal{A})$ has the homotopy type of a wedge of spheres. We also construct another chain complex whose homology is naturally isomorphic to $B(\mathcal{A})$, and show how these constructions are related. In Section 4.6 we generalize these two constructions to order complexes with arbitrary functor coefficient. This allows proof of an important technical result due to Yuzvinsky [254]. It is used in the proof of the Factorization Theorem, which asserts that if $\mathcal{A}$ is a free ℓ–arrangement with $\exp \mathcal{A} = \{b_1, \ldots, b_\ell\}$, then

$$\pi(\mathcal{A}, t) = (1 + b_1 t) \cdots (1 + b_\ell t).$$

Thus the exponents of a free arrangement are determined by combinatorial data.

In the first four chapters, the field $\mathbb{K}$ is arbitrary and the development of the material is essentially self-contained. In the last two chapters, we work mostly over the complex numbers and use more results from the literature.

Topology

In Chapter 5 we return to the convention that an arrangement is not necessarily central. The subject of this chapter is the topology of the complement of a complex arrangement, $M(\mathcal{A})$. In Section 5.1 we prove some elementary facts about $M = M(\mathcal{A})$ and discuss a few examples. In particular, if $\mathbf{c}\mathcal{A}$ is the cone over $\mathcal{A}$, then

(7) $$M(\mathbf{c}\mathcal{A}) \approx M(\mathcal{A}) \times \mathbb{C}^*,$$

where $\mathbb{C}^*$ denotes the nonzero complex numbers and $\approx$ denotes homeomorphism. We also review fundamental work of Arnold, Brieskorn, Deligne and Hattori. The rest of the chapter does not follow the chronology of discovery. In Section 5.2 we construct a finite simplicial complex M of the homotopy type of M. The construction uses an embedding in V of the order complex of the face poset of a real arrangement. In principle, M contains all information about the homotopy type of M. In the special case of a complexified real arrangement, Salvetti [203] constructed a smaller complex W of the homotopy type of M. Arvola [15] constructed a simplicial map $\mathsf{M} \to \mathsf{W}$, which is a homotopy equivalence. In practice, M and W are very large and unsuited for explicit calculations. It is therefore desirable to find simple algorithms to compute various topological invariants of M.

Arvola's presentation of the fundamental group of M is in Section 5.3. It generalizes Randell's presentation of the fundamental group of the complexification of a real arrangement. In Section 5.4 we consider the cohomology groups of $M(\mathcal{A})$ with integer coefficients. We use our results on $R(\mathcal{A})$ from Section 3.5 to prove that given a triple $(\mathcal{A}, \mathcal{A}', \mathcal{A}'')$, there are split short exact sequences for all $k \geq 0$

(8) $$0 \to H^k(M(\mathcal{A}')) \to H^k(M(\mathcal{A})) \to H^{k-1}(M(\mathcal{A}'')) \to 0.$$

It follows that there is an algebra isomorphism $R(\mathcal{A}) \to H^*(M(\mathcal{A}))$ induced by $\omega_H \mapsto [(1/2\pi i)\omega_H]$. Together with the algebra isomorphism $R(\mathcal{A}) \simeq A(\mathcal{A})$ established in Section 3.5, this provides a presentation of the cohomology algebra in terms of generators and relations. This is the topological interpretation of $A(\mathcal{A})$. Thus the cohomology algebra of $M(\mathcal{A})$ depends only on $L(\mathcal{A})$. These results have several consequences. They provide an elementary proof of Brieskorn's Lemma [41]. They show that the Poincaré polynomial of the complement is

$$\text{Poin}(M(\mathcal{A}), t) = \pi(\mathcal{A}, t).$$

Thus the coefficients of $\pi(\mathcal{A}, t)$ are also the Betti numbers of the complement. They show the common origin of formulas (2), (4), (8), and of (1), (5), (7). They also show that if $\mathcal{A}$ is a real arrangement, then $M(\mathcal{A})$ has the M–property. In Section 5.5 we prove that the complement of a supersolvable arrangement admits a strictly linear fibration. This is the topological interpretation of (6). In Section 5.6 we describe some related recent results: work of Falk and Kohno on minimal models, Manin and Schechtman's work on discriminantal arrangements, Falk's geometric linking, the cohomology of the Milnor fiber of a generic arrangement, and the results of Goresky and MacPherson on arrangements of subspaces of arbitrary codimension.

Reflection Arrangements

In Chapter 6 we study the equivariant theory of arrangements. Suppose $G \subseteq GL(V)$ is a finite linear group and $\mathcal{A}$ is an arrangement which is stable under the action of G. *We assume that the order of G is not divisible by the characteristic of the field* $\mathbb{K}$. Section 6.1 contains definitions, examples, and generalities about the action of G on the polynomial algebra S and on the S–module $\text{Der}_{\mathbb{K}}(S)$. We also study the action of G on the algebra $A(\mathcal{A})$ and on the cohomology ring $H^*(M)$ in this section.

There is a collection of linear groups and associated arrangements with particularly nice properties. The groups G are generated by reflections and the arrangements $\mathcal{A}(G)$ are the fixed hyperplanes of the reflections, called reflection arrangements. These are the objects we study in the rest of the chapter. In Section 6.2 we discuss some basic results. Chevalley's theorem asserts that finite groups generated by reflections are distinguished by the fact that their algebra of G–invariant polynomials is a polynomial algebra. This result is the key to their nice properties. Irreducible complex reflection groups were classified by Shephard and Todd [210]. It is customary to call a proof "case free" if it does not depend on this classification. In Section 6.3 we prove that every reflection arrangement $\mathcal{A}(G)$ is free. The argument here is case free. We show that the exponents of $\mathcal{A}(G)$ have an interpretation in the invariant theory of G. This allows determination of the exponents of $\mathcal{A}(G)$ using character tables for G. The results are listed in Appendix B. Construction of a basis for $D(\mathcal{A}(G))$ involves additional work. We also present these results in Appendix B. The group G acts by permutation on the poset $L(\mathcal{A})$. We discuss this action in Section 6.4. A complete set of orbit types and related information is presented for the irreducible groups in Appendix C. We study the structure of $L(\mathcal{A})$ and determine for all $X \in L(\mathcal{A})$ the structure of the restriction $L(\mathcal{A}^X)$. This work on the restrictions $L(\mathcal{A}^X)$ shows that if $\mathcal{A} = \mathcal{A}(G)$ is a reflection arrangement and $X \in \mathcal{A}$ with $p = \dim X$, then there exist integers $\{b_1^X, \ldots, b_p^X\}$ such that

$$\pi(\mathcal{A}^X, t) = (1 + b_1^X) \cdots (1 + b_p^X).$$

The values of b_i^X are tabulated in Appendix C. In Section 6.5 we consider restrictions of reflection arrangements. We show that if $\mathcal{A}$ is a Coxeter arrangement and $H \in \mathcal{A}$, then $\mathcal{A}^H$ is free. Unfortunately, induction may not be used in combination with this result because, in general, $\mathcal{A}^H$ is not the arrangement of any Coxeter group. In Section 6.6 we consider the orbit space V/G. It follows from Chevalley's theorem that V/G has the structure of affine space. The image N/G of the variety of $\mathcal{A}$ is the discriminant locus. We study the stratification of V/G induced by the stratification of V by the elements of $L(\mathcal{A})$. We also define a subclass of complex reflection groups called Shephard groups and associate to each Shephard group G an irreducible Coxeter group W. We show that with suitable choices, the discriminant loci of G and W are equal. It follows from Deligne's theorem that the complement of the discriminant of W is a $K(\pi, 1)$ space. This identification proves that the complement of the discriminant of G is a $K(\pi, 1)$ space. It follows that $\mathcal{A}(G)$ is a $K(\pi, 1)$–arrangement for all Shephard groups.

In Appendix A we collect certain facts from commutative algebra which are needed in the text. In Appendix B we construct basic derivations for all complex reflection groups. Detailed information on the intersection lattices of irreducible complex reflection groups is presented in Appendix C. Appendix D provides additional information on restrictions of reflection arrangements to 3–dimensional subspaces.

2. Combinatorics

The intersection poset $L(\mathcal{A})$ is an important combinatorial invariant of the arrangement $\mathcal{A}$. We study its properties in this chapter. In Section 2.1 we give $L(\mathcal{A})$ a partial order by reverse inclusion and show that it is a geometric lattice when $\mathcal{A}$ is a central arrangement. We construct the face poset of a real arrangement and show its connection with oriented matroids. We also define supersolvable arrangements here, and a generalization called arrangements with a nice partition. In Section 2.2 we define the Möbius function and study its properties. We also present notes on the interesting history of this function dating back to Euler. In Section 2.3 we define the Poincaré polynomial $\pi(\mathcal{A}, t)$, which is related to another combinatorial function called the characteristic polynomial. A fundamental technical tool in this book is the method of **deletion and restriction**, which allows induction on the number of hyperplanes in the arrangement. It uses the triple $(\mathcal{A}, \mathcal{A}', \mathcal{A}'')$ of Definition 1.14. The Deletion–Restriction Theorem states:

$$\pi(\mathcal{A}, t) = \pi(\mathcal{A}', t) + t\pi(\mathcal{A}'', t).$$

We prove a theorem of Stanley [218], which asserts that if $L(\mathcal{A})$ is supersolvable, then

$$\pi(\mathcal{A}, t) = (1 + b_1 t) \cdots (1 + b_\ell t),$$

where the b_i are nonnegative integers. We close the chapter with a section on the connections between arrangements and graph theory. This section includes a discussion of the chromatic polynomial, a precursor of the characteristic polynomial and the Poincaré polynomial. Many of the definitions and results may be extended to a larger class of objects. We use Aigner's book [1] as a general reference for undefined terms.

2.1 The Poset $L(\mathcal{A})$

Definitions

Definition 2.1 *Let $\mathcal{A}$ be an arrangement and let $L = L(\mathcal{A})$ be the set of nonempty intersections of elements of $\mathcal{A}$. Define a **partial order** on L by*

$$X \leq Y \iff Y \subseteq X.$$

Note that this is **reverse** inclusion. Thus V is the unique minimal element. Ordinary inclusion also gives a partial order preferred by many authors. With our convention the intersection poset of a central arrangement has all the properties of a **geometric lattice** shown in Lemma 2.3.

Definition 2.2 *Define a rank function on L by $r(X) = \mathrm{codim} X$. Thus $r(V) = 0$ and $r(H) = 1$ for $H \in \mathcal{A}$. Call H an* **atom** *of L. Let $X, Y \in L$. Define their* **meet** *by $X \wedge Y = \cap \{Z \in L \mid X \cup Y \subseteq Z\}$. If $X \cap Y \neq \emptyset$, we define their* **join** *by $X \vee Y = X \cap Y$.*

Lemma 2.3 *Let $\mathcal{A}$ be an arrangement and let $L = L(\mathcal{A})$. Then:*
(1) *Every element of $L \setminus \{V\}$ is a join of atoms.*
(2) *For every $X \in L$ all maximal linearly ordered subsets*
$$V = X_0 < X_1 < \ldots < X_p = X$$
have the same cardinality. Thus $L(\mathcal{A})$ is a geometric poset.
(3) *If $\mathcal{A}$ is central, then all joins exist, so L is a lattice. For all $X, Y \in L$ the rank function satisfies*
$$r(X \wedge Y) + r(X \vee Y) \leq r(X) + r(Y).$$
Thus for a central arrangement, $L(\mathcal{A})$ is a geometric lattice.

Proof. Assertion (1) follows from the definition. Assertion (2) is a consequence of the fact that the maximal number of linearly independent hyperplanes which can contain a subspace is its codimension. To see (3) recall that
$$\dim(X + Y) + \dim(X \cap Y) = \dim(X) + \dim(Y)$$
and $\dim(X + Y) \leq \dim(X \wedge Y)$. □

Lemma 2.4 *Maximal elements of $L(\mathcal{A})$ have the same rank.*

Proof. This is clear if $\mathcal{A}$ is a central arrangement, since $L(\mathcal{A})$ has a unique maximal element. If $\mathcal{A}$ is centerless, then it may have several maximal elements. Observe that $T \in L(\mathcal{A})$ is a maximal element if and only if for every $H \in \mathcal{A}$, either $T \subset H$ or $T \cap H = \emptyset$. Since this condition is invariant under affine linear transformations, maximal elements are affine linear images of each other, hence of the same dimension. □

Definition 2.5 *The* **rank** *of $\mathcal{A}$, $r(\mathcal{A})$, is the rank of a maximal element of $L(\mathcal{A})$. Call the ℓ-arrangement $\mathcal{A}$* **essential** *if $r(\mathcal{A}) = \ell$. If $\mathcal{A}$ is a central arrangement, let $T(\mathcal{A}) = \cap_{H \in \mathcal{A}} H$ be the* **unique maximal element** *of $L(\mathcal{A})$.*

Definition 2.6 *Call the arrangements $\mathcal{A} = (\mathcal{A}, V)$ and $\mathcal{B} = (\mathcal{B}, W)$* **lattice equivalent**, *or L-equivalent, if there is an order preserving bijection $\pi : L(\mathcal{A}) \to L(\mathcal{B})$.*

2.1 The Poset $L(\mathcal{A})$ 25

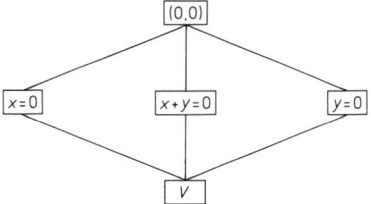

Fig. 2.1. The Hasse diagram of $Q(\mathcal{A}) = xy(x+y)$

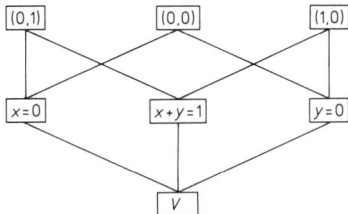

Fig. 2.2. The Hasse diagram of $Q(\mathcal{A}) = xy(x+y-1)$

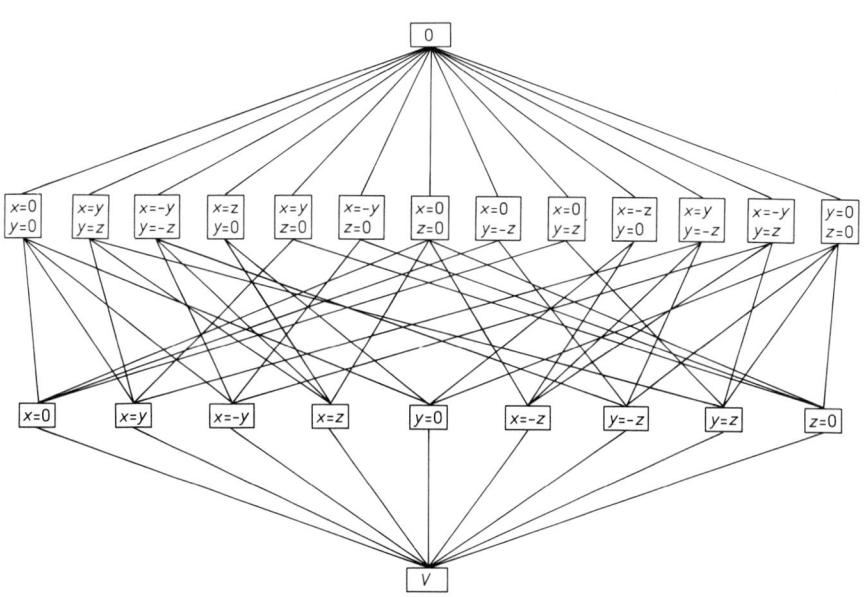

Fig. 2.3. The Hasse diagram of the B_3–arrangement

Definition 2.7 Let $L_p(\mathcal{A}) = \{X \in L(\mathcal{A}) \mid r(X) = p\}$. The **Hasse diagram** of $L(\mathcal{A})$ has vertices labeled by the elements of $L(\mathcal{A})$ and arranged on levels L_p for $p \geq 0$. Suppose $X \in L_p$ and $Y \in L_{p+1}$. An edge in the Hasse diagram connects X with Y if $X < Y$.

If $\mathcal{A}$ is defined by a polynomial $Q(\mathcal{A})$, it is sometimes convenient to label elements of $L(\mathcal{A})$ by the equations they satisfy. The Hasse diagrams of Examples 1.5, 1.6, 1.7 appear in Figures 2.1, 2.2, 2.3.

Examples

Example 2.8 The lattice $L(\mathcal{A})$ of the Boolean arrangement. Let $H_i = \ker(x_i)$. Let $I = \{i_1, \ldots, i_p\}$ where $1 \leq i_1 < \cdots < i_p \leq \ell$. Let $H_I = H_{i_1} \cap \cdots \cap H_{i_p}$. The lattice $L(\mathcal{A})$ consists of the 2^ℓ subspaces H_I for all subsets I.

Proposition 2.9 *The lattice $L(\mathcal{A})$ of the braid arrangement is isomorphic to the **partition lattice**.*

Proof. Let $I = \{1, \ldots, \ell\}$. Let $\mathcal{P}(\ell)$ be the set of partitions of I. An element of $\mathcal{P}(\ell)$ is a collection $\Lambda = \{\Lambda_1, \ldots, \Lambda_r\}$ of nonempty pairwise disjoint subsets of I, called the blocks of Λ, whose union is I. There is a natural partial order on $\mathcal{P}(\ell)$ given by $\Lambda \leq \Gamma$ if Λ is finer than Γ. Thus blocks of Γ are unions of blocks of Λ. In order to find a lattice isomorphism from the braid lattice $L(\mathcal{A})$ to $\mathcal{P}(\ell)$, it is convenient to define $H_{i,i} = V$ for all i. Let $X \in L(\mathcal{A})$. Define a relation $\sim_X$ on I by $i \sim_X j$ if and only if $X \subseteq H_{i,j}$. Since $H_{i,i} = V$, $H_{i,j} = H_{j,i}$ and $H_{i,j} \cap H_{j,k} \subseteq H_{i,k}$, this is an equivalence relation. Let Λ_X be the partition of I defined by $\sim_X$. The map $\pi : L(\mathcal{A}) \to \mathcal{P}(\ell)$ given by $\pi(X) = \Lambda_X$ is a lattice isomorphism. It is injective because

$$X = \bigcap_{k=1}^{r} \left(\bigcap_{i,j \in \Lambda_k} H_{i,j} \right)$$

is determined by the blocks of Λ. It is surjective because given any partition $\Lambda = \{\Lambda_1, \ldots, \Lambda_r\}$ we may define X by the intersection above and we get $\Lambda_X = \Lambda$. Note also that $X \leq Y$ if and only if every block of Λ_Y is a union of blocks of Λ_X. □

Definition 2.10 *Given a poset L and $X, Y \in L$ with $X < Y$, define the following subposets and segments*

$$L_X = \{Z \in L \mid Z \leq X\}, \quad L^X = \{Z \in L \mid Z \geq X\}.$$

$$[X, Y] = \{Z \in L \mid X \leq Z \leq Y\}, \quad [X, Y) = \{Z \in L \mid X \leq Z < Y\}.$$

Lemma 2.11 *Let $\mathcal{A}$ be an arrangement and let $X \in L(\mathcal{A})$. Then*

(1) $L(\mathcal{A})_X = L(\mathcal{A}_X)$,
(2) $L(\mathcal{A})^X = L(\mathcal{A}^X)$,
(3) if $Y \in L$ and $X \leq Y$, then $L((\mathcal{A}_Y)^X) = L(\mathcal{A}_Y)^X = [X, Y]$.

Example 2.12 The lattice $L(\mathcal{A})$ of Example 1.10 consists of all subspaces of V. If $X \in L(\mathcal{A})$ is p-dimensional, then $L(\mathcal{A}^X)$ is the lattice of all hyperplanes in X and $L(\mathcal{A}_X)$ is isomorphic to the lattice of all hyperplanes in the $(\ell - p)$-dimensional space V/X. If $X < Y$, then $[X, Y]$ is isomorphic to the lattice of all subspaces of X/Y.

Recall that $\mathcal{A}$ is essential if and only if it contains ℓ linearly independent hyperplanes. For a central arrangement, this is equivalent to the condition $T(\mathcal{A}) = \{0\}$. The braid arrangement is not essential; $T(\mathcal{A})$ is the line $x_1 = x_2 = \ldots = x_\ell$. All the other arrangements considered so far are essential.

Definition 2.13 Let $(\mathcal{A}_1, V_1)$ and $(\mathcal{A}_2, V_2)$ be arrangements and let $V = V_1 \oplus V_2$. Define the **product** arrangement $(\mathcal{A}_1 \times \mathcal{A}_2, V)$ by

$$\mathcal{A}_1 \times \mathcal{A}_2 = \{H_1 \oplus V_2 \mid H_1 \in \mathcal{A}_1\} \cup \{V_1 \oplus H_2 \mid H_2 \in \mathcal{A}_2\}.$$

Proposition 2.14 Let $\mathcal{A}_1, \mathcal{A}_2$ be arrangements. Define a partial order on the set $L(\mathcal{A}_1) \times L(\mathcal{A}_2)$ of pairs (X_1, X_2) with $X_i \in L(\mathcal{A}_i)$ by

$$(X_1, X_2) \leq (Y_1, Y_2) \iff X_1 \leq Y_1 \text{ and } X_2 \leq Y_2.$$

There is a natural isomorphism of lattices

$$\pi : L(\mathcal{A}_1) \times L(\mathcal{A}_2) \to L(\mathcal{A}_1 \times \mathcal{A}_2)$$

given by the map $\pi(X_1, X_2) = X_1 \oplus X_2$.

Definition 2.15 Call the arrangement $(\mathcal{A}, V)$ **reducible** if after a change of coordinates $(\mathcal{A}, V) = (\mathcal{A}_1 \times \mathcal{A}_2, V_1 \oplus V_2)$. Otherwise call $(\mathcal{A}, V)$ **irreducible**.

Example 2.16 The B_3-arrangement is irreducible. The Boolean arrangement is the product of ℓ copies of the 1-arrangement $(\{0\}, \mathbb{K})$. The braid arrangement is the product of the empty 1-arrangement with an irreducible arrangement.

Let $\mathcal{A}$ be an affine ℓ-arrangement. Recall the coning construction from Definition 1.15. Let $H \in \mathcal{A}$ be the kernel of the degree 1 polynomial $\alpha_H \in \mathbb{K}[x_1, \ldots, x_\ell]$. Then H corresponds to $\mathbf{c}H \in \mathbf{c}\mathcal{A}$, the kernel of the linear form obtained by homogenizing α_H in $\mathbb{K}[x_0, x_1, \ldots, x_\ell]$. Note that $\dim H = \ell - 1$ and $\dim \mathbf{c}H = \ell$. Recall that $\mathbf{c}\mathcal{A}$ also contains the additional hyperplane $K_0 = \ker(x_0)$.

Proposition 2.17 Let $\mathcal{A}$ be an affine arrangement with cone $\mathbf{c}\mathcal{A}$. Let $\mathcal{B} = (\{0\}, \mathbb{K})$ be the nonempty central 1-arrangement. Define the bijection $\phi : \mathcal{A} \times$

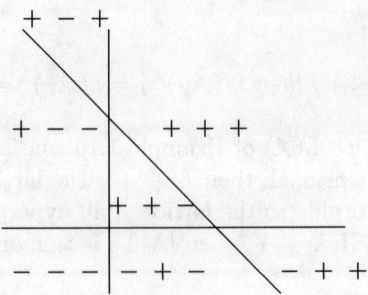

Fig. 2.4. The chambers of $Q(\mathcal{A}) = xy(x+y-1)$

$\mathcal{B} \to \mathbf{c}\mathcal{A}$ by $\phi(H \oplus \mathbb{K}) = \mathbf{c}H$ and $\phi(V \oplus \{0\}) = K_0$. Then ϕ induces a rank preserving surjective map of posets $\phi : L(\mathcal{A} \times \mathcal{B}) \to L(\mathbf{c}\mathcal{A})$.

Proof. Let $X \in L(\mathcal{A})$. Write $X = H_1 \cap \cdots \cap H_p$. Define $\mathbf{c}X = \mathbf{c}H_1 \cap \cdots \cap \mathbf{c}H_p$. A direct argument shows that $\mathbf{c}X$ is independent of the representation of X as an intersection of hyperplanes. We define $\phi(X \oplus \mathbb{K}) = \mathbf{c}X$ and $\phi(X \oplus \{0\}) = \mathbf{c}X \cap K_0$. It is easy to see that the map is rank preserving and surjective. In general, it is not injective. □

Oriented Matroids

Next we consider the special case of a real arrangement $\mathcal{A}$. Recall that $\mathcal{C}(\mathcal{A})$ is the set of chambers of $\mathcal{A}$. Thus $M(\mathcal{A}) = \cup_{C \in \mathcal{C}(\mathcal{A})} C$.

Definition 2.18 *Let $\mathcal{A}$ be a real arrangement. Let*

$$\mathcal{L}(\mathcal{A}) = \bigcup_{X \in L(\mathcal{A})} \mathcal{C}(\mathcal{A}^X).$$

View $\mathcal{L}(\mathcal{A})$ as a collection of subsets of V. An element $P \in \mathcal{L}(\mathcal{A})$ is a **face**. *The support $|P|$ of a face P is its affine linear span. Each face is open in its support. Let $\bar{P}$ denote the closure of P in V. The set $\mathcal{L}(\mathcal{A})$ is partially ordered by reverse inclusion: $P \leq Q$ if $Q \subseteq \bar{P}$. We call $\mathcal{L}(\mathcal{A})$ the* **face poset** *of $\mathcal{A}$.*

Definition 2.19 *Let $\mathcal{A}$ be a real arrangement. The poset map $\zeta : \mathcal{L}(\mathcal{A}) \to L(\mathcal{A})$ defined by $\zeta(P) = |P|$ is order preserving.*

There is a particularly efficient way to store the information in the face poset by using the associated oriented matroid; see [30]. Choose linear polynomials α_H so that $H = \ker \alpha_H$. Let $J = \{+,-,0\}$. We may view each face $P \in \mathcal{L}(\mathcal{A})$ as a map $P : \mathcal{A} \to J$ defined by $P(H) = \operatorname{sign}\alpha_H(p)$ for any $p \in P$. Note that $P(H) = 0$ if and only if $P \subseteq H$, and if $P(H) \neq 0$, then the sign indicates

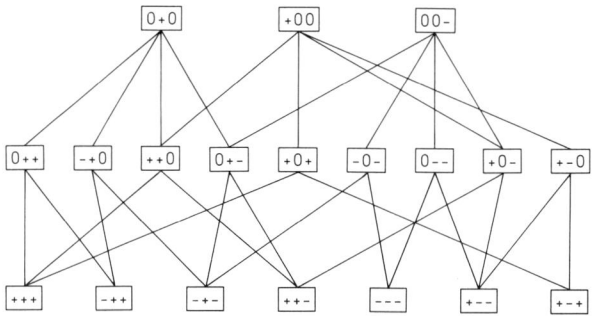

Fig. 2.5. The face poset of $Q(\mathcal{A}) = xy(x+y-1)$

whether P is in the positive or negative half-space determined by H. If we choose a linear order in $\mathcal{A}$, then we may write $\mathcal{A} = \{H_1, \ldots, H_n\}$ and let $H_k = \ker \alpha_k$. Let $W = J^n$, and let $\pi_k : W \to J$ be the projection onto the k-th coordinate. Define a map $\sigma : V \to W$ by

$$\pi_k \sigma(v) = \begin{cases} + & \text{if } \alpha_k(v) > 0, \\ 0 & \text{if } \alpha_k(v) = 0, \\ - & \text{if } \alpha_k(v) < 0. \end{cases}$$

We illustrate this concept in Example 1.6. In Figure 2.4 we labeled the chambers only. The reader is invited to label the remaining faces.

Definition 2.20 Let $\mathcal{G}(\mathcal{A}) = \sigma(V) \subseteq W$. Define a partial order in J by $+ < 0$, $- < 0$, while $+$ and $-$ are incomparable. This induces a partial order in W and in $\mathcal{G}(\mathcal{A})$. The poset $\mathcal{G}(\mathcal{A})$ is called the **oriented matroid** of $\mathcal{A}$.

Thus an element $P \in \mathcal{G}(\mathcal{A})$ is an ordered n-tuple $P = (P(1), \ldots, P(n))$ where each $P(k) \in J$ is one of $+, -, 0$. Equivalently, we may view it as a map $P : \mathcal{A} \to J$ where $P(H_k) = P(k)$. Let G_0 be the set of all subsets of $\{1, 2, \ldots, n\}$ partially ordered by inclusion. Define $\rho : \mathcal{G}(\mathcal{A}) \to G_0$ by $\rho(P) = \{k \mid P(k) = 0\}$. Let $G(\mathcal{A}) = \rho(\mathcal{G}(\mathcal{A}))$. Define $\tau : L(\mathcal{A}) \to G_0$ by $\tau(X) = \{k \mid H_k \leq X\}$. Then the following diagram of posets is commutative, and the vertical maps are poset isomorphisms:

$$\begin{array}{ccc} \mathcal{L}(\mathcal{A}) & \xrightarrow{\zeta} & L(\mathcal{A}) \\ \sigma \downarrow & & \downarrow \tau \\ \mathcal{G}(\mathcal{A}) & \xrightarrow{\rho} & G(\mathcal{A}) \end{array}$$

Definition 2.21 Identify $\mathcal{G}(\mathcal{A})$ and $\mathcal{L}(\mathcal{A})$. Define the **vector product** $\mathcal{L} \times \mathcal{L} \to \mathcal{L}$ by

$$(PQ)(i) = \begin{cases} P(i) & \text{if } P(i) \neq 0, \\ Q(i) & \text{if } P(i) = 0. \end{cases}$$

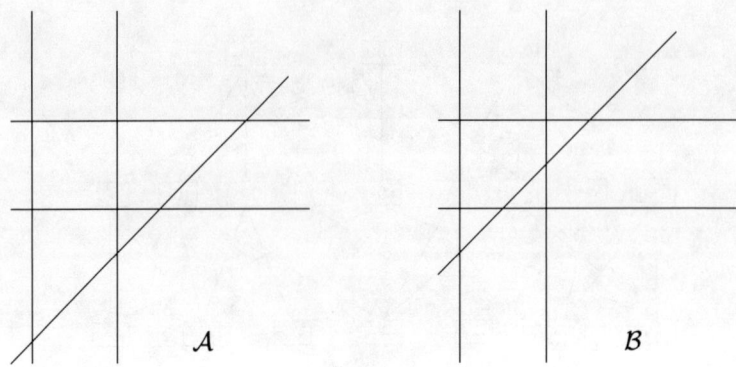

Fig. 2.6. Different face posets

Proposition 2.22 *The vector product is associative but not commutative. For every $Q \in \mathcal{L}$ we have $P \geq PQ$. For fixed $P \in \mathcal{L}$ the map $\mathcal{L} \to \mathcal{L}$ defined by $Q \mapsto PQ$ is order preserving. In particular, it carries chambers to chambers.*

The face poset of Example 1.6 is in Figure 2.5. It is identified with the oriented matroid. The face poset is a sharper invariant than the intersection poset. Arrangements $\mathcal{A}$ and $\mathcal{B}$ in Figure 2.6 are L–equivalent, but they have different face posets.

Supersolvable Arrangements

The next definitions are standard for lattices in general; see [1, 25, 217]. We will use them only for central arrangements. It simpifies notation to assume that $L = L(\mathcal{A})$ and $\mathcal{A}$ is central and essential. Let $r(\mathcal{A}) = \ell$ and write $T = T(\mathcal{A})$.

Definition 2.23 *A pair $(X,Y) \in L \times L$ is called a **modular pair** if for all Z with $Z \leq Y$*
$$Z \vee (X \wedge Y) = (Z \vee X) \wedge Y.$$

Lemma 2.24 *The following statements are equivalent:*
(1) the pair $(X,Y) \in L \times L$ is modular,
(2) $r(X) + r(Y) = r(X \vee Y) + r(X \wedge Y)$,
(3) $X \wedge Y = X + Y$,
(4) $X + Y \in L$.

Proof. The conditions (3) and (4) are obviously equivalent.
(1) $\Rightarrow$ (2) The order preserving map $\tau_X : [X \wedge Y, Y] \to [X, X \vee Y]$ defined by $\tau_X(Z) = X \vee Z$ is injective, because $\tau_X(Z) \wedge Y = (Z \vee X) \wedge Y = Z \vee (X \wedge Y) = Z$.

It shows
$$r(Y) - r(X \wedge Y) \leq r(X \vee Y) - r(X).$$
Combining with Lemma 2.3 (3), we have the desired result. (2) $\Rightarrow$ (3) Recall that $X + Y \subseteq X \wedge Y$. If (2) holds, then $\operatorname{codim}(X \wedge Y) = \operatorname{codim}(X + Y)$ so the spaces are equal. Finally, (3) $\Rightarrow$ (1) follows from
$$Z \vee (X \wedge Y) = Z \cap (X + Y) = (Z \cap X) + Y = (Z \vee X) \wedge Y$$
if $Y \subseteq Z$. □

Definition 2.25 *An element $X \in L$ is called* **modular** *if (X, Y) is a modular pair for all $Y \in L$.*

Corollary 2.26 *An element $X \in L$ is modular if and only if $X + Y \in L$ for all $Y \in L$.*

Lemma 2.27 *Let $X \in L$ be a modular element. For $Y \in L$ the map $\sigma_X : [Y, X \vee Y] \to [X \wedge Y, X]$ defined by $\sigma_X(Z) = X \wedge Z$ is an isomorphism with inverse $\tau_Y(Z) = Y \vee Z$.*

Proof. Both maps are clearly order preserving. Since X is modular, if $Z \in [X \wedge Y, X]$, then $\sigma_X \tau_Y(Z) = X \wedge (Y \vee Z) = (X \wedge Y) \vee Z = Z$. Similarly, if $Z \in [Y, X \vee Y]$, then $\tau_Y \sigma_X(Z) = Y \vee (X \wedge Z) = (Y \vee X) \wedge Z = Z$. □

Example 2.28 *For any central arrangement $\mathcal{A}$, the elements V, $T = T(\mathcal{A})$, and all atoms are modular.*

Example 2.29 *Consider $Q(\mathcal{A}) = xyz(x + y - z)$. This is the cone over the 2–arrangement of Example 1.6. In $L(\mathcal{A})$ every $H \in \mathcal{A}$ is modular, but no element of rank 2 is modular.*

Lemma 2.30 *An element $X \in L$ is modular if and only if (X, Y) is a modular pair for every $Y \in L$ such that $X \wedge Y = V$.*

Proof. Fix $X \in L$ and assume that (X, Y) is a modular pair for every $Y \in L$ such that $X \wedge Y = V$. We want to show that (X, Z) is a modular pair for every $Z \in L$. Set $a = r(Z) - r(X \wedge Z)$. There exist linearly independent hyperplanes $H_1, \ldots, H_a \in \mathcal{A}$ such that
$$Z = (X \wedge Z) \vee H_1 \vee \cdots \vee H_a.$$
Let $Y = H_1 \vee \cdots \vee H_a$. Then $r(Y) = a$ and $Z = (X \wedge Z) \vee Y$. We have
$$\begin{aligned} r(X \wedge Y) &= r(X \wedge Z \wedge Y) \leq r(X \wedge Z) + r(Y) - r((X \wedge Z) \vee Y) \\ &= r(X \wedge Z) + a - r(Z) = 0. \end{aligned}$$

Therefore $X \wedge Y = V$. By assumption, (X, Y) is a modular pair and hence $X + Y = V$. Then

$$\begin{aligned} X + Z &= X + ((X \wedge Z) \vee Y) = X + ((X \wedge Z) \cap Y) \\ &= (X \wedge Z) \cap (X + Y) = X \wedge Z. \end{aligned}$$

This shows that (X, Z) is a modular pair. □

Lemma 2.31 *If Y is a modular element in L and X is a modular element in L_Y, then X is a modular element in L.*

Proof. Let $Z \in L$. By Lemma 2.24, it is sufficient to show that $X + Z = X \wedge Z$. We have

$$\begin{aligned} X + Z &= (X + Y) + Z = X + (Y + Z) = X + (Y \wedge Z) \\ &= X \wedge (Y \wedge Z) = (X \wedge Y) \wedge Z = X \wedge Z \end{aligned}$$

as required. □

Definition 2.32 *Let $\mathcal{A}$ be an arrangement with $r(\mathcal{A}) = \ell$. We call $\mathcal{A}$ **supersolvable** if $L(\mathcal{A})$ has a maximal chain of modular elements*

$$V = X_0 < X_1 < \ldots < X_\ell = T.$$

Example 2.33 The arrangement in Example 2.29 is not supersolvable because it has no modular element of rank 2. The B_3-arrangement, the braid arrangement, and the Boolean arrangement are supersolvable. In the Boolean arrangement, every element is modular, so we may take any maximal chain. In the B_3-arrangement, the only modular elements of rank 2 are $x = y = 0$, $x = z = 0$, and $y = z = 0$; see Figure 2.3. Thus a maximal chain of modular elements is given by

$$V < \{x = 0\} < \{x = y = 0\} < \{0\}.$$

Not all elements of the braid arrangement are modular, but

$$V < \{x_1 = x_2\} < \{x_1 = x_2 = x_3\} < \ldots < \{x_1 = x_2 = \ldots = x_\ell\} = T$$

is a maximal chain of modular elements.

2.2 The Möbius Function

The Möbius Function

Definition 2.34 *Let $\mathcal{A}$ be an arrangement and let $L = L(\mathcal{A})$. Define the Möbius function $\mu_\mathcal{A} = \mu : L \times L \to \mathbb{Z}$ as follows:*

2.2 The Möbius Function   33

$$\begin{aligned}\mu(X,X) &= 1 & &\text{if } X \in L,\\ \sum_{X \leq Z \leq Y} \mu(X,Z) &= 0 & &\text{if } X,Y,Z \in L \text{ and } X < Y,\\ \mu(X,Y) &= 0 & &\text{otherwise.}\end{aligned}$$

Note that for fixed X the values of $\mu(X,Y)$ may be computed recursively. It follows that if ν is any other function which satisfies the defining properties of μ, then $\nu = \mu$. There is a useful reformulation of $\mu(X,Y)$.

Lemma 2.35 *Let $\mathcal{A}$ be an arrangement. For $X, Y \in L$ with $X \leq Y$, let $S(X,Y)$ be the set of central subarrangements $\mathcal{B} \subseteq \mathcal{A}$ such that $\mathcal{A}_X \subseteq \mathcal{B}$ and $T(\mathcal{B}) = Y$. Then*

$$\mu(X,Y) = \sum_{\mathcal{B} \in S(X,Y)} (-1)^{|\mathcal{B} \setminus \mathcal{A}_X|}.$$

Proof. Let $\nu(X,Y)$ denote the right side of the expression. Note that

$$\bigcup_{X \leq Z \leq Y} S(X,Z) = \{\mathcal{B} \subseteq \mathcal{A} \mid \mathcal{A}_X \subseteq \mathcal{B} \subseteq \mathcal{A}_Y\}$$

where the union is disjoint. Thus

$$\sum_{X \leq Z \leq Y} \nu(X,Z) = \sum_{\mathcal{A}_X \subseteq \mathcal{B} \subseteq \mathcal{A}_Y} (-1)^{|\mathcal{B} \setminus \mathcal{A}_X|} = \sum_{\mathcal{C}} (-1)^{|\mathcal{C}|}.$$

The last sum is over all subsets $\mathcal{C}$ of $\mathcal{A}_Y \setminus \mathcal{A}_X$. If $X = Y$, the sum is 1. If $X < Y$, then $\mathcal{A}_X$ is a proper subset of $\mathcal{A}_Y$, so the sum is zero. □

There is another interesting formula for $\mu(X,Y)$.

Definition 2.36 *Let $\mathcal{A}$ be an arrangement and let $L = L(\mathcal{A})$. Let $ch(L)$ be the set of all chains in L:*

$$ch(L) = \{(X_1, \ldots, X_p) \mid X_1 < \cdots < X_p\}.$$

Denote the first element of $c \in ch(L)$ by $\underline{c}$, the last element of c by $\bar{c}$, and the cardinality of c by $|c|$. Let $ch[X,Y] = \{c \in ch(L) \mid \underline{c} = X,\ \bar{c} = Y\}$ and $ch(X,Y) = \{c \in ch(L) \mid \underline{c} = X,\ \bar{c} < Y\}$.

Proposition 2.37 *For all $X, Y \in L$*

$$\mu(X,Y) = \sum_{c \in ch[X,Y]} (-1)^{|c|-1}.$$

Proof. We prove that the right side satisfies the defining properties of μ. This is clear for $X = Y$ and when X, Y are incomparable. Suppose $X < Y$. Then

$$\sum_{Z\in[X,Y]}\sum_{c\in ch[X,Z]}(-1)^{|c|-1} = \sum_{Z\in[X,Y)}\sum_{c\in ch[X,Z]}(-1)^{|c|-1} + \sum_{c\in ch[X,Y]}(-1)^{|c|-1}$$
$$= \sum_{c\in ch[X,Y)}(-1)^{|c|-1} + \sum_{c\in ch[X,Y]}(-1)^{|c|-1}$$
$$= \sum_{c\in ch[X,Y]}(-1)^{|c|-2} + \sum_{c\in ch[X,Y]}(-1)^{|c|-1} = 0,$$

because the map $(X, X_2, \ldots, X_p) \mapsto (X, X_2, \ldots, X_p, Y)$ from $ch[X, Y)$ to $ch[X, Y]$ is a bijection. $\square$

Möbius Inversion

Lemma 2.38 *Let $\mathcal{A}$ be an arrangement and let $L = L(\mathcal{A})$. Then*

$$\mu(X, X) = 1 \quad \text{if } X \in L,$$
$$\sum_{X \leq Z \leq Y} \mu(Z, Y) = 0 \quad \text{if } X, Y \in L \text{ and } X < Y.$$

Proof. Write $L = \{X_1, \ldots, X_r\}$ where the numbering is chosen so that $X_i \leq X_j$ implies $i \leq j$. Let A be the $r \times r$ matrix with (i, j) entry $\mu(X_i, X_j)$. Let B be the $r \times r$ matrix with (i, j) entry 1 if $X_i \leq X_j$ and 0 otherwise. Both A and B are upper unitriangular. It follows from the definition of μ that $\mathsf{AB} = \mathsf{I}_r$ is the identity matrix. Thus $\mathsf{BA} = \mathsf{I}_r$, which implies the assertions. $\square$

The next result is the Möbius inversion formula.

Proposition 2.39 *Let f, g be functions on $L(\mathcal{A})$ with values in an abelian group. Then*

$$g(Y) = \sum_{X \in L_Y} f(X) \iff f(Y) = \sum_{X \in L_Y} \mu(X, Y) g(X)$$

$$g(X) = \sum_{Y \in L^X} f(Y) \iff f(X) = \sum_{Y \in L^X} \mu(X, Y) g(Y).$$

Proof. Each of the four implications is based on an interchange of summation and the properties of μ given in the definition and in Lemma 2.38. We prove left-to-right implication in the first formula.

$$\sum_{Z \in L_Y} \mu(Z, Y) g(Z) = \sum_{Z \in L_Y} \mu(Z, Y) \sum_{X \in L_Z} f(X)$$
$$= \sum_{X \in L_Y} (\sum_{X \leq Z \leq Y} \mu(Z, Y)) f(X)$$
$$= f(Y)$$

as required. $\square$

The next result is due to Weisner [248].

Lemma 2.40 *Let $\mathcal{A}$ be an arrangement and let $L = L(\mathcal{A})$.*
(1) Suppose $Y \in L$ and $Y \neq V$. Then for all $Z \in L$

$$\sum_{X \vee Y = Z} \mu(V, X) = 0.$$

(2) Suppose $Y \in L$ and $T \in L$ is a maximal element such that $Y < T$. Then for all $Z \in L$

$$\sum_{X \wedge Y = Z} \mu(X, T) = 0.$$

Proof. We prove (1). The proof of (2) is similar. Note that $X \vee Y = Z$ implies $X \leq Z$, $Y \leq Z$, and $r(Z) \geq r(Y)$. We argue by induction on $r(Z)$. If $Z = Y$, then the sum to be computed is $\sum_{X \leq Y} \mu(V, X) = 0$, since $Y \neq V$. If $Z > Y$, then

$$\sum_{X \vee Y = Z} \mu(V, X) = \sum_{X \vee Y \leq Z} \mu(V, X) - \sum_{X \vee Y < Z} \mu(V, X)$$
$$= \sum_{X \leq Z} \mu(V, X) - \sum_{W < Z} \left(\sum_{X \vee Y = W} \mu(V, X) \right).$$

The first term is zero by definition. The second term is zero by induction. □

Lemma 2.41 *Let $(\mathcal{A}, V) = (\mathcal{A}_1, V_1) \times (\mathcal{A}_2, V_2)$ be the direct product of two arrangements. Let $\mu_i = \mu_{\mathcal{A}_i}$ and let $\mu = \mu_{\mathcal{A}}$. Let $X, Y \in L(\mathcal{A})$ where $X = (X_1, X_2)$ and $Y = (Y_1, Y_2)$ with $X_i, Y_i \in L(\mathcal{A}_i)$. Then*

$$\mu(X, Y) = \mu_1(X_1, Y_1)\mu_2(X_2, Y_2).$$

Proof. Define ν on $L(\mathcal{A}_1 \times \mathcal{A}_2)$ by $\nu(X, Y) = \mu_1(X_1, Y_1)\mu_2(X_2, Y_2)$. Then ν satisfies the defining conditions of μ, hence they are equal. □

The Function $\mu(X)$

Definition 2.42 *For $X \in L$ define $\mu(X) = \mu(V, X)$.*

Clearly $\mu(V) = 1$, $\mu(H) = -1$, for all $H \in L$ and if $r(X) = 2$, then $\mu(X) = |\mathcal{A}_X| - 1$. In general it is not possible to give a formula for $\mu(X)$. Recall the map $\phi : L(\mathcal{A} \times \mathcal{B}) \to L(\mathbf{c}\mathcal{A})$ of Proposition 2.17.

Proposition 2.43 *Let $\phi : L(\mathcal{A} \times \mathcal{B}) \to L(\mathbf{c}\mathcal{A})$. Let μ be the Möbius function of $L(\mathcal{A} \times \mathcal{B})$ and let μ_c be the Möbius function of $L(\mathbf{c}\mathcal{A})$. For all $Z \in L(\mathbf{c}\mathcal{A})$ we have*

$$\mu_c(Z) = \sum_{Y \in \phi^{-1}(Z)} \mu(Y).$$

Proof. It follows from Lemma 2.41 that $\mu(X \oplus \mathbb{K}) = -\mu(X \oplus \{0\})$ for $X \in L(\mathcal{A})$. If $K_0 \not\leq Z$, then there is a unique $Y \in L(\mathcal{A})$ such that $Z = \mathbf{c}Y$. Thus $\phi^{-1}(Z) = \{Y \oplus \mathbb{K}\}$ and $\mu(Y \oplus \mathbb{K}) = \mu_c(\mathbf{c}Y)$. If $K_0 \leq Z$, we argue by induction on $r(Z)$. If $r(Z) = 1$, then $Z = K_0$. Since $\phi^{-1}(K_0) = \{V \oplus \{0\}\}$, we have $\mu_c(K_0) = -1 = \mu(V \oplus \{0\})$. Now suppose $r(Z) \geq 2$.

$$\begin{aligned}
\mu_c(Z) &= -\sum_{\substack{Y \in L(\mathbf{c}\mathcal{A}) \\ Y < Z}} \mu_c(Y) \\
&= -\sum_{\substack{Y \in L(\mathbf{c}\mathcal{A}) \\ Y < Z}} \sum_{X \in \phi^{-1}(Y)} \mu(X) \\
&= -\sum_{\substack{X \in L(\mathcal{A} \times \mathcal{B}) \\ \phi(X) < Z}} \mu(X) \\
&= -\sum_{\substack{X \in L(\mathcal{A}) \\ \mathbf{c}X < Z}} \mu(X \oplus \mathbb{K}) - \sum_{\substack{X \in L(\mathcal{A}) \\ \mathbf{c}X \cap K_0 < Z}} \mu(X \oplus \{0\}) \\
&= \sum_{\substack{X \in L(\mathcal{A}) \\ \mathbf{c}X \cap K_0 \leq Z}} \mu(X \oplus \{0\}) - \sum_{\substack{X \in L(\mathcal{A}) \\ \mathbf{c}X \cap K_0 < Z}} \mu(X \oplus \{0\}) \\
&= \sum_{\substack{X \in L(\mathcal{A}) \\ \mathbf{c}X \cap K_0 = Z}} \mu(X \oplus \{0\}) \\
&= \sum_{\substack{Y \in L(\mathcal{A} \times \mathcal{B}) \\ \phi(Y) = Z}} \mu(Y).
\end{aligned}$$

The second equality is by the induction assumption. The fifth is by the fact that if $K_0 \leq Z$, then $\{X \in L(\mathcal{A}) \mid \mathbf{c}X < Z\} = \{X \in L(\mathcal{A}) \mid \mathbf{c}X \cap K_0 \leq Z\}$. $\square$

Proposition 2.44 *Define $\mathcal{A}$ by $Q(\mathcal{A}) = x_1 x_2 \cdots x_\ell$. Then for $X \in L$*

$$\mu(X) = (-1)^{r(X)}.$$

Proof. Define $\nu(X) = (-1)^{r(X)}$. It suffices to show that ν satisfies the defining properties of μ. Clearly $\nu(V) = (-1)^0 = 1$. If $X \neq V$, then $X = H_I$ for some I with $|I| = p = r(X) > 0$. If $V \leq Y \leq X$, then $Y = H_J$ where $J \subseteq I$ and

$$q = r(Y) = |J| \leq |I| = r(X) = p.$$

Thus

$$\sum_{Z \leq X} \nu(Z) = \sum_{J \subseteq I} (-1)^q = \sum_{q=0}^{p} (-1)^q \binom{p}{q} = 0$$

as required. $\square$

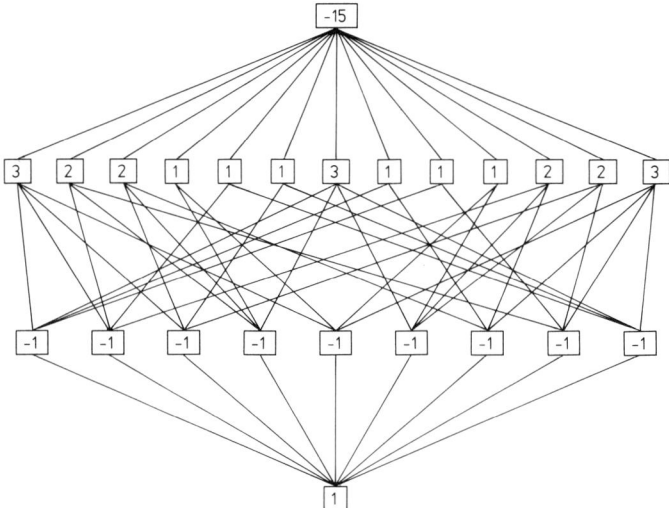

Fig. 2.7. The values of $\mu(X)$ for the B_3–arrangement

Here are the results of a hand calculation. Figure 2.7 gives the values of $\mu(X)$ for all $X \in L$ in Example 1.7.

Definition 2.45 *For a central arrangement* $\mathcal{A}$, *let* $\mu(\mathcal{A}) = \mu(T(\mathcal{A}))$.

It follows from Lemma 2.41 that when $\mathcal{A}_1$ and $\mathcal{A}_2$ are central arrangements, we have $\mu(\mathcal{A}_1 \times \mathcal{A}_2) = \mu_1(\mathcal{A}_1)\mu_2(\mathcal{A}_2)$.

Proposition 2.46 *Let* $\mathcal{A}$ *be the arrangement of* Example 1.10. *If* $X \leq Y$, *then* $\mu(X, Y) = (-1)^k q^{k(k-1)/2}$ *where* $k = \dim(X/Y)$. *In particular,* $\mu(\mathcal{A}) = (-1)^\ell q^{\ell(\ell-1)/2}$.

Proof. Since the segment $[X, Y]$ is isomorphic to the lattice of all subspaces of X/Y, it suffices to prove the second formula. We argue by induction on ℓ. If $\ell = 1$, then $L = \{V, T\}$, so $\mu(\mathcal{A}) = -1$. Suppose $\ell \geq 2$ and recall that the number of lines in V is $1+q+\cdots+q^{\ell-1}$. Apply Lemma 2.40.1 with $Z = T = \{0\}$ and $Y \in \mathcal{A}$. Then $X = T$ or $X \in R$ where R is the set of lines not contained in the hyperplane Y. Thus we have

$$0 = \mu(T) + \sum_{X \in R} \mu(X) = \mu(\mathcal{A}) + \sum_{X \in R} \mu(\mathcal{A}_X).$$

Clearly
$$|R| = (1 + q + \cdots + q^{\ell-1}) - (1 + q + \cdots + q^{\ell-2}) = q^{\ell-1}.$$
Since each $\mathcal{A}_X$ is isomorphic to the arrangement of all hyperplanes in V/X, we may assume by induction that $\mu(\mathcal{A}_X) = (-1)^{\ell-1} q^{(\ell-1)(\ell-2)/2}$. Thus

$$\mu(\mathcal{A}) = -q^{\ell-1}(-1)^{\ell-1} q^{(\ell-1)(\ell-2)/2}$$

as required. □

Theorem 2.47 *If $X \leq Y$, then $\mu(X,Y) \neq 0$ and $\operatorname{sign}\mu(X,Y) = (-1)^{r(X)-r(Y)}$.*

Proof. Since the the Möbius function of the segment $[X,Y]$ is the restriction of the Möbius function $\mu_{\mathcal{A}}$, we have $\mu(X,Y) = \mu((\mathcal{A}_Y)^X)$. Thus it suffices to show the assertion for a central arrangement $\mathcal{A}$, and there it suffices to prove that $\mu(\mathcal{A}) \neq 0$ and $\operatorname{sign}\mu(\mathcal{A}) = (-1)^{r(\mathcal{A})}$. We argue by induction on $r(\mathcal{A})$. The assertion is clear if $r(\mathcal{A}) = 0$. Suppose $r(\mathcal{A}) \geq 1$. Choose $H \in \mathcal{A}$ and apply Lemma 2.40.1 with $Y = H$ and $Z = T(\mathcal{A})$. We get

$$\begin{aligned} 0 &= \mu(\mathcal{A}) + \sum_{X \in M} \mu(X) \\ &= \mu(\mathcal{A}) + \sum_{X \in M} \mu(\mathcal{A}_X) \end{aligned}$$

where M is the set of all $X \in L$ such that $X \neq T(\mathcal{A})$ but $X \vee H = T(\mathcal{A})$. If $X \in M$, then

$$\begin{aligned} r(\mathcal{A}) &= r(X \vee H) \\ &\leq r(X \vee H) + r(X \wedge H) \\ &\leq r(X) + r(H) \\ &= r(X) + 1. \end{aligned}$$

Thus $r(X) = r(\mathcal{A}) - 1$ and therefore $r(\mathcal{A}_X) = r(\mathcal{A}) - 1$. By induction, $\mu(\mathcal{A}_X) \neq 0$ and $\operatorname{sign}\mu(\mathcal{A}_X) = (-1)^{r(\mathcal{A}_X)} = (-1)^{r(\mathcal{A})-1}$. The assertion follows from the equation $\mu(\mathcal{A}) = -\sum_{X \in M} \mu(\mathcal{A}_X)$. □

Historical Notes

Next we describe some of the history of the Möbius function. Inversion problems and considerations in number theory led to the basic definitions. The extension to partially ordered sets dates from around 1935. A revival started in 1964 with Rota's paper [195]. There are excellent surveys of the modern theory by C. Greene [95] and by R. Stanley [222].

Recall the Möbius function of elementary number theory. Let $\mathbb{N} = \{1, 2, \ldots\}$ denote the natural numbers. Define $\mu : \mathbb{N} \to \{0, \pm 1\}$ by

$$\begin{aligned} \mu(1) &= 1 \\ \mu(p_1 \ldots p_r) &= (-1)^r \quad \text{if } p_1, \ldots, p_r \text{ are distinct primes} \\ \mu(n) &= 0 \quad \text{otherwise.} \end{aligned}$$

Suppose $n > 1$ and write $n = p_1^{a_1} \ldots p_r^{a_r}$, where $p_1, \ldots, p_r$ are distinct primes. Since $r > 0$, we have

$$\sum_{d|n} \mu(d) = \sum_{k=0}^{r} \binom{r}{k} (-1)^k = 0.$$

2.2 The Möbius Function

The condition $\mu(1) = 1$ and this formula determine μ recursively.

The first known result which may be viewed as a precursor of the Möbius function appeared in the work of Euler. He started with the formula

$$\frac{\pi^2}{6} = 1 + \frac{1}{2^2} + \frac{1}{3^2} + \frac{1}{4^2} + \frac{1}{5^2} + \frac{1}{6^2} + \frac{1}{7^2} + \cdots$$

and formally inverted this infinite series to obtain

$$\frac{6}{\pi^2} = 1 - \frac{1}{2^2} - \frac{1}{3^2} - \frac{1}{5^2} + \frac{1}{6^2} - \frac{1}{7^2} + \frac{1}{10^2} \pm \cdots$$

In modern notation Euler's formula reads

$$\frac{6}{\pi^2} = \sum_{n=1}^{\infty} \frac{\mu(n)}{n^2}.$$

In 1832, Möbius [158] considered the following problem. Given a function

$$f(x) = \sum_{j=1}^{\infty} a_j x^j,$$

find coefficients b_k such that

$$x = \sum_{k=1}^{\infty} b_k f(x^k).$$

Since $f(x^k) = \sum_{j=1}^{\infty} a_j x^{jk}$, we have

$$x = \sum_{k=1}^{\infty} b_k \sum_{j=1}^{\infty} a_j x^{jk} = \sum_{n=1}^{\infty} (\sum_{jk=n} a_j b_k) x^n.$$

Thus the solution is given recursively by

$$a_1 b_1 = 1$$
$$\sum_{jk=n} a_j b_k = 0.$$

Möbius solved the problem explicitly for several functions. For

$$f(x) = \frac{x}{1-x} = x + x^2 + x^3 + \cdots$$

the $a_j = 1$ for all j. Thus $b_1 = 1$ and for $n > 1$ we have $\sum_{d|n} b_d = 0$. In modern notation he proved that $b_n = \mu(n)$. As a second example, let

$$f(x) = -\ln(1-x) = x + \frac{x^2}{2} + \frac{x^3}{3} + \cdots$$

so $a_j = 1/j$. He showed that $b_k = \mu(k)/k$. To see this, note that $b_1 = 1$ and for $n > 1$ we have

2. Combinatorics

$$\sum_{jk=n} a_j b_k = \sum_{jk=n} \frac{1}{j}\frac{\mu(k)}{k} = \frac{1}{n}\sum_{k|n}\mu(k) = 0.$$

This gives

$$x = \sum_{k=1}^{\infty}\frac{\mu(k)}{k}(-\ln(1-x^k)).$$

Formal exponentiation gives the remarkable formula

$$e^x = \prod_{k=1}^{\infty}(1-x^k)^{\frac{-\mu(k)}{k}}.$$

Work of Dedekind and Liouville is also relevant to these developments. Their problem concerned a function f defined on $\mathbb{N}$ and a second function g defined by the formula

$$g(n) = \sum_{d|n} f(d).$$

Can $f(n)$ be expressed in terms of g? Their formula

$$f(n) = g(n) - \sum g(\frac{n}{a}) + \sum g(\frac{n}{ab}) + \cdots$$

is written today as

(1) $$f(n) = \sum_{d|n}\mu(d)g(\frac{n}{d}).$$

For an application, recall Euler's function $\phi(n)$, which counts the number of integers k relatively prime to n such that $1 \leq k \leq n$. Let $f(n) = \phi(n)$. It is well known that $n = \sum_{d|n}\phi(d)$, thus $g(n) = n$. This gives the expression

$$\phi(n) = \sum_{d|n}\mu(d)\frac{n}{d}.$$

Let R be a commutative ring and let A denote the set of all functions from $\mathbb{N} \to R$. Dirichlet convolution

$$f * g(n) = \sum_{d|n} f(d)g(\frac{n}{d})$$

makes A an R–algebra. It is commutative and associative with identity function δ defined by $\delta(1) = 1$ and $\delta(n) = 0$ for $n > 1$. Let $U(R)$ denote the group of units in R and let $U(A)$ be the group of units in A. If $f(1) \in U(R)$, then $f \in U(A)$, since we can define its inverse, g, recursively by

$$f(1)g(1) = 1,$$
$$f(1)g(n) + \sum_{d>1} f(d)g(\frac{n}{d}) = 0.$$

Define $\zeta \in A$ by $\zeta(n) = 1$ for all $n \in \mathbf{N}$. Clearly $\zeta \in U(A)$. Let μ be its inverse. This is again the Möbius function. If

2.2 The Möbius Function

$$g(n) = \sum_{d|n} f(d),$$

then $g = f * \zeta$. It follows that $f = g * \mu$, which is formula (1).

In case $R = \mathbb{C}$, we may associate with $f \in A$ the **formal Dirichlet series**

$$\hat{f}(s) = \sum_{n=1}^{\infty} \frac{f(n)}{n^s}$$

where $s \in \mathbb{C}$ and there is no assumption of convergence. Let $\hat{A}$ denote the set of formal Dirichlet series. We may define addition and multiplication in $\hat{A}$ pointwise. Note that

$$\begin{aligned}
\hat{f}(s)\hat{g}(s) &= \left(\sum_{a=1}^{\infty} \frac{f(a)}{a^s}\right)\left(\sum_{b=1}^{\infty} \frac{g(b)}{b^s}\right) \\
&= \sum_{n=1}^{\infty} \left(\sum_{ab=n} f(a)g(b)\right)\frac{1}{n^s} \\
&= \widehat{(f * g)}(s).
\end{aligned}$$

Thus the map $A \to \hat{A}$ given by $f \to \hat{f}$ preserves addition and multiplication. In particular, the image of ζ is the Riemann zeta function.

$$\begin{aligned}
\hat{\zeta}(s) &= \sum_{n=1}^{\infty} \frac{\zeta(n)}{n^s} = \sum_{n=1}^{\infty} \frac{1}{n^s} \\
\hat{\mu}(s) &= \sum_{n=1}^{\infty} \frac{\mu(n)}{n^s} \\
\hat{\delta}(s) &= 1.
\end{aligned}$$

From $\zeta * \mu = \delta$ we get

$$\hat{\zeta}(s)\left(\sum_{n=1}^{\infty} \frac{\mu(n)}{n^s}\right) = 1.$$

Work of P. Hall [104, 105], L. Weisner [248], and M. Ward [247] extended these considerations to locally finite partially ordered sets. Let L be a poset and let $A(L)$ be the set of functions $f : L \times L \to R$ such that $f(x,y) = 0$ unless $x \leq y$. If we define addition pointwise and multiplication by

$$f * g(x,y) = \sum_{x \leq z \leq y} f(x,z)g(z,y),$$

then $A(L)$ forms an associative algebra, called the **incidence algebra** of L. It has an identity

$$\delta(x,y) = \begin{cases} 1 & x = y \\ 0 & otherwise. \end{cases}$$

If $f(x,x) \in U(R)$ for all $x \in L$, then f is a unit of $A(L)$ because we can define its inverse, g, recursively by

2. Combinatorics

$$f(x,x)g(x,x) = 1,$$
$$f(x,x)g(x,y) + \sum_{x<z\leq y} f(x,z)g(z,y) = 0.$$

Define
$$\zeta(x,y) = \begin{cases} 1 & x \leq y \\ 0 & otherwise. \end{cases}$$

Clearly ζ is a unit in $A(L)$. Let μ be its inverse so $\zeta * \mu = \delta = \mu * \zeta$. We call μ the **Möbius function** of L. Note that $\mu(x,x) = 1$ for all $x \in L$ and if $x < y$, then $\mu * \zeta = \delta$ implies

$$\mu * \zeta(x,y) = \sum_{x \leq z \leq y} \mu(x,z)\zeta(z,y)$$
$$= \sum_{x \leq z \leq y} \mu(x,z)$$
$$= 0.$$

In order to recover the number theoretic Möbius function from the Möbius function of a poset, recall that there is a natural partial order on the set $\mathbb{N}$ defined by $m \leq n \Leftrightarrow m$ divides n. Let $L(\mathbb{N})$ denote this poset. Its unique minimal element is 1. Let $\mu_L : L \times L \to \mathbb{Z}$ denote the Möbius function of this poset. Then $\mu(n) = \mu_L(1,n)$.

2.3 The Poincaré Polynomial

In this section we define one of the most important combinatorial invariants of an arrangement, its Poincaré polynomial, and study its properties.

Definition 2.48 *Let $\mathcal{A}$ be an arrangement with intersection poset L and Möbius function μ. Let t be an indeterminate. Define the **Poincaré polynomial** of $\mathcal{A}$ by*

$$\pi(\mathcal{A},t) = \sum_{X \in L} \mu(X)(-t)^{r(X)}.$$

It follows from Theorem 2.47 that $\pi(\mathcal{A},t)$ has nonnegative coefficients. In some cases it is easy to compute the values of μ directly in order to obtain the Poincaré polynomial.

Examples

Example 2.49 If $\mathcal{A} = \Phi_\ell$ is empty, then $\pi(\mathcal{A},t) = 1$. The Poincaré polynomial in Example 1.5 is $\pi(\mathcal{A},t) = 1+3t+2t^2 = (1+t)(1+2t)$. The Poincaré polynomial in Example 1.6 is $\pi(\mathcal{A},t) = 1+3t+3t^2$. The Poincaré polynomial in Example 1.7 is

$$\pi(\mathcal{A},t) = 1 + 9t + 23t^2 + 15t^3 = (1+t)(1+3t)(1+5t).$$

The Poincaré polynomial of the Boolean arrangement is

$$\pi(\mathcal{A},t) = \sum_{p=0}^{\ell} \binom{\ell}{p} t^p = (1+t)^{\ell}.$$

These examples may give the false impression that the Poincaré polynomial of every central arrangement is a product of linear terms $(1+bt) \in \mathbb{Z}[t]$. The reader is invited to check that in Example 2.29 $\pi(\mathcal{A},t) = 1 + 4t + 6t^2 + 3t^3 = (1+t)(1+3t+3t^2)$. Proposition 2.51 shows that $(1+t)$ divides the Poincaré polynomial of every central arrangement. More factors of the form $(1+bt) \in \mathbb{Z}[t]$ do not exist in general.

Lemma 2.50 *Let $\mathcal{A}_1$ and $\mathcal{A}_2$ be arrangements and let $\mathcal{A} = \mathcal{A}_1 \times \mathcal{A}_2$. Then*

$$\pi(\mathcal{A},t) = \pi(\mathcal{A}_1,t)\pi(\mathcal{A}_2,t).$$

Proof. Let $V = V_1 \oplus V_2$. Since $L(\mathcal{A}) = L(\mathcal{A}_1) \times L(\mathcal{A}_2)$, we have

$$\begin{aligned}
\pi(\mathcal{A},t) &= \sum_{X \in L(\mathcal{A})} \mu(X)(-t)^{r(X)} \\
&= \sum_{X_1 \oplus X_2 \in L(\mathcal{A}_1) \times L(\mathcal{A}_2)} \mu(X_1 \oplus X_2)(-t)^{r(X_1 \oplus X_2)} \\
&= \sum_{X_1 \in L(\mathcal{A}_1), X_2 \in L(\mathcal{A}_2)} \mu(X_1)\mu(X_2)(-t)^{r(X_1)}(-t)^{r(X_2)} \\
&= \pi(\mathcal{A}_1,t)\pi(\mathcal{A}_2,t)
\end{aligned}$$

as required. □

Proposition 2.51 *Let $\mathcal{A}$ be an affine arrangement with cone $\mathbf{c}\mathcal{A}$. Then*

$$\pi(\mathbf{c}\mathcal{A},t) = (1+t)\pi(\mathcal{A},t).$$

Proof. If $\mathcal{B} = (\{0\},\mathbb{K})$, then $\pi(\mathcal{B},t) = 1+t$. It follows from Lemma 2.50 that $\pi(\mathcal{A} \times \mathcal{B},t) = (1+t)\pi(\mathcal{A},t)$. Since $\phi : L(\mathcal{A} \times \mathcal{B}) \to L(\mathbf{c}\mathcal{A})$ is surjective and rank preserving, it follows from Proposition 2.43 that $\pi(\mathbf{c}\mathcal{A},t) = \pi(\mathcal{A} \times \mathcal{B},t)$. □

Definition 2.52 *Define the **characteristic polynomial** of $\mathcal{A}$ by*

$$\chi(\mathcal{A},t) = t^{\ell}\pi(\mathcal{A},-t^{-1}) = \sum_{X \in L} \mu(X) t^{\dim(X)}.$$

Note that $\chi(\mathcal{A},t)$ is a monic polynomial of degree ℓ. Our characteristic polynomial is slightly different from the usual definition of the characteristic polynomial of the lattice L. The definitions agree if $\mathcal{A}$ has rank ℓ. In some cases it is natural to compute the characteristic polynomial. Our first nontrivial computations obtain the characteristic polynomials of the arrangements of Examples 1.9 and 1.10. These computations have two interesting features: they use the combinatorial technique of proving an identity by expressing the cardinality of a set in two different ways and they use Möbius inversion to find $\chi(\mathcal{A}, t)$ without computing the individual values of $\mu(X)$.

Proposition 2.53 *Let $\mathcal{A}$ be the arrangement of Example 1.10. Then*

$$\pi(\mathcal{A}, t) = (1+t)(1+qt)\cdots(1+q^{\ell-1}t).$$

Proof. We prove the equivalent formula

$$\chi(\mathcal{A}, t) = (t-1)(t-q)\cdots(t-q^{\ell-1}).$$

Let W be a vector space of finite dimension over $\mathbf{F}_q$. Let $w = |W|$ be the cardinality of W. Then $|\hom(W, V)| = w^\ell$. We classify elements of $\hom(W, V)$ according to their images. If $X \in L$, define subsets P_X, Q_X of $\hom(W, V)$ by

$$\begin{aligned} P_X &= \{\phi \in \hom(W, V) \mid \operatorname{im}\phi = X\}, \\ Q_X &= \{\phi \in \hom(W, V) \mid \operatorname{im}\phi \subseteq X\}. \end{aligned}$$

If $\operatorname{im}\phi \subseteq X$, then $\operatorname{im}\phi = Y$ for some $Y \subseteq X$. Thus we have a disjoint union

$$Q_X = \bigcup_{Y \geq X} P_Y.$$

By Möbius inversion,

$$|P_Y| = \sum_{X \geq Y} \mu(Y, X)|Q_X|.$$

Since $Q_X = \hom(W, X)$, we have $|Q_X| = w^{\dim X}$. Taking $Y = V$ we get

$$|P_V| = \sum_{X \in L} \mu(X) w^{\dim X}.$$

If $\phi \in \hom(W, V)$, let $\phi^* \in \hom(V^*, W^*)$ be the transpose map. Since $\ker \phi^* = (\operatorname{im}\phi)^\circ$ is the annihilator of $(\operatorname{im}\phi)$, ϕ is surjective if and only if ϕ^* is injective. Thus $|P_V|$ is the number of injective maps in $\hom(V^*, W^*)$. This number is zero if $\dim W < \dim V$. Suppose $\dim W \geq \dim V$. Let $x_1, \ldots, x_\ell$ be a basis for V^*. If ϕ is injective, there are $w - 1$ possibilities for $\phi(x_1)$. Since $\phi(x_2)$ must lie outside the one-dimensional subspace spanned by $\phi(x_1)$, there are $w - q$ possibilities for $\phi(x_2)$. Similarly, there are $w - q^2$ possibilities for $\phi(x_3)$ and so on. Thus

$$|P_V| = (w-1)(w-q)\cdots(w-q^{\ell-1}).$$

Since the formulas hold for infinitely many integers w, the proposition follows by equating the two expressions for $|P_V|$. $\square$

2.3 The Poincaré Polynomial

Proposition 2.54 *Let $\mathcal{A}$ be the braid arrangement. Then*

$$\pi(\mathcal{A}, t) = (1+t)(1+2t)\cdots(1+(\ell-1)t).$$

Proof. We prove the equivalent formula

$$\chi(\mathcal{A}, t) = t(t-1)(t-2)\cdots(t-(\ell-1)).$$

Let $I = \{1, \ldots, \ell\}$. Let W be a set with cardinality $|W| = w$. Let $M = W^I$ denote the set of all maps from I into W so $|M| = w^\ell$. Each $\phi \in M$ determines an equivalence relation $\sim_\phi$ on I by $i \sim_\phi j$ if and only if $\phi(i) = \phi(j)$. Let Λ_ϕ be the corresponding partition. We classify the elements of M using the partitions Λ_ϕ. Given $X \in L(\mathcal{A})$ define subsets P_X and Q_X of M by

$$P_X = \{\phi \in M \mid \Lambda_\phi = \Lambda_X\}, \qquad Q_X = \{\phi \in M \mid \Lambda_\phi \geq \Lambda_X\}.$$

If $\Lambda_\phi \geq \Lambda_X$, then $\Lambda_\phi = \Lambda_Y$ for some $Y \geq X$ so we have a disjoint union

$$Q_X = \bigcup_{Y \geq X} P_Y.$$

Thus by Möbius inversion

$$|P_Y| = \sum_{X \geq Y} \mu(Y, X) |Q_X|.$$

Let $B(X)$ be the set of blocks of Λ_X and let $b(X) = |B(X)|$. If $\phi \in Q_X$, then ϕ is constant on the blocks of Λ_X. Thus there is a bijection from Q_X to $W^{B(X)}$. In particular, $|Q_X| = w^{b(X)}$.

Next note that $b(X) = \dim X$. We see this by choosing a basis for X consisting of vectors v^k defined by

$$v_i^k = \begin{cases} 1 & \text{if } i \in \Lambda_k \\ 0 & \text{otherwise.} \end{cases}$$

In case $Y = V$ this gives

$$|P_V| = \sum_{X \in L} \mu(X) w^{\dim X}.$$

Since Λ_V is the partition where each block is a singleton, P_V is the set of one-to-one maps from I into W. Therefore we have

$$|P_V| = w(w-1)\cdots(w-(\ell-1)).$$

Since these formulas hold for all positive integers w, we are done. □

The Deletion–Restriction Theorem

The formula for $\mu(X, Y)$ obtained in Lemma 2.35 provides a useful expression for $\pi(\mathcal{A}, t)$.

Lemma 2.55 *Let $\mathcal{A}$ be an arrangement. Then*
$$\pi(\mathcal{A}, t) = \sum_{\mathcal{B} \subseteq \mathcal{A}} (-1)^{|\mathcal{B}|}(-t)^{r(\mathcal{B})},$$
where the sum is over all central subarrangements $\mathcal{B}$ of $\mathcal{A}$.

Proof. Let $S(X) = S(V, X)$. From Lemma 2.35 we get
$$\pi(\mathcal{A}, t) = \sum_{X \in L} \mu(X)(-t)^{r(X)} = \sum_{X \in L} \Big(\sum_{\mathcal{B} \in S(X)} (-1)^{|\mathcal{B}|}(-t)^{r(X)} \Big).$$

If $\mathcal{B} \in S(X)$, then $T(\mathcal{B}) = X$, so $r(\mathcal{B}) = r(X)$. The result follows since every central subarrangement $\mathcal{B}$ of $\mathcal{A}$ occurs in a unique $S(X)$. □

We are now prepared for the main result of this section, the deletion–restriction theorem. A similar result was first proved by Brylawski [43] for central arrangements and by Zaslavsky [256] in general.

Theorem 2.56 (Deletion–Restriction) *If $(\mathcal{A}, \mathcal{A}', \mathcal{A}'')$ is a triple of arrangements, then*
$$\pi(\mathcal{A}, t) = \pi(\mathcal{A}', t) + t\pi(\mathcal{A}'', t).$$

Proof. We use the formula in Lemma 2.55. It is convenient here to let $H = H_0$ be the distinguished hyperplane. Separate the sum over $\mathcal{B} \subseteq \mathcal{A}$ into two sums: R' and R''. Here R' is the sum over those $\mathcal{B}$ which do not contain H and R'' is the sum over those $\mathcal{B}$ which contain H. It follows from Lemma 2.55 with $\mathcal{A}'$ in place of $\mathcal{A}$ that
$$R' = \pi(\mathcal{A}', t).$$
In order to compute R'' recall the definition of $S(X, Y)$ from Lemma 2.35. Since $H \in \mathcal{B}$, $\mathcal{A}_H \subseteq \mathcal{B}$. Thus if $T(\mathcal{B}) = Y$, then $\mathcal{B} \in S(H, Y)$. Let $L'' = L(\mathcal{A}'')$. Then
$$\begin{aligned}
R'' &= \sum_{H \in \mathcal{B} \subseteq \mathcal{A}} (-1)^{|\mathcal{B}|}(-t)^{r(\mathcal{B})} \\
&= \sum_{Y \in L''} \sum_{\mathcal{B} \in S(H,Y)} (-1)^{|\mathcal{B}|}(-t)^{r(Y)} \\
&= -\sum_{Y \in L''} \sum_{\mathcal{B} \in S(H,Y)} (-1)^{|\mathcal{B} \setminus \mathcal{A}_H|}(-t)^{r(Y)} \\
&= -\sum_{Y \in L''} \mu(H, Y)(-t)^{r(Y)} \\
&= t\pi(\mathcal{A}'', t).
\end{aligned}$$

The next-to-last equality follows from Lemma 2.35. The last equality follows because the Möbius function μ'' of L'' is the restriction of μ to L'' so $\mu''(Y) = \mu(H, Y)$, and the rank function r'' of L'' satisfies $r(Y) = r''(Y) + 1$. □

Corollary 2.57 *Let $(\mathcal{A}, \mathcal{A}', \mathcal{A}'')$ be a triple of arrangements. Then*

$$\chi(\mathcal{A}, t) = \chi(\mathcal{A}', t) - \chi(\mathcal{A}'', t).$$

Definition 2.58 *Let $\mathcal{A}$ be a central arrangement and let $(\mathcal{A}, \mathcal{A}', \mathcal{A}'')$ be a triple with respect to the hyperplane $H \in \mathcal{A}$. Call H a **separator** if $T(\mathcal{A}) \notin L(\mathcal{A}')$.*

Corollary 2.59 *Let $\mathcal{A}$ be a central arrangement and let $(\mathcal{A}, \mathcal{A}', \mathcal{A}'')$ be a triple with respect to $H \in \mathcal{A}$.*
  (1) If H is a separator, then $\mu(\mathcal{A}) = -\mu(\mathcal{A}'')$ and hence $|\mu(\mathcal{A})| = |\mu(\mathcal{A}'')|$.
  (2) If H is not a separator, then $\mu(\mathcal{A}) = \mu(\mathcal{A}') - \mu(\mathcal{A}'')$ and $|\mu(\mathcal{A})| = |\mu(\mathcal{A}')| + |\mu(\mathcal{A}'')|$.

Proof. It follows from Theorem 2.47 that $\pi(\mathcal{A}, t)$ has leading term

$$(-1)^{r(\mathcal{A})} \mu(\mathcal{A}) t^{r(\mathcal{A})}.$$

The conclusion follows by comparing coefficients of the leading terms on both sides of the equation in Theorem 2.56. If H is a separator, then $r(\mathcal{A}') < r(\mathcal{A})$ and there is no contribution from $\pi(\mathcal{A}', t)$. □

Definition 2.60 *Call the arrangements $\mathcal{A}$ and $\mathcal{B}$ π–**equivalent** if $\pi(\mathcal{A}, t) = \pi(\mathcal{B}, t)$.*

Example 2.61 It is clear that L–equivalent arrangements are π–equivalent. The converse is false. Consider the arrangements:

$$Q(\mathcal{A}) = xyz(x - z)(x + z)(y - z)(y + z).$$

$$Q(\mathcal{B}) = xyz(x + y + z)(x + y - z)(x - y + z)(x - y - z).$$

The reader should check that

$$\pi(\mathcal{A}, t) = (1 + t)(1 + 3t)(1 + 3t) = \pi(\mathcal{B}, t).$$

However, these arrangements are not L–equivalent. Figure 2.8 shows that $\mathcal{A}$ has two lines which are contained in four hyperplanes. These appear in the picture as the two common points on the line at infinity of the two sets of three parallel lines. Figure 2.8 also shows that $\mathcal{B}$ has no such lines. The factorization of these Poincaré polynomials is remarkable. Next we prove some general results to explain their factorization.

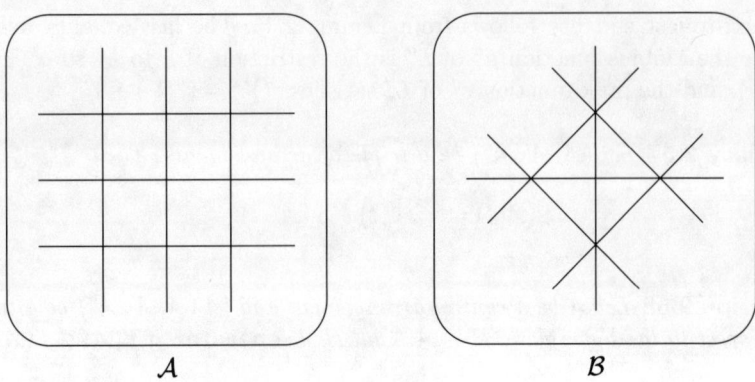

Fig. 2.8. π-equivalent but not L-equivalent

Supersolvable Arrangements

We noted in Example 2.33 that the braid arrangement is supersolvable. Our next aim is to prove a theorem due to Stanley [218]. It gives a factorization of $\pi(\mathcal{A}, t)$ for supersolvable arrangements and serves as a model for several results in this book. Recall that a supersolvable arrangement is central.

Lemma 2.62 *Let $\mathcal{A}$ be a supersolvable ℓ-arrangement with a maximal chain of modular elements*
$$V = X_0 < X_1 < \ldots < X_\ell = T.$$
Let $H \in \mathcal{A}$ be a complement of $X_{\ell-1}$ and let $(\mathcal{A}, \mathcal{A}', \mathcal{A}'')$ be the triple with respect to H. Then

(1) $\mathcal{A}_{X_{\ell-1}}$ is a supersolvable $(\ell-1)$-arrangement with a maximal chain of modular elements
$$V = X_0 < X_1 < \ldots < X_{\ell-1}.$$
The map $\tau_H : L(\mathcal{A}_{X_{\ell-1}}) \to L(\mathcal{A}'')$ defined by $\tau_H(Z) = H \vee Z$ is a lattice isomorphism. Thus $\mathcal{A}''$ is supersolvable.

(2) If H is not a separator, then
$$V = X_0 < X_1 < \ldots < X_\ell = T$$
is a maximal chain of modular elements in $\mathcal{A}'$. If H is a separator, then
$$V = X_0 < X_1 < \ldots < X_{\ell-1}$$
is a maximal chain of modular elements in $\mathcal{A}'$. Thus $\mathcal{A}'$ is supersolvable.

Proof. If X_p is modular in $L(\mathcal{A})$, then it is modular in $L(\mathcal{A}_{X_{\ell-1}})$. It follows from Lemma 2.27 that τ_H is an isomorphism. This implies (1).

To show that X_p is modular in $L(\mathcal{A}')$ note that $H \not\leq X_{\ell-1}$ and hence $H \not\leq X_p$. If we denote join and meet in $L(\mathcal{A}')$ by $\vee'$ and $\wedge'$, then for $Y \in L(\mathcal{A}')$

we have $Y \vee' X_p = Y \vee X_p$ and $Y \wedge' X_p = Y \wedge X_p$. Since $r(\mathcal{A}') = \ell$ if H is not a separator and $r(\mathcal{A}') = \ell - 1$ if H is a separator, this proves (2). □

Theorem 2.63 *Let $\mathcal{A}$ be a supersolvable ℓ-arrangement with a maximal chain of modular elements*

$$V = X_0 < X_1 < \ldots < X_\ell = T.$$

Let $b_i = |\mathcal{A}_{X_i} \setminus \mathcal{A}_{X_{i-1}}|$. Then

$$\pi(\mathcal{A}, t) = \prod_{i=1}^{\ell}(1 + b_i t).$$

Proof. We argue by induction on $|\mathcal{A}|$. The assertion is clear for $|\mathcal{A}| = 1$. For the induction step, choose $H \in \mathcal{A}$ as in Lemma 2.62. It follows from Lemma 2.62.1 that $\mathcal{A}''$ is supersolvable. Since $|\mathcal{A}''| < |\mathcal{A}|$, the induction hypothesis applies to $\mathcal{A}''$. It follows from the maximal chain of modular elements given in Lemma 2.62.1 that

$$\pi(\mathcal{A}'', t) = \prod_{i=1}^{\ell-1}(1 + b_i t).$$

It follows from Lemma 2.62.2 that $\mathcal{A}'$ is supersolvable. Since $|\mathcal{A}'| < |\mathcal{A}|$, the induction hypothesis applies to $\mathcal{A}'$. If H is not a separator, then it follows from the maximal chain of modular elements given in Lemma 2.62.2 that

$$\pi(\mathcal{A}', t) = \prod_{i=1}^{\ell-1}(1 + b_i t)(1 + (b_\ell - 1)t).$$

If H is a separator, then $\mathcal{A}' = \mathcal{A}_{X_{\ell-1}}$. It follows from the maximal chain of modular elements given in Lemma 2.62.2 that

$$\pi(\mathcal{A}', t) = \prod_{i=1}^{\ell-1}(1 + b_i t).$$

Note that in this case $b_\ell = 1$. The conclusion follows from Theorem 2.56. □

The arrangement $\mathcal{A}$ defined by

$$Q(\mathcal{A}) = xyz(x - z)(x + z)(y - z)(y + z)$$

in Example 2.61 is supersolvable. The following is a maximal chain of modular elements: $V < \{x = 0\} < \{x = z = 0\} < \{0\}$. Thus Theorem 2.63 explains the factorization $\pi(\mathcal{A}, t) = (1 + t)(1 + 3t)(1 + 3t)$. The arrangement $\mathcal{B}$ in the same example is not supersolvable. We prove a result next which explains the factorization of its Poincaré polynomial.

Nice Partitions

In Theorem 2.63 we considered the sets $\pi_i = \mathcal{A}_{X_i} \setminus \mathcal{A}_{X_{i-1}}$. These sets provide a partition $(\pi_1, \ldots, \pi_\ell)$ of $\mathcal{A}$. Next we generalize supersolvable arrangements by defining the notion of a nice partition for a central arrangement, which was introduced in [236]. We show in Proposition 2.67 that the partition which arises from a maximal chain of modular elements is nice. Moreover, we show in Corollary 3.88 that the Poincaré polynomial of an arrangement with a nice partition has a factorization similar to Theorem 2.63. Nice partition of an affine arrangement is introduced in Definition 3.91.

Definition 2.64 *A partition $\pi = (\pi_1, \ldots, \pi_s)$ of $\mathcal{A}$ is called* **independent** *if for every choice of hyperplanes $H_i \in \pi_i$ for $1 \leq i \leq s$, the resulting s hyperplanes are independent, $r(H_1 \vee \cdots \vee H_s) = s$.*

Definition 2.65 *Let $X \in L$. Let $\pi = (\pi_1, \ldots, \pi_s)$ be a partition of $\mathcal{A}$. Then the* **induced partition** *π_X is a partition of $\mathcal{A}_X$. Its blocks are the nonempty subsets $\pi_i \cap \mathcal{A}_X$.*

Definition 2.66 *A partition $\pi = (\pi_1, \ldots, \pi_s)$ of $\mathcal{A}$ is called* **nice** *if:*
 (1) *π is independent and*
 (2) *if $X \in L \setminus \{V\}$, then the induced partition π_X contains a block which is a singleton.*

Proposition 2.67 *Let $\mathcal{A}$ be a supersolvable ℓ-arrangement with a maximal chain of modular elements $V = X_0 < X_1 < \ldots < X_\ell = T$. Let $\pi_i = \mathcal{A}_{X_i} \setminus \mathcal{A}_{X_{i-1}}$. Then the partition $(\pi_1, \ldots, \pi_\ell)$ is nice.*

Proof. Choose $H_i \in \pi_i$ for each i. First we use induction on i to prove that $r(H_1 \vee \cdots \vee H_i) = i$. This is clear when $i = 1$. Let $Y = H_1 \vee \cdots \vee H_{i-1}$. Then $Y \leq X_{i-1}$. Since $H_i \not\leq X_{i-1}$, we have $H_i \not\leq Y$. Thus $H_i \wedge Y = V$. By the inductive assumption, we have $r(Y) = i - 1$. Therefore we have

$$\begin{aligned} r(H_1 \vee \cdots \vee H_i) &= r(Y \vee H_i) = r(Y) + r(H_i) - r(Y \wedge H_i) \\ &= (i-1) + 1 - r(V) = i. \end{aligned}$$

This shows that the partition $(\pi_1, \ldots, \pi_\ell)$ is independent. Next let $X \in L \setminus \{V\}$. Let j be the largest integer such that $V = X \wedge X_j$. Then

$$\begin{aligned} 0 < r(X \wedge X_{j+1}) &= r(X) + r(X_{j+1}) - r(X \vee X_{j+1}) \\ &\leq r(X) + r(X_{j+1}) - r(X \vee X_j) \\ &= r(X) + r(X_{j+1}) - (r(X) + r(X_j) - r(X \wedge X_j)) \\ &= 1. \end{aligned}$$

This implies that $X \wedge X_{j+1}$ is a hyperplane belonging to $\mathcal{A}$. Thus $\mathcal{A}_X \cap \pi_{j+1} = \{X \wedge X_{j+1}\}$ is a singleton. $\square$

We will prove in Corollary 3.88 that if $\mathcal{A}$ has a nice partition $\pi = (\pi_1, \ldots, \pi_s)$ and $b_i = |\pi_i|$, then

$$\pi(\mathcal{A}, t) = \prod_{i=1}^{s}(1 + b_i t).$$

The arrangement $\mathcal{B}$ defined by

$$Q(\mathcal{B}) = xyz(x+y+z)(x+y-z)(x-y+z)(x-y-z)$$

in Example 2.61 has a nice partition $\pi = (\pi_1, \pi_2, \pi_3)$. We may take $\pi_1 = \{\ker(z)\}$, $\pi_2 = \{\ker(x), \ker(x-y+z), \ker(x-y-z)\}$, $\pi_3 = \{\ker(y), \ker(x+y-z), \ker(x+y+z)\}$. Thus Corollary 3.88 explains the factorization $\pi(\mathcal{B}, t) = (1+t)(1+3t)(1+3t)$.

Counting Functions

The Poincaré polynomial of an arrangement will appear repeatedly in this book. It will be shown to equal the Poincaré polynomial of several graded algebras which we are going to associate with $\mathcal{A}$. It is also the Poincaré polynomial of the complement $M(\mathcal{A})$ of a complex arrangement. Here we prove that the Poincaré polynomial is also a counting function. First suppose that $\mathcal{A}$ is a real arrangement. Recall from the introduction that the complement $M(\mathcal{A})$ is a disjoint union of chambers

$$M(\mathcal{A}) = \bigcup_{C \in \mathcal{C}(\mathcal{A})} C.$$

Zaslavsky [256] showed that the number of chambers is determined by the Poincaré polynomial as follows.

Theorem 2.68 (Zaslavsky) *Let $\mathcal{A}$ be a real arrangement. Then*

$$|\mathcal{C}(\mathcal{A})| = \pi(\mathcal{A}, 1).$$

Proof. If $\mathcal{A}$ is empty, then $|\mathcal{C}(\mathcal{A})| = 1 = \pi(\mathcal{A}, 1)$. We showed in the introduction that $|\mathcal{C}(\mathcal{A})| = |\mathcal{C}(\mathcal{A}')| + |\mathcal{C}(\mathcal{A}'')|$. Thus the result is a consequence of Theorem 2.56. □

Next assume that $\mathcal{A}$ is an ℓ-arrangement over the finite field $\mathbb{F}_q$. Then the complement $M(\mathcal{A})$ is a finite set of points. Its cardinality is also determined by the Poincaré polynomial.

Theorem 2.69 *Let $\mathcal{A}$ be an ℓ-arrangement over $\mathbb{F}_q$. Let $|M(\mathcal{A})|$ denote the cardinality of the complement. Then*

$$|M(\mathcal{A})| = q^\ell \pi(\mathcal{A}, -q^{-1}) = \chi(\mathcal{A}, q).$$

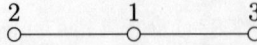

Fig. 2.9. A graph with three vertices

Proof. If $\mathcal{A}$ is empty, then $|M(\mathcal{A})| = q^\ell = \chi(\mathcal{A}, q)$. Suppose $\mathcal{A}$ is nonempty and let $(\mathcal{A}, \mathcal{A}', \mathcal{A}'')$ be a triple. Evidently

$$|M(\mathcal{A})| = |M(\mathcal{A}')| - |M(\mathcal{A}'')|.$$

Thus the functions $|M(\mathcal{A})|$ and $\chi(\mathcal{A}, q)$ agree on Φ_ℓ and by Corollary 2.57 they satisfy the same recursion for deletion and restriction. It follows that they are equal. □

2.4 Graphic Arrangements

In this section we study certain special central arrangements obtained from finite simple nonoriented graphs. They are called **graphic arrangements**. Let G be a finite simple nonoriented graph and let $\mathcal{A}(G)$ be the corresponding graphic arrangement. The correspondence $G \mapsto \mathcal{A}(G)$ gives a map from the set of finite simple nonoriented graphs into the set of arrangements. This map may be used to "pull back" results concerning arrangements to results concerning graphs. For example, Zaslavsky's chamber counting theorem 2.68 for arrangements can be translated into Stanley's negative one color theorem 2.94, which determines the number of acyclic orientations for graphs. The characteristic polynomial of the arrangement $\mathcal{A}(G)$ corresponds to the chromatic polynomial of the graph G.

Definitions

Definition 2.70 *A finite simple nonoriented* **graph** $G = (\mathcal{V}, \mathcal{E})$ *is an ordered pair consisting of the set $\mathcal{V}$ of vertices and the set $\mathcal{E}$ of edges. They satisfy the following two conditions:*
  (1) $\mathcal{V}$ *is a finite set,*
  (2) $\mathcal{E}$ *is a collection of 2–element subsets of $\mathcal{V}$.*
When it is convenient, we let $\mathcal{V} = \{1, 2, \ldots, \ell\}$.

Example 2.71 If $\mathcal{V} = \{1, 2, 3\}$ and $\mathcal{E} = \{\{1, 2\}, \{1, 3\}\}$, then $G = (\mathcal{V}, \mathcal{E})$ is a graph. This graph is shown in Figure 2.9.

Example 2.72 When $\mathcal{E}$ is the set of *all* 2–element subsets of $\mathcal{V}$, the graph $G = (\mathcal{V}, \mathcal{E})$ is called a **complete graph**. Figure 2.10 shows the complete graph for $\mathcal{V} = \{1, 2, 3, 4\}$.

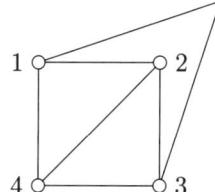

Fig. 2.10. A complete graph

Definition 2.73 *Let $\mathbb{K}$ be a field and let $V = \mathbb{K}^\ell$. Let $x_1, \ldots x_\ell$ be a basis for the dual space V^*. Given the graph $G = (\mathcal{V}, \mathcal{E})$, define an arrangement $\mathcal{A}(G)$ by*

$$\mathcal{A}(G) = \{\ker(x_i - x_j) \mid \{i,j\} \in \mathcal{E}\}.$$

*The arrangement $\mathcal{A}(G)$ is called a **graphic arrangement**.*

Example 2.74 If G is the graph in Figure 2.9, then a defining polynomial Q of the graphic arrangement is given by $Q = (x_1 - x_2)(x_1 - x_3)$. The arrangement consists of two planes in $\mathbb{K}^3$.

Example 2.75 Let $\mathbb{K} = \mathbb{R}$. For the complete graph with ℓ vertices, $\mathcal{A}(G)$ has a defining polynomial

$$Q(\mathcal{A}(G)) = \prod_{1 \leq i < j \leq \ell} (x_i - x_j).$$

Thus $\mathcal{A}(G)$ is equal to the braid arrangement $\mathcal{A}_\mathbb{R}$ of Definition 1.9.

It follows that an arrangement is graphic if and only if it is a subarrangement of the braid arrangement.

Definition 2.76 *Let C be a finite set of cardinality n. Let $G = (\mathcal{V}, \mathcal{E})$ be a graph. A **coloring** of G by C is a map $\varphi : \mathcal{V} \to C$ such that $\varphi(i) \neq \varphi(j)$ whenever $\{i,j\} \in \mathcal{E}$.*

Example 2.77 If G is the graph in Figure 2.9 and $C = \{a,b\}$, then the map $\varphi : \{1,2,3\} \to \{a,b\}$ defined by $\varphi(1) = a$, $\varphi(2) = b$, $\varphi(3) = b$ is a coloring. The map $\phi : \{1,2,3\} \to \{a,b\}$ defined by $\phi(1) = a$, $\phi(2) = a$, $\phi(3) = b$ is not a coloring.

The concept of coloring originates from the "map coloring problem." Visualize a map of Canada, the United States, and Mexico. We assign a vertex to each country and connect two vertices only when the corresponding countries are adjacent. This map yields the graph in Figure 2.9. Here 1 = United States, 2 = Canada, 3 = Mexico. In this way we can assign a (finite simple nonoriented) graph to any map of countries. A graph coloring corresponds to a way of coloring the map so that two adjacent countries are colored differently.

2. Combinatorics

Definition 2.78 *Let $G = (\mathcal{V}, \mathcal{E})$ be a graph. The* **chromatic function** *$\chi(G, t)$ is a function defined on the set of nonnegative integers by*

$$\chi(G, t) = \text{the number of colorings of } G \text{ with } t \text{ colors.}$$

The famous "four color theorem" is translated into the assertion that $\chi(G, 4) > 0$ for any planar graph G, or any graph G obtained from an arbitrary map on the plane. The chromatic function $\chi(G, t)$ was introduced by G. Birkhoff in his study [26] of the four color problem.

Example 2.79 If a graph G has ℓ vertices and no edges, then there is no restriction for its coloring. Thus we have

$$\chi(G, t) = t^\ell.$$

Example 2.80 If G is the graph in Figure 2.9, then there are t ways to color the vertex 1 and there are $t - 1$ ways to color the vertices 2 and 3 each. Thus we have

$$\chi(G, t) = t(t-1)^2.$$

Example 2.81 If G is the complete graph with ℓ vertices, then there are t ways to color the first vertex, $t - 1$ ways to color the second vertex, $t - 2$ ways to color the third vertex, and so on. Thus we have

$$\chi(G, t) = t(t-1)(t-2)\cdots(t-\ell+1).$$

It is easy to find the chromatic function directly in the examples above. It is not so easy to find the chromatic function of the graph consisting of the vertices and edges of a square. To compute its chromatic function we use a deletion–contraction technique, closely related to the method of deletion and restriction of arrangements studied in Section 2.3. We also prove that the chromatic function is a monic polynomial of degree ℓ.

Deletion–Contraction

Definition 2.82 *Let $G = (\mathcal{V}, \mathcal{E})$, $\mathcal{V} = \{1, 2, \ldots, \ell\}$, and $\mathcal{E} \neq \emptyset$. Fix an edge $e_0 = \{i, j\} \in \mathcal{E}$. The* **deletion** *$G' = (\mathcal{V}', \mathcal{E}')$ of G with respect to e_0 is defined by*

$$\mathcal{V}' = \mathcal{V}, \qquad \mathcal{E}' = \mathcal{E} \setminus \{e_0\}.$$

The **contraction** *G'' of G with respect to e_0 is the graph $G'' = (\mathcal{V}'', \mathcal{E}'')$. Here $\mathcal{V}''$ is the vertex set with cardinality $\ell - 1$ obtained by identifying i and j in $\mathcal{V}$. Write $\mathcal{V}'' = \{\bar{1}, \bar{2}, \ldots, \bar{\ell}\}$ where $\bar{p} = \bar{q}$ if and only if either $p = q$ or $\{p, q\} = \{i, j\}$. Define $\mathcal{E}''$ by*

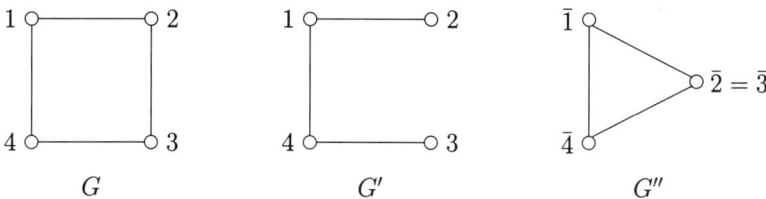

Fig. 2.11. Deletion and contraction

$$\mathcal{E}'' = \{\{\bar{p},\bar{q}\} \mid \{p,q\} \in \mathcal{E}'\}.$$

Example 2.83 Let G be the graph in Figure 2.11 and let $e_0 = \{2,3\}$. Figure 2.11 shows G together with the deletion G' and the contraction G'' with respect to e_0.

Proposition 2.84 Let $G = (\mathcal{V}, \mathcal{E})$ be a graph with $\mathcal{E} \neq \emptyset$. Let G' and G'' be the deletion and contraction with respect to the edge e_0. Then

$$\chi(G',t) = \chi(G,t) + \chi(G'',t).$$

Proof. Suppose $e_0 = \{1,2\}$. Every coloring of G induces a coloring of G'. Thus there is an injection from the set of all colorings of G to the set of all colorings of G'. The complement of the image is exactly equal to the set of colorings φ of G' such that $\varphi(1) = \varphi(2)$. This set is in one-to-one correspondence with the set of all colorings of G''. □

Example 2.85 This allows us to find the chromatic function of the graph in Figure 2.11. Let $e_0 = \{2,3\}$. The deletion G' and the contraction G'' are given in Figure 2.11. Proposition 2.84 gives

$$\begin{aligned} \chi(G,t) &= \chi(G',t) - \chi(G'',t) \\ &= t(t-1)^3 - t(t-1)(t-2) = t(t-1)(t^2 - 3t + 3). \end{aligned}$$

Corollary 2.86 Let G be a graph with ℓ vertices. Then the chromatic function $\chi(G,t)$ is a monic polynomial in t of degree ℓ.

Proof. We argue by induction on the number of edges. When there are no edges, we have $\chi(G,t) = t^\ell$. Suppose that G has at least one edge. Fix an edge e_0. Consider the deletion G' and the contraction G''. By the induction assumption, $\chi(G',t)$ is a monic polynomial of degree ℓ and $\chi(G'',t)$ is a polynomial of degree less than ℓ. The result follows from Proposition 2.84. □

This shows that the chromatic function $\chi(G,t)$ is a polynomial. From now on we will call it the **chromatic polynomial**. As we have just seen, the construction of the deletion G' and the contraction G'' is very useful. Next we consider the corresponding graphic arrangements $\mathcal{A}(G')$ and $\mathcal{A}(G'')$.

Proposition 2.87 *Let G be a graph with edge e_0. Let G' and G'' be the deletion and contraction of G with respect to e_0. Let $H_0 \in \mathcal{A}(G)$ be the hyperplane corresponding to e_0. Write $\mathcal{A} = \mathcal{A}(G)$. Let $\mathcal{A}'$ and $\mathcal{A}''$ denote the deletion and restriction of $\mathcal{A}$ with respect to H_0. Then $\mathcal{A}(G') = \mathcal{A}'$ and $\mathcal{A}(G'') = \mathcal{A}''$.*

Proof. Assume $e_0 = \{1, 2\}$. Then $H_0 = \ker(x_1 - x_2)$. Clearly
$$\mathcal{A}(G') = \mathcal{A}(G) \setminus \{H_0\} = \mathcal{A}'.$$
Denote the set of vertices of G'' by $\{\bar{1} = \bar{2}, \bar{3}, \ldots, \bar{\ell}\}$. Write the corresponding basis for $\mathbb{K}^{\ell-1}$ as $\bar{x}_1 = \bar{x}_2, \bar{x}_3, \ldots, \bar{x}_\ell$. It is naturally identified with a basis for the dual space H_0^* of H_0. We have
$$\begin{aligned}\mathcal{A}(G'') &= \{\ker(\bar{x}_i - \bar{x}_j) \mid \{\bar{i}, \bar{j}\} \in \mathcal{E}''\} = \{\ker(\bar{x}_i - \bar{x}_j) \mid \{i, j\} \in \mathcal{E}'\} \\ &= \{\ker(\bar{x}_i - \bar{x}_j) \mid \ker(x_i - x_j) \in \mathcal{A}'\} = \mathcal{A}''\end{aligned}$$
as required. $\square$

Recall the characteristic polynomial of an arrangement from Definition 2.52:
$$\chi(\mathcal{A}, t) = t^\ell \pi(\mathcal{A}, -t^{-1}) = \sum_{X \in L} \mu(X) t^{\dim(X)}.$$

The Greek words for characteristic and chromatic begin with χ. This may explain why the corresponding polynomials are called χ. It is a pleasant coincidence that these polynomials are equal.

Theorem 2.88 *Let G be a graph and let $\mathcal{A}(G)$ be the corresponding graphic arrangement. Then*
$$\chi(G, t) = \chi(\mathcal{A}(G), t).$$

Proof. We argue by induction on the number of edges in G. Equality holds when G has no edges: $\chi(G, t) = t^\ell = \chi(\mathcal{A}(G), t)$. The induction is completed by applying Proposition 2.84, Theorem 2.88, Proposition 2.87, and Corollary 2.57 to give
$$\begin{aligned}\chi(G, t) &= \chi(G', t) - \chi(G'', t) \\ &= \chi(\mathcal{A}(G'), t) - \chi(\mathcal{A}(G''), t) \\ &= \chi(\mathcal{A}', t) - \chi(\mathcal{A}'', t) \\ &= \chi(\mathcal{A}, t)\end{aligned}$$
as required. $\square$

2.4 Graphic Arrangements

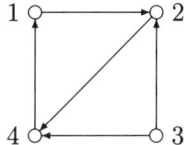

Fig. 2.12. An oriented graph

Theorem 2.68 on chamber counting gives

Corollary 2.89 *Let $\mathbb{K} = \mathbb{R}$. The number of chambers of the graphic arrangement $\mathcal{A}(G)$ is $(-1)^\ell \chi(G, -1)$.*

Proof. By Theorem 2.68 and Theorem 2.88 we have

$$|\mathcal{C}(\mathcal{A}(G))| = \pi(\mathcal{A}(G), 1) = (-1)^\ell \chi(\mathcal{A}(G), -1) = (-1)^\ell \chi(G, -1)$$

as required. □

Acyclic Orientations

Let $\mathcal{A}$ be an arrangement in $\mathbb{R}^\ell$. Denote the set of chambers of $\mathcal{A}$ by $\mathcal{C}(\mathcal{A})$ as before. Corollary 2.89 asserts that the value $\chi(G, -1)$ of the chromatic polynomial of the graph G has an interpretation as $(-1)^\ell |\mathcal{C}(\mathcal{A}(G))|$. If we give a graph theoretic meaning to the number $|\mathcal{C}(\mathcal{A}(G))|$ of chambers of the arrangement $\mathcal{A}(G)$, we obtain a result in graph theory called the negative one color theorem. It was first proved by R. Stanley in [219]. The original proof is not directly related to arrangements. The first argument using graphic arrangements is due to C. Greene [94, 96].

Definition 2.90 *Let G be a graph. An **orientation** of G is an assignment of a direction to each edge $\{i, j\}$, denoted by $i \to j$ or $j \to i$. An orientation is called **acyclic** if it has no directed cycles.*

Example 2.91 The orientation in Figure 2.12 is not acyclic because it contains the directed cycle $1 \to 2 \to 4 \to 1$.

Example 2.92 The complete graph on three vertices has $8 = 2^3$ orientations. Six of these orientations are acyclic. The two shown in Figure 2.13 are not.

Lemma 2.93 *Let G be a graph. Denote the set of all acyclic orientations by $AO(G)$. Let $\mathcal{A} = \mathcal{A}(G)$. There exists a bijection from $AO(G)$ to $\mathcal{C}(\mathcal{A})$.*

Proof. Let $\omega \in AO(G)$. For $i \in \mathcal{V}$ let $p_i(\omega)$ denote the number of vertices which can be reached along the directions of the orientation from the vertex i.

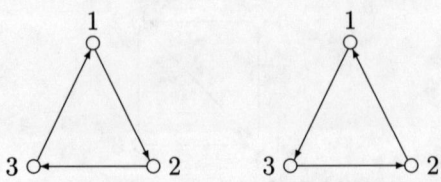

Fig. 2.13. Not acyclic orientations

Consider a point $p(\omega) = (p_1(\omega), \ldots, p_\ell(\omega)) \in \mathbb{R}^\ell$. Let $\{i,j\} \in \mathcal{E}$. Define half spaces
$$H_{ij}^+ = \{(x_1, \ldots, x_\ell) \in \mathbb{R}^\ell \mid x_i > x_j\},$$
$$H_{ij}^- = \{(x_1, \ldots, x_\ell) \in \mathbb{R}^\ell \mid x_i < x_j\}.$$

If $i \to j$, then every vertex which can be reached from j can also be reached from i. Thus $p_i(\omega) \geq p_j(\omega)$. Since the orientation is acyclic, it is impossible to reach i from j. Thus we have

$$i \to j \text{ in } \omega \quad \Leftrightarrow \quad p_i(\omega) > p_j(\omega) \quad \Leftrightarrow \quad p(\omega) \in H_{ij}^+,$$
$$j \to i \text{ in } \omega \quad \Leftrightarrow \quad p_j(\omega) > p_i(\omega) \quad \Leftrightarrow \quad p(\omega) \in H_{ij}^-.$$

Therefore $p(\omega) \notin \ker(x_i - x_j)$. Thus $p(\omega) \in M(\mathcal{A}) = \mathbb{R}^\ell \setminus \bigcup_{H \in \mathcal{A}} H$ and there exists a unique chamber $C(\omega) \in \mathcal{C}(\mathcal{A})$ which contains $p(\omega)$. We have

$$C(\omega) \subseteq H_{ij}^+ \quad \Leftrightarrow \quad i \to j \text{ in } \omega$$
$$C(\omega) \subseteq H_{ij}^- \quad \Leftrightarrow \quad j \to i \text{ in } \omega.$$

We show next that the correspondence $\omega \mapsto C(\omega)$ gives a bijection from $AO(G)$ to $\mathcal{C}(\mathcal{A})$. It is obvious from the last equations that this map is injective. In order to prove surjectivity let $C \in \mathcal{C}(\mathcal{A})$. Choose a point $p = (p_1, \ldots, p_\ell) \in C$. Define a direction on each edge $\{i,j\}$ by

$$i \to j \quad \Leftrightarrow \quad p_i > p_j \quad \Leftrightarrow \quad p \in H_{ij}^+$$
$$j \to i \quad \Leftrightarrow \quad p_i < p_j \quad \Leftrightarrow \quad p \in H_{ij}^-$$

These formulas show that

$$C(\omega) \subseteq H_{ij}^+ \quad \Leftrightarrow \quad p \in H_{ij}^+$$
$$C(\omega) \subseteq H_{ij}^- \quad \Leftrightarrow \quad p \in H_{ij}^-.$$

It follows that $p \in C(\omega)$ and thus $C = C(\omega)$. □

By combining Lemma 2.93 with Corollary 2.89, we obtain the following theorem due to R. P. Stanley [219]:

Theorem 2.94 *The number of acyclic orientations of G is $(-1)^\ell \chi(G, -1)$.* □

3. Algebras

Let $\mathcal{K}$ be a commutative ring. We construct certain algebras over $\mathcal{K}$ associated with $\mathcal{A}$. We construct the graded algebra $A(\mathcal{A})$ for a central arrangement $\mathcal{A}$ in Section 3.1. This construction is generalized to affine arrangements in Section 3.2. The algebra $A(\mathcal{A})$ is the quotient of the exterior algebra $E(\mathcal{A})$ based on $\mathcal{A}$ by a homogeneous ideal $I(\mathcal{A})$, $A(\mathcal{A}) = E(\mathcal{A})/I(\mathcal{A})$. This algebra is constructed using only $L(\mathcal{A})$. It will reappear in Chapter 5 with a topological significance. We prove that the $\mathcal{K}$-algebra $A(\mathcal{A})$ is a free graded $\mathcal{K}$-module and that its Poincaré polynomial is equal to $\pi(\mathcal{A}, t)$. This gives an interpretation of the coefficients of $\pi(\mathcal{A}, t)$. We construct a $\mathcal{K}$-basis for $A(\mathcal{A})$ as a free graded $\mathcal{K}$-module using broken circuits. We also show that given a triple $(\mathcal{A}, \mathcal{A}', \mathcal{A}'')$, there is an exact sequence of $\mathcal{K}$-modules

$$0 \to A(\mathcal{A}') \to A(\mathcal{A}) \to A(\mathcal{A}'') \to 0.$$

We prove some algebra factorization theorems in Section 3.3. If $\mathbf{c}\mathcal{A}$ is the cone over $\mathcal{A}$, then

$$A(\mathbf{c}\mathcal{A}) \simeq (\mathcal{K} + \mathcal{K}a_0) \otimes A(\mathcal{A}).$$

Recall from Theorem 2.63 that if the central arrangement $\mathcal{A}$ is supersolvable, then

(1) $$\pi(\mathcal{A}, t) = (1 + b_1 t) \cdots (1 + b_\ell t),$$

where the b_i are nonnegative integers. In Section 3.3 we prove that the algebra $A(\mathcal{A})$ has a tensor product decomposition

(2) $$(\mathcal{K} + B_1) \otimes \cdots \otimes (\mathcal{K} + B_\ell)$$

as graded $\mathcal{K}$-module with $b_i = \mathrm{rank} B_i$. It follows that (1) is a consequence of (2). This decomposition of $A(\mathcal{A})$ is generalized to arrangements with a nice partition. In particular, the Poincaré polynomial of an arrangement with a nice partition has a factorization like (1). In Section 3.4 we define another graded algebra, $B(\mathcal{A})$, whose multiplication is a shuffle product. We prove that $B(\mathcal{A})$ is algebra isomorphic to $A(\mathcal{A})$. In Section 3.5 we assume that $\mathcal{K}$ is a subring of $\mathbb{K}$. We associate to the arrangement $\mathcal{A}$ the $\mathcal{K}$-algebra $R(\mathcal{A})$ generated by the differential forms $\omega_H = d\alpha_H/\alpha_H$. (Note that this algebra is not a purely combinatorial object, since the defining polynomials α_H enter the definition.)

The main result of Section 3.5 is that there exists an isomorphism of algebras $A(\mathcal{A}) \simeq R(\mathcal{A})$, which shows that $R(\mathcal{A})$ depends only on $L(\mathcal{A})$. The argument uses the fact that there is a short exact sequence of $\mathcal{K}$–modules

$$0 \to R(\mathcal{A}') \to R(\mathcal{A}) \to R(\mathcal{A}'') \to 0.$$

3.1 $A(\mathcal{A})$ for Central Arrangements

In this section we assume that $\mathcal{A}$ is a central arrangement. We associate to the arrangement $\mathcal{A}$ a graded anticommutative algebra $A(\mathcal{A})$ over $\mathcal{K}$. In the literature, this algebra is sometimes called the **Orlik–Solomon algebra**. The algebra $A(\mathcal{A})$ was first defined in [171], where it was used to prove that for a complex arrangement, $A(\mathcal{A})$ is isomorphic as a graded algebra to the cohomology algebra of the complement $M(\mathcal{A})$. We show this in Section 5.4. The algebra $A(\mathcal{A})$ has since been used by several authors in work on hypergeometric functions.

Construction of $A(\mathcal{A})$

Definition 3.1 *Let $\mathcal{A}$ be an arrangement over $\mathbb{K}$. Let $\mathcal{K}$ be a commutative ring. Let*

$$E_1 = \bigoplus_{H \in \mathcal{A}} \mathcal{K} e_H$$

and let

$$E = E(\mathcal{A}) = \Lambda(E_1)$$

be the exterior algebra of E_1.

Note that E_1 has a $\mathcal{K}$–basis consisting of elements e_H in one-to-one correspondence with the hyperplanes of $\mathcal{A}$. Write $uv = u \wedge v$ and note that $e_H^2 = 0$, $e_H e_K = -e_K e_H$ for $H, K \in \mathcal{A}$. The algebra E is graded. If $|\mathcal{A}| = n$, then

$$E = \bigoplus_{p=0}^{n} E_p,$$

where $E_0 = \mathcal{K}$, E_1 agrees with its earlier definition and E_p is spanned over $\mathcal{K}$ by all $e_{H_1} \cdots e_{H_p}$ with $H_k \in \mathcal{A}$.

Definition 3.2 *Define a $\mathcal{K}$–linear map $\partial_E = \partial : E \to E$ by $\partial 1 = 0$, $\partial e_H = 1$ and for $p \geq 2$*

$$\partial(e_{H_1} \cdots e_{H_p}) = \sum_{k=1}^{p} (-1)^{k-1} e_{H_1} \cdots \widehat{e_{H_k}} \cdots e_{H_p}$$

for all $H_1, \ldots, H_p \in \mathcal{A}$.

3.1 $A(\mathcal{A})$ for Central Arrangements

Definition 3.3 *Given a p-tuple of hyperplanes, $S = (H_1, \ldots, H_p)$, write $|S| = p$,*
$$e_S = e_{H_1} \cdots e_{H_p} \in E, \qquad \cap S = H_1 \cap \cdots \cap H_p.$$
Since $\mathcal{A}$ is central, $\cap S \in L$ for all S.

If $p = 0$, we agree that $S = (\ )$ is the empty tuple, $e_S = 1$, and $\cap S = V$. Since the rank function on L is codimension, it is clear that $r(\cap S) \le |S|$.

Definition 3.4 *Call S **independent** if $r(\cap S) = |S|$ and **dependent** if $r(\cap S) < |S|$.*

The terminology has geometric significance. The tuple S is independent if the corresponding linear forms $\alpha_1, \ldots, \alpha_p$ are linearly independent. Equivalently, the hyperplanes of S are in general position. Let $\mathbf{S}_p$ denote the set of all p-tuples $(H_1, \ldots, H_p)$ and let $\mathbf{S} = \cup_{p \ge 0} \mathbf{S}_p$.

Definition 3.5 *Let $\mathcal{A}$ be an arrangement. Let $I = I(\mathcal{A})$ be the ideal of E generated by ∂e_S for all dependent $S \in \mathbf{S}$.*

Since I is generated by homogeneous elements, it is a graded ideal. Let $I_p = I \cap E_p$. Then
$$I = \bigoplus_{p=0}^{n} I_p.$$

Definition 3.6 *Let $\mathcal{A}$ be an arrangement. Let $A = A(\mathcal{A}) = E/I$. Let $\varphi : E \to A$ be the natural homomorphism and let $A_p = \varphi(E_p)$. If $H \in \mathcal{A}$, let $a_H = \varphi(e_H)$ and if $S \in \mathbf{S}$, let $a_S = \varphi(e_S)$.*

Lemma 3.7 *If $S \in \mathbf{S}$ and $H \in S$, then $e_S = e_H \partial e_S$.*

Proof. If $H \in S$, then $e_H e_S = 0$. Thus $0 = \partial(e_H e_S) = e_S - e_H \partial e_S$. □

Since both E and I are graded, A is a graded anticommutative algebra. Since the elements of $\mathbf{S}_1$ are independent, we have $I_0 = 0$ and hence $A_0 = \mathcal{K}$. The only dependent elements of $\mathbf{S}_2$ are of the form $S = (H, H)$. Since $e_S = e_H^2 = 0$, we have $I_1 = 0$. Thus the elements a_H are linearly independent over $\mathcal{K}$ and $A_1 = \bigoplus_{H \in \mathcal{A}} \mathcal{K} a_H$. If $p > \ell$, then every element of $\mathbf{S}_p$ is dependent and it follows from Lemma 3.7 that $A_p = 0$. Thus
$$A = \bigoplus_{p=0}^{\ell} A_p.$$

Example 3.8 Suppose $\ell = 2$ and $\mathcal{A} = \{H_1, \ldots, H_n\}$. Write $a_k = a_{H_k}$. Then
$$A(\mathcal{A}) = \mathcal{K} \oplus \bigoplus_{p=1}^{n} \mathcal{K} a_p \oplus \bigoplus_{k=1}^{n-1} \mathcal{K} a_k a_n.$$

We have computed A_0, A_1 and we know that $A_p = 0$ for $p > 2$. It remains to compute A_2. Since $\dim V = 2$, (H_i, H_j, H_k) is dependent for all (i,j,k). Thus I_2 contains the element

$$\partial(e_i e_j e_k) = e_j e_k - e_i e_k + e_i e_j = e_i e_j + e_j e_k + e_k e_i.$$

It follows that A_2 is spanned by $a_p a_q$ subject to the relations

$$a_i a_j + a_j a_k + a_k a_i = 0$$

for all (i,j,k). This shows that A_2 is spanned by $a_k a_n$ for $1 \leq k < n$. It remains to show that the sum is direct. Suppose $\sum_{k=1}^{n-1} c_k a_k a_n = 0$ with $c_k \in \mathcal{K}$. Then $\sum_{k=1}^{n-1} c_k e_k e_n \in I_2$. Recall that I_2 is spanned by the elements $\partial(e_i e_j e_k)$. Since $\partial\partial = 0$, we have $\partial I_2 = 0$ and hence

$$\partial\left(\sum_{k=1}^{n-1} c_k e_k e_n\right) = \sum_{k=1}^{n-1} c_k (e_n - e_k) = 0.$$

Since $e_1, \ldots, e_n$ are linearly independent over $\mathcal{K}$, we get that $c_k = 0$ for all k.

Example 3.9 If $\mathcal{A}$ is the Boolean arrangement, then $S = (H_1, \ldots, H_p)$ is independent if and only if $H_1, \ldots, H_p$ are distinct hyperplanes. Hence if S is dependent, then $e_S = 0$. Thus $I = 0$ and $A = E$.

An Acyclic Complex

It is convenient to introduce some more notation. If $S = (H_1, \ldots, H_p)$, we say that $H_i \in S$. If T is a subsequence of S, we write $T \subseteq S$. If $T = (K_1, \ldots, K_q)$, we write $(S,T) = (H_1, \ldots, H_p, K_1, \ldots, K_q)$. Thus $e_{(S,T)} = e_S e_T$ and in particular for $H \in \mathcal{A}$ we have $e_{(H,S)} = e_H e_S$.

Lemma 3.10 *The map $\partial : E \to E$ satisfies*
 (1) $\partial^2 = 0$,
 (2) *if $u \in E_p$ and $v \in E$, then $\partial(uv) = (\partial u)v + (-1)^p u(\partial v)$.*

Proof. Part (1) is the standard boundary formula. It suffices to check (2) for $u = e_S$ and $v = e_T$ for $S, T \in \mathbf{S}$, where it follows by direct computation. □

Note that this lemma has nothing to do with arrangements. It states two familiar properties of the exterior algebra. Since the map ∂ is homogeneous of degree -1, we see from (1) that (E, ∂) is a chain complex. Part (2) says that ∂ is a derivation of the exterior algebra. It may be characterized as the unique derivation of E with $\partial e_H = 1$.

Lemma 3.11 $\partial_E I \subseteq I$.

Proof. Recall that I is a $\mathcal{K}$–linear combination of elements of the form $e_T \partial e_S$ where $T, S \in \mathbf{S}$ and S is dependent. We have

$$\partial(e_T \partial e_S) = (\partial e_T)(\partial e_S) \pm e_T(\partial^2 e_S) = (\partial e_T)(\partial e_S) \in I$$

as required. □

Definition 3.12 *Since $\partial_E I \subseteq I$, we may define $\partial_A : A \to A$ by $\partial_A \varphi u = \varphi \partial_E u$ for $u \in E$.*

Lemma 3.13 *The map $\partial_A : A \to A$ satisfies*
 (1) $\partial_A^2 = 0$,
 (2) if $a \in A_p$ and $b \in A$, then $\partial_A(ab) = (\partial_A a)b + (-1)^p a(\partial_A b)$,
 (3) if $\mathcal{A}$ is not empty, then the chain complex (A, ∂_A) is acyclic.

Proof. Parts (1) and (2) follow from the corresponding facts for ∂_E. Since ∂_A is homogeneous of degree -1, (A, ∂_A) is a chain complex. It follows from (1) that $\text{im}\,\partial_A \subseteq \ker \partial_A$. To prove that the complex is acyclic we must show the reverse inclusion. Since $\mathcal{A}$ is not empty, we may choose $H \in \mathcal{A}$. Let $v = e_H$. Then $\partial_E v = 1$. Let $b = \varphi v$ and let $a \in A$. Choose $u \in E$ with $\varphi u = a$. Then $\partial_E(vu) = (\partial_E v)u - v(\partial_E u) = u - v(\partial_E u)$. Since $\varphi \partial_E = \partial_A \varphi$ and φ is a $\mathcal{K}$–algebra homomorphism, applying φ to the first and last terms gives $a = \partial_A(ba) + b\partial_A a$ for all $a \in A$. Thus $\text{im}\,\partial_A \supseteq \ker \partial_A$. □

Next we study the ideal I and return to the notation $\partial = \partial_E$.

Definition 3.14 *Let $J = J(\mathcal{A})$ be the submodule of E spanned over $\mathcal{K}$ by all e_S such that $S \in \mathbf{S}$ is dependent.*

Lemma 3.15 *J is an ideal of E and $I = J + \partial J$.*

Proof. If $T \in \mathbf{S}$ is dependent, then (S, T) is dependent for all $S \in \mathbf{S}$. Thus $e_S e_T = e_{(S,T)} \in J$ and hence J is an ideal. The formula $e_S = e_H \partial e_S$ when $H \in S$ applied to a dependent S shows that $J \subseteq I$. The definitions of J and I imply that $\partial J \subseteq I$. Thus $J + \partial J \subseteq I$. For the reverse inclusion note that $J + \partial J$ contains the generators of I. It suffices to show that $J + \partial J$ is an ideal. Since J is an ideal, it is enough to show that $e_H \partial e_S \in J + \partial J$ when $H \in \mathcal{A}$ and $S \in \mathbf{S}$ is dependent. Since (H, S) is also dependent, this follows from the formula

$$e_H \partial e_S = e_S - \partial(e_H e_S) = e_S - \partial e_{(H,S)}.$$

This completes the proof. □

The Structure of $A(\mathcal{A})$

We decompose the algebra E into a direct sum indexed by elements of L.

Definition 3.16 *For $X \in L$ let $\mathbf{S}_X = \{S \in \mathbf{S} \mid \cap S = X\}$ and let*
$$E_X = \sum_{S \in \mathbf{S}_X} \mathcal{K}e_S.$$

Note that $e_S \in E_{\cap S}$ for all $S \in \mathbf{S}$.

Lemma 3.17 *Since $\mathbf{S} = \bigcup_{X \in L} \mathbf{S}_X$ is a disjoint union, $E = \bigoplus_{X \in L} E_X$ is a direct sum.*

Our next aim is to show that the algebra $\mathcal{A}$ has an analogous direct sum decomposition.

Definition 3.18 *Let $\pi_X : E \to E_X$ be the projection. Thus*
$$\pi_X e_S = \begin{cases} e_S & \text{if } \cap S = X \\ 0 & \text{otherwise.} \end{cases}$$

The next result follows from Lemma 3.17.

Lemma 3.19 *If F is a submodule of E, write $F_X = F \cap E_X$. If $\pi_X(F) \subseteq F$ for all $X \in L$, then $\pi_X(F) = F_X$ and $F = \bigoplus_{X \in L} F_X$.*

Lemma 3.20 $J = \bigoplus_{X \in L} J_X$.

Proof. Since J is spanned by elements e_S where $S \in \mathbf{S}$ is dependent, it follows from the definition of π_X that $\pi_X(J) \subseteq J$. The result follows from Lemma 3.19. □

Definition 3.21 *Let $J' = J'(\mathcal{A})$ be the submodule of E spanned by all e_S where $S \in \mathbf{S}$ is independent. Thus $E = J \oplus J'$. Let $\pi = \pi_\mathcal{A} : E \to J'$ be the projection which annihilates J. Let $K = K(\mathcal{A}) = \pi(\partial J)$.*

Lemma 3.22 $I = J \oplus K$.

Proof. The map $1 - \pi : E \to J$ is the projection which annihilates J'. Since $J \subseteq I$, we have $(1 - \pi)I = J$. It follows from Lemma 3.15 that $\pi(I) = \pi(J + \partial J) = \pi(\partial J) = K$. From $(1 - \pi)I = J \subseteq I$ we get $\pi I \subseteq I$ and hence $I = (1 - \pi)I \oplus \pi I = J \oplus K$. □

Lemma 3.23 $K = \bigoplus_{X \in L} K_X$.

Proof. By Lemma 3.19, it suffices to show that $\pi_X K \subseteq K$ for all $X \in L$. By Lemma 3.20, we have $K = \pi(\partial J) = \sum_{Y \in L} \pi(\partial J_Y)$. The module ∂J_Y is spanned by elements ∂e_S where S is dependent and $\cap S = Y$. Let $S = (H_1, \ldots, H_p)$

and let $S_k = (H_1, \ldots, \hat{H}_k, \ldots, H_p)$. Then $\partial e_S = \sum_{k=1}^{p}(-1)^{k-1} e_{S_k}$. If S_k is dependent, then $\pi e_{S_k} = 0$. If S_k is independent, then $\cap S_k = \cap S = Y$ because S is dependent. This gives $\pi(\partial e_S) \in E_Y$, so $\pi(\partial J_Y) \subseteq E_Y$. For $Y \neq X$ we have $\pi_X(E_Y) = 0$. Thus $\pi_X(K) = \pi(\partial J_X) \subseteq \pi(\partial J) = K$. □

Proposition 3.24 $I = \bigoplus_{X \in L} I_X$.

Proof. Recall that $I = J \oplus K$. We showed $\pi_X(J) \subseteq J$ in Lemma 3.20 and $\pi_X(K) \subseteq K$ in Lemma 3.23. Thus $\pi_X I \subseteq I$. The conclusion follows from Lemma 3.19. □

Definition 3.25 *If $X \in L$, let $A_X = \varphi(E_X)$.*

Theorem 3.26 *Let $\mathcal{A}$ be a central arrangement and let $A = A(\mathcal{A})$. Then*
$$A = \bigoplus_{X \in L} A_X.$$

Proof. Since $E = \oplus E_X$, $A = \sum A_X$. Proposition 3.24 shows that the sum is direct. □

Corollary 3.27 *Let $\mathcal{A}$ be a central arrangement. Then*
$$A_p = \bigoplus_{X \in L_p} A_X.$$

Proof. Suppose $a \in A_X$ where $X \in L_p$. Write $a = \varphi(u)$ where $u \in E_X$. Write $u = \sum_{S \in \mathbf{S}_X} c_S e_S$ where $c_S \in K$. If $S \in \mathbf{S}_X$ is dependent, then $e_S \in I$ and $\varphi(e_S) = 0$. If S is independent, then $r(\cap S) = r(X) = p$ implies $e_S \in E_p$ and $\varphi(e_S) \in A_p$. Thus $a = \varphi(u) \in A_p$ and hence
$$\sum_{X \in L_p} A_X \subseteq A_p.$$

Conversely, suppose $a \in A_p$ and write $a = \varphi(u)$ where $u \in E_p$. Write $u = \sum_{S \in \mathbf{S}_p} c_S e_S$ where $c_S \in K$. If $S \in \mathbf{S}_p$ is dependent, then $\varphi(e_S) = 0$. If $S \in \mathbf{S}_p$ is independent, let $X = \cap S$. Then $r(X) = p$ and $e_S \in E_X$ implies $\varphi(e_S) \in A_X$. Thus
$$A_p \subseteq \sum_{X \in L_p} A_X.$$

The sum is direct by Theorem 3.26. □

The Injective Map $A(\mathcal{A}_X) \to A(\mathcal{A})$

If $\mathcal{B}$ is a subarrangement of $\mathcal{A}$, then we view $E(\mathcal{B})$ as a subalgebra of $E(\mathcal{A})$ and $L(\mathcal{B})$ as a sublattice of $L(\mathcal{A})$. Note that $\mathbf{S}(\mathcal{B}) \subseteq \mathbf{S}(\mathcal{A})$ and an element $S \in \mathbf{S}(\mathcal{B})$

is dependent viewed in $\mathbf{S}(\mathcal{B})$ if and only if it is dependent in $\mathbf{S}(\mathcal{A})$. The map $\partial_{E(\mathcal{B})}$ is the restriction of $\partial_{E(\mathcal{A})}$ to $E(\mathcal{B})$. Since $J(\mathcal{B}) \subseteq J(\mathcal{A})$ and $J'(\mathcal{B}) \subseteq J'(\mathcal{A})$, the projection $\pi_\mathcal{B}$ of $E(\mathcal{B})$ onto $J'(\mathcal{B})$ is the restriction to $E(\mathcal{B})$ of the projection $\pi_\mathcal{A}$ of $E(\mathcal{A})$ onto $J'(\mathcal{A})$. For simplicity we let ∂ denote both $\partial_{E(\mathcal{A})}$ and $\partial_{E(\mathcal{B})}$ and we let π denote both $\pi_\mathcal{A}$ and $\pi_\mathcal{B}$. Thus $K(\mathcal{B}) = \pi(\partial J(\mathcal{B})) \subseteq \pi(\partial J(\mathcal{A})) = K(\mathcal{A})$. It is convenient to agree that undefined modules are zero. It follows that $K_X(\mathcal{B}) \subseteq K_X(\mathcal{A})$ for all $X \in L(\mathcal{A})$, since by this convention $K_X(\mathcal{B}) = 0$ if $X \notin L(\mathcal{B})$.

Clearly, $I(\mathcal{B}) \subseteq I(\mathcal{A}) \cap E(\mathcal{B})$ for any subarrangement $\mathcal{B}$ of $\mathcal{A}$. In [171, Lemma 2.14] it was asserted that

(1) $$I(\mathcal{B}) = I(\mathcal{A}) \cap E(\mathcal{B})$$

for any subarrangement $\mathcal{B}$ of $\mathcal{A}$. The proof given there is correct if $\mathcal{B} = \mathcal{A}_X$ for $X \in L(\mathcal{A})$ and this is the only case used in the rest of [171]. The mistake in [171, Lemma 2.14] lies in the claim that $E_X(\mathcal{A}) \subseteq E(\mathcal{B})$ if $X \in L(\mathcal{B})$. This was pointed out by George Glauberman, W. A. M. Janssen, and Nguyễn Viêt Dũng. Although this claim is false, equation (1) holds for all subarrangements $\mathcal{B}$ of $\mathcal{A}$. We will prove (1) in Proposition 3.66. Here we establish it in case $\mathcal{B} = \mathcal{A}_X$.

Lemma 3.28 *If $X \in L(\mathcal{A})$, then $I(\mathcal{A}_X) = I(\mathcal{A}) \cap E(\mathcal{A}_X)$.*

Proof. The inclusion $\subseteq$ is obvious. Suppose $Y \in L(\mathcal{A}_X)$. If $S \in \mathbf{S}_Y(\mathcal{A})$, then $S = (H_1, \ldots, H_p)$ with $\cap S = Y$. Thus $Y \subseteq H_k$ for $1 \leq k \leq p$. Since $Y \in L(\mathcal{A}_X)$, we have $X \subseteq Y$ and hence $X \subseteq H_k$ for $1 \leq k \leq p$. Thus $S \in \mathbf{S}_Y(\mathcal{A}_X)$. This gives $\mathbf{S}_Y(\mathcal{A}) \subseteq \mathbf{S}_Y(\mathcal{A}_X)$. Since $\mathcal{A}_X \subseteq \mathcal{A}$, we have $\mathbf{S}_Y(\mathcal{A}) = \mathbf{S}_Y(\mathcal{A}_X)$. Thus $E_Y(\mathcal{A}) = E_Y(\mathcal{A}_X)$ and $J_Y(\mathcal{A}) = J_Y(\mathcal{A}_X)$. It follows from the argument of Lemma 3.23 that $K_Y(\mathcal{A}) = \pi(\partial J_Y(\mathcal{A})) = \pi(\partial J_Y(\mathcal{A}_X)) = K_Y(\mathcal{A}_X)$. Since $K_Y(\mathcal{A}_X) \subseteq E(\mathcal{A}_X)$, this shows

$$I(\mathcal{A}) \cap E(\mathcal{A}_X) \subseteq J(\mathcal{A}_X) \oplus \Big(\bigoplus_{Y \in L(\mathcal{A}_X)} K_Y(\mathcal{A}_X) \Big).$$

If we apply Lemmas 3.22 and 3.23 to $\mathcal{A}_X$, we see that the right side is $I(\mathcal{A}_X)$. □

Definition 3.29 *Let $\mathcal{B}$ be a subarrangement of $\mathcal{A}$. Since $I(\mathcal{B}) \subseteq I(\mathcal{A}) \cap E(\mathcal{B})$, the inclusion $E(\mathcal{B}) \subseteq E(\mathcal{A})$ induces a $\mathcal{K}$-algebra homomorphism $i : A(\mathcal{B}) \to A(\mathcal{A})$ such that for $H \in \mathcal{A}$*

$$i(e_H + I(\mathcal{B})) = e_H + I(\mathcal{A}).$$

Note that i is a monomorphism precisely when (1) holds.

The next result follows from Lemma 3.28.

Proposition 3.30 *The map i is a monomorphism for $\mathcal{B} = \mathcal{A}_X$.*

Proposition 3.31 *Let $\mathcal{A}$ be a central arrangement. If $Y \leq X$, then $A_Y(\mathcal{A}_X) \simeq A_Y(\mathcal{A})$.*

Proof. Let $i : A(\mathcal{A}_X) \to A(\mathcal{A})$ be the homomorphism of Definition 3.29. It is a monomorphism by Lemma 3.28. The module $A_Y(\mathcal{A}) = \varphi(E_Y(\mathcal{A}))$ is spanned over $\mathcal{K}$ by all elements $e_S + I(\mathcal{A})$ with $S \in \mathbf{S}_Y(\mathcal{A})$. Similarly $A_Y(\mathcal{A}_X)$ is spanned over $\mathcal{K}$ by all elements $e_S + I(\mathcal{A}_X)$ with $S \in \mathbf{S}_Y(\mathcal{A}_X)$. Since $\mathbf{S}_Y(\mathcal{A}) = \mathbf{S}_Y(\mathcal{A}_X)$, we have $i(A_Y(\mathcal{A}_X)) = A_Y(\mathcal{A})$. Since i is a monomorphism, this completes the proof. □

The Broken Circuit Basis

Next we show that the $\mathcal{K}$-algebra $A(\mathcal{A})$ is a free $\mathcal{K}$-module by constructing a standard $\mathcal{K}$-basis for $A(\mathcal{A})$. These results can be extended to geometric lattices. For the more general results see [44] and [121].

We introduce an arbitrary linear order $\prec$ in $\mathcal{A}$. Call a p–tuple $S = (H_1, \ldots, H_p)$ **standard** if $H_1 \prec \ldots \prec H_p$. Note that $E = E(\mathcal{A})$ has a $\mathcal{K}$–basis consisting of all e_S with standard S.

Definition 3.32 *A p–tuple $S = (H_1, \ldots, H_p)$ is a **circuit** if it is minimally dependent. Thus $(H_1, \ldots, H_p)$ is dependent, but for $1 \leq k \leq p$ the $(p-1)$-tuple $(H_1, \ldots, \hat{H}_k, \ldots, H_p)$ is independent.*

Definition 3.33 *Given $S = (H_1, \ldots, H_p)$, let $\max S$ be the maximal element of S in the linear order $\prec$ in $\mathcal{A}$.*

Definition 3.34 *A standard p–tuple $S \in \mathbf{S}$ is a **broken circuit** if there exists $H \in \mathcal{A}$ such that $\max S \prec H$ and (S, H) is a circuit.*

Definition 3.35 *A standard p–tuple S is called χ-**independent** if it does not contain any broken circuit. Define*

$$\mathcal{C}_p = \{S \in \mathbf{S}_p \mid S \text{ is standard and } \chi\text{-independent}\}.$$

Let $\mathcal{C} = \cup_{p \geq 0} \mathcal{C}_p$.

Definition 3.36 *The **broken circuit module** $C = C(\mathcal{A})$ is defined as follows. Let $C_0 = \mathcal{K}$, and for $p \geq 1$ let C_p be the free $\mathcal{K}$–module with basis $\{e_S \in E \mid S \in \mathcal{C}_p\}$. Let $C = C(\mathcal{A}) = \oplus_{p \geq 0} C_p$. Then $C(\mathcal{A})$ is a free graded $\mathcal{K}$-module.*

It is clear that every broken circuit is obtained by deleting the maximal element in a standard circuit. Note that if S is χ–independent, then S is independent. Thus every $S \in \mathcal{C}$ is independent. By definition $C(\mathcal{A})$ is a submodule of $E(\mathcal{A})$. In general $C(\mathcal{A})$ is not closed under multiplication in $E(\mathcal{A})$, so $C(\mathcal{A})$ is not a subalgebra. Recall the natural projection $\varphi : E(\mathcal{A}) \to A(\mathcal{A})$ and let

$\psi : C(\mathcal{A}) \to A(\mathcal{A})$ be its restriction. Our aim is to show that ψ is an isomorphism of graded modules.

Example 3.37 Define $\mathcal{A}$ by $Q(\mathcal{A}) = xyz(x+y)(x+y-z)$. Let $H_0 = \ker(x+y-z)$, $H_1 = \ker(x)$, $H_2 = \ker(y)$, $H_3 = \ker(z)$, and $H_4 = \ker(x+y)$. Define the linear order on $\mathcal{A}$ by $H_i \prec H_j \iff i < j$. Here (H_0, H_1, H_2, H_3), (H_0, H_3, H_4), and (H_1, H_2, H_4) are the standard circuits. Thus the broken circuits are (H_0, H_1, H_2), (H_0, H_3), and (H_1, H_2). Writing $e_i = e_{H_i}$ we get the following basis for $C(\mathcal{A})$:

$$1$$
$$e_0, \ e_1, \ e_2, \ e_3, \ e_4$$
$$e_0 e_1, \ e_0 e_2, \ e_0 e_4, \ e_1 e_3, \ e_1 e_4, \ e_2 e_3, \ e_2 e_4, \ e_3 e_4$$
$$e_0 e_1 e_4, \ e_0 e_2 e_4, \ e_1 e_3 e_4, \ e_2 e_3 e_4$$

Definition 3.38 Recall that for $S = (H_1, \ldots, H_p)$ we write $\cap S = H_1 \cap \cdots \cap H_p$ and that $E_X = \sum_{\cap S = X} \mathcal{K} e_S$. Let $C_X(\mathcal{A}) = C_X = C \cap E_X$. Then each C_X is a free $\mathcal{K}$-module.

Lemma 3.39 For $p \geq 0$ we have $C_p = \oplus_{X \in L_p} C_X$ and hence $C = \oplus_{X \in L} C_X$.

Proof. If $S \in \mathbf{S}$ is χ-independent with $|S| = p$, then S is independent. If $Y = \cap S$, then $r(Y) = p$ and $e_S \in \oplus_{X \in L_p} C_X$. Conversely, if $e_S \in C_X$ and $r(X) = p$, then $|S| = p$, so $e_S \in C_p$. $\square$

Lemma 3.40 If $Y \leq X$, then $C_Y(\mathcal{A}_X) = C_Y(\mathcal{A})$.

Proof. Let $S \in \mathbf{S}(\mathcal{A}_X) \subseteq \mathbf{S}(\mathcal{A})$. It suffices to show that S is a broken circuit of $\mathcal{A}_X$ if and only if S is a broken circuit of $\mathcal{A}$. First observe that S is dependent in $\mathcal{A}_X$ if and only if S is dependent in $\mathcal{A}$. Thus S is a circuit of $\mathcal{A}_X$ if and only if S is a circuit of $\mathcal{A}$. Suppose S is a broken circuit of $\mathcal{A}_X$. Then S is obtained by removing the maximal element of a standard circuit of $\mathcal{A}_X$. Since the latter is also a standard circuit of $\mathcal{A}$, S is a broken circuit of $\mathcal{A}$. Conversely, suppose S is a broken circuit of $\mathcal{A}$. Then there exists $H \in \mathcal{A}$ such that $\max S \prec H$ and (S, H) is a circuit. Since S is independent and (S, H) is dependent, we have $\cap S = \cap S \cap H$. Thus $X \geq \cap S \geq H$ and $(S, H) \in \mathbf{S}_X$. It follows that (S, H) is a circuit of $\mathcal{A}_X$ and S is a broken circuit of $\mathcal{A}_X$. $\square$

Lemma 3.41 Let H_n be the maximal element of $\mathcal{A}$ under $\prec$ and write $e_n = e_{H_n}$. Then $e_n C \subseteq C$, so C is closed under multiplication by e_n.

Proof. Since a broken circuit is obtained from a standard circuit by deleting the maximal element, no broken circuit has the form (S, H_n). $\square$

Lemma 3.42 Suppose $\mathcal{A}$ is not empty. Let ∂_C denote the restriction of the map $\partial : E \to E$ to C. Then $\partial_C(C) \subseteq C$ and (C, ∂_C) is an acyclic complex.

Proof. Deleting an element of a χ-independent p-tuple results in a χ-independent $(p-1)$-tuple. This shows that $\partial_C(C) \subseteq C$. Suppose $c \in C$ and $\partial_C c = 0$. By Lemma 3.41, $e_n c \in C$ and

$$c = c - e_n(\partial_C c) = \partial_C(e_n c) \in \partial_C C.$$

This shows that the complex is acyclic. □

Theorem 3.43 *For each $X \in L$, the restriction $\psi_X : C_X(\mathcal{A}) \to A_X(\mathcal{A})$ is an isomorphism. The map $\psi : C(\mathcal{A}) \to A(\mathcal{A})$ is an isomorphism of graded $\mathcal{K}$-modules. The set*

$$\{e_S + I \in A(\mathcal{A}) \mid S \text{ is standard and } \chi\text{-independent}\}$$

is a basis for $A(\mathcal{A})$ as a graded $\mathcal{K}$-module.

Proof. Clearly, $\psi(C_X) \subseteq A_X$, so ψ induces a map $\psi_X : C_X \to A_X$. It suffices to show that this map is an isomorphism for all $X \in L(\mathcal{A})$. We use induction on $r = r(\mathcal{A})$. The assertion holds for the empty arrangement with $r = 0$ and $C(\mathcal{A}) = \mathcal{K} = A(\mathcal{A})$. Suppose $r > 0$. Let $X \in L(\mathcal{A})$ with $r(X) < r$. Then $r(\mathcal{A}_X) < r$, so by the induction hypothesis $\psi_X : C_X(\mathcal{A}_X) \to A_X(\mathcal{A}_X)$ is an isomorphism. We see from Proposition 3.31 that $A_X(\mathcal{A}_X) \simeq A_X(\mathcal{A})$ and from Lemma 3.40 that $C_X(\mathcal{A}_X) = C_X(\mathcal{A})$. It follows from the commutativity of the diagram

$$\begin{array}{ccc} C_X(\mathcal{A}_X) & \xrightarrow{\psi_X(\mathcal{A}_X)} & A_X(\mathcal{A}_X) \\ \downarrow & & \downarrow \\ C_X(\mathcal{A}) & \xrightarrow{\psi_X(\mathcal{A})} & A_X(\mathcal{A}) \end{array}$$

that $\psi_X(\mathcal{A})$ is an isomorphism for $X \in L$ with $r(X) < r$. Since $\mathcal{A}$ is central, it has a unique maximal element $T = T(\mathcal{A})$ of rank r. It remains to prove the isomorphism for $X = T$. In the commutative diagram

$$\begin{array}{ccccccccc} 0 & \to & C_r & \to & C_{r-1} & \to & \cdots & \to & C_0 & \to & 0 \\ & & \psi_r \downarrow & & \psi_{r-1} \downarrow & & & & \psi_0 \downarrow & & \\ 0 & \to & A_r & \to & A_{r-1} & \to & \cdots & \to & A_0 & \to & 0 \end{array}$$

the horizontal maps are the respective boundary operators in the two acyclic complexes, so the sequences are exact. Since $A_p = \bigoplus_{X \in L_p} A_X$ and $C_p = \bigoplus_{X \in L_p} C_X$, the first part of the argument shows that all vertical maps except ψ_r are isomorphisms. It follows from the diagram that ψ_r is an isomorphism. This completes the argument because $C_r = C_T(\mathcal{A})$, $A_r = A_T(\mathcal{A})$, and $\psi_r = \psi_T(\mathcal{A})$. The remaining assertions follow from Theorem 3.25 and Lemma 3.39. □

Corollary 3.44 *The algebra $A(\mathcal{A})$ is a free graded $\mathcal{K}$-module. The $\mathcal{K}$-modules $A_X(\mathcal{A})$ for $X \in L$ and $A_p(\mathcal{A})$ for $p \geq 0$ are also free.*

Proof. The $\mathcal{K}$-modules $C_X(\mathcal{A})$ are free by definition. It follows from Theorem 3.43 that $C_X(\mathcal{A}) \simeq A_X(\mathcal{A})$. Thus $A_X(\mathcal{A})$ is a free $\mathcal{K}$-module. We showed in Corollary 3.27 that $A_p = \bigoplus_{X \in L_p} A_X$. Thus A_p is also free. □

3.2 $A(\mathcal{A})$ for Affine Arrangements

In this section we generalize the constructions and results of the last section to affine arrangements. Our main tool is the interplay between the affine ℓ-arrangement $\mathcal{A}$ and the central $(\ell+1)$-arrangement $\mathbf{c}\mathcal{A}$. We recall the basic notation and properties of the coning construction from Definition 1.15 and Proposition 2.17. Let $Q(\mathcal{A}) \in S = \mathbb{K}[x_1, \ldots, x_\ell]$ be a defining polynomial of $\mathcal{A}$ and let $Q' \in \mathbb{K}[x_0, x_1, \ldots, x_\ell]$ be the polynomial $Q(\mathcal{A})$ homogenized. Then $\mathbf{c}\mathcal{A}$ is a central $(\ell+1)$-arrangement with defining polynomial $Q(\mathbf{c}\mathcal{A}) = x_0 Q'$. The cone $\mathbf{c}\mathcal{A}$ consists of the hyperplane $K_0 = \ker(x_0)$ together with $\{\mathbf{c}H \mid H \in \mathcal{A}\}$ where $\mathbf{c}H$ is the cone over the affine hyperplane H. If $H \in \mathcal{A}$ is the kernel of the degree 1 polynomial $\alpha_H \in \mathbb{K}[x_1, \ldots, x_\ell]$, then $\mathbf{c}H \in \mathbf{c}\mathcal{A}$ is the kernel of the linear form $\alpha_{\mathbf{c}H}$ obtained by homogenizing α_H in $\mathbb{K}[x_0, x_1, \ldots, x_\ell]$. For example, if $\alpha_H = x_1 + x_2 - 1$, then $\alpha_{\mathbf{c}H} = x_1 + x_2 - x_0$.

Construction of $A(\mathcal{A})$

Let $\mathcal{K}$ be a commutative ring. The first definitions are the same as in the central case. Define a $\mathcal{K}$-module $E_1(\mathcal{A})$ which has a $\mathcal{K}$-basis consisting of elements e_H in one-to-one correspondence with the hyperplanes of $\mathcal{A}$. Let

$$E(\mathcal{A}) = \Lambda(E_1(\mathcal{A}))$$

be the exterior algebra of E_1. Let $\mathbf{S}_p(\mathcal{A})$ denote the set of all p-tuples $(H_1, \ldots, H_p)$ of hyperplanes in $\mathcal{A}$. Define $\mathbf{S}(\mathcal{A}) = \cup_{p \geq 0} \mathbf{S}_p(\mathcal{A})$. For $S = (H_1, \ldots, H_p) \in \mathbf{S}(\mathcal{A})$, define $e_S = e_{H_1} \cdots e_{H_p} \in E(\mathcal{A})$. For $S \in \mathbf{S}(\mathcal{A})$, define the p-tuple $\mathbf{c}S \in \mathbf{S}(\mathbf{c}\mathcal{A})$ of hyperplanes in $\mathbf{c}\mathcal{A}$ by $\mathbf{c}S = (\mathbf{c}H_1, \ldots, \mathbf{c}H_p)$. Write $e_0 = e_{K_0} \in E(\mathbf{c}\mathcal{A})$. Then $E(\mathbf{c}\mathcal{A})$ has a $\mathcal{K}$-basis

$$\{e_0 e_{\mathbf{c}S} \mid S \in \mathbf{S}(\mathcal{A})\} \cup \{e_{\mathbf{c}S} \mid S \in \mathbf{S}(\mathcal{A})\}.$$

Given $S = (H_1, \ldots, H_p) \in \mathbf{S}(\mathcal{A})$, recall that $\cap S = H_1 \cap \cdots \cap H_p$. The crucial difference between central and affine arrangements is that here $\cap S$ may be empty. Since K_0 is sent to infinity in the deconing, $\cap S = \emptyset$ if and only if $\cap(\mathbf{c}S) \subseteq K_0$.

Definition 3.45 *Let $\mathcal{A}$ be an affine arrangement. We say that S is* **dependent** *if $\cap S \neq \emptyset$ and $r(\cap S) = \text{codim}(\cap S) < |S|$. Let $I(\mathcal{A})$ be the ideal of $E(\mathcal{A})$ generated by*

$$\{e_S \mid \cap S = \emptyset\} \cup \{\partial e_S \mid S \text{ is dependent}\}.$$

Define the algebra $A(\mathcal{A})$ by

$$A(\mathcal{A}) = E(\mathcal{A})/I(\mathcal{A}).$$

Example 3.46 Recall the affine 2-arrangement $\mathcal{A}$ defined by $Q(\mathcal{A}) = xy(x+y-1)$ in Example 1.6. Let $H_1 = \ker(x)$, $H_2 = \ker(y)$, and $H_3 = \ker(x+y-1)$.

Note that $H_1 \cap H_2 \cap H_3 = \emptyset$. Write $e_i = e_{H_i}$ and $a_i = e_i + I(\mathcal{A}) \in A(\mathcal{A})$. Then $e_1 e_2 e_3 \in I(\mathcal{A})$ and thus $a_1 a_2 a_3 = 0$. The ideal $I(\mathcal{A})$ is generated by $e_1 e_2 e_3$. We have

$$A(\mathcal{A}) = \mathcal{K} \oplus (\mathcal{K}a_1 \oplus \mathcal{K}a_2 \oplus \mathcal{K}a_3) \oplus (\mathcal{K}a_1 a_2 \oplus \mathcal{K}a_2 a_3 \oplus \mathcal{K}a_3 a_1).$$

Lemma 3.47 *Let $S \in \mathbf{S}(\mathcal{A})$.*
(1) Assume $\cap S \neq \emptyset$. Then S is dependent if and only if $\mathbf{c}S$ is dependent.
(2) The $(p+1)$-tuple $(K_0, \mathbf{c}S)$ is dependent if and only if either $\cap S = \emptyset$ or S is dependent.

Proof. If $\cap S \neq \emptyset$, then $r(\cap S) = r(\cap(\mathbf{c}S))$. This proves (1). If $\cap S = \emptyset$, then $\cap(\mathbf{c}S) \subseteq K_0$. Thus $(K_0, \mathbf{c}S)$ is dependent. If S is dependent, then $\mathbf{c}S$ is dependent and so is $(K_0, \mathbf{c}S)$. For the converse, suppose that $(K_0, \mathbf{c}S)$ is dependent. If we assume that $\cap S \neq \emptyset$ and that S is independent, we derive a contradiction as follows. Since S is independent, $\mathbf{c}S$ is independent by (1). Since $(K_0, \mathbf{c}S)$ is dependent, $\cap(\mathbf{c}S) \subseteq K_0$ and hence $\cap S = \emptyset$. □

We define maps in both directions between $E(\mathcal{A})$ and $E(\mathbf{c}\mathcal{A})$.

Definition 3.48 *Let $S \in \mathbf{S}(\mathcal{A})$. Define a $\mathcal{K}$-algebra homomorphism*

$$s : E(\mathbf{c}\mathcal{A}) \to E(\mathcal{A}) \quad \text{by} \quad s(e_0 e_{\mathbf{c}S}) = 0, \quad s(e_{\mathbf{c}S}) = e_S.$$

Define a $\mathcal{K}$-linear homomorphism

$$t : E(\mathcal{A}) \to E(\mathbf{c}\mathcal{A}) \quad \text{by} \quad t(e_S) = e_0 e_{\mathbf{c}S}.$$

Lemma 3.49 *We have $s(I(\mathbf{c}\mathcal{A})) \subseteq I(\mathcal{A})$. It follows that s induces a $\mathcal{K}$-algebra homomorphism $s : A(\mathbf{c}\mathcal{A}) \to A(\mathcal{A})$.*

Proof. Let $S = (H_1, \ldots, H_p) \in \mathbf{S}(\mathcal{A})$. It follows from Definition 3.5 that the ideal $I(\mathbf{c}\mathcal{A})$ is generated by

$$\{\partial(e_0 e_{\mathbf{c}S}) \mid (K_0, \mathbf{c}S) \text{ is dependent}\} \cup \{\partial(e_{\mathbf{c}S}) \mid \mathbf{c}S \text{ is dependent}\}.$$

Case 1. If $(K_0, \mathbf{c}S)$ is dependent and $\cap S = \emptyset$, then $s(\partial(e_0 e_{\mathbf{c}S})) = s(e_{\mathbf{c}S}) = e_S \in I(\mathcal{A})$.
Case 2. If $(K_0, \mathbf{c}S)$ is dependent and $\cap S \neq \emptyset$, then S is dependent by Lemma 3.47.2. Thus we have $s(\partial(e_0 e_{\mathbf{c}S})) = e_S = e_{H_1}(\partial e_S) \in I(\mathcal{A})$.
Case 3. If $\mathbf{c}S$ is dependent and $\cap S \neq \emptyset$, then S is dependent by Lemma 3.47.1. Thus we have $s(\partial e_{\mathbf{c}S}) = \partial e_S \in I(\mathcal{A})$.
Case 4. Assume that $\mathbf{c}S$ is dependent and $\cap S = \emptyset$. Let $S_k = (H_1, \ldots, \hat{H}_k, \ldots, H_p)$ for $k = 1, \ldots, p$. If $\cap S_k = \emptyset$, then $e_{S_k} \in I(\mathcal{A})$. If $\cap S_k \neq \emptyset$, then $\cap(\mathbf{c}S_k) \not\subseteq K_0 \supseteq \cap(\mathbf{c}S)$. Thus $\cap(\mathbf{c}S)$ is a proper subspace of $\cap(\mathbf{c}S_k)$. So $\mathbf{c}S_k$

is dependent and $e_{S_k} \in I(\mathcal{A})$ by Case 2. Then we have $s(\partial e_{\mathbf{c}S}) = \partial e_S = \sum_k (-1)^{k-1} e_{S_k} \in I(\mathcal{A})$. □

Lemma 3.50 *We have $t(I(\mathcal{A})) \subseteq I(\mathbf{c}\mathcal{A})$. It follows that t induces a $\mathcal{K}$-linear homomorphism $t : A(\mathcal{A}) \to A(\mathbf{c}\mathcal{A})$.*

Proof. Case 1. If S satisfies $\cap S = \emptyset$, then $(K_0, \mathbf{c}S)$ is dependent by Lemma 3.47.2. So we have $t(e_S) = e_0 e_{\mathbf{c}S} \in I(\mathbf{c}\mathcal{A})$.

Case 2. If S is dependent, then $\mathbf{c}S$ is dependent by Lemma 3.47.1. We have $t(\partial e_S) = e_0 \partial(e_{\mathbf{c}S}) \in I(\mathbf{c}\mathcal{A})$. □

Note that $st = 0$. Thus we have a complex

$$0 \to A(\mathcal{A}) \xrightarrow{t} A(\mathbf{c}\mathcal{A}) \xrightarrow{s} A(\mathcal{A}) \to 0.$$

We will prove that this is a short exact sequence by using the broken circuit basis.

The Broken Circuit Basis

We extend the construction of the broken circuit basis from central arrangements to affine arrangements. We introduce an arbitrary linear order $\prec$ in the affine arrangement $\mathcal{A}$. Call a p-tuple $S = (H_1, \ldots, H_p)$ **standard** if $H_1 \prec \ldots \prec H_p$. Note that $E = E(\mathcal{A})$ has a $\mathcal{K}$-basis consisting of all e_S with standard S. Recall that S is dependent if $\cap S \neq \emptyset$ and $r(\cap S) < |S|$. Next we generalize the notions of circuit and broken circuit of Definitions 3.32 and 3.34. A p-tuple is a **circuit** if it is minimally dependent. A standard p-tuple S is a **broken circuit** if there exists $H \in \mathcal{A}$ such that $\max S \prec H$ and (S, H) is a circuit.

Definition 3.51 *A standard p-tuple S is called χ-independent if $\cap S \neq \emptyset$ and it does not contain any broken circuit.*

Define a linear order $\prec$ in the central arrangement $\mathbf{c}\mathcal{A}$ by
(1) $\mathbf{c}H_1 \prec \mathbf{c}H_2$ if $H_1 \prec H_2$ for $H_1, H_2 \in \mathcal{A}$,
(2) K_0 is the maximal element in $\mathbf{c}\mathcal{A}$.

Lemma 3.52 *Let $S \in \mathbf{S}(\mathcal{A})$. The following three conditions are equivalent:*
(1) S is χ-independent,
(2) $\mathbf{c}S$ is χ-independent,
(3) $(\mathbf{c}S, K_0)$ is χ-independent.

Proof. Since K_0 is the maximal element in $\mathbf{c}\mathcal{A}$, (2) and (3) are equivalent. We show first that if (1) is false, then either (2) or (3) must fail. Suppose that S is not χ-independent. Then either $\cap S = \emptyset$ or S contains a broken circuit. If

3.2 $A(\mathcal{A})$ for Affine Arrangements

$\cap S = \emptyset$, then $(\mathbf{c}S, K_0)$ is dependent by Lemma 3.47.2. This contradicts (3). If S contains a broken circuit, then there exists $H \in \mathcal{A}$ with $\max S \prec H$ such that (S, H) is dependent. So $(\cap S) \cap H \neq \emptyset$. By Lemma 3.47.1, $(\mathbf{c}S, \mathbf{c}H)$ is dependent. Since $\max(\mathbf{c}S) \prec \mathbf{c}H$, $\mathbf{c}S$ is not χ–independent. This contradicts (2).

Next we show that (1) implies (2). Suppose that S is χ–independent. Then $\cap S \neq \emptyset$ and S is independent. It follows from Lemma 3.47.1 that $\mathbf{c}S$ is independent. Since $\cap S \neq \emptyset$, we have $\cap(\mathbf{c}S) \not\subseteq K_0$. If $\mathbf{c}S$ is not χ–independent, then at least one of the following two conditions must be true:

(a) $(\mathbf{c}S, K_0)$ is dependent,
(b) $(\mathbf{c}S, \mathbf{c}H)$ is dependent for some $H \in \mathcal{A}$ with $\max S \prec H$.

Since $\mathbf{c}S$ is independent, (a) implies that $\cap(\mathbf{c}S) = \cap(\mathbf{c}S) \cap K_0 \subseteq K_0$. Thus $\cap S = \emptyset$, which is a contradiction. In case (b) we have $\cap(\mathbf{c}S) \cap \mathbf{c}H = \cap(\mathbf{c}S) \not\subseteq K_0$. Thus $(\cap S) \cap H \neq \emptyset$. It follows from Lemma 3.47.1 that (S, H) is dependent and thus S is not χ–independent. This is a contradiction. □

Definition 3.53 *The* **broken circuit module** $C(\mathcal{A})$ *is defined as follows. Let $C(\mathcal{A})$ be the free $\mathcal{K}$–module with basis $\{1\} \cup \{e_S \in E(\mathcal{A}) \mid S$ is χ–independent$\}$. Then $C(\mathcal{A})$ is a free graded $\mathcal{K}$–module.*

Proposition 3.54 *The following sequence is exact:*

$$0 \to C(\mathcal{A}) \xrightarrow{t} C(\mathbf{c}\mathcal{A}) \xrightarrow{s} C(\mathcal{A}) \to 0.$$

Proof. It is clear that t is injective. The implication (1) $\Rightarrow$ (3) in Lemma 3.52 shows that $t(C(\mathcal{A})) \subseteq C(\mathbf{c}\mathcal{A})$. The implication (2) $\Rightarrow$ (1) in Lemma 3.52 shows that $s(C(\mathbf{c}\mathcal{A})) \subseteq C(\mathcal{A})$. The implication (1) $\Rightarrow$ (2) in Lemma 3.52 shows the surjectivity of s. The implication (3) $\Rightarrow$ (1) in Lemma 3.52 shows $\ker(s) = t(C(\mathcal{A}))$. □

Theorem 3.55 *Let $\varphi : E(\mathcal{A}) \to A(\mathcal{A})$ be the natural homomorphism and let $\psi : C(\mathcal{A}) \to A(\mathcal{A})$ be its restriction. The map $\psi : C(\mathcal{A}) \to A(\mathcal{A})$ is an isomorphism of graded $\mathcal{K}$–modules. The set*

$$\{e_S + I \in A(\mathcal{A}) \mid S \text{ is standard and } \chi\text{–independent}\}$$

is a basis for $A(\mathcal{A})$ as a graded $\mathcal{K}$–module.

Proof. We have a commutative diagram

$$\begin{array}{ccccccccc}
0 & \to & C(\mathcal{A}) & \xrightarrow{t} & C(\mathbf{c}\mathcal{A}) & \xrightarrow{s} & C(\mathcal{A}) & \to & 0 \\
 & & \psi \downarrow & & \psi \downarrow & & \psi \downarrow & & \\
0 & \to & A(\mathcal{A}) & \xrightarrow{t} & A(\mathbf{c}\mathcal{A}) & \xrightarrow{s} & A(\mathcal{A}) & \to & 0.
\end{array}$$

Note that $s : A(\mathbf{c}\mathcal{A}) \to A(\mathcal{A})$ is surjective and it follows from Theorem 3.43 that $\psi : C(\mathbf{c}\mathcal{A}) \to A(\mathbf{c}\mathcal{A})$ is an isomorphism. The top row is exact by Proposition 3.54. Diagram chasing shows that $\psi : C(\mathcal{A}) \to A(\mathcal{A})$ is an isomorphism. □

Corollary 3.56 *The algebra $A(\mathcal{A})$ is a free graded $\mathcal{K}$-module.*

Corollary 3.57 *The following sequence is exact:*
$$0 \to A(\mathcal{A}) \xrightarrow{t} A(\mathbf{c}\mathcal{A}) \xrightarrow{s} A(\mathcal{A}) \to 0.$$

Let $\text{Poin}(M,t)$ be the Poincaré polynomial of the free graded $\mathcal{K}$-module $M = \oplus_{p=0}^{\ell} M_p$:
$$\text{Poin}(M,t) = \sum_{p=0}^{\ell} (\text{rank} M_p) t^p.$$
In the commutative diagram above, t is a homogeneous map of degree 1 and s is of degree 0 in each row. This provides an analog of Proposition 2.51.

Corollary 3.58 $\text{Poin}(A(\mathbf{c}\mathcal{A}), t) = (1+t)\text{Poin}(A(\mathcal{A}), t).$

Deletion and Restriction

Next we consider properties of the algebra A under deletion and restriction. Suppose that $\mathcal{A}$ is a nonempty affine arrangement. Let $H_0 \in \mathcal{A}$ be the distinguished hyperplane. Write $L = L(\mathcal{A})$, $L' = L(\mathcal{A}')$, $L'' = L(\mathcal{A}'')$ for the corresponding posets, and $A = A(\mathcal{A})$, $A' = A(\mathcal{A}')$, $A'' = A(\mathcal{A}'')$. We use similar notation for E, E', E'', I, I', I'', etc.

It is easy to see that $I' \subseteq I''$. Let $i : A' \to A$ be the $\mathcal{K}$-algebra homomorphism induced by the inclusion $E' \subseteq E$. If $H \in \mathcal{A}$, we write $a_H = e_H + I$. If $H \in \mathcal{A}'$, it is important to distinguish between a_H and $e_H + I'$. We cannot identify the two because we do not know that i is a monomorphism. If $S = (H_1, \ldots, H_p) \in \mathbf{S}$, write $a_S = a_{H_1} \ldots a_{H_p}$. If $S \in \mathbf{S}'$, then $a_S \in i A'$. The hyperplanes of $\mathcal{A}''$ have the form $H_0 \cap H$ where $H \in \mathcal{A}'$. We write the corresponding generators of E'' and A'' as $e_{H_0 \cap H}$ and $a_{H_0 \cap H}$. If $S = (H_1, \ldots, H_p) \in \mathbf{S}$ and σ is a permutation of $1, \ldots, p$, let $\sigma S = (H_{\sigma 1}, \ldots, H_{\sigma p})$. To define a $\mathcal{K}$-linear map θ from E to some module over $\mathcal{K}$, it suffices to prescribe the values $\theta(e_S)$ for $S \in \mathbf{S}$ and check that $\theta(e_{\sigma S}) = \text{sign}(\sigma) \theta(e_S)$. For convenience we agree that if $H_0 \in S$, then H_0 is the first element of the tuple S and write $S = (H_0, H_1, \ldots, H_p)$ where $H_1, \ldots, H_p \in \mathcal{A}'$.

Lemma 3.59 *There exists a surjective $\mathcal{K}$-linear map $\theta : E \to E''$ such that*
$$\begin{aligned}
\theta(e_{H_1} \cdots e_{H_p}) &= 0, \\
\theta(e_{H_0} e_{H_1} \cdots e_{H_p}) &= e_{H_0 \cap H_1} \cdots e_{H_0 \cap H_p}
\end{aligned}$$
for all $(H_1, \ldots, H_p) \in \mathbf{S}'$. This map satisfies $\theta(I) \subseteq I''$.

Proof. Since $E = E' \oplus e_{H_0} E'$, we may define θ by the formulas in the lemma. It is understood that $\theta(1) = 0$ and $\theta(e_{H_0}) = 1$. Define $\lambda : \mathcal{A}' \to \mathcal{A}''$ by $\lambda H = H_0 \cap H$

for $H \in \mathcal{A}'$. Extend this map to $\lambda : \mathbf{S}' \to \mathbf{S}''$ by $\lambda(H_1, \ldots, H_p) = (\lambda H_1, \ldots, \lambda H_p)$. In case $S = (\ )$ we agree that $\lambda S = (\ )$. In terms of this notation, θ is defined by $\theta(e_S) = 0$ and $\theta(e_{H_0}e_S) = e_{\lambda S}$ for $S \in \mathbf{S}'$. Since θ is surjective, it suffices to show that $\theta(\partial e_T) \in I''$ for any dependent $T \in \mathbf{S}$. If $T \in \mathbf{S}'$, then $\theta(e_T) = 0$. Thus we may assume that $T = (H_0, S)$ is dependent. Note that $\cap(\lambda S) = H_0 \cap (\cap S) \neq \emptyset$. Thus λS is dependent and $\theta(\partial(e_{H_0}e_S)) = \theta(e_S - e_{H_0}\partial e_S) = -\partial(e_{\lambda T}) \in I''$. □

Corollary 3.60 *There exists a surjective $\mathcal{K}$–linear map $j : A \to A''$ such that the diagram*

$$\begin{array}{ccc} E & \xrightarrow{\theta} & E'' \\ \varphi \downarrow & & \downarrow \varphi'' \\ A & \xrightarrow{j} & A'' \end{array}$$

commutes. In particular, for all $(H_1, \ldots, H_p) \in \mathbf{S}'$

$$j(a_{H_1} \cdots a_{H_p}) = 0,$$
$$j(a_{H_0}a_{H_1} \cdots a_{H_p}) = a_{H_0 \cap H_1} \cdots a_{H_0 \cap H_p}.$$

In order to prove that the sequence

$$0 \to A' \xrightarrow{i} A \xrightarrow{j} A'' \to 0$$

is exact, we utilize broken circuits. Fix linear orders on $\mathcal{A}$, $\mathcal{A}'$, and $\mathcal{A}''$ so that

(1) H_0 is the minimal element in $\mathcal{A}$,
(2) the linear order on $\mathcal{A}'$ is induced by the linear order of $\mathcal{A}$,
(3) if $H, K \in \mathcal{A}'$, then $\lambda H \prec \lambda K$ implies $H \prec K$.

Write $C = C(\mathcal{A})$, $C' = C(\mathcal{A}')$, and $C'' = C(\mathcal{A}'')$.

Lemma 3.61 $C' \subseteq C$.

Proof. Let $S' \in \mathbf{S}'$ be χ–independent. Note that $\cap S' \neq \emptyset$. Assume that $e_{S'} \notin C$. Then S' contains a broken circuit of $\mathcal{A}$, so there exists $H \in \mathcal{A}$ such that $(S', H) \in \mathbf{S}$ contains a circuit and $\max S' \prec H$. Since H_0 is the minimal element of the linear order, $H \in \mathcal{A}'$, implying that S' contains a broken circuit of $\mathcal{A}'$, which is a contradiction. □

Let $i : C' \to C$ be the inclusion map. Recall the definition of $\theta : E \to E''$ from Lemma 3.59.

Lemma 3.62 $\theta(C) = C''$.

Proof. We claim first that $\theta(C) \subseteq C''$. Otherwise there exists $S \in \mathbf{S}$ such that $(H_0, S) \in \mathbf{S}$ is χ–independent but $\lambda S \in \mathbf{S}''$ is not. Since $\cap(\lambda S) = H_0 \cap (\cap S) \neq \emptyset$, λS contains a broken circuit. So there exists $K \in \mathcal{A}'$ such that $\max \lambda S \prec \lambda K$ and

$(\lambda S, \lambda K)$ is dependent. It follows from our choice of linear orders on $\mathcal{A}, \mathcal{A}', \mathcal{A}''$ that $\max S \prec K$ and (H_0, S, K) is dependent. Thus (H_0, S) contains a broken circuit, which is a contradiction.

Next we show that $\theta(C) \supseteq C'''$. Let $S'' \in \mathbf{S}''$ be χ–independent. For each $H'' \in S''$, let $\mu H'' = \max\{\lambda^{-1}(H'')\} \in \mathcal{A}$. Arrange the $\mu H''$ for $H'' \in S''$ into a standard tuple $S \in \mathbf{S}$. Then obviously $\lambda S = S''$. Suppose that (H_0, S) is not χ-independent. Since $H_0 \cap (\cap S) = \cap(\lambda S) = \cap S'' \neq \emptyset$, (H_0, S) contains a broken circuit. So there exists $K \in \mathcal{A}'$ such that $\max S \prec K$ and (H_0, S, K) is dependent. It follows from the definition of S that $\max \lambda S \prec \lambda K$ and $(\lambda S, \lambda K)$ is dependent. Thus λS contains a broken circuit, which is a contradiction. $\square$

Lemma 3.63 *Let $S_1 \in \mathbf{S}'$ and $S_2 \in \mathbf{S}'$. If (H_0, S_1) and (H_0, S_2) are χ-independent with $\lambda S_1 = \lambda S_2$, then $S_1 = S_2$.*

Proof. Suppose that $S_1 \neq S_2$. Then there exist $H_i \in \mathbf{S}_i$ ($i = 1, 2$) such that $H_1 \neq H_2$ and $\lambda H_1 = \lambda H_2$. We may assume that $H_1 \prec H_2$. So $(H_0, H_1, H_2) \in \mathbf{S}$ is dependent, and (H_0, H_1) is a broken circuit. This contradicts the χ-independence of (H_0, S_1). $\square$

Proposition 3.64 *Let $j : C \to C''$ be the restriction of θ. The following sequence is exact:*
$$0 \to C' \xrightarrow{i} C \xrightarrow{j} C'' \to 0.$$

Proof. By Lemmas 3.61 and 3.62, it is sufficient to show that $\ker(j) \subseteq \operatorname{im}(i)$. Suppose
$$j(\sum c_S e_{H_0} e_S) = \sum c_S e_{\lambda S} = 0,$$
where the sum is over $\{S \in \mathbf{S}' \mid (H_0, S) \text{ is } \chi\text{-independent}\}$ and $c_S \in \mathcal{K}$. By Lemma 3.63, we have $c_S = 0$ for all S. $\square$

Theorem 3.65 *Let $\mathcal{A}$ be an affine arrangement. Let $H_0 \in \mathcal{A}$ and let $(\mathcal{A}, \mathcal{A}', \mathcal{A}'')$ be the corresponding triple. Let $i : A(\mathcal{A}') \to A(\mathcal{A})$ be the natural homomorphism and let $j : A(\mathcal{A}) \to A(\mathcal{A}'')$ be the $\mathcal{K}$-linear map defined by*
$$j(a_{H_1} \cdots a_{H_p}) = 0,$$
$$j(a_{H_0} a_{H_1} \cdots a_{H_p}) = a_{H_0 \cap H_1} \cdots a_{H_0 \cap H_p}$$
for $(H_1, \ldots, H_p) \in \mathbf{S}(\mathcal{A}')$. Then the following sequence is exact:
$$0 \to A(\mathcal{A}') \xrightarrow{i} A(\mathcal{A}) \xrightarrow{j} A(\mathcal{A}'') \to 0.$$

Proof. This follows from the commutative diagram

$$\begin{array}{ccccccccc} 0 & \to & C' & \xrightarrow{i} & C & \xrightarrow{j} & C'' & \to & 0 \\ & & \psi' \downarrow & & \psi \downarrow & & \psi'' \downarrow & & \\ 0 & \to & A' & \xrightarrow{i} & A & \xrightarrow{j} & A'' & \to & 0, \end{array}$$

Theorem 3.55, and Proposition 3.64. $\square$

Proposition 3.66 *Let $\mathcal{A}$ be an arrangement and let $\mathcal{B}$ be a subarrangement. The natural homomorphism $i : A(\mathcal{B}) \to A(\mathcal{A})$ is a monomorphism.*

Proof. The assertion is true if $|\mathcal{A}| - |\mathcal{B}| = 1$ by Theorem 3.65. The conclusion follows by induction on $|\mathcal{A}| - |\mathcal{B}|$. $\square$

In Theorem 3.65 the map i has degree 0 and j has degree -1. Thus we get the following results.

Corollary 3.67 *Let $(\mathcal{A}, \mathcal{A}', \mathcal{A}'')$ be a triple and let A, A', A'' be the corresponding algebras. Then*
(1) $\mathrm{Poin}(A,t) = \mathrm{Poin}(A',t) + t\mathrm{Poin}(A'',t)$,
(2) $\mathrm{rank}A = \mathrm{rank}A' + \mathrm{rank}A''$.

Theorem 3.68 $\mathrm{Poin}(A(\mathcal{A}), t) = \pi(\mathcal{A}, t)$.

Proof. We argue by induction on the cardinality of $|\mathcal{A}|$. When $\mathcal{A}$ is the empty arrangement, $A(\mathcal{A}) = \mathcal{K}$. So $\mathrm{Poin}(A(\mathcal{A}), t) = 1 = \pi(\mathcal{A}, t)$. We proved the formula $\pi(\mathcal{A}, t) = \pi(\mathcal{A}', t) + t\pi(\mathcal{A}'', t)$ in Theorem 2.56. This recursion combined with Corollary 3.67.1 completes the proof. $\square$

The Structure of $A(\mathcal{A})$

As in the central case, we define for $X \in L$

$$\mathbf{S}_X = \mathbf{S}_X(\mathcal{A}) = \{S \in \mathbf{S}(\mathcal{A}) \mid \cap S = X\}, \quad E_X = E_X(\mathcal{A}) = \sum_{S \in \mathbf{S}_X} \mathcal{K}e_S,$$

$$A_X(\mathcal{A}) = \phi(E_X), \quad C_X = C_X(\mathcal{A}) = C(\mathcal{A}) \cap E_X(\mathcal{A}).$$

We can generalize Lemmas 3.39 and 3.40 without changing their proofs.

Lemma 3.69 *For $p \geq 0$, we have $C_p = \oplus_{X \in L_p} C_X$ and hence $C = \oplus_{X \in L} C_X$.*

Lemma 3.70 *If $Y \leq X$, then $C_Y(\mathcal{A}_X) = C_Y(\mathcal{A})$.*

Let $X \in L$. Since the natural map $i : A(\mathcal{A}_X) \to A(\mathcal{A})$ is injective by Proposition 3.66, we have the affine version of Proposition 3.31.

Proposition 3.71 *If $Y \leq X$, then $A_Y(\mathcal{A}_X) \simeq A_Y(\mathcal{A})$.*

Theorem 3.72 *Let $\mathcal{A}$ be an affine arrangement and let $A = A(\mathcal{A})$. Then*

$$A = \bigoplus_{X \in L} A_X.$$

Proof. Let $X \in L$. Note that $\mathcal{A}_X$ is a central arrangement. By Lemma 3.70, Theorem 3.43, and Proposition 3.71, we have

$$C_X(\mathcal{A}) = C_X(\mathcal{A}_X) \simeq A_X(\mathcal{A}_X) \simeq A_X(\mathcal{A}).$$

Since $C(\mathcal{A}) = \oplus_{X \in L} C_X(\mathcal{A})$ by Lemma 3.69 and $C(\mathcal{A}) \simeq A(\mathcal{A})$ by Theorem 3.55, we have the desired result. □

Recall Brieskorn's Lemma from the introduction. The next result is its algebraic analog. Its proof is the same as the proof of the central version, Corollary 3.27. Corollary 3.73 will be used in Theorem 5.91 to give an elementary proof of Brieskorn's Lemma.

Corollary 3.73 *Let $\mathcal{A}$ be an affine arrangement. Then $A_p = \oplus_{X \in L_p} A_X$.*

These results are summarized in the next corollary.

Corollary 3.74 *The algebra $A(\mathcal{A})$ is a free graded $\mathcal{K}$–module. The $\mathcal{K}$–modules $A_X(\mathcal{A})$ for $X \in L$ and $A_p(\mathcal{A})$ for $p \geq 0$ are also free.*

Proof. The $\mathcal{K}$–modules $C_X(\mathcal{A})$ are free by definition. It follows from Lemma 3.70 that $C_X(\mathcal{A}) \simeq A_X(\mathcal{A})$. Thus $A_X(\mathcal{A})$ is a free $\mathcal{K}$–module. We showed in Corollary 3.73 that $A_p = \oplus_{X \in L_p} A_X$. Thus A_p is also free. □

Proposition 3.75 *If $X \in L(\mathcal{A})$, then the rank of the free $\mathcal{K}$–module $A_X(\mathcal{A})$ is equal to $(-1)^{r(X)}\mu(X)$.*

Proof. The leading coefficient of $\pi(\mathcal{A}_X, t)$ is equal to $(-1)^{r(X)}\mu(X)$. Since $\mathrm{Poin}(A(\mathcal{A}_X), t) = \pi(\mathcal{A}_X, t)$, it is also equal to $\mathrm{rank} A_X(\mathcal{A}_X) = \mathrm{rank} A_X(\mathcal{A})$. □

A–equivalence

Definition 3.76 *The arrangements $\mathcal{A}$ and $\mathcal{B}$ are $\mathcal{K}$–algebra equivalent, or A–equivalent, if there is an isomorphism of graded $\mathcal{K}$–algebras $\phi : A(\mathcal{A}) \to A(\mathcal{B})$.*

Clearly, L–equivalent arrangements are A–equivalent and A–equivalent arrangements are π–equivalent. Example 2.61 showed that π–equivalent arrangements are not L–equivalent. In fact, these three notions of equivalence are distinct. Falk [77] used his work on minimal models to find an invariant of the algebra A which is different for the two arrangements of Example 2.61. He conjectured and L. Rose and H. Terao showed that the arrangements in Figure 3.1 are algebra equivalent.

Example 3.77 The 3–arrangements $\mathcal{A}$ and $\mathcal{B}$ in Figure 3.1 are A–equivalent but not L–equivalent. To see that $L(\mathcal{A})$ and $L(\mathcal{B})$ are not isomorphic, note that

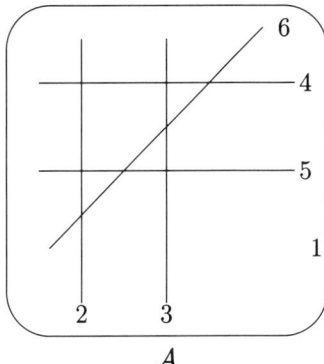 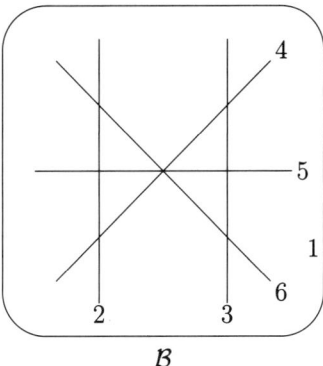

Fig. 3.1. $\mathcal{A}$–equivalent but not L–equivalent

the two triple points of $\mathcal{A}$ are on the same line, while the two triple points of $\mathcal{B}$ are on different lines. In order to show that $A(\mathcal{A}) \simeq A(\mathcal{B})$, label the hyperplanes as in Figure 3.1. Let $E(\mathcal{A})$ have generators e_i for $1 \leq i \leq 6$, and let $E(\mathcal{B})$ have generators f_i for $1 \leq i \leq 6$. Define $\phi : E(\mathcal{A}) \to E(\mathcal{B})$ by $\phi(e_i) = f_i$ for $i = 1, 2, 3, 6$ and $\phi(e_4) = f_5 - f_4 + f_1$, $\phi(e_5) = f_6 - f_5 + f_1$. Note that $I_2(\mathcal{A})$ is generated by $e_{1,2} - e_{1,3} + e_{2,3}$ and $e_{1,4} - e_{1,5} + e_{4,5}$. We have

$$\phi(e_{1,2} - e_{1,3} + e_{2,3}) = f_{1,2} - f_{1,3} + f_{2,3}$$
$$\in I_2(\mathcal{B}),$$
$$\phi(e_{1,4} - e_{1,5} + e_{4,5}) = \phi((e_1 - e_4)(e_1 - e_5))$$
$$= \phi(e_1 - e_4)\phi(e_1 - e_5)$$
$$= (f_4 - f_5)(f_5 - f_6)$$
$$= f_{4,5} - f_{4,6} + f_{5,6}$$
$$\in I_2(\mathcal{B}).$$

Since these images generate $I_2(\mathcal{B})$, it follows that ϕ induces an isomorphism $A(\mathcal{A}) \simeq A(\mathcal{B})$.

More recently, Falk [79] showed that the complements of the complexified arrangements $M(\mathcal{A})$ and $M(\mathcal{B})$ are homotopy equivalent.

3.3 Algebra Factorizations

Let $\mathcal{A}$ be an affine arrangement. Let $\mathbf{c}\mathcal{A}$ be the cone over $\mathcal{A}$. Recall that K_0 is the additional hyperplane and we write $e_0 = e_{K_0}$. Let $\phi : E(\mathbf{c}\mathcal{A}) \to A(\mathbf{c}\mathcal{A})$ be the natural surjection and let $a_0 = \phi(e_0)$. It follows from Corollary 3.56 that $A(\mathcal{A})$ and $A(\mathbf{c}\mathcal{A})$ are free $\mathcal{K}$–modules. Thus the short exact sequence

$$0 \to A(\mathcal{A}) \xrightarrow{t} A(\mathbf{c}\mathcal{A}) \xrightarrow{s} A(\mathcal{A}) \to 0$$

of Corollary 3.57 splits. This yields the following result.

Theorem 3.78 *Let $c\mathcal{A}$ be the cone over the affine arrangement $\mathcal{A}$. Let $a_0 \in A(c\mathcal{A})$ correspond to the additional hyperplane. There is an isomorphism of graded $\mathcal{K}$-modules*

$$(\mathcal{K} + \mathcal{K}a_0) \otimes A(\mathcal{A}) \simeq A(c\mathcal{A}).$$

It follows from Theorem 3.68 that Proposition 2.51 is a consequence of Theorem 3.78. The topological interpretation of this algebra factorization follows from Proposition 5.1 and Theorem 5.90.

In this section we prove algebra factorization in two more cases. The factorization for supersolvable arrangements also has a topological interpretation, which is described after the Fibration Theorem 5.113. We prove here that the existence of a nice partition is equivalent to algebra factorization. We do not know of a topological interpretation in this case.

Supersolvable Arrangements

Recall that a supersolvable arrangement is central. Thus we may assume that $\mathcal{A}$ is a central arrangement.

Lemma 3.79 *Suppose there exists a modular element $Y \in L(\mathcal{A})$ with $r(Y) = r(\mathcal{A}) - 1$. For every $H \in \mathcal{A} \setminus \mathcal{A}_Y$ there is a $\mathcal{K}$-algebra isomorphism $\rho : A(\mathcal{A}_Y) \to A(\mathcal{A}^H)$ defined by $\rho(a_K) = a_{H \cap K}$ for all $K \in \mathcal{A}_Y$.*

Proof. Since $H \in \mathcal{A} \setminus \mathcal{A}_Y$, we have $L_Y = L(\mathcal{A}_Y) = [H \wedge Y, Y]$ and $L^H = L(\mathcal{A}^H) = [H, H \vee Y]$. It follows from Lemma 2.27 that the map $\tau : L_Y \to L^H$ given by $\tau(Z) = Z \vee H = Z \cap H$ is a lattice isomorphism. If $S = (H_1, \ldots, H_p) \in \mathbf{S}(\mathcal{A}_Y)$, define $\tau S = (\tau H_1, \ldots, \tau H_p) \in \mathbf{S}(\mathcal{A}^H)$. The $\mathcal{K}$-algebra isomorphism from $E(\mathcal{A}_Y)$ to $E(\mathcal{A}^H)$ which sends e_S to $e_{\tau S}$ maps $I(\mathcal{A}_Y)$ to $I(\mathcal{A}^H)$ and induces a $\mathcal{K}$-algebra homomorphism $\rho : A(\mathcal{A}_Y) \to A(\mathcal{A}^H)$ such that $\rho a_S = a_{\tau S}$. The inverse of ρ is constructed using σ, the inverse of τ in Lemma 2.27. □

Lemma 3.80 *Suppose there exists a modular element $Y \in L(\mathcal{A})$ with $r(Y) = r(\mathcal{A}) - 1$. Let $\mathcal{B} = \mathcal{A} \setminus \mathcal{A}_Y$. Then*

$$A(\mathcal{A}) = A(\mathcal{A}_Y) \oplus \big(\bigoplus_{H \in \mathcal{B}} A(\mathcal{A}_Y) a_H \big).$$

Proof. It follows from Proposition 3.66 that we may identify $A(\mathcal{A}_Y)$ with the $\mathcal{K}$-subalgebra of $A(\mathcal{A})$ generated by the elements $a_K = e_K + I(\mathcal{A})$ for $K \in \mathcal{A}_Y$. Let $U = \sum_{H \in \mathcal{B}} A(\mathcal{A}_Y) a_H$. Note first that if $H, K \in \mathcal{B}$, then $a_H a_K \in U$. This is clear if $H = K$, since $a_H^2 = 0$. Suppose $H \neq K$. Recall again that the map $\tau : L_Y \to L^H$ given by $\tau Z = Z \cap H$ is an isomorphism. It follows that there exists $M \in L_Y$ such that $M \cap H = K \cap H$. Since $K \neq H$, we have $r(K \cap H) = 2$, so $r(M \cap H) = 2$, and thus $M \in \mathcal{A}_Y$. Since $r(M \cap H \cap K) = 2$, the 3–tuple

(M, H, K) is dependent, and thus $a_H a_K - a_M a_K + a_M a_H = 0$. This shows that $a_H a_K \in A(\mathcal{A}_Y) a_H + A(\mathcal{A}_Y) a_K \subseteq U$.

Since $A(\mathcal{A}_Y)$ is a $\mathcal{K}$-subalgebra of $A(\mathcal{A})$ containing the identity, it follows that U is closed under multiplication and $A(\mathcal{A}_Y)U \subseteq U$. Thus $A(\mathcal{A}_Y) + U$ is a $\mathcal{K}$- subalgebra of $A(\mathcal{A})$. But $\mathcal{A} = \mathcal{A}_Y \cup \mathcal{B}$, so $a(\mathcal{A}_Y) + U$ contains all the generators a_H of $A(\mathcal{A})$ where $H \in \mathcal{A}$. It follows that

$$A(\mathcal{A}) = A(\mathcal{A}_Y) + U = A(\mathcal{A}_Y) + (\sum_{H \in \mathcal{B}} A(\mathcal{A}_Y) a_H).$$

We show next that this is a direct sum. We need $A(\mathcal{A}_Y) \cap U = 0$. Recall that $A = \oplus_{X \in L} A_X$ where $A_X = \phi(E_X) = \sum_{\cap S = X} \mathcal{K} a_S$. Let $\pi_X : A \to A_X$ be the natural projection. Thus

$$\pi_X a_S = \begin{cases} a_S & \text{if } \cap S = X \\ 0 & \text{otherwise.} \end{cases}$$

We show that $\pi_X(A(\mathcal{A}_Y) \cap U) = 0$ for all $X \in L$ by showing that (i) $\pi_X(A(\mathcal{A}_Y)) = 0$ if $X \not\leq Y$, and (ii) $\pi_X(U) = 0$ if $X \leq Y$. Assertion (i) is immediate from $A(\mathcal{A}_Y) = \sum_{\cap S \leq Y} \mathcal{K} a_S$. To prove (ii) we observe that if $H \in \mathcal{B}$ and $X \leq Y$, then $\pi_X(a_S a_H) = \pi_X(a_{(S,H)}) = 0$ because $(\cap S) \cap H \not\leq Y$. Thus $A(\mathcal{A}_Y) \cap U = 0$.

It remains to show that the sum $\sum_{H \in \mathcal{B}} A(\mathcal{A}_Y) a_H$ is also direct. Fix $H_0 \in \mathcal{B}$ and let $(\mathcal{A}, \mathcal{A}', \mathcal{A}'')$ be the inductive triple with respect to H_0. Let $\lambda : \mathbf{S}' \to \mathbf{S}''$ and $j : A \to A''$ be the maps defined in Lemma 3.59 and Corollary 3.60. Since $H_0 \in \mathcal{B}$, we have $\mathbf{S}(\mathcal{A}_Y) \subseteq \mathbf{S}'$. It follows from Corollary 3.60 that for $S \in \mathbf{S}(\mathcal{A}_Y)$ and $H \in \mathcal{B}$, we have $j(a_H a_S) = a_{\lambda S}$ if $H = H_0$, and $j(a_H a_S) = 0$ otherwise. By Lemma 3.79 with H replaced by H_0, there exists a $\mathcal{K}$-algebra isomorphism $\rho : A(\mathcal{A}_Y) \to A''$ with $\rho(a_S) = a_{\lambda S}$. Thus $j(a_H a_S) = \rho(a_S)$ if $H = H_0$, and $j(a_H a_S) = 0$ otherwise. It follows that for $u \in A(\mathcal{A}_Y)$ we have

$$j(a_H u) = \begin{cases} \rho(u) & \text{if } H = H_0 \\ 0 & \text{if } H \in \mathcal{B} \setminus \{H_0\}. \end{cases}$$

Suppose $\sum_{H \in \mathcal{B}} a_H u_H = 0$ where $u_H \in A(\mathcal{A}_Y)$. Then

$$0 = j(\sum_{H \in \mathcal{B}} a_H u_H) = \rho(u_{H_0})$$

and thus $u_{H_0} = 0$. Since H_0 was an arbitrary hyperplane of $\mathcal{B}$, it follows that $u_H = 0$ for all $H \in \mathcal{B}$. □

Theorem 3.81 *Let $\mathcal{A}$ be a supersolvable arrangement with $r = r(\mathcal{A})$ and maximal chain of modular elements*

$$V = Y_0 < Y_1 < \ldots < Y_r = T.$$

Let $A = A(\mathcal{A})$. For $1 \leq i \leq r$ let $\mathcal{B}_i = \mathcal{A}_{Y_i} \setminus \mathcal{A}_{Y_{i-1}}$ and let $B_i = \sum_{H \in \mathcal{B}_i} \mathcal{K} a_H$. The $\mathcal{K}$-linear map

$$(\mathcal{K}+\mathcal{B}_1) \otimes \ldots \otimes (\mathcal{K}+\mathcal{B}_r) \to A$$

defined by multiplication in A is an isomorphism of graded $\mathcal{K}$-modules. In particular, with $b_i = |\mathcal{B}_i|$

$$\mathrm{Poin}(A,t) = (1+b_1)\ldots(1+b_r).$$

Proof. Let $Y = Y_{r-1}$. Then Y is a modular element of $L(\mathcal{A})$ with $r(Y) = r-1$. Let $\mathcal{B} = \mathcal{A} \setminus \mathcal{A}_Y$ and let $B = \sum_{H \in \mathcal{B}} \mathcal{K} a_H$. Lemma 3.80 shows that the $\mathcal{K}$-linear map $A(\mathcal{A}_Y) \otimes (\mathcal{K}+B) \to A$ defined by the multiplication in A is an isomorphism of modules. The result follows by induction on r. □

Nice Partitions of Central Arrangements

Assume first that $\mathcal{A}$ is a central arrangement. Generalizations of supersolvable arrangements were suggested in [117, 33]. The present notion was introduced in [236]. Let $\pi = (\pi_1, \ldots, \pi_s)$ be a partition of $\mathcal{A}$. Let (π_i) denote the free $\mathcal{K}$-module with basis 1 and the elements of π_i. It is graded by $\deg 1 = 0$ and $\deg H = 1$. Define the graded $\mathcal{K}$-module

$$(\pi) = (\pi_1) \otimes (\pi_2) \otimes \cdots \otimes (\pi_s).$$

We agree that $(\pi) = \mathcal{K}$ when $\mathcal{A} = \Phi_\ell$. Since $\mathrm{Poin}((\pi_i), t) = (1 + |\pi_i| t)$, we obtain

$$\mathrm{Poin}((\pi), t) = \prod_{i=1}^{s} (1 + |\pi_i| t).$$

Definition 3.82 *Let $S = (H_1, \ldots, H_k) \in \mathbf{S}_k$. Call S a k-section of π if for $1 \le i \le k$*

$$H_i \in \pi_{n(i)}, \quad 1 \le n(1) < n(2) < \ldots < n(k) \le s.$$

We agree that the 0-section is $S = ( )$. Let $\mathcal{S}_k \subset \mathbf{S}_k$ be the set of k-sections of π, and let $\mathcal{S} = \cup_{k=0}^s \mathcal{S}_k$. Given $S \in \mathcal{S}_k$, let $p_S = x_1 \otimes \cdots \otimes x_s \in (\pi)$ where

$$x_j = \begin{cases} H_i & \text{if } j = n(i) \\ 1 & \text{if } j \notin \{n(1), \ldots, n(k)\}. \end{cases}$$

Note that $p_{()} = 1$ and p_S is homogeneous of degree k. The graded $\mathcal{K}$-module (π) is free with basis $\{p_S \mid S \in \mathcal{S}\}$. Recall the notation $e_S \in E(\mathcal{A})$ and $a_S \in A(\mathcal{A})$. Each element of the algebra $A(\mathcal{A})$ may be expressed as a linear combination of elements $\{a_S \mid S \in \mathbf{S}\}$, but this expression is not necessarily unique.

Definition 3.83 *Define $\kappa : (\pi) \to A(\mathcal{A})$ as follows. For $S \in \mathcal{S}$ assign $\kappa(p_S) = a_S$ and let κ be the unique homogeneous $\mathcal{K}$-linear map of degree 0 which extends this assignment.*

3.3 Algebra Factorizations

We will show that the map κ is an isomorphism of graded $\mathcal{K}$–modules if and only if the partition π is nice. Denote the homogeneous part of degree k of (π) by $(\pi)_k$. Then
$$(\pi) = \bigoplus_{k=0}^{s} (\pi)_k.$$
Here $(\pi)_0 = \mathcal{K}$. Given $S = (H_1, \ldots, H_k) \in \mathbf{S}_k$, recall that $\cap S = H_1 \cap \cdots \cap H_k \in L$. For $X \in L$, define a free submodule $(\pi)_X$ of (π) with basis $\{p_S \mid S \in \mathcal{S}, \cap S = X\}$. It follows that $(\pi)_V = \mathcal{K}$.

Lemma 3.84 *Suppose that π is an independent partition. For each $k \geq 0$, we have*
$$(\pi)_k = \bigoplus_{X \in L_k} (\pi)_X.$$

Proof. Note that $\{p_S \mid S \in \mathcal{S}_k\}$ is a basis for $(\pi)_k$. If $\cap S = X$, then $p_S \in (\pi)_X$. We have $X \in L_k$ because π is independent. $\square$

Lemma 3.85 *For $X, Y \in L$ with $Y \leq X$, the natural map $(\pi_X)_Y \to (\pi)_Y$ is an isomorphism.*

Proof. If $S \in \mathcal{S}$ with $\cap S = Y$, then $S \subseteq \mathcal{A}_Y \subseteq \mathcal{A}_X$. Thus S is also a section of π_X:
$$\{S \mid S \in \mathcal{S}, \cap S = Y\} = \{S \mid S \text{ is a section of } \pi_X, \cap S = Y\}.$$
The isomorphism $p_S \in (\pi_X)_Y \mapsto p_S \in (\pi)_Y$ is obtained by inserting "$1 \otimes$" the required number of times. $\square$

Let $S = (H_1, \ldots, H_k) \in \mathbf{S}_k$. Recall that S_j denotes the tuple with H_j deleted. Define a $\mathcal{K}$–linear map $\partial : (\pi)_k \to (\pi)_{k-1}$ by $\partial(p_{()}) = 0$, $\partial(p_H) = 1$, and for $k \geq 2$ and $S \in \mathcal{S}_k$
$$\partial(p_S) = \sum_{j=1}^{k} (-1)^{j-1} p_{S_j}.$$
Then $\partial\partial = 0$ and $((\pi)_*, \partial)$ is a chain complex.

Lemma 3.86 *If the partition π contains a block which is a singleton, then the complex $((\pi)_*, \partial)$ is acyclic.*

Proof. We may assume that π_1 is a singleton, $\pi_1 = \{H_1\}$. Suppose that $x \in (\pi)_k$ is a cycle, $\partial x = 0$. Write x as $x = H_1 \otimes x_1 + 1 \otimes x_2$, where $x_1, x_2 \in (\pi_2) \otimes \cdots \otimes (\pi_s)$. Then
$$0 = \partial x = 1 \otimes x_1 - H_1 \otimes (\partial x_1) + 1 \otimes (\partial x_2) = 1 \otimes (x_1 + \partial x_2) - H_1 \otimes (\partial x_1).$$
This implies that $x_1 = -\partial x_2$. Define $y = H_1 \otimes x_2 \in (\pi)_{k+1}$. Then

$$\partial y = 1 \otimes x_2 - H_1 \otimes (\partial x_2) = 1 \otimes x_2 + H_1 \otimes x_1 = x$$

as required. □

Theorem 3.87 *Let $\mathcal{A}$ be a central arrangement and let π be a partition of $\mathcal{A}$. The homogeneous $\mathcal{K}$-linear map κ of Definition 3.83 is an isomorphism if and only if the partition π is nice.*

Proof. Assume that π is a nice partition. We argue by induction on $r = r(\mathcal{A})$. If $r(\mathcal{A}) = 0$, then $\mathcal{A} = \Phi_\ell$. Thus $(\pi) = \mathcal{K} = A(\mathcal{A})$. Assume that $r = r(\mathcal{A}) > 0$. Note $s \leq r$ because π is independent. Consider the diagram

$$\begin{array}{ccccccccccc} 0 & \to & (\pi)_r & \xrightarrow{\partial} & (\pi)_{r-1} & \xrightarrow{\partial} & \cdots & \xrightarrow{\partial} & (\pi)_1 & \xrightarrow{\partial} & (\pi)_0 & \to & 0 \\ & & \downarrow \kappa_r & & \downarrow \kappa_{r-1} & & & & \downarrow \kappa_1 & & \downarrow \kappa_0 & & \\ 0 & \to & A_r(\mathcal{A}) & \xrightarrow{\partial} & A_{r-1}(\mathcal{A}) & \xrightarrow{\partial} & \cdots & \xrightarrow{\partial} & A_1(\mathcal{A}) & \xrightarrow{\partial} & A_0(\mathcal{A}) & \to & 0. \end{array}$$

The vertical maps are induced by $\kappa : (\pi) \to A(\mathcal{A})$. The top row is exact by Lemma 3.86. The bottom row is exact by Lemma 3.13. Note that

$$(\pi)_k = \bigoplus_{Y \in L_k} (\pi)_Y \simeq \bigoplus_{Y \in L_k} (\pi_Y)_Y$$

by Lemmas 3.84 and 3.85. Also note that

$$A_k(\mathcal{A}) = \bigoplus_{Y \in L_k} A_Y(\mathcal{A}) \simeq \bigoplus_{Y \in L_k} A_Y(\mathcal{A}_Y)$$

by Corollary 3.27 and Proposition 3.31. By applying the induction assumption to L_Y for $r(Y) < r$, we obtain that κ_i is an isomorphism for $1 \leq i < r$. It follows from the commutative diagram that κ_r is also an isomorphism. Thus $\kappa : (\pi) \to A(\mathcal{A})$ is an isomorphism.

For the converse, suppose κ is an isomorphism. First we show that π is independent. Let $S \in \mathcal{S}$. Then $p_S \neq 0$ and $a_S = \kappa(p_S) \neq 0$. This shows that S is independent. Next we show that if $X \neq V$, then π_X contains a block which is a singleton. Since

$$(\pi) = \bigoplus_{Y \in L} (\pi)_Y, \quad A(\mathcal{A}) = \bigoplus_{Y \in L} A_Y(\mathcal{A}),$$

κ induces isomorphisms $(\pi)_Y \to A_Y(\mathcal{A})$. By Lemma 3.85 and Proposition 3.31, we obtain

$$\begin{aligned} (\pi_X) &= \bigoplus_{Y \in L_X} (\pi_X)_Y \simeq \bigoplus_{Y \in L_X} (\pi)_Y \\ &\simeq \bigoplus_{Y \in L_X} A_Y(\mathcal{A}) \simeq \bigoplus_{Y \in L_X} A_Y(\mathcal{A}_X) = A(\mathcal{A}_X). \end{aligned}$$

Let $X \neq V$. Then

$$0 = \sum_{Y \in L_X} \mu(Y) = \text{Poin}(A(\mathcal{A}_X), -1) = \text{Poin}((\pi_X), -1)$$
$$= \prod_i (1 - |\pi_i \cap \mathcal{A}_X|).$$

This implies that π_X contains a block which is a singleton. □

Corollary 3.88 *If $\mathcal{A}$ has a nice partition $\pi = (\pi_1, \ldots, \pi_s)$, then $s = r$ and*

$$\text{Poin}(A(\mathcal{A}), t) = \prod_{i=1}^r (1 + |\pi_i| t).$$

Corollary 3.89 *If $\mathcal{A}$ has a nice partition $\pi = (\pi_1, \ldots, \pi_s)$, then the multiset $\{|\pi_1|, \ldots, |\pi_s|\}$ depends only on $\mathcal{A}$.*

Corollary 3.90 *If $\mathcal{A}$ has a nice partition $\pi = (\pi_1, \ldots, \pi_s)$, then for all $X \in L$*

$$r(X) = |\{i \mid \pi_i \cap \mathcal{A}_X \neq \emptyset\}|.$$

Proof. We showed in the proof of Theorem 3.87 that the isomorphism κ induces isomorphisms $\kappa_X : (\pi_X) \to A(\mathcal{A}_X)$ for all $X \in L$. Thus π_X is a nice partition of $\mathcal{A}_X$. By Corollary 3.88, we have

$$r(X) = r(\mathcal{A}_X) = |\pi_X| = |\{i \mid \pi_i \cap \mathcal{A}_X \neq \emptyset\}|.$$

This completes the proof. □

Nice Partitions of Affine Arrangements

Next assume that $\mathcal{A}$ is an affine arrangement. We generalize Theorem 3.87 to affine arrangements. Recall that $\mathcal{A}_X$ is central for each $X \in L$.

Definition 3.91 *Let $\pi = (\pi_1, \ldots, \pi_s)$ be a partition of the affine arrangement $\mathcal{A}$. It is called **nice** if*
(1) for every choice of hyperplanes $H_i \in \pi_i$ for $1 \leq i \leq s$, the intersection $H_1 \cap H_2 \cap \cdots \cap H_s$ is not empty, and
(2) if $X \in L \setminus \{V\}$, then the induced partition π_X of the central arrangement $\mathcal{A}_X$ is a nice partition in the sense of Definition 2.66.

For any partition of $\mathcal{A}$, we can define the graded free $\mathcal{K}$–module $(\pi) = \bigoplus_k (\pi)_k$, the submodule $(\pi)_X$ for $X \in L$, and the map $\kappa : (\pi) \to A(\mathcal{A})$ as before. If the first condition of Definition 3.91 is satisfied, then

$$(\pi) = \bigoplus_{X \in L} (\pi)_X.$$

As in Lemma 3.85 for the central case, the natural map $(\pi_X)_Y \to (\pi)_Y$ is an isomorphism for $X, Y \in L$ with $Y \leq X$.

Theorem 3.92 *Let $\mathcal{A}$ be an affine arrangement and let π be a partition of $\mathcal{A}$. Then κ is an isomorphism if and only if the partition π is nice.*

Proof. Suppose π is nice. Theorem 3.87 and Proposition 3.71 give

$$(\pi) \simeq \bigoplus_{X \in L} (\pi)_X \simeq \bigoplus_{X \in L} (\pi_X)_X$$
$$\simeq \bigoplus_{X \in L} A_X(\mathcal{A}_X) \simeq \bigoplus_{X \in L} A_X(\mathcal{A}) \simeq A(\mathcal{A}).$$

Conversely, assume that the map κ is an isomorphism. If we have $S = (H_1, \ldots, H_s)$ with $H_i \in \pi_i$ for $1 \leq i \leq s$, then e_S is not zero in $A(\mathcal{A})$ because κ is injective. It follows that $\cap S \neq \emptyset$. This is the first condition. The isomorphism κ induces isomorphisms $(\pi)_Y \simeq A_Y(\mathcal{A})$ for all $Y \in L$. Let $X \in L$. Proposition 3.71 gives

$$(\pi_X) \simeq \bigoplus_{Y \in L_X} (\pi)_Y \simeq \bigoplus_{Y \in L_X} A_Y(\mathcal{A}) \simeq A(\mathcal{A}_X).$$

Theorem 3.87 implies that each partition π_X is nice. □

3.4 The Algebra $B(\mathcal{A})$

Let $\mathcal{A}$ be an affine arrangement. In this section we construct a $\mathcal{K}$–algebra $B(\mathcal{A})$ whose elements are certain $\mathcal{K}$–linear combinations of ordered subsets of $L(\mathcal{A})$ with multiplication defined using a shuffle product. Thus $B(\mathcal{A})$ depends only on $L(\mathcal{A})$. We prove that the algebras $A(\mathcal{A})$ and $B(\mathcal{A})$ are isomorphic. This algebra will reappear in Section 4.5 as the homology of a chain complex based on $L(\mathcal{A})$.

The Shuffle Product

Definition 3.93 *Let $\mathcal{A}$ be an arrangement with intersection poset $L = L(\mathcal{A})$. For $p \geq 0$ define free $\mathcal{K}$–modules $\mathcal{T}_p$ as follows: $\mathcal{T}_0 = \mathcal{K}$ and for $p > 0$, $\mathcal{T}_p$ has a basis consisting of all p-tuples $(X_1, \ldots, X_p)$ where $X_i \in L \setminus \{V\}$. Let*

$$\mathcal{T} = \bigoplus_{p \geq 0} \mathcal{T}_p.$$

Let $Sym(p)$ be the symmetric group on the letters $1, \ldots, p$. If $\pi \in Sym(p)$ and $u = (X_1, \ldots, X_p)$, let $\pi u = (X_{\pi^{-1}1}, \ldots, X_{\pi^{-1}p})$. This makes $\mathcal{T}_p$ a $Sym(p)$–module.

3.4 The Algebra $B(\mathcal{A})$

Definition 3.94 *Define a product $\mathcal{T} \times \mathcal{T} \to \mathcal{T}$, written $*$, as follows. If $u = (X_1, \ldots, X_p)$ and $v = (Y_1, \ldots, Y_q)$, let*

$$w = (Z_1, \ldots, Z_{p+q}) = (X_1, \ldots, X_p, Y_1, \ldots, Y_q).$$

Define
$$u * v = \sum \mathrm{sign}\pi(\pi w)$$

where the sum is over all (p,q)-shuffles π of $1, \ldots, p+q$.

Recall [147, p.243] that a (p,q)-shuffle of $1, \ldots, p+q$ is a permutation $\pi \in \mathrm{Sym}(p+q)$ such that $\pi i < \pi j$ whenever $i < j \leq p$ or $p < i < j$. This makes $\mathcal{T}$ into an associative graded anticommutative $\mathcal{K}$-algebra with identity.

Definition 3.95 *Let $\eta : \mathcal{T} \to \mathcal{T}$ be the antisymmetrizer defined for $u = (X_1, \ldots, X_p)$ by*

$$\eta u = \sum \mathrm{sign}\, \pi\, (\pi u) = \sum \mathrm{sign}\, \pi\, (\pi^{-1} u)$$

summed over all $\pi \in \mathrm{Sym}(p)$. Define a $\mathcal{K}$-linear map $\lambda : \mathcal{T} \to \mathcal{T}$ by $\lambda 1 = 1$ and

$$\lambda(X_1, \ldots, X_p) = \begin{cases} (X_1, X_1 \cap X_2, \ldots, X_1 \cap X_2 \cap \ldots \cap X_p) & \text{if } Z \neq \emptyset \\ 0 & \text{if } Z = \emptyset \end{cases}$$

where $Z = X_1 \cap X_2 \cap \ldots \cap X_p$.

Lemma 3.96 *We have*
(1) $\eta(X_1, \ldots, X_p) = (X_1) * \ldots * (X_p)$,
(2) if $u, v \in \mathcal{T}$, then $\lambda(\lambda u * \lambda v) = \lambda(u * v)$.

Proof. Assertion (1) follows by induction. In (2), if one side is zero, so is the other. Otherwise it suffices to check (2) for $u = (X_1, \ldots, X_p)$ and $v = (Y_1, \ldots, Y_q)$. Then $\lambda u = (X'_1, \ldots, X'_p)$ and $\lambda v = (Y'_1, \ldots, Y'_q)$ where $X'_i = X_1 \cap \ldots \cap X_i$ and $Y'_j = Y_1 \cap \ldots \cap Y_j$. Write $(Z_1, \ldots, Z_{p+q}) = (X_1, \ldots, X_p, Y_1, \ldots, Y_q)$ and $(Z'_1, \ldots, Z'_{p+q}) = (X'_1, \ldots, X'_p, Y'_1, \ldots, Y'_q)$. It follows from the idempotence $Z \cap Z = Z$ that $Z'_{\pi 1} \cap \ldots \cap Z'_{\pi i} = Z_{\pi 1} \cap \ldots \cap Z_{\pi i}$ for all $1 \leq i \leq p+q$, and all permutations π of $1, \ldots, p+q$. Thus

$$\begin{aligned} \lambda(\lambda u * \lambda v) &= \sum (\mathrm{sign}\pi) \lambda(Z'_{\pi 1}, \ldots, Z'_{\pi(p+q)}) \\ &= \sum (\mathrm{sign}\pi) \lambda(Z_{\pi 1}, \ldots, Z_{\pi(p+q)}) \\ &= \lambda(u * v) \end{aligned}$$

as required. □

Definition 3.97 *Let $\mathcal{U} = \lambda(\mathcal{T})$. Then $\mathcal{U}$ inherits a grading from $\mathcal{T}$. Since λ is idempotent, $\mathcal{U}$ is spanned by the identity and all $(X_1, \ldots, X_p)$ with $X_1 \leq \ldots \leq X_p$. Define a product in $\mathcal{U}$ by $uv = \lambda(u * v)$ for $u, v \in \mathcal{U}$.*

88   3. Algebras

The multiplication in $\mathcal{U}$ is associative. To see this, let $u, v, w \in \mathcal{U}$. Since $\lambda w = w$, it follows from Lemma 3.96.2 that

$$(uv)w = \lambda(uv * w) = \lambda(\lambda(u * v) * \lambda w) = \lambda((u * v) * w).$$

The conclusion follows since $*$ is associative. Thus $\mathcal{U}$ is an associative anticommutative algebra with identity.

The Algebra $B(\mathcal{A})$

Recall the notation $S = (H_1, \ldots, H_p) \in \mathbf{S}$. We may view each element $S \in \mathbf{S}$ as an element of $\mathcal{T}$.

Definition 3.98 *For $S \in \mathbf{S}$ define an element $b_S \in \mathcal{U}$ as follows: if $S = (\ )$, let $b_S = 1$, and for $S = (H_1, \ldots, H_p)$, let $b_S = \lambda(\eta S)$. Thus $b_S = 0$ if $\cap S = \emptyset$, and if $\cap S \neq \emptyset$, then*

$$b_S = \sum_{\pi \in Sym(p)} \mathrm{sign}\pi (H_{\pi 1}, H_{\pi 1} \cap H_{\pi 2}, \ldots, H_{\pi 1} \cap H_{\pi 2} \cap \ldots \cap H_{\pi p}).$$

Lemma 3.99 *Let $S, T \in \mathbf{S}$. Then $b_S b_T = b_{(S,T)}$.*

Proof. Let $S = (H_1, \ldots, H_p)$ and $T = (K_1, \ldots, K_q)$ where $H_i, K_j \in \mathcal{A}$. Using Lemma 3.96 we get

$$\begin{aligned} b_S b_T &= \lambda(b_S * b_T) \\ &= \lambda(\lambda(\eta S) * \lambda(\eta T)) \\ &= \lambda(\eta S * \eta T) \\ &= \lambda((H_1) * \ldots * (H_p) * (K_1) * \ldots * (K_q)) \\ &= \lambda\eta(H_1, \ldots, H_p, K_1, \ldots, K_q) \\ &= b_{(S,T)}. \end{aligned}$$

This completes the proof. $\square$

Definition 3.100 *Let*

$$B = B(\mathcal{A}) = \sum_{S \in \mathbf{S}} \mathbb{K} b_S.$$

Define $B_p(\mathcal{A}) = B_p = B \cap \mathcal{T}_p$. It follows from Lemma 3.99 that $B = \oplus_{p \geq 0} B_p$ is a graded subalgebra of $\mathcal{U}$.

Lemma 3.101 *If $S \in \mathbf{S}$ is dependent, then $b_S = 0$. In particular, $b_S = 0$ if $|S| > \ell$, so*

$$B = \bigoplus_{p=0}^{\ell} B_p.$$

Proof. Since S is dependent, $\cap S \neq \emptyset$. Let $S = (H_1, \ldots, H_p)$. If $S_k = (H_1, \ldots, \hat{H}_k, \ldots, H_p)$ is dependent for some k, then we have $b_S = (-1)^{k-1} b_{(H_k, S_k)} = (-1)^{k-1} b_{H_k} b_{S_k}$ and we are done by induction. Thus we may assume that S_k is independent for each k. It follows that $\cap S_k = \cap S$ for all k. If $\pi \in Sym(p)$, let ζ be the permutation defined by $\zeta k = \pi k$ for $1 \leq k \leq p-2$, $\zeta(p-1) = \pi(p)$ and $\zeta(p) = \pi(p-1)$. Then $\text{sign}\zeta = -\text{sign}\pi$, and the terms corresponding to π and ζ in Definition 3.98 cancel. □

Example 3.102 Let $\mathcal{A} = \{H_1, \ldots, H_n\}$ be a central 2–arrangement. Write $b_k = b_{H_k} = (H_k)$. Then we have

$$B(\mathcal{A}) = \mathcal{K} \oplus \bigoplus_{p=1}^{n} \mathcal{K} b_p \oplus \bigoplus_{k=1}^{n-1} \mathcal{K} b_k b_n.$$

We know B_0, B_1 and that $B_p = 0$ for $p > 2$. By definition

$$b_i b_j = b_{(H_i, H_j)} = (H_i, H_i \cap H_j) - (H_j, H_j \cap H_i).$$

Thus it is clear that B_2 is spanned by $b_k b_n$ for $1 \leq k < n$. It is equally clear that these generators are linearly independent and hence the sum is direct. The reader should compare this example with Example 3.8.

Example 3.103 Recall the affine 2–arrangement $\mathcal{A}$ defined by $Q(\mathcal{A}) = xy(x+y-1)$ in Example 1.6. Let $H_1 = \ker(x)$, $H_2 = \ker(y)$, and $H_3 = \ker(x+y-1)$. Note that $H_1 \cap H_2 \cap H_3 = \emptyset$. Write $e_i = e_{H_i}$ and $b_i = b_{H_i}$. Then $b_1 b_2 b_3 = 0$. We have

$$B(\mathcal{A}) = \mathcal{K} \oplus (\mathcal{K} b_1 \oplus \mathcal{K} b_2 \oplus \mathcal{K} b_3) \oplus (\mathcal{K} b_1 b_2 \oplus \mathcal{K} b_2 b_3 \oplus \mathcal{K} b_3 b_1).$$

The reader should compare this example with Example 3.46.

The Isomorphism of B and A

Recall the algebra $E(\mathcal{A})$. Note that for $S = (H_1, \ldots, H_p)$ and $\pi \in Sym(p)$ we have $\eta \pi S = (\text{sign}\pi)\eta S$ and hence $b_{\pi S} = (\text{sign}\pi)b_S$. This allows us to define the following map.

Definition 3.104 Define a $\mathcal{K}$–linear map $\psi : E \to B$ by $\psi e_S = b_S$. Since $e_S e_T = e_{(S,T)}$, the map ψ is a homomorphism of algebras.

Definition 3.105 Define a $\mathcal{K}$–linear map $\tau : T \to T$ by $\tau 1 = 0$, $\tau(X) = 1$ for $X \in L \setminus \{V\}$, and for $p \geq 2$ and $X_i \in L \setminus \{V\}$

$$\tau(X_1, \ldots, X_p) = (-1)^{p-1}(X_1, \ldots, X_{p-1}).$$

Lemma 3.106 *If $S \in \mathbf{S}$ and $\cap S \neq \emptyset$, then $\psi \partial e_S = \tau \psi e_S$.*

Proof. Suppose $S \in \mathbf{S}_p$. Then

$$\psi \partial e_S = \lambda(\sum_{k=1}^{p}(-1)^{k-1}\eta S_k)$$
$$= \lambda(\sum_{k=1}^{p}\sum_{\zeta \in W_k}(-1)^{k-1}(\text{sign}\zeta)(H_{\zeta 1},\ldots,\hat{H}_{\zeta k},\ldots,H_{\zeta p}))$$

where W_k is the group of permutations of $1,\ldots,\hat{k},\ldots,p$. On the other hand, since $\cap S \neq \emptyset$, we have $\tau\lambda = \lambda\tau$, and hence

$$\tau\psi e_S = \lambda(\sum \text{sign}\pi(H_{\pi 1},\ldots,H_{\pi(p-1)}))$$

where π ranges over $Sym(p)$. If $\pi \in Sym(p)$ and $\pi(p) = k$, define $\zeta \in W_k$ by $\zeta i = \pi i$ for $1 \leq i \leq k-1$, $\zeta i = \pi(i-1)$ for $i > k$. Then $\text{sign}\pi = (-1)^{p-k}\text{sign}\zeta$, and the sums $\psi \partial e_S$ and $\tau\psi e_S$ are equal term for term. □

Lemma 3.107 *The map $\psi : E \to B$ induces a surjection of algebras $\theta : A \to B$ such that $\theta a_S = b_S$.*

Proof. If $\cap S = \emptyset$, then $\psi(e_S) = b_S = 0$. If S is dependent, then $\psi\partial e_S = \tau\psi e_S = \tau b_S = 0$, so $\partial e_S \in \ker\psi$. Thus $I \subseteq \ker\psi$ and ψ induces a surjective map $\theta : A \to B$ such that $\theta a_S = b_S$. Since ψ is an algebra homomorphism, so is θ. □

Lemma 3.108 *If $X \in L$, let $B_X = \sum_{S \in \mathbf{S}_X} \mathcal{K}b_S$. Then*
  (1) $B(\mathcal{A}) = \bigoplus_{X \in L(\mathcal{A})} B_X(\mathcal{A})$,
  (2) if $Y \leq X$, then $B_Y(\mathcal{A}_X) = B_Y(\mathcal{A})$,
  (3) $B_p(\mathcal{A}) = \bigoplus_{X \in L_p(\mathcal{A})} B_X(\mathcal{A})$.

Proof. Assertion (1) is immediate from the definition of B. To prove (2) note that there is a natural inclusion $\mathcal{T}(\mathcal{A}_X) \to \mathcal{T}(\mathcal{A})$, and because intersections in L_X are the same as in L, there is a natural inclusion $\mathcal{U}(\mathcal{A}_X) \to \mathcal{U}(\mathcal{A})$. Hence $B(\mathcal{A}_X) \to B(\mathcal{A})$ is an inclusion. Assertion (1) and Lemma 3.107 prove (3). □

Lemma 3.109 *Suppose $\mathcal{A}$ is a central arrangement. Then τ induces a map $\tau : B(\mathcal{A}) \to B(\mathcal{A})$ which satisfies*
  (1) $\tau^2 = 0$,
  (2) if $b \in B_p$ and $u \in B$, then $\tau(bu) = \tau(b)u + (-1)^p b\tau(u)$,
  (3) if $\mathcal{A}$ is not empty, then the complex (B,τ) is acyclic.

Proof. Properties (1) and (2) follow from the corresponding facts for ∂_E. We argue (3) as in Lemma 3.13.3. □

Theorem 3.110 *Let $\mathcal{A}$ be an arrangement. Then $\theta : A(\mathcal{A}) \to B(\mathcal{A})$ is an isomorphism of graded $\mathcal{K}$-algebras.*

Proof. Assume first that $\mathcal{A}$ is a central arrangement of rank $r = r(\mathcal{A})$ with $T = T(\mathcal{A})$. Clearly, $\theta(A_X) \subseteq B_X$, so θ induces a map $\theta_X : A_X \to B_X$. It suffices to show that this map is an isomorphism for all $X \in L(\mathcal{A})$. We use induction on r. The assertion holds for the empty arrangement with $r = 0$ and $A(\mathcal{A}) = \mathcal{K} = B(\mathcal{A})$. Suppose $r > 0$. Let $X \in L(\mathcal{A})$ with $r(X) < r$. Then $r(\mathcal{A}_X) < r$, so by the induction hypothesis $\theta_X : A_X(\mathcal{A}_X) \to B_X(\mathcal{A}_X)$ is an isomorphism. We see from Proposition 3.31 that $A_X(\mathcal{A}_X) \simeq A_X(\mathcal{A})$ and from Lemma 3.108 that $B_X(\mathcal{A}_X) = B_X(\mathcal{A})$. It follows from the commutativity of the diagram

$$\begin{array}{ccc} A_X(\mathcal{A}_X) & \stackrel{\theta_X(\mathcal{A}_X)}{\longrightarrow} & B_X(\mathcal{A}_X) \\ \downarrow & & \downarrow \\ A_X(\mathcal{A}) & \stackrel{\theta_X(\mathcal{A})}{\longrightarrow} & B_X(\mathcal{A}) \end{array}$$

that $\theta_X(\mathcal{A})$ is an isomorphism for $X \in L$ with $r(X) < r$. It remains to prove the isomorphism for $X = T$. In the commutative diagram

$$\begin{array}{ccccccccc} 0 & \to & A_r & \to & A_{r-1} & \to & \cdots & \to & A_0 & \to & 0 \\ & & \theta_r \downarrow & & \theta_{r-1} \downarrow & & & & \theta_0 \downarrow & & \\ 0 & \to & B_r & \to & B_{r-1} & \to & \cdots & \to & B_0 & \to & 0 \end{array}$$

the horizontal maps are the respective boundary operators in the two acyclic complexes, so the sequences are exact. Since $B_p = \oplus_{X \in L_p} B_X$ and $A_p = \oplus_{X \in L_p} A_X$, the first part of the argument shows that all vertical maps except θ_r are isomorphisms. It follows from the diagram that θ_r is an isomorphism. This completes the argument because $A_r = A_T(\mathcal{A})$, $B_r = B_T(\mathcal{A})$ and $\theta_r = \theta_T(\mathcal{A})$.

Now assume that $\mathcal{A}$ is an affine arrangement. By Proposition 3.71, Corollary 3.73, and Lemma 3.108 we have

$$A(\mathcal{A}) \simeq \bigoplus_{X \in L} A_X(\mathcal{A}) \simeq \bigoplus_{X \in L} A_X(\mathcal{A}_X)$$
$$\simeq \bigoplus_{X \in L} B_X(\mathcal{A}_X) = \bigoplus_{X \in L} B_X(\mathcal{A}) = B(\mathcal{A})$$

as required. □

The following results are consequences of Corollary 3.74, Theorem 3.68, and Proposition 3.75 respectively.

Corollary 3.111 *The algebra $B(\mathcal{A})$ is a free graded $\mathcal{K}$-module. The $\mathcal{K}$-modules $B_X(\mathcal{A})$ for $X \in L$ and $B_p(\mathcal{A})$ for $p \geq 0$ are also free.*

Corollary 3.112 *The Poincaré polynomial of $B(\mathcal{A})$ is*

$$\mathrm{Poin}(B(\mathcal{A}), t) = \pi(\mathcal{A}, t).$$

Corollary 3.113 *If $X \in L(\mathcal{A})$, then $\mathrm{rank} B_X = (-1)^{r(X)} \mu(X)$.*

3.5 Differential Forms

In this section we study the algebra of differential forms generated by 1 and the differential forms $\omega_H = d\alpha_H/\alpha_H$ for $H \in \mathcal{A}$. This algebra was first computed by Arnold [6] for the braid arrangement. Brieskorn [41] defined it for all arrangements and showed that it is isomorphic to the cohomology algebra. Its isomorphism with $A(\mathcal{A})$ was established in [171] for central arrangements. In all these topological considerations the field was $\mathbb{C}$. Our presentation is based on [179], where the properties of $R(\mathcal{A})$ over an arbitrary field $\mathbb{K}$ were first studied. The results here extend the results of [179] to affine arrangements. It is important to note that the definition of $R(\mathcal{A})$ involves the polynomials α_H of degree 1, and thus this algebra is not obviously a combinatorial invariant of $\mathcal{A}$. Its combinatorial nature is a consequence of the main theorem of this section, which establishes an algebra isomorphism between $A(\mathcal{A})$ and $R(\mathcal{A})$. We also study the properties of $R(\mathcal{A})$ with respect to deletion and restriction. Let $(\mathcal{A}, \mathcal{A}', \mathcal{A}'')$ be a triple of arrangements with distinguished hyperplane $H_0 \in \mathcal{A}$. We construct linear maps $i : R(\mathcal{A}') \to R(\mathcal{A})$ and $j : R(\mathcal{A}) \to R(\mathcal{A}'')$ and prove that there is an exact sequence

$$0 \to R(\mathcal{A}') \xrightarrow{i} R(\mathcal{A}) \xrightarrow{j} R(\mathcal{A}'') \to 0.$$

The corresponding exact sequence for $A(\mathcal{A})$ will allow us to prove the isomorphism $A(\mathcal{A}) \simeq R(\mathcal{A})$ by induction.

The de Rham Complex

Let $(\mathcal{A}, V)$ be an affine arrangement. Let S be the symmetric algebra of V^* and let F be the quotient field of S. Sometimes it will be convenient to indicate the dependence of S and F on V. In this case we write $S = \mathbb{K}[V]$ and $F = \mathbb{K}(V)$. We view $F \otimes_\mathbb{K} V^*$ as a vector space over F by defining $f(g \otimes \alpha) = fg \otimes \alpha$ where $f, g \in F$ and $\alpha \in V^*$. There exists a unique $\mathbb{K}$–linear map $d : F \to F \otimes V^*$ such that $d(fg) = f(dg) + g(df)$ for $f, g \in F$ and $d\alpha \in \mathbb{K}$ for $\alpha \in V^*$. Recall that we have chosen a basis $x_1, \ldots, x_\ell$ for V^* so we may identify the symmetric algebra of V^* with the polynomial algebra $S = \mathbb{K}[x_1, \ldots, x_\ell]$ and its quotient field with the field of rational functions $F = \mathbb{K}(x_1, \ldots, x_\ell)$. In terms of this basis, the differential df is given by the usual formula

$$df = \sum_{i=1}^{\ell} \frac{\partial f}{\partial x_i} \otimes x_i = \sum_{i=1}^{\ell} \frac{\partial f}{\partial x_i} dx_i.$$

Note that $F \otimes V^* = Fdx_1 \oplus \ldots \oplus Fdx_\ell$.

Definition 3.114 *Let $\Omega(V)$ be the exterior algebra of the F–vector space $F \otimes V^*$ graded by $\Omega(V) = \oplus_{p=0}^{\ell} \Omega^p(V)$ where*

$$\Omega^p(V) = \bigoplus_{1 \leq i_1 < \ldots < i_p \leq \ell} Fdx_{i_1} \wedge \ldots \wedge dx_{i_p}.$$

For simplicity of notation, we write $\omega\eta = \omega\wedge\eta$ for $\omega, \eta \in \Omega(V)$. In particular, we write $dx_1\ldots dx_p = dx_1\wedge\ldots\wedge dx_p$. We identify Ω^0 with F. The elements of $\Omega^p(V)$ are called rational differential p–forms on V. We list some well-known properties of d for future reference.

Proposition 3.115 *The map $d: F \to F \otimes V^*$ may be extended in a unique way to a $\mathbb{K}$–linear map $d: \Omega(V) \to \Omega(V)$ with the following properties:*
(1) $d^2 = 0$,
(2) if $\omega \in \Omega^p(V)$ and $\eta \in \Omega(V)$, then $d(\omega\eta) = (d\omega)\eta + (-1)^p\omega(d\eta)$,
(3) if $\omega = \sum f_{i_1\ldots i_p} dx_{i_1}\ldots dx_{i_p}$ where $1 \leq i_1 < \ldots < i_p \leq \ell$ and $f_{i_1\ldots i_p} \in F$, then
$$d\omega = \sum_{j=1}^{\ell}\sum (\partial f_{i_1\ldots i_p}/\partial x_j)dx_j dx_{i_1}\ldots dx_{i_p}.$$

The Algebra $R(\mathcal{A})$

Let $\mathcal{K}$ be a commutative subring of the field $\mathbb{K}$.

Definition 3.116 *Let $\mathcal{A}$ be an affine arrangement. For $H \in \mathcal{A}$, let $\alpha_H \in S$ be a polynomial of degree 1 with $H = \ker(\alpha_H)$ and let $\omega_H = d\alpha_H/\alpha_H \in \Omega^1(V)$. Let $R = R(\mathcal{A})$ be the $\mathcal{K}$–subalgebra of $\Omega(V)$ generated by 1 and ω_H for $H \in \mathcal{A}$.*

Let $R_p = R \cap \Omega^p(V)$. Since R is generated by 1 and the 1–forms ω_H, it is naturally graded $R = \oplus_{p=0}^{\ell} R_p$.

To give the reader some intuitive idea why this algebra is again isomorphic to $A(\mathcal{A})$, we work out the analog of Examples 3.8 and 3.102.

Example 3.117 Let $\mathcal{A} = \{H_1,\ldots,H_n\}$ be a central 2–arrangement. Write $\omega_i = \omega_{H_i}$. Then
$$R(\mathcal{A}) = \mathcal{K} \oplus \bigoplus_{p=1}^{n}\mathcal{K}\omega_p \oplus \bigoplus_{k=1}^{n-1}\mathcal{K}\omega_k\omega_n.$$

We know that $R_0 = \mathcal{K}$ and that $R_p = 0$ for $p > 2$. By definition $\omega_1,\ldots,\omega_n$ span R_1 over $\mathcal{K}$. These 1–forms are linearly independent over $\mathcal{K}$ because the rational functions $1/\alpha_1,\ldots,1/\alpha_n$ are linearly independent over $\mathcal{K}$. Since $\omega_i^2 = 0$ and $\omega_i\omega_j = -\omega_j\omega_i$, the space R_2 is spanned over $\mathcal{K}$ by the $\omega_i\omega_j$ with $i < j$. In order to discover the remaining relations among these generators, let x, y be a basis for V^* and write $\alpha_i = a_i x + b_i y$ with $a_i, b_i \in \mathbb{K}$. Then $\omega_i = (a_i/\alpha_i)dx + (b_i/\alpha_i)dy$ and we have
$$d\alpha_i d\alpha_j = (a_i b_j - b_i a_j)dxdy.$$
Thus for any i, j, k we have
$$\alpha_k d\alpha_i d\alpha_j + \alpha_i d\alpha_j d\alpha_k + \alpha_j d\alpha_k d\alpha_i =$$

$$\det \begin{bmatrix} a_i & a_j & a_k \\ b_i & b_j & b_k \\ \alpha_i & \alpha_j & \alpha_k \end{bmatrix} dxdy = 0$$

because the third row is a linear combination of the first two. If we multiply this equation by $1/(\alpha_i\alpha_j\alpha_k)$ we get:

$$\omega_i\omega_j + \omega_j\omega_k + \omega_k\omega_i = 0.$$

In particular, we have $\omega_i\omega_j = \omega_i\omega_n - \omega_j\omega_n$ if $1 \leq i < j \leq n$, so R_2 is spanned by the elements $\omega_k\omega_n$ for $1 \leq k < n$. It remains to show that these elements are linearly independent over $\mathcal{K}$. Define an F-linear map $\partial : \Omega^2(V) \to \Omega^1(V)$ by $\partial(f dx dy) = f_x dy - f_y dx$. Then $\partial(\omega_i\omega_j) = \omega_j - \omega_i$. If $\sum_{k=1}^{n-1} c_k \omega_k \omega_n = 0$ with $c_k \in \mathcal{K}$, then applying ∂ gives $\sum_{k=1}^{n-1} c_k (\omega_n - \omega_k) = 0$. Since $\omega_1, \ldots, \omega_n$ are linearly independent over $\mathcal{K}$, we get $c_1 = \ldots = c_{n-1} = 0$. This proves the assertion.

Next consider the analog of Examples 3.46 and 3.103.

Example 3.118 Recall the affine 2–arrangement $\mathcal{A}$ defined by $Q(\mathcal{A}) = xy(x + y - 1)$ in Example 1.6. Let $H_1 = \ker(x)$, $H_2 = \ker(y)$, and $H_3 = \ker(x + y - 1)$. Note that $H_1 \cap H_2 \cap H_3 = \emptyset$. Write $\omega_i = \omega_{H_i}$. Then

$$\omega_1 = \frac{dx}{x}, \quad \omega_2 = \frac{dy}{y}, \quad \omega_3 = \frac{d(x+y-1)}{x+y-1}.$$

Note the relation $\omega_1\omega_2\omega_3 = 0$. We have

$$R(\mathcal{A}) = \mathcal{K} \oplus (\mathcal{K}\omega_1 \oplus \mathcal{K}\omega_2 \oplus \mathcal{K}\omega_3) \oplus (\mathcal{K}\omega_1\omega_2 \oplus \mathcal{K}\omega_2\omega_3 \oplus \mathcal{K}\omega_3\omega_1).$$

Lemma 3.119 *There exists a surjective homomorphism* $\gamma : A(\mathcal{A}) \to R(\mathcal{A})$ *of graded $\mathcal{K}$-algebras such that* $\gamma(a_H) = \omega_H$ *for all* $H \in \mathcal{A}$.

Proof. Define a $\mathcal{K}$-algebra homomorphism $\nu : E \to R$ by $\nu(e_H) = \omega_H$. To prove that ν induces a homomorphism $\gamma : A \to R$, we must show that $\nu(I) = 0$. Thus we need to show that if $\cap S = \emptyset$, then $\nu(e_S) = 0$, and that if $S = (H_1, \ldots, H_p)$ is dependent, then $\nu(\partial e_S) = 0$. In the first case it is easy to see that there exist $c_i \in \mathcal{K}$, not all zero, with $\sum_{i=1}^p c_i \alpha_i = 1$. Thus $\sum_{i=1}^p c_i(d\alpha_i) = 0$ and hence $d\alpha_1, \ldots, d\alpha_p$ are linearly dependent. Thus we have

$$\nu(e_S) = \omega_1 \cdots \omega_p = (d\alpha_1 \cdots d\alpha_p)/(\alpha_1 \cdots \alpha_p) = 0.$$

In the second case, since $\alpha_1, \ldots, \alpha_p$ is a linearly dependent set, there exist $c_i \in \mathcal{K}$, not all zero, with $\sum_{i=1}^p c_i\alpha_i = 0$. The following argument, suggested by M. Kervaire, is a simplification of the proof in [171]. We may assume that $c_p = -1$, so we have $\alpha_p = \sum_{k=1}^{p-1} c_k\alpha_k$ and $d\alpha_p = \sum_{k=1}^{p-1} c_k d\alpha_k$. Thus

(1) $$\omega_p = \sum_{k=1}^{p-1} \frac{c_k\alpha_k}{\alpha_p}\omega_k.$$

We get

$$\nu(\partial e_S) = \sum_{k=1}^{p}(-1)^{k-1}\omega_1\ldots\widehat{\omega_k}\ldots\omega_p$$
$$= \sum_{k=1}^{p-1}(-1)^{k-1}\omega_1\ldots\widehat{\omega_k}\ldots\omega_p + (-1)^{p-1}\omega_1\ldots\omega_{p-1}.$$

Substitute (1) to get

$$\nu(\partial e_S) = (\sum_{k=1}^{p-1}(-1)^{k-1}(-1)^{p-(k-1)}\frac{c_k\alpha_k}{\alpha_p})\omega_1\ldots\omega_{p-1}$$
$$+ (-1)^{p-1}\omega_1\ldots\omega_{p-1}$$
$$= ((-1)^p\frac{\sum_{k=1}^{p-1}c_k\alpha_k}{\alpha_p} + (-1)^{p-1})\omega_1\ldots\omega_{p-1}$$
$$= ((-1)^p + (-1)^{p-1})\omega_1\ldots\omega_{p-1}$$
$$= 0.$$

Thus $\nu(I) = 0$ and ν induces a surjection $\gamma : A \to R$ such that $\gamma(a_H) = \omega_H$. □

Deletion and Restriction

Let $\mathcal{A}$ be a nonempty arrangement, let $H_0 \in \mathcal{A}$, and let $(\mathcal{A}, \mathcal{A}', \mathcal{A}'')$ be the inductive triple with respect to H_0. Note that $R(\mathcal{A}')$ and $R(\mathcal{A})$ are both subalgebras of $\Omega(V)$ and that $R(\mathcal{A}') \subseteq R(\mathcal{A})$. We prove next that there is a short exact sequence of $\mathcal{K}$–modules

$$0 \to R(\mathcal{A}') \xrightarrow{i} R(\mathcal{A}) \xrightarrow{j} R(\mathcal{A}'') \to 0.$$

We define j with the help of the Leray residue map on differential forms. This definition is analogous to Pham's definition [185, Chap.III] in case $\mathbb{K} = \mathbb{C}$ and the forms are holomorphic. Let $\alpha_0 = \alpha_{H_0}$ and let S_0 be the localization of S at α_0. By definition, S_0 is the subring of F consisting of all f/g such that $f, g \in S$ and g is prime to α_0. Let $\rho : V^* \to H_0^*$ be the restriction map and let $y_i = \rho(x_i)$. We may extend ρ uniquely to a $\mathbb{K}$–algebra homomorphism $\rho : S_0 \to \mathbb{K}(H_0)$. Both existence and uniqueness follow from the formula

$$\rho(f/g) = f(y_1,\ldots,y_\ell)/g(y_1,\ldots,y_\ell).$$

Note that $g(y_1,\ldots,y_\ell) \neq 0$ because g is prime to α_0. Define a $\mathbb{K}$–subalgebra Ω_0 of $\Omega(V)$ by

$$\Omega_0 = \oplus_{p=0}^{\ell}\bigoplus_{i_1<\ldots<i_p}S_0 dx_{i_1}\ldots dx_{i_p}.$$

This subalgebra does not depend on the basis for V^*.

Lemma 3.120 *The map $\rho : S_0 \to \mathbb{K}(H_0)$ may be extended in a unique way to a $\mathbb{K}$–linear map $\rho : \Omega_0 \to \Omega(H_0)$ such that for $\omega, \eta \in \Omega_0$, $f \in S_0$, and $\beta \in V^*$ we have*

(1) $\rho(\omega\eta) = \rho(\omega)\rho(\eta)$,
(2) $\rho(f\omega) = \rho(f)\rho(\omega)$,
(3) $\rho(d\beta) = d\rho(\beta)$,
(4) *if $\omega = \sum f_{i_1\ldots i_p} dx_{i_1} \ldots dx_{i_p}$, then*

$$\rho(\omega) = \sum f_{i_1\ldots i_p}(y_1, \ldots, y_\ell) dy_{i_1} \ldots dy_{i_p}.$$

Proof. If $\omega = \sum f_{i_1\ldots i_p} dx_{i_1} \ldots dx_{i_p}$ and ρ has the properties (1)–(3), then

$$\begin{aligned}\rho(\omega) &= \sum \rho(f_{i_1\ldots i_p})\rho(dx_{i_1})\ldots\rho(dx_{i_p})\\ &= \sum \rho(f_{i_1\ldots i_p}) dy_{i_1}\ldots dy_{i_p}.\end{aligned}$$

This shows that $\rho(\omega)$ is given by (4) and proves uniqueness. To prove existence, define $\rho(\omega)$ by (4) and then (1)–(3) are clear. □

Lemma 3.121 *Suppose $\beta \in V^*$ and $\beta \neq 0$. If $\omega \in \Omega_0$ and $(d\beta)\omega = 0$, then there exists $\psi \in \Omega_0$ with $\omega = (d\beta)\psi$.*

Proof. Choose a basis $x_1, \ldots, x_\ell$ for V^* such that $\beta = x_1$. We may assume that ω is a p-form. Write $\omega = \sum f_{i_1\ldots i_p} dx_{i_1} \ldots dx_{i_p}$ where $f_{i_1\ldots i_p} \in S_0$ and the sum is over all $1 \leq i_1 < \ldots < i_p \leq \ell$. Then

$$0 = (dx_1)\omega = \sum f_{i_1\ldots i_p} dx_1 dx_{i_1} \ldots dx_{i_p}$$

where the sum is over all $2 \leq i_1 < \ldots < i_p \leq \ell$. Thus $f_{i_1\ldots i_p} = 0$ if $i_1 \geq 2$. □

Definition 3.122 *Say that $\phi \in \Omega(V)$ has at most a **simple pole** along H_0 if $\alpha_0 \phi \in \Omega_0$.*

Lemma 3.123 *Suppose $\phi \in \Omega(V)$ has at most a simple pole along H_0 and that $d\phi = 0$. Then there exist $\psi, \theta \in \Omega_0$ such that*

$$\phi = (d\alpha_0/\alpha_0)\psi + \theta.$$

The form $\rho(\psi) \in \Omega(H_0)$ is uniquely determined by ϕ.

Proof. For simplicity, write $\alpha = \alpha_0$. Let $\beta \in V^*$ be the degree 1 homogeneous part of α. Then $d\alpha = d\beta$. Since $d\phi = 0$, it follows from Lemma 3.115.2 that $d(\alpha\phi) = (d\alpha)\phi - \alpha(d\phi) = (d\alpha)\phi = (d\beta)\phi$. Since $\alpha\phi \in \Omega_0$ by hypothesis and $d\Omega_0 \subseteq \Omega_0$, it follows from Lemma 3.121 that there exists $\theta \in \Omega_0$ such that $d(\alpha\phi) = (d\beta)\theta$. Thus $(d\beta)\phi = (d\beta)\theta$, which implies $(d\beta)\alpha(\phi - \theta) = 0$. Since $\alpha(\phi - \theta) \in \Omega_0$, it follows from Lemma 3.121 that there exists $\psi \in \Omega_0$ such that $\alpha(\phi - \theta) = (d\beta)\psi = (d\alpha)\psi$. This proves the existence of θ and ψ.

3.5 Differential Forms

To prove the uniqueness of $\rho(\psi)$, it suffices to show that if $\psi, \theta \in \Omega_0$ and $(d\alpha/\alpha)\psi + \theta = 0$, then $\rho(\psi) = 0$. First note that $(d\beta)\theta = (d\alpha)\theta = 0$. It follows from Lemma 3.121 that there exists $\theta' \in \Omega_0$ such that $\theta = (d\beta)\theta'$. Now $(d\beta)(\psi + \alpha\theta') = (d\beta)\psi + \alpha\theta = (d\alpha)\psi + \alpha\theta = 0$. Since $\psi + \alpha\theta' \in \Omega_0$, we may apply Lemma 3.121 again to conclude that there exists $\theta'' \in \Omega_0$ with $\psi + \alpha\theta' = (d\beta)\theta'' = (d\alpha)\theta''$. Since $\rho(\alpha) = 0$, it follows from Lemma 3.120 that $\rho(\alpha\theta') = 0$ and $\rho((d\alpha)\theta'') = 0$. Thus $\rho(\psi) = 0$. □

Definition 3.124 *The uniquely determined form $\rho(\psi)$ is called the* **residue** *of ϕ along H_0. We denote it* $\mathrm{res}(\phi)$.

If $H \in \mathcal{A}$, then $d\omega_H = 0$, so $d(\omega_{H_1} \cdots \omega_{H_p}) = 0$ for all $H_1, \ldots, H_p \in \mathcal{A}$. Thus $d\phi = 0$ for all $\phi \in R(\mathcal{A})$. It is clear from the definition that each $\phi \in R(\mathcal{A})$ has at most a simple pole along H_0. Thus $\mathrm{res}(\phi)$ is defined for all $\phi \in R(\mathcal{A})$.

Lemma 3.125 *Suppose $H_1, \ldots, H_p \in \mathcal{A}'$. Then*
(1) $\mathrm{res}(\omega_{H_1} \cdots \omega_{H_p}) = 0$,
(2) $\mathrm{res}(\omega_{H_0}\omega_{H_1} \cdots \omega_{H_p}) = \omega_{H_0 \cap H_1} \cdots \omega_{H_0 \cap H_p}$,
(3) $\mathrm{res} R(\mathcal{A}) \subseteq R(\mathcal{A}'')$.

Proof. In case $p = 0$, formulas (1) and (2) are interpreted as $\mathrm{res}(1) = 0$ and $\mathrm{res}(\omega_{H_0}) = 1$. Let $\phi = \omega_{H_1} \cdots \omega_{H_p}$. We may choose $\psi = 0$ and $\theta = \phi$ in Lemma 3.123. This shows that $\mathrm{res}(\phi) = 0$ and proves (1). Now let $\phi = \omega_{H_0}\omega_{H_1} \cdots \omega_{H_p}$. We may choose $\psi = \omega_{H_1} \cdots \omega_{H_p}$ and $\theta = 0$ in Lemma 3.123. This shows that $\mathrm{res}(\phi) = \rho(\omega_{H_1} \cdots \omega_{H_p})$. By Lemma 3.120.1, we have $\rho(\omega_{H_1} \cdots \omega_{H_p}) = \rho(\omega_{H_1}) \cdots \rho(\omega_{H_p})$. It remains to show that $\rho(\omega_{H_i}) = \omega_{H_0 \cap H_i}$. If $H \in \mathcal{A}'$, then it follows from Lemma 3.120 that $\rho(\omega_H) = \rho(d\alpha_H/\alpha_H) = d\rho(\alpha_H)/\rho(\alpha_H)$. Since $\rho(\alpha_H)$ is a polynomial function on H_0 which defines the hyperplane $H_0 \cap H \in \mathcal{A}''$, we have $\rho(\omega_H) = \omega_{H_0 \cap H}$. This proves (2). To prove (3), note that since $\omega_{H_0}^2 = 0$, it follows from the definition of $R(\mathcal{A})$ and $R(\mathcal{A}')$ that $R(\mathcal{A}) = R(\mathcal{A}') + \omega_{H_0} R(\mathcal{A}')$. Thus (3) follows from (1) and (2). □

The Isomorphism of R and A

Theorem 3.126 *Let $\mathcal{A}$ be an arrangement and let $R(\mathcal{A})$ be the algebra of differential forms generated by 1 and $\omega_H = d\alpha_H/\alpha_H$. The map $\gamma : A(\mathcal{A}) \to R(\mathcal{A})$ induces an isomorphism of graded $\mathcal{K}$-algebras such that $\gamma(a_H) = \omega_H$.*

Theorem 3.127 *Let $(\mathcal{A}, \mathcal{A}', \mathcal{A}'')$ be a triple of arrangements with respect to $H_0 \in \mathcal{A}$. Let $i : R(\mathcal{A}') \to R(\mathcal{A})$ be the inclusion map and define $j : R(\mathcal{A}) \to R(\mathcal{A}'')$ by $j(\phi) = \mathrm{res}(\phi)$ for $\phi \in R(\mathcal{A})$, where $\mathrm{res}(\phi)$ is the residue of ϕ along H_0. Then there is an exact sequence:*

$$0 \to R(\mathcal{A}') \xrightarrow{i} R(\mathcal{A}) \xrightarrow{j} R(\mathcal{A}'') \to 0.$$

Proof. We prove Theorems 3.126 and 3.127 simultaneously by induction on $|\mathcal{A}|$. If $\mathcal{A}$ is empty, then $A(\mathcal{A}) = \mathcal{K} = R(\mathcal{A})$ and the first result holds. The second assumes that $\mathcal{A}$ is nonempty. If $|\mathcal{A}| = 1$, then $\mathcal{A}'$ and $\mathcal{A}''$ are empty arrangements. Let $\mathcal{A} = \{H\}$. Then $R(\mathcal{A}) = \mathcal{K} + \mathcal{K}\omega_H$ and $R(\mathcal{A}') = \mathcal{K} = R(\mathcal{A}'')$, so both statements are clear. If $|\mathcal{A}| > 1$, then we see from Lemma 3.125.3 that $jR(\mathcal{A}) \subseteq R(\mathcal{A}'')$ and from Lemma 3.125.2 that j is surjective. It follows from Lemma 3.125.1 that $ji = 0$, so $\mathrm{im}(i) \subseteq \mathrm{ker}(j)$. To prove that $\mathrm{ker}(j) \subseteq \mathrm{im}(i)$ consider the following diagram.

$$\begin{array}{ccccccccc} 0 & \to & A(\mathcal{A}') & \stackrel{i_A}{\to} & A(\mathcal{A}) & \stackrel{j_A}{\to} & A(\mathcal{A}'') & \to & 0 \\ & & \gamma' \downarrow & & \gamma \downarrow & & \gamma'' \downarrow & & \\ 0 & \to & R(\mathcal{A}') & \stackrel{i}{\to} & R(\mathcal{A}) & \stackrel{j}{\to} & R(\mathcal{A}'') & \to & 0 \end{array}$$

The diagram is commutative. This is clear for the left square by the definitions of i_A and i. For the right square it follows from Lemma 3.125. The top row is exact by Theorem 3.65. We may assume by the induction hypothesis in Theorem 3.126 that γ' and γ'' are isomorphisms. A diagram chase shows that $\mathrm{ker}(j) \subseteq \mathrm{im}(i)$. This proves that the second row of the diagram is exact. Thus Theorem 3.127 holds for $\mathcal{A}$. It follows from the Five Lemma that γ is an isomorphism, so Theorem 3.126 is also established for $\mathcal{A}$. □

The next results follow from Corollary 3.74 and Theorem 3.68.

Corollary 3.128 *The algebra $R(\mathcal{A})$ is a free graded $\mathcal{K}$-module. The $\mathcal{K}$-module $R_p(\mathcal{A})$ is free for $p \geq 0$.*

Corollary 3.129 *Let $\mathcal{A}$ be an arrangement and let $R(\mathcal{A})$ be the algebra of differential forms generated by 1 and $\omega_H = d\alpha_H/\alpha_H$. The Poincaré polynomial of $R(\mathcal{A})$ is*

$$\mathrm{Poin}(R(\mathcal{A}), t) = \pi(\mathcal{A}, t).$$

Definition 3.130 *For $X \in L$ let $R_X = R_X(\mathcal{A}) = \sum \mathcal{K} \omega_{H_1} \cdots \omega_{H_p}$ where the sum is over all $(H_1, \ldots, H_p) \in \mathbf{S}_X$.*

Proposition 3.131 *We have*

$$R_p = \bigoplus_{X \in L_p} R_X.$$

Proof. The sum is direct because $R_X = \gamma(A_X)$, γ is an isomorphism, and $A_p = \bigoplus_{X \in L_X} A_X$ by Corollary 3.73. □

4. Free Arrangements

In this chapter we assume that all arrangements are central and use "arrangement" in place of "central hyperplane arrangement." Section 4.1 contains basic definitions. In Section 4.2 we define free arrangements and establish their fundamental properties. If $\mathcal{A}$ is free, then we can associate with it a collection of nonnegative integers, called its exponents, $\exp \mathcal{A} = \{b_1, \ldots, b_\ell\}$. These integers are unique up to order, but they are not necessarily distinct. In Section 4.3 we prove the Addition–Deletion Theorem 4.51 following [226]. It asserts that if $(\mathcal{A}, \mathcal{A}', \mathcal{A}'')$ is a triple, then any two of the following statements imply the third:

$$\begin{aligned} \mathcal{A} \text{ is free with } \exp \mathcal{A} &= \{b_1, \ldots, b_{\ell-1}, b_\ell\}, \\ \mathcal{A}' \text{ is free with } \exp \mathcal{A}' &= \{b_1, \ldots, b_{\ell-1}, b_\ell - 1\}, \\ \mathcal{A}'' \text{ is free with } \exp \mathcal{A}'' &= \{b_1, \ldots, b_{\ell-1}\}. \end{aligned}$$

This result leads to the definition of inductively free arrangements. We give several examples and prove that a supersolvable arrangement is inductively free. In Section 4.4 we define the module $\Omega^p(\mathcal{A})$ of logarithmic p-forms with poles on the hypersurface $N(\mathcal{A})$. We show that the complex $\Omega^\cdot(\mathcal{A})$ is closed under exterior product and that $\Omega^1(\mathcal{A})$ is the dual of $D(\mathcal{A})$. We also study several lattice homology theories in this chapter. In Section 4.5 we construct a simplicial complex $\mathsf{F}(\mathcal{A})$ associated to $L(\mathcal{A})$ by Folkman [85]. We compute its homology groups and show that $\mathsf{F}(\mathcal{A})$ has the homotopy type of a wedge of spheres. We also construct another chain complex whose homology is naturally isomorphic to the algebra $B(\mathcal{A})$, defined in Section 3.4. We show how these constructions are related. These lattice homology theories are part of a more general theory essentially due to K. Baclawski [16]. In Section 4.6 we generalize these constructions to order complexes with arbitrary functor coefficient. This allows proof of an important technical result, Theorem 4.128, due to Yuzvinsky [254]. It is used in the proof of Theorem 4.136, first obtained in [212], which gives a formula for the characteristic polynomial of any arrangement.

When this formula is applied to a free arrangement, it yields the Factorization Theorem 4.137 of [228], which asserts that if $\mathcal{A}$ is a free ℓ-arrangement with $\exp \mathcal{A} = \{b_1, \ldots, b_\ell\}$, then

$$\pi(\mathcal{A}, t) = (1 + b_1 t) \cdots (1 + b_\ell t).$$

Thus the exponents of a free arrangement are determined by combinatorial data. The class of free arrangements contains the important class we call **reflection arrangements**, which is the subject of Chapter 6.

4.1 The Module $D(\mathcal{A})$

Derivations

Recall that $S = S(V^*)$ is the symmetric algebra of the dual space V^* of V. If $x_1, \ldots, x_\ell$ is a basis for V^*, then $S \simeq \mathbb{K}[x_1, \ldots, x_\ell]$. We identify S with the polynomial ring $\mathbb{K}[x_1, \ldots, x_\ell]$ by this isomorphism.

Definition 4.1 *Let $\mathrm{Der}_\mathbb{K}(S)$ be the set of $\mathbb{K}$–linear maps $\theta : S \to S$ such that*
$$\theta(fg) = f\theta(g) + g\theta(f) \qquad f, g \in S.$$
*An element of $\mathrm{Der}_\mathbb{K}(S)$ is called a **derivation** of S over $\mathbb{K}$.*

For $f \in S$ and $\theta_1, \theta_2 \in \mathrm{Der}_\mathbb{K}(S)$, define $f\theta_1 \in \mathrm{Der}_\mathbb{K}(S)$ and $\theta_1 + \theta_2 \in \mathrm{Der}_\mathbb{K}(S)$ by $f\theta_1(g) = f(\theta_1(g))$ and $(\theta_1 + \theta_2)(g) = \theta_1(g) + \theta_2(g)$ for any $g \in S$. Any $\mathbb{K}$–linear map from V^* to S can be extended uniquely to a derivation of S over $\mathbb{K}$. In particular, for any $v \in V$ there exists a unique $D_v \in \mathrm{Der}_\mathbb{K}(S)$ such that $D_v(\alpha) = \alpha(v)$ for any $\alpha \in V^*$. Let $e_1, \ldots, e_\ell \in V$ be the dual basis of $x_1, \ldots, x_\ell$. Define
$$D_i = D_{e_i} \qquad 1 \le i \le \ell.$$
Then D_i is the usual derivation $\partial/\partial x_i$:
$$D_i(f) = \partial f/\partial x_i, \qquad f \in S.$$
It is easy to see that $D_1, \ldots, D_\ell$ is a basis for $\mathrm{Der}_\mathbb{K}(S)$ over S. Thus any derivation θ of S over $\mathbb{K}$ is expressed uniquely as
$$\theta = f_1 D_1 + \cdots + f_\ell D_\ell, \qquad f_1, \ldots, f_\ell \in S.$$
It follows that $\mathrm{Der}_\mathbb{K}(S)$ is a free S–module of rank ℓ.

Let S_p denote the $\mathbb{K}$–vector subspace of S consisting of 0 and the homogeneous polynomials of degree p for $p \ge 0$. For $p < 0$ define $S_p = 0$. Then
$$S = \bigoplus_{p \in \mathbb{Z}} S_p$$
is a graded $\mathbb{K}$–algebra. It follows that $\deg x = 1$ for $x \in V^*$ and $x \ne 0$.

Definition 4.2 *A nonzero element $\theta \in \mathrm{Der}_\mathbb{K}(S)$ is **homogeneous of polynomial degree** p if $\theta = \sum_{k=1}^\ell f_k D_k$ and $f_k \in S_p$ for $1 \le k \le \ell$. In this case we write $\mathrm{pdeg}\,\theta = p$. Note that $\mathrm{pdeg}\,D_i = 0$. Let $\mathrm{Der}_\mathbb{K}(S)_p$ denote the vector space* consisting of all homogeneous elements of pdegree p for $p \geq 0$. Let $\mathrm{Der}_{\mathbb{K}}(S)_p = 0$ if $p < 0$.

With this pdegree function $\mathrm{Der}_{\mathbb{K}}(S)$ is a graded S–module

$$\mathrm{Der}_{\mathbb{K}}(S) = \bigoplus_{p \in \mathbb{Z}} \mathrm{Der}_{\mathbb{K}}(S)_p.$$

If we view derivations as a subset of the set of $\mathbb{K}$–linear endomorphisms of S, then there is another natural grading of $\mathrm{Der}_{\mathbb{K}}(S)$.

Definition 4.3 *A nonzero element $\theta \in \mathrm{Der}_{\mathbb{K}}(S)$ is* **homogeneous of total degree** r *if $\theta(S_q) \subseteq S_{r+q}$. In this case we write $\mathrm{tdeg}\,\theta = r$. Note that $\mathrm{tdeg}\,\theta = \mathrm{pdeg}\,\theta - 1$. In particular, $\mathrm{tdeg}\, D_i = -1$.*

In this chapter, polynomial degree is the natural grading in all formulas and proofs. In Chapter 6, functoriality requires the use of total degree. In our earlier work we sometimes used degree to denote total degree, and called the polynomial degree of θ its exponent. This led to some confusion. We hope this new terminology will clarify the issue. See also Definitions 4.62 and 4.63.

Definition 4.4 *For any $f \in S$, define*

$$D(f) = \{\theta \in \mathrm{Der}_{\mathbb{K}}(S) \mid \theta(f) \in fS\}.$$

Note that $D(f)$ is an S–submodule of $\mathrm{Der}_{\mathbb{K}}(S)$.

Definition 4.5 *Let $\mathcal{A}$ be an arrangement in V with defining polynomial*

$$Q(\mathcal{A}) = \prod_{H \in \mathcal{A}} \alpha_H$$

where $H = \ker(\alpha_H)$. Define the **module of $\mathcal{A}$–derivations** *by*

$$D(\mathcal{A}) = D(Q(\mathcal{A})).$$

Clearly, $D(\mathcal{A})$ does not depend on the choice of $Q(\mathcal{A})$. In particular, $D(\Phi_\ell) = \mathrm{Der}_{\mathbb{K}}(S)$ because $Q(\Phi_\ell) = 1$. An element of $D(\mathcal{A})$ is called a **derivation tangent to $\mathcal{A}$**. This terminology is justified by the topological significance of the module $D(\mathcal{A})$ in case $\mathbb{K} = \mathbb{C}$; see Proposition 5.17.

Example 4.6 Let $\mathcal{A}$ be the Boolean arrangement defined by $Q(\mathcal{A}) = x_1 \cdots x_\ell$. Then

$$\sum_{i=1}^{\ell} f_i D_i \in D(\mathcal{A})$$

$$\Leftrightarrow \sum_{i=1}^{\ell} f_i(\partial(x_1\cdots x_\ell)/\partial x_i) \in x_1\cdots x_\ell S$$

$$\Leftrightarrow (x_1\cdots x_\ell)\sum_{i=1}^{\ell}(f_i/x_i) \in x_1\cdots x_\ell S$$

$$\Leftrightarrow f_i \in x_i S \qquad (1 \leq i \leq \ell).$$

This implies that $D(\mathcal{A})$ is a free S-module with basis $\{x_1 D_1, \ldots, x_\ell D_\ell\}$.

Basic Properties

Definition 4.7 *The **Euler derivation** $\theta_E \in \mathrm{Der}_{\mathbb{K}}(S)$ is defined by*

$$\theta_E = \sum_{i=1}^{\ell} x_i D_i.$$

For any homogeneous $f \in S$,

$$\theta_E(f) = (\deg f)f.$$

Thus θ_E is independent of the choice of $\{x_1, \ldots, x_\ell\}$. Taking $f = Q = Q(\mathcal{A})$, we get $\theta_E(Q) = |\mathcal{A}|Q \in QS$. Thus $\theta_E \in D(\mathcal{A})$ for any arrangement $\mathcal{A}$.

Proposition 4.8

$$D(\mathcal{A}) = \bigcap_{H \in \mathcal{A}} D(\alpha_H) = \{\theta \in \mathrm{Der}_{\mathbb{K}}(S) \mid \theta(\alpha_H) \in \alpha_H S \text{ for all } H \in \mathcal{A}\}.$$

Proof. It is sufficient to prove

$$D(f_1 f_2) = D(f_1) \cap D(f_2)$$

for any $f_1, f_2 \in S$ such that f_1 and f_2 are coprime. If $\theta \in \mathrm{Der}_{\mathbb{K}}(S)$, then

$$\begin{aligned}
\theta &\in D(f_1 f_2) \\
&\Leftrightarrow \theta(f_1 f_2) \in f_1 f_2 S \\
&\Leftrightarrow f_1 \theta(f_2) + f_2 \theta(f_1) \in f_1 f_2 S \\
&\Leftrightarrow \theta(f_i) \in f_i S \quad (i = 1, 2) \\
&\Leftrightarrow \theta \in D(f_1) \cap D(f_2)
\end{aligned}$$

as required. $\square$

Corollary 4.9 *Let $\mathcal{A}_1$ and $\mathcal{A}_2$ be two arrangements in V such that $\mathcal{A}_1 \subseteq \mathcal{A}_2$. Then $D(\mathcal{A}_1) \supseteq D(\mathcal{A}_2)$.*

4.1 The Module $D(\mathcal{A})$

Proposition 4.10 *Let $D(\mathcal{A})_p = D(\mathcal{A}) \cap \mathrm{Der}_{\mathbb{K}}(S)_p$. Then*

$$D(\mathcal{A}) = \bigoplus_{p \in \mathbb{Z}} D(\mathcal{A})_p.$$

Thus $D(\mathcal{A})$ is a graded S-submodule of $\mathrm{Der}_{\mathbb{K}}(S)$.

Proof. Decompose $\theta \in D(\mathcal{A})$ into homogeneous components

$$\theta = \theta_0 + \theta_1 + \cdots$$

where θ_p is zero or homogeneous of pdegree $p \geq 0$. Since the ideal QS is generated by the homogeneous polynomial Q, each homogeneous component $\theta_p(Q)$ of $\theta(Q)$ also lies in QS. This shows that $\theta_p \in D(\mathcal{A})$ for $p \geq 0$. □

Definition 4.11 *If $\theta \in \mathrm{Der}_{\mathbb{K}}(S)$, then $\theta = \sum \theta(x_i) D_i$. Given derivations $\theta_1, \ldots, \theta_\ell \in D(\mathcal{A})$, define the **coefficient matrix** $\mathsf{M}(\theta_1, \ldots, \theta_\ell)$ by $\mathsf{M}_{i,j} = \theta_j(x_i)$.*

Thus

$$\mathsf{M}(\theta_1, \ldots, \theta_\ell) = \begin{bmatrix} \theta_1(x_1) & \cdots & \theta_\ell(x_1) \\ \cdot & \cdot & \cdot \\ \cdot & \cdot & \cdot \\ \cdot & \cdot & \cdot \\ \theta_1(x_\ell) & \cdots & \theta_\ell(x_\ell) \end{bmatrix}$$

and $\theta_j = \sum \mathsf{M}_{i,j} D_i$.

Proposition 4.12 *If $\theta_1, \ldots, \theta_\ell \in D(\mathcal{A})$, then $\det \mathsf{M}(\theta_1, \ldots, \theta_\ell) \in QS$.*

Proof. This is clear for $\mathcal{A} = \Phi_\ell$ since $Q(\mathcal{A}) = 1$. Let $H \in \mathcal{A}$ and let $H = \ker(\alpha_H)$, where $\alpha_H = \sum_{i=1}^{\ell} c_i x_i \in V^*$. We may assume that $c_i = 1$ for some i. Then

$$\det \mathsf{M}(\theta_1, \ldots, \theta_\ell) = \det \begin{bmatrix} \theta_1(x_1) & \cdots & \theta_\ell(x_1) \\ \cdot & \cdot & \cdot \\ \theta_1(\alpha_H) & \cdots & \theta_\ell(\alpha_H) \\ \cdot & \cdot & \cdot \\ \theta_1(x_\ell) & \cdots & \theta_\ell(x_\ell) \end{bmatrix} \in \alpha_H S.$$

Since H is arbitrary, $\det \mathsf{M}(\theta_1, \ldots, \theta_\ell)$ is divisible by all α_H, and hence by Q. □

Let $(\mathcal{A}_1, V_1)$ and $(\mathcal{A}_2, V_2)$ be two arrangements. Let $S_i = S(V_i^*)$ for $i = 1, 2$ and let $V = V_1 \oplus V_2$. Then S_1 and S_2 may be regarded as $\mathbb{K}$-subalgebras of $S = S(V^*)$. An element $\theta \in \mathrm{Der}(S_1)$ is uniquely extended to an element ϑ of $\mathrm{Der}(S)$ such that $\vartheta|_{S_2} = 0$. By this extension, $\mathrm{Der}(S_i)$ may be regarded as a subset of $\mathrm{Der}(S)$ for $i = 1, 2$. The following is easy to see.

Proposition 4.13 $\mathrm{Der}(S) = S\mathrm{Der}(S_1) \oplus S\mathrm{Der}(S_2)$.

Proposition 4.14 *For two arrangements $(\mathcal{A}_1, V_1)$ and $(\mathcal{A}_2, V_2)$, we have*

$$D(\mathcal{A}_1 \times \mathcal{A}_2) = SD(\mathcal{A}_1) \oplus SD(\mathcal{A}_2).$$

Proof. Write $\mathcal{A} = \mathcal{A}_1 \times \mathcal{A}_2$. For $i = 1, 2$ let $S_i = S(V_i^*)$ and let $Q_i \in S_i$ be defining polynomials for $\mathcal{A}_i$. Then $Q_1 Q_2$ is a defining polynomial for $\mathcal{A}$. Since an element of $\mathrm{Der}(S_1)$ annihilates every element in S_2, we get

$$SD(\mathcal{A}_1) \subseteq D(\mathcal{A}).$$

Thus we obtain

$$SD(\mathcal{A}_1) \oplus SD(\mathcal{A}_2) \subseteq D(\mathcal{A}).$$

By Proposition 4.13, any element $\theta \in D(\mathcal{A})$ can be written as $\theta = \theta_1 + \theta_2$ for some $\theta_i \in S\mathrm{Der}(S_i)$ with $i = 1, 2$. By symmetry we only have to prove $\theta_1 \in SD(\mathcal{A}_1)$. For that purpose we may assume that $\theta = \theta_1$. Since

$$Q_1 Q_2 S \ni \theta(Q_1 Q_2) = Q_2 \theta(Q_1) + Q_1 \theta(Q_2) = Q_2 \theta(Q_1),$$

we have $\theta(Q_1) \in Q_1 S$. Let $G = \{g_1, g_2, \ldots\}$ be a $\mathbb{K}$-basis for S_2. For example, take G to be the set of all monomials. Note that G is linearly independent over S_1 also. There is a unique expression

$$\theta = \sum_{i \geq 1} g_i \eta_i$$

with $\eta_i \in \mathrm{Der}(S_1)$. There is also a unique expression

$$\theta(Q_1) = Q_1 \sum_{i \geq 1} g_i h_i$$

with $h_i \in S_1$. Thus we have

$$\sum_{i \geq 1} g_i (h_i Q_1) = \theta(Q_1) = \sum_{i \geq 1} g_i \eta_i(Q_1).$$

By the uniqueness of the expression, we have for $i \geq 1$

$$\eta_i(Q_1) = h_i Q_1 \in Q_1 S_1,$$

so $\eta_i \in D(\mathcal{A}_1)$. Thus $\theta = \sum_i g_i \eta_i \in SD(\mathcal{A}_1)$. □

4.2 Free Arrangements

Saito's Criterion

Definition 4.15 *An arrangement $\mathcal{A}$ is called a **free arrangement** if $D(\mathcal{A})$ is a free module over S.*

Example 4.16 Let $\mathcal{A} = \Phi_\ell$. Then $Q(\mathcal{A}) = 1$ and $D_1, \ldots, D_\ell$ is a basis for $D(\mathcal{A}) = \mathrm{Der}_\mathbb{K}(S)$. Thus the empty arragement is free.

Example 4.17 Let $\mathcal{A}$ be the Boolean arragement defined by $Q(\mathcal{A}) = x_1 \cdots x_\ell$. It follows from the calculation in Example 4.6 that $\mathcal{A}$ is a free arrangement.

Proposition 4.18 *If $\mathcal{A}$ is a free arrangement, then $D(\mathcal{A})$ has a basis consisting of ℓ homogeneous elements.*

Proof. Let r be the rank of the free S-module $D(\mathcal{A})$; see Definition A.4. Note that
$$Q\mathrm{Der}_\mathbb{K}(S) \subseteq D(\mathcal{A}) \subseteq \mathrm{Der}_\mathbb{K}(S).$$
Since $\mathrm{Der}_\mathbb{K}(S)$ contains the ℓ linearly independent elements $D_1, \ldots, D_\ell$, and $Q\mathrm{Der}_\mathbb{K}(S)$ contains the ℓ linearly independent elements $QD_1, \ldots, QD_\ell$, it follows from Proposition A.3.1 that $\ell \leq r \leq \ell$. By Theorem A.20, we can choose a basis consisting of ℓ homogeneous elements in $D(\mathcal{A})$. □

Let $\theta_1, \ldots, \theta_\ell \in D(\mathcal{A})$. Recall that the (i,j)-entry of the $\ell \times \ell$ matrix $\mathsf{M}(\theta_1, \ldots, \theta_\ell)$ is $\theta_j(x_i)$. The following criterion is very useful.

Theorem 4.19 (Saito's criterion) *Given $\theta_1, \ldots, \theta_\ell \in D(\mathcal{A})$, the following two conditions are equivalent:*
(1) $\det \mathsf{M}(\theta_1, \ldots, \theta_\ell) \doteq Q(\mathcal{A})$,
(2) $\theta_1, \ldots, \theta_\ell$ *form a basis for $D(\mathcal{A})$ over S.*

Proof. $(1) \Rightarrow (2)$ First note that the derivations $\theta_1, \ldots, \theta_\ell$ are linearly independent over S because $\det \mathsf{M}(\theta_1, \ldots, \theta_\ell) \neq 0$. We may assume that $\det \mathsf{M}(\theta_1, \ldots, \theta_\ell) = Q$. It suffices to show that $\theta_1, \ldots, \theta_\ell$ generate $D(\mathcal{A})$ over S. Let $\eta \in D(\mathcal{A})$. We shall show that $\eta \in S\theta_1 + \cdots + S\theta_\ell$. Since
$$\theta_i = \sum_{j=1}^\ell \theta_i(x_j) D_j,$$
Cramer's rule implies that
$$QD_j \in S\theta_1 + \cdots + S\theta_\ell.$$
Write
$$Q\eta = \sum_{j=1}^\ell f_j \theta_j.$$
By Proposition 4.12, $\det \mathsf{M}(\theta_1, \ldots, \theta_{i-1}, \eta, \theta_{i+1}, \ldots, \theta_\ell) \in QS$. Thus
$$\begin{aligned}
Q \det \mathsf{M}(\theta_1, \ldots, \theta_{i-1}, \eta, \theta_{i+1}, \ldots, \theta_\ell) &= \det \mathsf{M}(\theta_1, \ldots, \theta_{i-1}, Q\eta, \theta_{i+1}, \ldots, \theta_\ell) \\
&= \det \mathsf{M}(\theta_1, \ldots, \theta_{i-1}, f_i \theta_i, \theta_{i+1}, \ldots, \theta_\ell) \\
&= f_i \det \mathsf{M}(\theta_1, \ldots, \theta_\ell) \\
&= f_i Q \\
&\in Q^2 S.
\end{aligned}$$

Thus $f_i \in QS$ for each i. This shows that

$$\eta = \sum_{i=1}^{\ell}(f_i/Q)\theta_i \in S\theta_1 + \cdots + S\theta_\ell.$$

(2) $\Rightarrow$ (1) By Proposition 4.12, we can write $\det \mathsf{M}(\theta_1,\ldots,\theta_\ell) = fQ$ for some $f \in S$. Fix $H \in \mathcal{A}$. We can assume that $H = \ker(x_1)$. Then $Q_H = Q/x_1$ is a defining polynomial for $\mathcal{A} \setminus \{H\}$. Define $\eta_1 = QD_1$ and for $2 \leq i \leq \ell$ let $\eta_i = Q_H D_i$. These derivations are in $D(\mathcal{A})$. Since each η_i is an S–linear combination of $\theta_1,\ldots,\theta_\ell$, there exists an $\ell \times \ell$ matrix N with entries in S such that $\mathsf{M}(\eta_1,\ldots,\eta_\ell) = \mathsf{M}(\theta_1,\ldots,\theta_\ell)\mathsf{N}$. Thus we have

$$QQ_H^{\ell-1} = \det \mathsf{M}(\eta_1,\ldots,\eta_\ell) \in \det \mathsf{M}(\theta_1,\ldots,\theta_\ell)S = fQS.$$

Therefore f divides $Q_H^{\ell-1}$. This is true for all $H \in \mathcal{A}$. Since the polynomials $\{Q_H^{\ell-1}\}_{H\in\mathcal{A}}$ have no common factor, $f \in \mathbb{K}^*$. □

Example 4.20 Let $\ell = 2$. Assume that $\mathcal{A} \neq \Phi_\ell$ and let $Q = Q(\mathcal{A})$. Choose coordinates so that $\ker(x_1) \in \mathcal{A}$ and let $Q = x_1 Q_0$. Let $\theta_E = x_1 D_1 + x_2 D_2$ be the Euler derivation 4.7 and let $\theta = Q_0 D_2$. Then $\theta_E, \theta \in D(\mathcal{A})$. Moreover

$$\det \mathsf{M}(\theta_E, \theta) = \det \begin{bmatrix} x_1 & 0 \\ x_2 & Q_0 \end{bmatrix} = Q.$$

It follows from Saito's criterion 4.19 that θ_E and θ form a basis for $D(\mathcal{A})$. This implies that all 2–arrangements are free.

Example 4.21 Let $\mathcal{A}$ be the Boolean arrangement defined by $Q(\mathcal{A}) = x_1 \cdots x_\ell$. We showed in Example 4.6 that $\theta_i = x_i D_i$ for $1 \leq i \leq \ell$ form a basis for $D(\mathcal{A})$. This can be verifed by applying Saito's criterion 4.19:

$$\det \mathsf{M}(\theta_1,\ldots,\theta_\ell) = \det \begin{bmatrix} x_1 & & 0 \\ & \ddots & \\ 0 & & x_\ell \end{bmatrix} = x_1 \cdots x_\ell = Q.$$

Example 4.22 Recall the braid arrangement of Example 1.10 defined by

$$Q(\mathcal{A}) = \prod_{1 \leq i < j \leq \ell}(x_i - x_j).$$

For $1 \leq k \leq \ell$ define

$$\theta_k = \sum_{n=1}^{\ell} x_n^{k-1} D_n.$$

Then

$$\theta_k(x_i - x_j) = x_i^{k-1} - x_j^{k-1} \in (x_i - x_j)S$$

for $1 \leq k \leq \ell$ and $1 \leq i < j \leq \ell$. Thus $\theta_1, \ldots, \theta_\ell \in D(\mathcal{A})$ by Proposition 4.8. Moreover

$$\det \mathsf{M}(\theta_1, \ldots, \theta_\ell) = \det \begin{bmatrix} 1 & x_1 & \cdots & x_1^{\ell-1} \\ \cdot & \cdot & & \cdot \\ \cdot & \cdot & & \cdot \\ \cdot & \cdot & & \cdot \\ 1 & x_\ell & \cdots & x_\ell^{\ell-1} \end{bmatrix}$$

is the Vandermonde determinant and thus equals $\prod_{1 \leq i < j \leq \ell}(x_i - x_j) = Q$. It follows from Saito's criterion 4.19 that $\theta_1, \ldots, \theta_\ell$ is a basis for $D(\mathcal{A})$. Thus $\mathcal{A}$ is a free arrangement.

In fact, $\mathcal{A}$ is a reflection arrangement and we will show in Chapter 6 that all reflection arrangements are free.

Exponents

Theorem 4.23 *Let $\theta_1, \ldots, \theta_\ell \in D(\mathcal{A})$ be homogeneous and linearly independent over S. Then $\mathcal{A}$ is free with basis $\theta_1, \ldots, \theta_\ell$ if and only if*

$$\sum_{i=1}^{\ell} \mathrm{pdeg}\,\theta_i = |\mathcal{A}|.$$

Proof. Since $\theta_1, \ldots, \theta_\ell$ are linearly independent, $\det \mathsf{M}(\theta_1, \ldots, \theta_\ell) \neq 0$. By Proposition 4.12, we may write $\det \mathsf{M}(\theta_1, \ldots, \theta_\ell) = fQ$ with some nonzero homogeneous polynomial $f \in S$. Since $\deg \det \mathsf{M}(\theta_1, \ldots, \theta_\ell) = \sum_{i=1}^{\ell} \mathrm{pdeg}\,\theta_i = |\mathcal{A}| = \deg Q$, we see that $f \in \mathbb{K}^*$. The conclusion follows from Saito's criterion 4.19. $\square$

Example 4.24 Let $\mathcal{A}$ be the arrangement consisting of all hyperplanes through the origin in an ℓ–dimensional vector space over a finite field of q elements, $\mathbb{K} = \mathbb{F}_q$. We showed after Definition 1.11 that $|\mathcal{A}| = 1 + q + q^2 + \cdots + q^{\ell-1}$. For $1 \leq i \leq \ell$ define

$$\theta_i = \sum_{j=1}^{\ell} x_j^{q^{i-1}} D_j.$$

We use the fact that $c = c^q$ for any $c \in \mathbb{K}$ to show that $\theta_i \in D(\mathcal{A})$:

$$\begin{aligned}
\theta_i(\textstyle\sum_{j=1}^{\ell} c_j x_j) &= \textstyle\sum_{j=1}^{\ell} c_j x_j^{q^{i-1}} \\
&= \textstyle\sum_{j=1}^{\ell} c_j^{q^{i-1}} x_j^{q^{i-1}} \\
&= (\textstyle\sum_{j=1}^{\ell} c_j x_j)^{q^{i-1}} \in (\textstyle\sum_{j=1}^{\ell} c_j x_j)S
\end{aligned}$$

for any $c_1, \ldots, c_\ell \in \mathbb{K}$. In order to prove that $\det \mathsf{M}(\theta_1, \ldots, \theta_\ell) \neq 0$ by induction on ℓ, it is sufficient to consider the coefficient of $x_\ell^{q^{\ell-1}}$ in $\det \mathsf{M}(\theta_1, \ldots, \theta_\ell)$. Note that each θ_i is homogeneous with pdegree q^{i-1}. Thus

$$\sum_{i=1}^{\ell} \operatorname{pdeg}\theta_i = 1 + q + q^2 + \cdots + q^{\ell-1} = |\mathcal{A}|.$$

It follows from Theorem 4.23 that $\mathcal{A}$ is a free arrangement with basis $\theta_1, \ldots, \theta_\ell$.

If $\mathcal{A}$ is free, then by Proposition 4.18 there exists a homogeneous basis $\{\theta_1, \ldots, \theta_\ell\}$ for $D(\mathcal{A})$. It follows from the general result in Proposition A.24 that the pdegrees

$$\{\operatorname{pdeg}\theta_1, \ldots, \operatorname{pdeg}\theta_\ell\}$$

(with multiplicity but neglecting the order) depend only on $\mathcal{A}$.

Definition 4.25 Let $\mathcal{A}$ be a free arrangement and let $\{\theta_1, \ldots, \theta_\ell\}$ be a homogeneous basis for $D(\mathcal{A})$. We call $\operatorname{pdeg}\theta_1, \ldots, \operatorname{pdeg}\theta_\ell$ the **exponents** of $\mathcal{A}$ and write

$$\exp \mathcal{A} = \{\operatorname{pdeg}\theta_1, \ldots, \operatorname{pdeg}\theta_\ell\}.$$

Note that $\exp \mathcal{A}$ may have repetitions and that the order should be neglected. If the integer m occurs $e \geq 0$ times in the multi-set $\exp \mathcal{A}$, we write $m^e \in \exp \mathcal{A}$. Using this notation there is a unique expression

$$\exp \mathcal{A} = \{0^{e_0}, 1^{e_1}, 2^{e_2}, \ldots\},$$

where $e_i \geq 0$. If $e_i = 0$, then it is understood that $i \notin \exp \mathcal{A}$. For example, $\{0, 3, 1, 3, 5\} = \{0^1, 1^1, 2^0, 3^2, 4^0, 5^1\}$. The next result follows from Theorem 4.23.

Proposition 4.26 If $\mathcal{A}$ is a free ℓ-arrangement with

$$\exp \mathcal{A} = \{0^{e_0}, 1^{e_1}, 2^{e_2}, \ldots\},$$

then

$$\sum_{k \geq 0} e_k = \ell, \qquad \sum_{k \geq 0} k e_k = |\mathcal{A}|.$$

Proposition 4.27 If $\mathcal{A} \neq \Phi$ is free, then there exists a homogeneous basis $\{\theta_1, \ldots, \theta_\ell\}$ for $D(\mathcal{A})$ such that $\theta_1 = \theta_E$ is the Euler derivation.

Proof. Let $H \in \mathcal{A}$ and let $\alpha = \alpha_H$. Then $\theta_E(\alpha) = \alpha$. Define

$$\operatorname{Ann}(H) = \{\theta \in D(\mathcal{A}) \mid \theta(\alpha) = 0\}.$$

Then $\operatorname{Ann}(H)$ is a graded submodule of $D(\mathcal{A})$. For any $\theta \in D(\mathcal{A})$

$$\theta - \frac{\theta(\alpha)}{\alpha}\theta_E \in \mathrm{Ann}(H).$$

Also
$$S\theta_E \cap \mathrm{Ann}(H) = 0.$$

Thus
$$D(\mathcal{A}) = S\theta_E \oplus \mathrm{Ann}(H).$$

Let G be a minimal system of homogeneous generators for $\mathrm{Ann}(H)$. Then $\{\theta_E\} \cup G$ is a minimal system of generators for $D(\mathcal{A})$, and by Theorem A.19 it is a homogeneous basis for $D(\mathcal{A})$. □

Proposition 4.28 *Let $(\mathcal{A}_1, V_1)$ and $(\mathcal{A}_2, V_2)$ be two arrangements. The product arrangement $(\mathcal{A}_1 \times \mathcal{A}_2, V_1 \oplus V_2)$ is free if and only if both $(\mathcal{A}_1, V_1)$ and $(\mathcal{A}_2, V_2)$ are free. In this case*

$$\exp(\mathcal{A}_1 \times \mathcal{A}_2) = \{\exp \mathcal{A}_1, \exp \mathcal{A}_2\}.$$

Proof. Recall from Proposition 4.14 that

$$D(\mathcal{A}_1 \times \mathcal{A}_2) = SD(\mathcal{A}_1) \oplus SD(\mathcal{A}_2).$$

Assume that both $(\mathcal{A}_1, V_1)$ and $(\mathcal{A}_2, V_2)$ are free. Let θ_i for $1 \leq i \leq \ell_1$ and η_j for $1 \leq j \leq \ell_2$ be homogeneous bases for $D(\mathcal{A}_1)$ and $D(\mathcal{A}_2)$ respectively. Then $\{\theta_1, \ldots, \theta_{\ell_1}, \eta_1, \ldots, \eta_{\ell_2}\}$ is linearly independent over S. By Theorem 4.23

$$\sum_{i=1}^{\ell_1} \mathrm{pdeg}\,\theta_i + \sum_{j=1}^{\ell_2} \mathrm{pdeg}\,\eta_j = |\mathcal{A}_1| + |\mathcal{A}_2| = |\mathcal{A}_1 \times \mathcal{A}_2|.$$

It follows from Theorem 4.23 that $\{\theta_1, \ldots, \theta_{\ell_1}, \eta_1, \ldots, \eta_{\ell_2}\}$ is a basis for $D(\mathcal{A}_1 \times \mathcal{A}_2)$.

Conversely, assume that $D(\mathcal{A}_1 \times \mathcal{A}_2)$ is free. Let $\{\theta_i \mid 1 \leq i \leq m_1\}$ and $\{\eta_j \mid 1 \leq j \leq m_2\}$ be minimal sets of homogeneous generators for $D(\mathcal{A}_1)$ and $D(\mathcal{A}_2)$ respectively. Then by Proposition 4.14

$$\{\theta_i \mid 1 \leq i \leq m_1\} \cup \{\eta_j \mid 1 \leq j \leq m_2\}$$

is also a minimal set of homogeneous generators for $D(\mathcal{A}_1 \times \mathcal{A}_2)$ over S, and by Theorem A.19 it is a basis. In particular, the basis elements are independent over S. Thus $\{\theta_i \mid 1 \leq i \leq m_1\}$ and $\{\eta_j \mid 1 \leq j \leq m_2\}$ are linearly independent over $S(V_1^*)$ and $S(V_2^*)$, respectively. □

Proposition 4.29
(1) *If $\mathcal{A} = \Phi_\ell$, then $\exp \mathcal{A} = \{0^\ell\}$.*
(2) *If $\mathcal{A}$ is free of rank $r(\mathcal{A})$, then $\exp \mathcal{A} = \{0^{\ell - r(\mathcal{A})}, 1^{e_1}, 2^{e_2}, \ldots\}$.*
(3) *If $\mathcal{A} \neq \Phi_\ell$ is free and $\exp \mathcal{A} = \{0^{e_0}, 1^{e_1}, 2^{e_2}, \ldots\}$, then $\mathcal{A}$ is a direct product of e_1 nonempty irreducible arrangements.*

(4) If $\ell = 2$ and $\mathcal{A} \neq \Phi_2$, then $\exp \mathcal{A} = \{1, |\mathcal{A}| - 1\}$.

Proof. (1) Let $\mathcal{A} = \Phi_\ell$. We showed in Example 4.16 that the exponents of $\mathcal{A}$ are all zero.

(2) Since $\mathcal{A} = \Phi_{\ell - r(\mathcal{A})} \times \mathcal{A}_0$ with an essential free arrangement $\mathcal{A}_0$, it follows from Proposition 4.28 and from (1) that

$$\exp \mathcal{A} = \{\exp \Phi_{\ell - r(\mathcal{A})}, \exp \mathcal{A}_0\} = \{0^{\ell - r(\mathcal{A})}, \exp \mathcal{A}_0\}.$$

Thus it suffices to show that if $\mathcal{A}$ is essential and free, then the integer 0 does not appear in $\exp \mathcal{A}$. Assume that $\mathcal{A}$ is essential and free. Let $0 \neq \theta \in D(\mathcal{A})$ be homogeneous with pdeg$\theta = 0$. Then for every $H \in \mathcal{A}$, $\theta(\alpha_H) \in \alpha_H S$ and $\deg \theta(\alpha_H) = 0$. This implies that $\theta(\alpha_H) = 0$ for every $H \in \mathcal{A}$. Write $\theta = D_v$ for some $v \in V$. Therefore

$$0 = \theta(\alpha_H) = D_v(\alpha_H) = \alpha_H(v).$$

Thus $v \in \bigcap_{H \in \mathcal{A}} H = 0$. Therefore $v = 0$ and $\theta = D_v = 0$, which is a contradiction.

(3) In light of Proposition 4.28, we may assume that $\mathcal{A}$ is irreducible. It follows from Proposition 4.27 that $e_1 \geq 1$. Suppose $\theta \in D(\mathcal{A})$ and pdeg$\theta = 1$. Then there exists $c_H \in \mathbb{K}$ such that $\theta(\alpha_H) = c_H \alpha_H$. We see from Proposition 4.8 that α_H is an eigenvector of the linear transformation $\theta|_{V^*}$. Since $V^* = \sum_{H \in \mathcal{A}} \mathbb{K} \alpha_H$, we see that $\theta|_{V^*}$ is semisimple. For $\lambda \in \mathbb{K}$, let

$$W_\lambda = \{\alpha \in V^* \mid \theta(\alpha) = \lambda \alpha\}.$$

Since $\theta|_{V^*}$ is semisimple, we have

$$V^* = \bigoplus_{\lambda \in \mathbb{K}} W_\lambda.$$

Choose $\lambda \in \mathbb{K}$ so that $W_\lambda \neq 0$ and let

$$W'_\lambda = \bigoplus_{\mu \neq \lambda} W_\mu.$$

Then $V^* = W_\lambda \oplus W'_\lambda$. If $H \in \mathcal{A}$, then $\alpha_H \in W_\mu$ for some $\mu \in \mathbb{K}$ so $\alpha_H \in W_\lambda \cup W'_\lambda$. Choose a basis $x_1, \ldots, x_k$ for W_λ and a basis $x_{k+1}, \ldots, x_\ell$ for W'_λ. Define

$$Q_1 = \prod_{\alpha_H \in W_\lambda} \alpha_H \in \mathbb{K}[x_1, \ldots, x_k], \quad Q_2 = \prod_{\alpha_H \in W'_\lambda} \alpha_H \in \mathbb{K}[x_{k+1}, \ldots, x_\ell].$$

Then $Q = Q_1 Q_2$. This implies that $\mathcal{A}$ is a direct product of two nonempty arrangements. This is a contradiction. Therefore $W'_\lambda = 0$ and $V^* = W_\lambda$. Thus $\theta(\alpha) = \lambda \alpha$ for all $\alpha \in V^*$, so $\theta = \lambda \theta_E$ where θ_E is the Euler derivation. This implies that $e_1 = 1$.

(4) A homogeneous basis for $D(\mathcal{A})$ is given in Example 4.20. □

Corollary 4.30 *A free arrangement $\mathcal{A}$ is irreducible if and only if $e_0 = 0$ and $e_1 = 1$*

Examples

Example 4.31 Let $\mathcal{A}$ be a Boolean arrangement defined by $Q(\mathcal{A}) = x_1 \cdots x_\ell$. We showed in Example 4.6 that the derivations $\theta_i = x_i D_i$ for $1 \leq i \leq \ell$ form a basis for $D(\mathcal{A})$. Thus $\exp \mathcal{A} = \{1^\ell\}$.

Example 4.32 Let $\mathcal{A}$ be the braid arrangement. We showed in Example 4.22 that the derivations $\theta_k = \sum_{j=1}^\ell x_j^k D_j$ for $0 \leq k \leq (\ell-1)$ form a homogeneous basis for $D(\mathcal{A})$. Thus

$$\exp \mathcal{A} = \{0, 1, \ldots, \ell-1\}.$$

The integer 0 appears once here. On the other hand, $\ell - r(\mathcal{A}) = \ell - (\ell-1) = 1$. This is consistent with the assertion in Proposition 4.29.2. The sum of the exponents is equal to $\ell(\ell-1)/2$, which is consistent with Proposition 4.26.

Example 4.33 Let $\mathcal{A}$ be the arrangement consisting of all hyperplanes through the origin in an ℓ-dimensional vector space over a finite field of q elements, $\mathbb{K} = \mathbb{F}_q$. In Example 4.24 we showed that for $1 \leq i \leq \ell$

$$\theta_i = \sum_{j=1}^\ell x_j^{q^{i-1}} D_j$$

form a basis for $D(\mathcal{A})$. Thus $\exp \mathcal{A} = \{1, q, q^2, \ldots, q^{\ell-1}\}$. Note that the sum of the exponents $\sum_{i=1}^\ell q^{i-1}$ is equal to $|\mathcal{A}|$ by Proposition 4.26. This is consistent with the calculation following Definition 1.11.

Example 4.34 Consider 3–arrangements $\mathcal{A}$ with $|\mathcal{A}| = 4$. If $r(\mathcal{A}) < 3$, then $\mathcal{A}$ is free by Example 4.20. Assume that $\mathcal{A}$ is essential. After a linear change of coordinates, we may assume that $\mathcal{A}$ is defined by

$$Q(\mathcal{A}) = x_1 x_2 x_3 (a_1 x_1 + a_2 x_2 + a_3 x_3)$$

where at most one of a_1, a_2, a_3 is 0. There are two cases to consider.

(i) If one of the parameters is 0, say $a_1 = 0$, then $Q = Q_1 Q_2$. Here $Q_1 = x_1$ defines a 1–arrangement $\mathcal{A}_1$ and $Q_2 = x_2 x_3 (a_2 x_2 + a_3 x_3)$ defines a 2–arrangement such that $\mathcal{A} = \mathcal{A}_1 \times \mathcal{A}_2$. Clearly, $\mathcal{A}_1$ is free with $\exp \mathcal{A}_1 = \{1\}$. It follows from Proposition 4.29.4 that $\mathcal{A}_2$ is free with $\exp \mathcal{A}_2 = \{1, 2\}$. It follows from Proposition 4.28 that $\mathcal{A}$ is free with $\exp \mathcal{A} = \{1, 1, 2\}$.

(ii) If $a_1 a_2 a_3 \neq 0$, then $\mathcal{A}$ is not free. These are the simplest examples of nonfree arrangements. We argue by assuming that $\mathcal{A}$ is free. Let

$$\exp \mathcal{A} = \{0^{e_0}, 1^{e_1}, 2^{e_2}, \ldots\}.$$

Since $\mathcal{A}$ is irreducible and $r(\mathcal{A}) = 3$, it follows from Corollary 4.30 that $e_0 = 0$, $e_1 = 1$, and from Proposition 4.26 that $\sum_i e_i = 3$ and $\sum_i i e_i = 4$. This is impossible.

This example illustrates the fact that arrangements are in general not free. Consider the parameter space P spanned by a_1, a_2, a_3. Then P is a subset of $\mathbb{K}^3$ where at most one parameter is 0. Let S be the subset of P where one parameter is 0. Then S is the parameter space for free arrangements and $P \setminus S$ is the parameter space for nonfree arrangements. Note that S is a *thin* subset of P. In case $\mathbb{K} = \mathbb{R}$ or $\mathbb{K} = \mathbb{C}$, the set $P \setminus S$ is open and dense.

Example 4.35 Let $Q(\mathcal{A}) = x_1 x_2 x_3 (x_1 - x_2)(x_2 + x_3)(x_1 + x_3)(x_1 + x_2 + x_3)$. Then $\mathcal{A}$ is free for all $\mathbb{K}$, but its exponents depend on the characteristic of $\mathbb{K}$.

Assume $\mathrm{char}(\mathbb{K}) \neq 2$. Then the derivations

$$\begin{aligned}
\theta_1 &= x_1 D_1 + x_2 D_2 + x_3 D_3, \\
\theta_2 &= x_1(x_1 + x_3)(x_1 + x_2 + x_3) D_1 + x_2(x_2 + x_3)(x_1 + x_2 + x_3) D_2, \\
\theta_3 &= x_1(x_1 + x_3)(2x_2 + x_3) D_1 + x_2(x_2 + x_3)(2x_1 + x_3) D_2
\end{aligned}$$

are in $D(\mathcal{A})$ and $\det \mathsf{M}(\theta_1, \theta_2, \theta_3) = 2Q \in \mathbb{K}^* Q$. By Saito's criterion 4.19, they form a basis for $D(\mathcal{A})$. It follows that $\exp \mathcal{A} = \{1, 3, 3\}$.

For $\mathrm{char}(\mathbb{K}) = 2$ the derivations

$$\begin{aligned}
\theta_1 &= x_1 D_1 + x_2 D_2 + x_3 D_3, \\
\theta_2 &= x_1^2 D_1 + x_2^2 D_2 + x_3^2 D_3, \\
\theta_3 &= x_1^4 D_1 + x_2^4 D_2 + x_3^4 D_3
\end{aligned}$$

are in $D(\mathcal{A})$. By Saito's criterion 4.19 they form a basis for $D(\mathcal{A})$. It follows that $\exp \mathcal{A} = \{1, 2, 4\}$. Note that

$$|\mathcal{A}| = 7 = 1 + 3 + 3 = 1 + 2 + 4$$

in agreement with Proposition 4.26.

We close the section with a discussion of subarrangements. Restrictions are considered at the end of Section 4.6. Note first that a free arrangement may have a subarrangement which is not free, or may be the subarrangement of an arrangement which is not free.

Example 4.36 Define 3-arrangements $\mathcal{C} \subset \mathcal{B} \subset \mathcal{A}$ by

$$Q(\mathcal{A}) = xyz(x+y)(x+y-z), \quad Q(\mathcal{B}) = xyz(x+y-z), \quad Q(\mathcal{C}) = xyz.$$

We will show in Example 4.54 that $\mathcal{A}$ is free. It follows from Example 4.34 that $\mathcal{B}$ is not free. It follows from Example 4.6 that $\mathcal{C}$ is free.

The next result shows the special nature of the subarrangements $\mathcal{A}_X$ for $X \in L(\mathcal{A})$.

Theorem 4.37 *If $\mathcal{A}$ is free, then $\mathcal{A}_X$ is free for all $X \in L(\mathcal{A})$.*

Proof. Let $Q_X = Q(\mathcal{A}_X)$ and let $Q_0 = Q(\mathcal{A})/Q_X$. Choose $w \in M(\mathcal{A}^X)$ and note that by this choice $\alpha_H(w) = 0$ if and only if $X \subseteq H$. Thus $Q_0(w) \neq 0$. Define the affine linear map $\tau : V \to V$ by $\tau(v) = v + w$. If $e_1, \ldots, e_\ell$ is a basis for V dual to $x_1, \ldots, x_\ell$ and $w = \sum w_i e_i$, then τ induces a map, which we again call $\tau : V^* \to V^*$. This map extends to a $\mathbb{K}$–algebra isomorphism $\tau : S \to S$ given by $\tau(x_i) = x_i + w_i$ for $1 \leq i \leq \ell$. It maps the ideal $\mathcal{M}_w = (x_1 - w_1, \ldots, x_\ell - w_\ell)$ to the maximal ideal $\mathcal{M} = (x_1, \ldots, x_\ell)$. Define $\tau : \mathrm{Der}_{\mathbb{K}}(S) \to \mathrm{Der}_{\mathbb{K}}(S)$ by $\tau(\sum_{j=1}^\ell h_j D_j) = \sum_{j=1}^\ell \tau(h_j) D_j$. Note that $(\tau\theta)(\tau f) = \tau(\theta(f))$ and that $\tau Q_X = Q_X$. Since $Q_0(w) \neq 0$, we get $\tau Q_0(0) \neq 0$. Suppose that $\theta \in D(\mathcal{A})$. Then $(\tau\theta)(Q_X) = (\tau\theta)(\tau Q_X) = \tau(\theta(Q_X)) \in SQ_X$. Thus $\tau(D(\mathcal{A})) \subseteq D(\mathcal{A}_X)$. Let $\theta_1, \ldots, \theta_\ell$ be a basis for $D(\mathcal{A})$ and let $\mathsf{M} = \mathsf{M}(\theta_1, \ldots, \theta_\ell)$. We may assume that $\det \mathsf{M} = Q(\mathcal{A})$. Then

$$\begin{aligned}
\det \mathsf{M}(\tau\theta_1, \ldots, \tau\theta_\ell) &= \det[\tau \mathsf{M}_{i,j}] \\
&= \tau Q(\mathcal{A}) \\
&= (\tau Q_X)(\tau Q_0) \\
&= Q_X(\tau Q_0).
\end{aligned}$$

Write $\tau\theta_i = \sum_{k \geq 0} \phi_i^{(k)}$ where $\mathrm{pdeg}\,\phi_i^{(k)} = k$. It follows that $\phi_i^{(k)} \in D(\mathcal{A}_X)$. Since $\tau Q_0(0) \neq 0$, there exist $\phi_1^{(k_1)}, \ldots, \phi_\ell^{(k_\ell)}$ such that $\det \mathsf{M}(\phi_1^{(k_1)}, \ldots, \phi_\ell^{(k_\ell)}) = cQ_X$ with $c \in \mathbb{K}^*$. Saito's criterion 4.19 implies that $\phi_1^{(k_1)}, \ldots, \phi_\ell^{(k_\ell)}$ form a basis for $D(\mathcal{A}_X)$. □

4.3 The Addition–Deletion Theorem

In this section we study the properties of free arrangements for triples $(\mathcal{A}, \mathcal{A}', \mathcal{A}'')$. We may use Example 4.36 to construct a triple where $\mathcal{A}$ and $\mathcal{A}''$ are free but $\mathcal{A}'$ is not. The same example provides a triple where $\mathcal{A}'$ and $\mathcal{A}''$ are free but $\mathcal{A}$ is not. We have no example of a triple where $\mathcal{A}, \mathcal{A}'$ are free but $\mathcal{A}''$ is not. The main result of this section is the Addition–Deletion Theorem, which asserts that for a triple $(\mathcal{A}, \mathcal{A}', \mathcal{A}'')$, any two of the following statements imply the third:

$$\begin{aligned}
\mathcal{A} \text{ is free with } \exp \mathcal{A} &= \{b_1, \ldots, b_{\ell-1}, b_\ell\}, \\
\mathcal{A}' \text{ is free with } \exp \mathcal{A}' &= \{b_1, \ldots, b_{\ell-1}, b_\ell - 1\}, \\
\mathcal{A}'' \text{ is free with } \exp \mathcal{A}'' &= \{b_1, \ldots, b_{\ell-1}\}.
\end{aligned}$$

For an alternate proof see [267]. This result leads to the definition of inductively free arrangements, introduced in [226]. We give several examples and prove that a supersolvable arrangement is inductively free. This allows us to give an example of a free arrangement which is not inductively free. This example leads to the more general notion of recursively free arrangements, defined by Ziegler [264]. It is not known whether every free arrangement is recursively free. We show that every supersolvable arrangement is inductively free. See [120, 221, 267] for more work on supersolvable arrangements. We close this section with the result that the Poincaré polynomial of a recursively free arrangement factors. All derivations in this section are homogeneous.

4. Free Arrangements

Basis Extension

Let $\mathcal{A}$ be a nonempty arrangement defined by $Q = Q(\mathcal{A}) = \prod_{H \in \mathcal{A}} \alpha_H$. Let $H_0 \in \mathcal{A}$ be a distinguished hyperplane. Let $(\mathcal{A}, \mathcal{A}', \mathcal{A}'')$ be the corresponding triple. Recall the map $\lambda : \mathcal{A}' \to \mathcal{A}''$ defined in Lemma 3.59 by $\lambda(H) = H_0 \cap H$. Since λ is surjective, we may define a map $\nu : \mathcal{A}'' \to \mathcal{A}'$ by the property $\lambda\nu(X) = X$ for all $X \in \mathcal{A}''$.

Definition 4.38 *Define the polynomial*
$$b(\mathcal{A}, \nu) = \frac{Q}{\alpha_0 \prod_{X \in \mathcal{A}''} \alpha_{\nu(X)}}.$$

Lemma 4.39
(1) $\deg b(\mathcal{A}, \nu) = |\mathcal{A}'| - |\mathcal{A}''|$.
(2) The ideal $(\alpha_0, b(\mathcal{A}, \nu))$ is independent of the choice of ν. Thus we may write $(\alpha_0, b(\mathcal{A}))$.
(3) $(\alpha_0, b(\mathcal{A})) = (\alpha_0, \prod_{X \in \mathcal{A}''} \alpha_{\nu(X)}^{|\mathcal{A}_X|-2})$.

Proof. (1) $\deg b(\mathcal{A}) = |\mathcal{A}| - (1 + |\mathcal{A}''|) = |\mathcal{A}'| - |\mathcal{A}''|$. (2) Suppose $\rho : \mathcal{A}'' \to \mathcal{A}'$ such that $\lambda\rho(X) = X$ for all $X \in \mathcal{A}''$. Then $H_0 \cap \nu(X) = X = H_0 \cap \rho(X)$ and hence H_0, $\nu(X)$, and $\rho(X)$ are dependent. Thus $(\alpha_0, b(\mathcal{A}, \nu)) = (\alpha_0, b(\mathcal{A}, \rho))$. Part (3) follows from (2). □

Definition 4.40 *Let $D(\mathcal{A}')\alpha_0 = \{\theta(\alpha_0) \mid \theta \in D(\mathcal{A}')\}$.*

Proposition 4.41 *The ideal $D(\mathcal{A}')\alpha_0$ is contained in $(\alpha_0, b(\mathcal{A}))$.*

Proof. Since $D(\mathcal{A}')$ is an S–module, $D(\mathcal{A}')\alpha_0$ is an ideal. Let $X \in \mathcal{A}''$ and $\nu(X) \in \mathcal{A}'$. Let $\mathcal{A}'_X = \mathcal{A}_X \setminus \{H_0\}$. Then $r(\mathcal{A}'_X) \leq r(\mathcal{A}_X) = 2$, so $\mathcal{A}'_X$ is a nonempty free arrangement. If we choose coordinates so $\alpha_0 = x_1$ and $\alpha_{\nu(X)} = x_2$, then $Q(\mathcal{A}_X) \in \mathbb{K}[x_1, x_2]$ is divisible by $x_1 x_2$. Note that $b(\mathcal{A}_X) = Q(\mathcal{A}_X)/x_1 x_2$. Saito's criterion 4.19 shows that
$$\{\theta_E, b(\mathcal{A}_X)D_1, D_3, \ldots, D_\ell\}$$
is a basis for $D(\mathcal{A}'_X)$. Since $D(\mathcal{A}'_X)\alpha_0 = (x_1, b(\mathcal{A}_X))$ and $D(\mathcal{A}') \subseteq D(\mathcal{A}'_X)$, we conclude that $D(\mathcal{A}')\alpha_0 \subseteq (\alpha_0, \alpha_{\nu(X)}^{|\mathcal{A}_X|-2})$ for all $X \in \mathcal{A}''$. Since the polynomials $\alpha_{\nu(X)}^{|\mathcal{A}_X|-2}$ are coprime, we have
$$D(\mathcal{A}')\alpha_0 \subseteq \bigcap_{X \in \mathcal{A}''} (\alpha_0, \alpha_{\nu(X)}^{|\mathcal{A}_X|-2})$$
$$= (\alpha_0, \prod_{X \in \mathcal{A}''} \alpha_{\nu(X)}^{|\mathcal{A}_X|-2})$$
$$= (\alpha_0, b(X))$$

as required. □

Theorem 4.42 *Let $\mathcal{A}$ be a free arrangement with $\exp \mathcal{A} = \{b_1, \ldots, b_\ell\}$, where $b_1 \leq \cdots \leq b_\ell$. If $\theta_1, \ldots, \theta_k \in D(\mathcal{A})$ satisfy for $1 \leq i \leq k$,*
  (1) $\mathrm{pdeg}\,\theta_i = b_i$,
  (2) $\theta_i \notin S\theta_1 + \cdots + S\theta_{i-1}$,
then $\theta_1, \ldots, \theta_k$ may be extended to a basis for $D(\mathcal{A})$.

Proof. We argue by induction on k. The assertion is true for $k = 0$. For $k \geq 1$ we assume that $D(\mathcal{A})$ has a basis $\theta_1, \ldots, \theta_{k-1}, \phi_k, \ldots, \phi_\ell$ such that $\mathrm{pdeg}\,\phi_j = b_j$. Write
$$\theta_k = f_1\theta_1 + \cdots + f_{k-1}\theta_{k-1} + f_k\phi_k + \cdots + f_\ell\phi_\ell.$$
Compare homogeneous components of degree b_k. It follows from hypothesis (2) and $b_k \leq b_{k+1} \leq \cdots \leq b_\ell$ that there is a nonzero term $f_j\phi_j$ of pdegree b_k. We may assume that $f_k\phi_k \neq 0$. Since $\mathrm{pdeg}\,\phi_k = b_k$, this implies that $f_k \in \mathbb{K}^*$ and we may replace ϕ_k by θ_k. □

The Map from $D(\mathcal{A})$ to $D(\mathcal{A}'')$

Definition 4.43 *Let $\bar{S} = S/\alpha_0 S$. If $\theta \in D(\mathcal{A})$, then $\theta(\alpha_0 S) \subseteq \alpha_0 S$. Thus we may define $\bar{\theta} : \bar{S} \to \bar{S}$ by $\bar{\theta}(f + \alpha_0 S) = \theta(f) + \alpha_0 S$.*

Proposition 4.44 *If $\theta \in D(\mathcal{A})$, then $\bar{\theta} \in D(\mathcal{A}'')$. If $\bar{\theta} \neq 0$, then $\mathrm{pdeg}\,\bar{\theta} = \mathrm{pdeg}\,\theta$.*

Proof. It follows from the definition that $\bar{\theta} \in \mathrm{Der}_{\mathbb{K}}(\bar{S})$. Suppose $H \in \mathcal{A}'$ and $\alpha = \alpha_H$. Then $H \cap H_0 \in \mathcal{A}''$. If we identify $H_0^* = V^*/\alpha_0$, then $H \cap H_0 = \ker \bar{\alpha}$. It follows from $\theta(\alpha) \in \alpha S$ that $\bar{\theta}(\bar{\alpha}) \in \bar{\alpha}\bar{S}$ and therefore $\bar{\theta} \in D(\{H \cap H_0\})$. Since this holds for all $H \in \mathcal{A}'$, we get
$$\bar{\theta} \in \bigcap_{H \in \mathcal{A}'} D(\{H \cap H_0\}) = D(\mathcal{A}'').$$
This completes the argument. □

Proposition 4.45 *Define $p : D(\mathcal{A}') \to D(\mathcal{A})$ by $p(\theta) = \alpha_0\theta$ and $q : D(\mathcal{A}) \to D(\mathcal{A}'')$ by $q(\theta) = \bar{\theta}$. The sequence*
$$0 \to D(\mathcal{A}') \xrightarrow{p} D(\mathcal{A}) \xrightarrow{q} D(\mathcal{A}'')$$
is exact.

Proof. If $\alpha_0\theta = 0$, then $\theta = 0$, so p is a monomorphism. It is clear that $\mathrm{im}\,p \subseteq \ker q$. To show the reverse inclusion, let $\theta \in D(\mathcal{A})$ with $\bar{\theta} = 0$. Then $\theta(f) \in \alpha_0 S$ for all $f \in S$. Write $\theta = \sum h_j D_j$. Then $\theta(x_j) = h_j \in \alpha_0 S$ shows that $\theta = \alpha_0\eta$ for some $\eta \in \mathrm{Der}_{\mathbb{K}}(S)$. It remains to show that $\eta \in D(\mathcal{A}')$. Let $H \in \mathcal{A}'$. Since $H \in \mathcal{A}$, we have $\theta(\alpha_H) = \alpha_0\eta(\alpha_H) \in \alpha_H S$. Since α_0 and α_H are relatively prime, we have $\eta \in D(\alpha_H)$. Since this holds for all $H \in \mathcal{A}'$, we conclude that $\eta \in D(\mathcal{A}')$. □

4. Free Arrangements

We show in Example 4.56 that q is not always surjective. The next theorem provides necessary conditions.

Theorem 4.46 *Suppose $\mathcal{A}$ and $\mathcal{A}'$ are free arrangements. Then there is a basis $\{\theta_1, \ldots, \theta_\ell\}$ for $D(\mathcal{A}')$ such that*
  (1) $\{\theta_1, \ldots, \theta_{i-1}, \alpha_0 \theta_i, \theta_{i+1}, \ldots, \theta_\ell\}$ *is a basis for $D(\mathcal{A})$,*
  (2) $\{\bar{\theta}_1, \ldots, \bar{\theta}_{i-1}, \bar{\theta}_{i+1}, \ldots, \bar{\theta}_\ell\}$ *is a basis for $D(\mathcal{A}'')$.*

Proof. Let $\theta_1, \ldots, \theta_\ell$ be a basis for $D(\mathcal{A}')$. We may assume that $D(\mathcal{A})$ has a basis $\{\theta_1, \ldots, \theta_{i-1}, \phi_i, \phi_{i+1}, \ldots, \phi_\ell\}$ where $1 \leq i \leq \ell$. Thus if $i = 1$, the two bases have no common element. Let $d_j = \text{pdeg}\,\theta_j$, $e_j = \text{pdeg}\,\phi_j$, and assume $d_1 \leq \cdots \leq d_\ell$ and $d_1 \leq \cdots \leq d_{i-1} \leq e_i \leq \cdots \leq e_\ell$. Note that $\theta_i \in D(\mathcal{A}')$ implies that $\alpha_0 \theta_i \in D(\mathcal{A})$. Thus we can write

$$\alpha_0 \theta_i = \sum_{k=1}^{i-1} f_k \theta_k + \sum_{p=i}^{\ell} g_p \phi_p.$$

Since $\{\theta_1, \ldots, \theta_\ell\}$ is an S–independent set, some $g_p \neq 0$ for $p \geq i$. Thus $\text{pdeg}\,\alpha_0 \theta_i \geq \text{pdeg}\,\phi_p \geq \text{pdeg}\,\phi_i$ and we have $d_i + 1 \geq e_i$. On the other hand, $\mathcal{A}' \subset \mathcal{A}$ implies that $D(\mathcal{A}) \subset D(\mathcal{A}')$. Thus we have

$$\phi_i = \sum_{k=1}^{i-1} a_k \theta_k + \sum_{p=i}^{\ell} b_p \theta_p.$$

Since $\{\theta_1, \ldots, \theta_{i-1}, \phi_i\}$ is an S–independent set, some $b_p \neq 0$ for $p \geq i$. Thus $\text{pdeg}\,\phi_i \geq \text{pdeg}\,\theta_p \geq \text{pdeg}\,\theta_i$ and we have $e_i \geq d_i$. These two inequalities give

$$d_i \leq e_i \leq d_i + 1.$$

If $d_i = e_i$, then $\text{pdeg}\,\theta_i = e_i$ and we may apply Theorem 4.42 to $\theta_1, \ldots \theta_i$ and extend it to a basis for $D(\mathcal{A})$: $\theta_1, \ldots \theta_i, \phi_{i+1}, \ldots, \phi_\ell$. We can then repeat the first part of the proof with $i + 1$ in place of i. Since $\sum_{j=1}^\ell d_j = |\mathcal{A}'| < |\mathcal{A}| = \sum_{j=1}^\ell e_j$, we cannot have $d_i = e_i$ for $1 \leq i \leq \ell$. Thus we may assume that i is the index where $d_i + 1 = e_i$.

Since $d_i + 1 = e_i$, we may apply Theorem 4.42 to the S–independent set $\theta_1, \ldots, \theta_{i-1}, \alpha_0 \theta_i$ and extend it to a basis for $D(\mathcal{A})$ by choosing new $\theta_{i+1}, \ldots, \theta_\ell$. It follows from Saito's criterion 4.19 that $\theta_1, \ldots \theta_{i-1}, \theta_i, \theta_{i+1}, \ldots, \theta_\ell$ is a basis for $D(\mathcal{A}')$. This proves assertion (1).

To prove (2), recall that $\deg b(\mathcal{A}) = |\mathcal{A}'| - |\mathcal{A}''|$ and that $\theta_i(\alpha_0) \in (\alpha_0, b(\mathcal{A}))$. If $\text{pdeg}\,\theta_i < \deg b(\mathcal{A})$, then $\theta_i(\alpha_0) \in \alpha_0 S$, and hence $\theta_i \in D(\mathcal{A})$. This contradicts (1). It follows that $\text{pdeg}\,\theta_i \geq |\mathcal{A}'| - |\mathcal{A}''|$. Since $\bar{\theta}_j \neq 0$ for $j \neq i$, we have

$$\Big(\sum_{j \neq i} \text{pdeg}\,\bar{\theta}_j\Big) \leq |\mathcal{A}'| - (|\mathcal{A}'| - |\mathcal{A}''|) = |\mathcal{A}''|.$$

It remains to show that $\{\bar{\theta}_1, \ldots, \bar{\theta}_{i-1}, \bar{\theta}_{i+1}, \ldots, \bar{\theta}_\ell\}$ are independent over $\bar{S}$. Otherwise there is a dependence. Since $S \to \bar{S}$ is surjective, we may write this dependence as

$$\sum_{k=1}^{i-1} \bar{a}_k \bar{\theta}_k + \sum_{p=i+1}^{\ell} \bar{a}_p \bar{\theta}_p = 0$$

with some $\bar{a}_m \neq 0$. Thus a_m is not divisible by α_0. It follows that for some $\theta \in \mathrm{Der}_{\mathbb{K}}(S)$,

$$\sum_{k=1}^{i-1} a_k \theta_k + \sum_{p=i+1}^{\ell} a_p \theta_p = \alpha_0 \theta.$$

Since the left side is in $D(\mathcal{A}')$, so is $\alpha_0 \theta$. Since α_0 is coprime to α_H for all $H \in \mathcal{A}'$, it follows that $\theta \in D(\mathcal{A}')$. Since $\theta_1, \ldots, \theta_\ell$ is an S-independent set, it follows that α_0 divides a_m for all m. This is a contradiction. □

Corollary 4.47 *If $\mathcal{A}$ and $\mathcal{A}'$ are free, then $\mathcal{A}''$ is free and there exist nonnegative integers $b_1 \leq \cdots \leq b_\ell$ such that*

$$\begin{aligned}
\exp \mathcal{A} &= \{b_1, \ldots, b_{k-1}, b_k + 1, b_{k+1}, \ldots, b_\ell\}, \\
\exp \mathcal{A}' &= \{b_1, \ldots, b_{k-1}, b_k, b_{k+1}, \ldots, b_\ell\}, \\
\exp \mathcal{A}'' &= \{b_1, \ldots, b_{k-1}, b_{k+1}, \ldots, b_\ell\}.
\end{aligned}$$

The Addition–Deletion Theorem

Lemma 4.48 *Let $(\mathcal{A}, \mathcal{A}', \mathcal{A}'')$ be a triple with respect to H_0. Assume that $\mathcal{A}''$ is free with $\exp \mathcal{A}'' = \{b_1, \ldots, b_{\ell-1}\}$ where*

$$b_1 \leq \cdots \leq b_{k-1} < b_k \leq \cdots \leq b_{\ell-1}.$$

Suppose $\theta_1, \ldots, \theta_k \in D(\mathcal{A})$ such that $\mathrm{pdeg}\,\theta_j = b_j$ for $1 \leq j \leq k-1$ and $\mathrm{pdeg}\,\theta_k < b_k$. There exists p with $1 \leq p \leq k$ such that

(1) $$\theta_p \in S\theta_1 + \cdots + S\theta_{p-1} + \alpha_0 D(\mathcal{A}').$$

Proof. Assume that (1) is false for $1 \leq p \leq k$. It follows that for $1 \leq p < k$

$$\bar{\theta}_p \notin \bar{S}\bar{\theta}_1 + \cdots + \bar{S}\bar{\theta}_{p-1}.$$

By Theorem 4.42, we may extend $\bar{\theta}_1, \ldots, \bar{\theta}_{k-1}$ to a basis for $D(\mathcal{A}'')$. Since $\mathrm{pdeg}\,\bar{\theta}_k < b_k$, it follows that $\bar{\theta}_k \in \bar{S}\bar{\theta}_1 + \cdots + \bar{S}\bar{\theta}_{k-1}$, and hence $\theta_k \in S\theta_1 + \cdots + S\theta_{k-1} + \alpha_0 D(\mathcal{A}')$. This contradicts the assumption that (1) is false for $1 \leq p \leq k$. □

Theorem 4.49 (Deletion) *If $\mathcal{A}$ and $\mathcal{A}''$ are free and $\exp \mathcal{A}'' \subset \exp \mathcal{A}$, then $\mathcal{A}'$ is free.*

Proof. Let $\exp \mathcal{A} = \{b_1, \ldots, b_{i-1}, b_i, b_{i+1}, \ldots, b_\ell\}$ where $b_1 \leq \cdots \leq b_\ell$ and let $\exp \mathcal{A}'' = \{b_1, \ldots, b_{i-1}, b_{i+1}, \ldots, b_\ell\}$. We may assume that either $b_i < b_{i+1}$ or

$i = \ell$. Let $\theta_1, \ldots, \theta_\ell$ be a basis for $D(\mathcal{A})$ with pdeg$\theta_k = b_k$ for $1 \leq k \leq \ell$. Apply Lemma 4.48 to $\theta_1, \ldots, \theta_i$ to find p with $1 \leq p \leq i$ so that

$$\theta_p \in S\theta_1 + \cdots + S\theta_{p-1} + \alpha_0 D(\mathcal{A}').$$

We may assume that $\theta_p \in \alpha_0 D(\mathcal{A}')$. It follows from Saito's criterion 4.19 that

$$\theta_1, \ldots, \theta_{p-1}, \frac{\theta_p}{\alpha_0}, \theta_{p+1}, \ldots, \theta_\ell$$

is a basis for $D(\mathcal{A}')$. □

Theorem 4.50 (Addition) *If $\mathcal{A}'$ and $\mathcal{A}''$ are free and $\exp \mathcal{A}'' \subset \exp \mathcal{A}'$, then $\mathcal{A}$ is free.*

Proof. Let $\exp \mathcal{A}' = \{b_1, \ldots, b_{i-1}, b_i, b_{i+1}, \ldots, b_\ell\}$ where $b_1 \leq \cdots \leq b_\ell$ and let $\exp \mathcal{A}'' = \{b_1, \ldots, b_{i-1}, b_{i+1}, \ldots, b_\ell\}$. We may assume that either $b_i < b_{i+1}$ or $i = \ell$. Note that $\deg b(\mathcal{A}) = |\mathcal{A}'| - |\mathcal{A}''| = b_i$. Let $\theta_1, \ldots, \theta_\ell$ be a basis for $D(\mathcal{A}')$ with pdeg$\theta_k = b_k$ for $1 \leq k \leq \ell$. It follows from Proposition 4.41 that $\theta_k(\alpha_0) \in (\alpha_0, b(\mathcal{A}))$. This implies that if pdeg$\theta_k < b_i = \deg b(\mathcal{A})$, then $\theta_k(\alpha_0) \in \alpha_0 S$, and hence $\theta_k \in D(\mathcal{A})$.

We show first the existence of $\theta_r \in \{\theta_1, \ldots, \theta_\ell\}$ with pdeg$\theta_r = b_i$ so that $\theta_r \notin D(\mathcal{A})$. Suppose not. Then $\theta_1, \ldots, \theta_i \in D(\mathcal{A})$ and we conclude from Lemma 4.48 that there exists p with $1 \leq p \leq i$ so that

$$\theta_p \in S\theta_1 + \cdots + S\theta_{p-1} + \alpha_0 D(\mathcal{A}').$$

We may assume $\theta_p \in \alpha_0 D(\mathcal{A}')$. Thus α_0 divides $\det \mathsf{M}(\theta_1, \ldots, \theta_\ell) = Q(\mathcal{A}')$. This is a contradiction.

The existence of $\theta_r \in \{\theta_1, \ldots, \theta_\ell\}$ with pdeg$\theta_r = b_i$ so that $\theta_r \notin D(\mathcal{A})$ implies that $\theta_r(\alpha_0) \notin \alpha_0 S$. Thus $\theta_r(\alpha_0) = c_r b(\mathcal{A}) + f_r \alpha_0$ with $c_r \neq 0$. Since pdeg$\theta_r = b_i = \deg b(\mathcal{A})$, we may assume $c_r = 1$. Let $\eta_r = \alpha_0 \theta_r$. For $k \neq r$ write $\theta_k(\alpha_0) = c_k b(\mathcal{A}) + f_k \alpha_0$ and let $\eta_k = \theta_k - c_k \theta_r$. Then $\eta_k(\alpha_0) \in \alpha_0 S$ and therefore $\eta_k \in D(\mathcal{A})$. It follows from Saito's criterion 4.19 that $\eta_1, \ldots, \eta_\ell$ is a basis for $D(\mathcal{A})$. □

These results may be conveniently summarized in a single statement. Here we return to the convention that the exponents are unordered.

Theorem 4.51 (Addition–Deletion) *Suppose $\mathcal{A} \neq \Phi_\ell$. Let $(\mathcal{A}, \mathcal{A}', \mathcal{A}'')$ be a triple. Any two of the following statements imply the third:*

$$\begin{aligned}
\mathcal{A} \text{ is free with } \exp \mathcal{A} &= \{b_1, \ldots, b_{\ell-1}, b_\ell\}, \\
\mathcal{A}' \text{ is free with } \exp \mathcal{A}' &= \{b_1, \ldots, b_{\ell-1}, b_\ell - 1\}, \\
\mathcal{A}'' \text{ is free with } \exp \mathcal{A}'' &= \{b_1, \ldots, b_{\ell-1}\}.
\end{aligned}$$

It follows from Example 4.20 that every 2–arrangement is free. Thus for 3–arrangements the Addition–Deletion Theorem has a special form.

4.3 The Addition–Deletion Theorem

Table 4.1. Induction table

$\exp \mathcal{A}'$	α_H	$\exp \mathcal{A}''$
0,0,0	x	0,0
0,0,1	y	0,1
0,1,1	z	1,1
1,1,1	$x+y$	1,1
1,1,2	$x+y-z$	1,2
1,2,2		

Theorem 4.52 *Let $\mathcal{A}$ be a nonempty 3-arrangement. Let $(\mathcal{A}, \mathcal{A}', \mathcal{A}'')$ be a triple. Any two of the following statements imply the third:*

$$\mathcal{A} \text{ is free with } \exp \mathcal{A} = \{b_1, b_2, b_3\},$$
$$\mathcal{A}' \text{ is free with } \exp \mathcal{A}' = \{b_1, b_2, b_3 - 1\},$$
$$|\mathcal{A}''| = b_1 + b_2.$$

Inductively Free Arrangements

Definition 4.53 *The class $\mathcal{IF}$ of* **inductively free** *arrangements is the smallest class of arrangements which satisfies*
  (1) $\Phi_\ell \in \mathcal{IF}$ for $\ell \geq 0$,
  (2) if there exists $H \in \mathcal{A}$ such that $\mathcal{A}'' \in \mathcal{IF}$, $\mathcal{A}' \in \mathcal{IF}$, and $\exp \mathcal{A}'' \subset \exp \mathcal{A}'$, then $\mathcal{A} \in \mathcal{IF}$.

In order to show that a given arrangement $\mathcal{A}$ is inductively free, we must start with some inductively free arrangement and add hyperplanes one at a time satisfying (2). This process may be described conveniently in an **induction table**. Each row is one step in the process. The first column gives $\exp \mathcal{A}'$ of the arrangement which is the $\mathcal{A}'$ of that step. The second column gives α_H, where $H = \ker \alpha_H$ is the hyperplane added to $\mathcal{A}'$. The third column gives $\exp \mathcal{A}''$. The last row displays $\exp \mathcal{A}$. Since $Q(\mathcal{A}')$ is the product of the α_H in the rows above the row in consideration, it is easy to compute $Q(\mathcal{A}'')$. At each step the difficulty lies in showing that $\mathcal{A}''$ is free, and in computing $\exp \mathcal{A}''$.

Example 4.54 Table 4.1 shows that the arrangement $\mathcal{A}$ defined by $Q(\mathcal{A}) = xyz(x+y)(x+y-z)$ is inductively free. The most delicate problem is to determine in which order to add the hyperplanes. Even in this simple example the order of the last two hyperplanes could not be reversed.

This process shows only that $\mathcal{A}$ is free. Construction of a basis for $D(\mathcal{A})$ requires more work. Let $\alpha_0 = x+y-z$. Then $\mathcal{A}'$ defined by $Q(\mathcal{A}') = xyz(x+y)$ is the product of a 1-arrangement and a 2-arrangement and hence $\mathcal{A}'$ is free. Let

$$\theta_1 = xD_x + yD_y,$$
$$\theta_2 = y(x+y)D_y,$$
$$\theta_3 = zD_z.$$

It follows from Saito's criterion 4.19 that $\{\theta_1, \theta_2, \theta_3\}$ is a basis for $D(\mathcal{A}')$. Choose coordinates x', y' for H_0^*. Then $Q(\mathcal{A}'') = x'y'(x'+y')$. Let $\nu(x') = x$, $\nu(y') = y$, and $\nu(x'+y') = x+y$. Then $b(\mathcal{A}) = z$. Thus we have

$$\begin{aligned} \theta_1(\alpha_0) &= x+y &= b(\mathcal{A}) + \alpha_0, \\ \theta_2(\alpha_0) &= y(x+y) &= yb(\mathcal{A}) + y\alpha_0, \\ \theta_3(\alpha_0) &= -z &= -b(\mathcal{A}). \end{aligned}$$

It follows from the proof of the Addition Theorem 4.50 that we may choose the following basis for $D(\mathcal{A})$:

$$\begin{aligned} \eta_1 &= \alpha_0\theta_1 &= (x+y-z)(xD_x + yD_y), \\ \eta_2 &= \theta_2 - y\theta_1 &= xy(D_y - D_x), \\ \eta_3 &= \theta_3 - (-1)\theta_1 &= xD_x + yD_y + zD_z. \end{aligned}$$

Example 4.55 The braid arrangement is inductively free. This is argued by a double induction. For $k \leq \ell$ define an ℓ-arrangement $\mathcal{A}_\ell(k)$ by $Q_\ell(k) = \prod_{1 \leq i < j \leq k}(x_i - x_j)$. We will show that $\mathcal{A}_\ell(k)$ is inductively free with

$$\exp \mathcal{A}_\ell(k) = \{0^{\ell-k+1}, 1, 2, \ldots, k-1\}.$$

By induction we may assume that $\mathcal{A}_p(q)$ is inductively free with the appropriate exponents for $p < \ell$, $q \leq p$, and that $\mathcal{A}_\ell(q)$ is inductively free with the appropriate exponents for $q < k$. We want to show that $\mathcal{A}_\ell(k)$ is inductively free. We may start with $\mathcal{A}_\ell(k-1)$ and add the hyperplanes $H_{i,k}$ for $1 \leq i < k$. The crucial fact is that $\mathcal{A}'' = \mathcal{A}_{\ell-1}(k-1)$ independently of i. Thus we get Table 4.2. In particular, we recover the exponents of the braid arrangement

$$\exp \mathcal{A} = \{0, 1, 2, \ldots, (\ell-1)\}.$$

Example 4.56 The map $q : D(\mathcal{A}) \to D(\mathcal{A}'')$ is not always surjective. Let $Q(\mathcal{A}) = xyz(x+y)(x+y-z)$ define $\mathcal{A}$. In Example 4.54 we showed that $\mathcal{A}$ is free and $\exp \mathcal{A} = \{1, 2, 2\}$. It follows from Proposition 4.44 that the image under q of any basis for $D(\mathcal{A})$ has at most one element of pdegree 1. Let $\alpha_0 = x+y$. Choose coordinates x', z' for H_0^*. Then $Q(\mathcal{A}'') = x'z'$. In particular, $\mathcal{A}''$ is free with $\exp \mathcal{A}'' = \{1, 1\}$. Thus the map $q : D(\mathcal{A}) \to D(\mathcal{A}'')$ is not surjective.

Proposition 4.57 *Suppose $\mathcal{A}$ and $\mathcal{A}''$ are free. The map $q : D(\mathcal{A}) \to D(\mathcal{A}'')$ is surjective if and only if $\mathcal{A}'$ is free.*

Proof. Since $\mathcal{A}$ and $\mathcal{A}''$ are free, surjectivity of q implies that $\exp \mathcal{A}'' \subset \exp \mathcal{A}$. It follows from the Deletion Theorem 4.49 that $\mathcal{A}'$ is free. If $\mathcal{A}$ and $\mathcal{A}'$ are free, then the proof of Theorem 4.42 shows that q is surjective. □

Table 4.2. The braid arrangement

$\exp \mathcal{A}'$	α_H	$\exp \mathcal{A}''$
$0^{\ell-k+2}, 1, \ldots, k-2$	$x_1 - x_k$	$0^{\ell-k+1}, 1, \ldots, k-2$
$0^{\ell-k+1}, 1, 1, 2, \ldots, k-2$	$x_2 - x_k$	$0^{\ell-k+1}, 1, \ldots, k-2$
$\vdots$	$x_i - x_k$	$0^{\ell-k+1}, 1, \ldots, k-2$
$0^{\ell-k+1}, 1, \ldots, k-2, k-2$	$x_{k-1} - x_k$	$0^{\ell-k+1}, 1, \ldots, k-2$
$0^{\ell-k+1}, 1, \ldots, k-1$		

Supersolvable Arrangements

Theorem 4.58 *Let $\mathcal{A}$ be a supersolvable ℓ-arrangement with a maximal chain of modular elements*

$$V = X_0 < X_1 < \cdots < X_\ell = T.$$

Define $b_i = |\mathcal{A}_{X_i} \setminus \mathcal{A}_{X_{i-1}}|$ for $1 \leq i \leq \ell$. Then $\mathcal{A}$ is inductively free with

$$\exp \mathcal{A} = \{b_1, \ldots, b_\ell\}.$$

Proof. We argue by induction on $|\mathcal{A}|$. The assertion is clear for $|\mathcal{A}| = 1$. As in Lemma 2.27, we let $H \in \mathcal{A}$ be a complement of $X_{\ell-1}$ in $L(\mathcal{A})$. Then both $\mathcal{A}'$ and $\mathcal{A}''$ are supersolvable and the induction hypothesis applies to them. It follows from Lemma 2.27 that

$$\exp \mathcal{A}' = \{b_1, \ldots, b_\ell - 1\},$$
$$\exp \mathcal{A}'' = \{b_1, \ldots, b_{\ell-1}\}.$$

The conclusion follows from the Addition–Deletion Theorem 4.51. □

Example 4.59 The central 3–arrangement $\mathcal{A}$ whose projective image $\mathbf{d}\mathcal{A}$ contains the five sides and five diagonals of a regular pentagon together with the line at infinity is free but not inductively free.

The first example of a free arrangement which is not inductively free appeared in [226]. The example presented here is due to K. Brandt and J. Keaty. To see that $\mathcal{A}$ is free, we consider the arrangement $\mathcal{B}$ obtained by adding the two dotted lines in Figure 4.1. Then $\mathcal{B}$ is supersolvable. This can be seen directly from $L(\mathcal{B})$ or by using the Fibration Theorem 5.113. See the discussion after Proposition 5.114. It follows from Theorem 4.58 that $\mathcal{B}$ is inductively free with $\exp \mathcal{B} = \{1, 5, 7\}$. Apply the Deletion Theorem 4.49 to $\mathcal{B}$ and the horizontal dotted line to get the free arrangement $\mathcal{B}'$ with $\exp \mathcal{B}' = \{1, 5, 6\}$. Apply the Deletion Theorem to $\mathcal{B}'$ and the vertical dotted line to see that $\mathcal{A}$ is free with $\exp \mathcal{A} = \{1, 5, 5\}$.

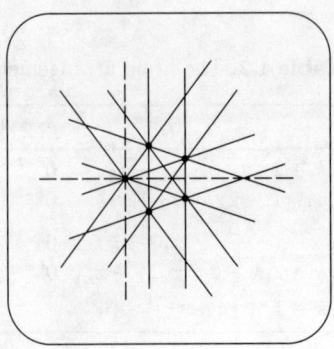

Fig. 4.1. Free but not inductively free

To see that $\mathcal{A}$ is not inductively free, we note that the last addition would require $\exp \mathcal{A}' = \{1, 4, 5\}$. It follows from Theorem 4.52 that $|\mathcal{A}''| = 6$, but each solid line contains five intersection points.

Definition 4.60 *The class $\mathcal{RF}$ of* **recursively free** *arrangements is the smallest class of arrangements which satisfies*
  (1) $\Phi_\ell \in \mathcal{RF}$ for $\ell \geq 0$,
  (2) if there exists $H \in \mathcal{A}$ such that $\mathcal{A}'' \in \mathcal{RF}$, $\mathcal{A}' \in \mathcal{RF}$ and $\exp \mathcal{A}'' \subset \exp \mathcal{A}'$, then $\mathcal{A} \in \mathcal{RF}$,
  (3) if there exists $H \in \mathcal{A}$ such that $\mathcal{A}'' \in \mathcal{RF}$, $\mathcal{A} \in \mathcal{RF}$, and $\exp \mathcal{A}'' \subset \exp \mathcal{A}$, then $\mathcal{A}' \in \mathcal{RF}$.

Example 4.59 shows that $\mathcal{IF} \subseteq \mathcal{RF}$ is a proper inclusion. It is not known whether all free arrangements are recursively free.

Factorization Theorem

Theorem 4.61 *Suppose $\mathcal{A} \in \mathcal{RF}$ with $\exp \mathcal{A} = \{b_1, \ldots, b_\ell\}$. Then*

$$\pi(\mathcal{A}, t) = \prod_{i=1}^{\ell}(1 + b_i t).$$

Proof. We argue by induction on $|\mathcal{A}|$. If $|\mathcal{A}| = 0$, then $\mathcal{A} = \Phi_\ell$. Since $\pi(\Phi_\ell, t) = 1$ and $\exp \Phi_\ell = \{0^\ell\}$, the assertion holds. The induction step consists of two cases, one for addition and one for deletion. Since they are similar, it suffices to show the argument for addition. Let $\mathcal{A}'$ and $\mathcal{A}''$ be recursively free with $\exp \mathcal{A}' = \{b_1, \ldots, b_{\ell-1}, b_\ell - 1\}$ and $\exp \mathcal{A}'' = \{b_1, \ldots, b_{\ell-1}\}$. By the induction hypothesis we have

$$\pi(\mathcal{A}', t) = \prod_{i=1}^{\ell-1}(1 + b_i t)(1 + (b_\ell - 1)t)$$

$$\pi(\mathcal{A}'', t) = \prod_{i=1}^{\ell-1}(1 + b_i t).$$

In Theorem 2.56 we proved the formula:

$$\pi(\mathcal{A}, t) = \pi(\mathcal{A}', t) + t\pi(\mathcal{A}'', t).$$

Thus

$$\begin{aligned}\pi(\mathcal{A}, t) &= \prod_{i=1}^{\ell-1}(1 + b_i t)(1 + (b_\ell - 1)t) + t\prod_{i=1}^{\ell-1}(1 + b_i t) \\ &= \prod_{i=1}^{\ell-1}(1 + b_i t)(1 + b_\ell t) \\ &= \prod_{i=1}^{\ell}(1 + b_i t)\end{aligned}$$

as required. □

4.4 The Modules $\Omega^p(\mathcal{A})$

In this section we study the S–modules $\Omega^p(\mathcal{A})$ of logarithmic differential p-forms with poles along $N = \bigcup_{H \in \mathcal{A}} H$. Modules of logarithmic differential forms with poles along a divisor with normal crossings were used by P. Deligne in [62] to define mixed Hodge structures. K. Saito [199, 201] generalized the definition to any divisor. His work is in the analytic category. Here we consider the special case when the divisor is a union of hyperplanes, and we work in the algebraic category.

Recall the module $\Omega^p(V)$ of rational differential p-forms on V from Definition 3.114. Let $x_1, x_2, \ldots, x_\ell$ be a basis for V^* and let p be a nonnegative integer throughout this section. Let

$$\Omega^p[V] = \bigoplus_{1 \leq i_1 < \cdots < i_p \leq \ell} S(dx_{i_1} \wedge \cdots \wedge dx_{i_p}).$$

We agree that $\Omega^0[V] = S$. The elements of $\Omega^p[V]$ are called **regular** differential p-forms on V.

Definition 4.62 *An element $\omega \in \Omega^p[V]$ is homogeneous of **polynomial degree** q if the coefficient of each $dx_{i_1} \wedge \cdots \wedge dx_{i_p}$ $(1 \leq i_1 < i_2 < \cdots < i_p \leq \ell)$ is homogeneous of degree q. In this case we write $\mathrm{pdeg}\,\omega = q$. Let $\Omega^p_q[V]$ denote the vector space consisting of all homogeneous p-forms of polynomial degree q for $q \in \mathbb{Z}$. Note that $\Omega^p_q[V] = 0$ if $q < 0$. This gives $\Omega^p[V]$ a graded S–module structure*

$$\Omega^p[V] = \bigoplus_{q \in \mathbb{Z}} \Omega_q^p[V].$$

This grading of $\Omega^p[V]$ by polynomial degree is similar to the grading of $\mathrm{Der}_{\mathbb{K}}(S)$ in Definition 4.2. For example, $\mathrm{pdeg}(dx_1) = 0$, $\mathrm{pdeg}(x_2 dx_1) = 1$. We may also define a total degree as in Definition 4.3.

Definition 4.63 *We say that $\omega \in \Omega_q^p[V]$ has **total degree** $p + q$ and write $\mathrm{tdeg}\,\omega = p + q$.*

Definition of $\Omega^p(\mathcal{A})$

Definition 4.64 *Let p be a nonnegative integer. Let Q be a defining polynomial for $\mathcal{A}$. The module $\Omega^p(\mathcal{A})$ of **logarithmic p-forms with poles along $\mathcal{A}$** is defined as*

$$\Omega^p(\mathcal{A}) = \{\omega \in \Omega^p(V) \mid Q\omega \in \Omega^p[V] \text{ and } Q(d\omega) \in \Omega^{p+1}[V]\}.$$

It is easy to see that $\Omega^p(\mathcal{A})$ is closed under addition and under multiplication by an element of S. Thus it is an S–module. Let

$$\Omega(\mathcal{A}) = \bigoplus_{p \geq 0} \Omega^p(\mathcal{A}).$$

Example 4.65 Let $Q = xyz$ define the Boolean 3–arrangement. Here $\omega_1 = dx/x \in \Omega^1(\mathcal{A})$ because $Q\omega_1 = yz\,dx \in \Omega^1[V]$ and $d\omega_1 = 0$. On the other hand, $\omega_2 = dx/y \notin \Omega^1(\mathcal{A})$ because

$$Q(d\omega_2) = Q(dx \wedge dy)/y^2 = (xz\,dx \wedge dy)/y \notin \Omega^2[V].$$

Proposition 4.66 *For any arrangement $\mathcal{A}$ with defining polynomial Q we have:*
(1) $dQ/Q \in \Omega^1(\mathcal{A})$.
(2) Let $\alpha \in V^*$. Then $d\alpha/\alpha \in \Omega^1(\mathcal{A}) \iff \ker(\alpha) \in \mathcal{A}$.

Basic Properties of $\Omega^p(\mathcal{A})$

Lemma 4.67 *Let Q define a nonempty arrangement. Then Q and the partial derivatives $\partial Q/\partial x_1$, $\partial Q/\partial x_2$, ..., $\partial Q/\partial x_\ell$ have no common factor of positive degree.*

Proof. Suppose f is a nonconstant common factor. Then there exists a linear form $\alpha \in V^*$ which divides Q and f. We may assume that for some i we have $\partial \alpha / \partial x_i = 1$. Write $Q = \alpha Q'$. Since Q is square free, Q' is not divisible by α. But

$$Q' = Q'(\partial\alpha/\partial x_i) = \partial(\alpha Q')/\partial x_i - \alpha(\partial Q'/\partial x_i) = \partial Q/\partial x_i - \alpha(\partial Q'/\partial x_i)$$

shows that Q' is divisible by α. This is a contradiction. $\square$

For the extreme values of p it is easy to compute $\Omega^p(\mathcal{A})$.

Proposition 4.68
(1) $\Omega^p(\mathcal{A}) = 0$ for $p > \ell$,
(2) $\Omega^\ell(\mathcal{A}) = (1/Q)\Omega^\ell[V]$,
(3) $\Omega^0(\mathcal{A}) = S$.

Proof. (1) and (2) are obvious from the definition. Let $f/Q \in \Omega^0(\mathcal{A})$ with $f \in S$. If $\mathcal{A}$ is the empty arrangement, then $Q = 1$ and there is nothing to prove. Suppose $\mathcal{A}$ is nonempty. We have

$$Qd(f/Q) = df - (fdQ/Q) \in \Omega^1[V].$$

Thus $f(\partial Q/\partial x_i) \in QS$ for $1 \le i \le \ell$ and $f \in QS$ by Lemma 4.67. $\square$

Proposition 4.69 *For $\omega \in \Omega^p(V)$, the following three conditions are equivalent:*
(1) $\omega \in \Omega^p(\mathcal{A})$,
(2) $Q\omega \in \Omega^p[V]$ and $dQ \wedge \omega \in \Omega^{p+1}[V]$,
(3) $Q\omega \in \Omega^p[V]$ and $Q(d\alpha/\alpha)\wedge\omega \in \Omega^{p+1}[V]$ for all $\alpha \in V^*$ with $\ker(\alpha) \in \mathcal{A}$.

Proof. The equivalence of (1) and (2) follows from the formula

$$d(Q\omega) = Q(d\omega) + (dQ) \wedge \omega.$$

Let $Q = \prod_{i=1}^n \alpha_i$ be a defining polynomial for $\mathcal{A}$. Then

$$dQ \wedge \omega = \sum_{i=1}^n Q(d\alpha_i/\alpha_i) \wedge \omega.$$

This shows that (3) implies (2). For the converse, assume (2) and let $\alpha = \alpha_1$. Write $Q = \alpha Q'$. We have

$$Q(dQ \wedge \omega) = Q'(d\alpha \wedge (Q\omega)) + \alpha(dQ' \wedge (Q\omega)).$$

Thus each coefficient of $Q'(d\alpha \wedge (Q\omega))$ is a polynomial divisible by α. Since Q' is not divisible by α, each coefficient of $d\alpha \wedge (Q\omega)$ is a polynomial divisible by α. This proves (3). $\square$

Corollary 4.70 *Let $\mathcal{A}_1$ and $\mathcal{A}_2$ be two arrangements in V with $\mathcal{A}_1 \subseteq \mathcal{A}_2$. Let $Q_3 \in S$ be a defining polynomial of the arrangement $\mathcal{A}_2 \setminus \mathcal{A}_1$. Then*

$$Q_3 \Omega^p(\mathcal{A}_2) \subseteq \Omega^p(\mathcal{A}_1) \subseteq \Omega^p(\mathcal{A}_2).$$

Proof. For $i = 1, 2$ let Q_i be a defining polynomial for $\mathcal{A}_i$. Then $Q_2 = Q_1 Q_3$. If $\omega \in \Omega^p(\mathcal{A}_1)$, then

$$dQ_2 \wedge \omega = d(Q_1 Q_3) \wedge \omega = dQ_3 \wedge (Q_1\omega) + Q_3(dQ_1 \wedge \omega)$$

is a regular differential form. It follows from Proposition 4.69 that $\omega \in \Omega^p(\mathcal{A}_2)$.

Suppose $\eta \in \Omega^p(\mathcal{A}_2)$. To show that $Q_3\eta \in \Omega^p(\mathcal{A}_1)$, we verify the conditions of Proposition 4.69.3. Clearly $Q_1(Q_3\eta) = Q_2\eta$ is regular. Now suppose that $\alpha \in V^*$ and $\ker(\alpha) \in \mathcal{A}_1$. Then

$$Q_1(d\alpha/\alpha) \wedge (Q_3\eta) = Q_2(d\alpha/\alpha) \wedge \eta.$$

Since $\eta \in \Omega^p(\mathcal{A}_2)$ and $\ker(\alpha) \in \mathcal{A}_2$, the last form is regular by Proposition 4.69. □

Definition 4.71 *An element $\omega \in \Omega^p(\mathcal{A})$ is homogeneous of* **polynomial degree** *q if the regular differential form $Q\omega \in \Omega^p[V]$ is homogeneous of polynomial degree $(q + \deg Q)$. Let $\Omega^p_q(\mathcal{A})$ denote the vector space consisting of all homogeneous p-forms of polynomial degree q for $q \in \mathbb{Z}$. Note that $\Omega^p_q(\mathcal{A}) = 0$ if $q < -\deg Q$.*

Thus, for example, $\mathrm{pdeg}(dx_1) = 0$, $\mathrm{pdeg}(dQ/Q) = -1$.

Proposition 4.72 *We have*

$$\Omega^p(\mathcal{A}) = \bigoplus_{q \in \mathbb{Z}} \Omega^p_q(\mathcal{A}).$$

Thus $\Omega^p(\mathcal{A})$ is a graded S-module.

Proof. Let $\omega \in \Omega^p(\mathcal{A})$. Decompose $Q\omega \in \Omega^p[V]$ into homogeneous components

$$Q\omega = \sum_{q \in \mathbb{Z}} \omega_q$$

where $\omega_q \in \Omega^p_q[V]$ for each q. We obtain

$$Q(dQ \wedge \omega) = \sum_{q \in \mathbb{Z}} (dQ \wedge \omega_q).$$

This implies that each coefficient on the left side belongs to the ideal QS. Since QS is generated by the homogeneous element Q, each coefficient of $dQ \wedge \omega_q$ belongs to QS. This shows that $(\omega_q/Q) \in \Omega^p(\mathcal{A})$. □

Let $1 \leq p \leq \ell$. There is an S-bilinear pairing $\langle\,,\,\rangle$ between $\mathrm{Der}_{\mathbb{K}}(S)$ and $\Omega^1(V)$ called the **interior product**. More generally, the interior product

$$\langle\,,\,\rangle : \mathrm{Der}_{\mathbb{K}}(S) \times \Omega^p(V) \longrightarrow \Omega^{p-1}(V)$$

is the unique S-bilinear map which satisfies

$$\left\langle \theta, dx_{i_1} \wedge \cdots \wedge dx_{i_p} \right\rangle = \sum_{k=1}^{p} (-1)^{k-1} \theta(x_{i_k}) dx_{i_1} \wedge \cdots \wedge \widehat{dx_{i_k}} \wedge \cdots \wedge dx_{i_p}$$

for $1 \leq i_1 < \cdots < i_p \leq \ell$. Here $\widehat{dx_{i_k}}$ denotes the deletion of dx_{i_k}. Restriction of the interior product gives an S–bilinear pairing

$$\langle\,,\,\rangle : \mathrm{Der}_{\mathbb{K}}(S) \times \Omega^p[V] \longrightarrow \Omega^{p-1}[V].$$

The following properties of the interior product are well known.

Lemma 4.73 *If $\theta \in \mathrm{Der}_{\mathbb{K}}(S)$, $f \in S$, $\omega_1 \in \Omega^p(V)$, and $\omega_2 \in \Omega^q(V)$, then*
(1) $\langle \theta, \omega_1 \wedge \omega_2 \rangle = \langle \theta, \omega_1 \rangle \wedge \omega_2 + (-1)^p \omega_1 \wedge \langle \theta, \omega_2 \rangle$,
(2) $\langle \theta, df \rangle = \theta(f)$.

Proof. Since both sides are S–linear with respect to ω_1 and ω_2, we may assume that $\omega_1 = dx_{i_1} \wedge \cdots \wedge dx_{i_p}$ and $\omega_2 = dx_{j_1} \wedge \cdots \wedge dx_{j_q}$, where $1 \leq i_1 < \cdots < i_p \leq \ell$ and $1 \leq j_1 < \cdots < j_q \leq \ell$. Then

$$\begin{aligned}
\langle \theta, \omega_1 \wedge \omega_2 \rangle &= \left\langle \theta, dx_{i_1} \wedge \cdots \wedge dx_{i_p} \wedge dx_{j_1} \wedge \cdots \wedge dx_{j_q} \right\rangle \\
&= \left\langle \theta, dx_{i_1} \wedge \cdots \wedge dx_{i_p} \right\rangle \wedge dx_{j_1} \wedge \cdots \wedge dx_{j_q} \\
&\quad + (-1)^p dx_{i_1} \wedge \cdots \wedge dx_{i_p} \wedge \left\langle \theta, dx_{j_1} \wedge \cdots \wedge dx_{j_q} \right\rangle \\
&= \langle \theta, \omega_1 \rangle \wedge \omega_2 + (-1)^p \omega_1 \wedge \langle \theta, \omega_2 \rangle.
\end{aligned}$$

$$\begin{aligned}
\langle \theta, df \rangle &= \left\langle \theta, \sum_{i=1}^{\ell} (\partial f/\partial x_i) dx_i \right\rangle \\
&= \sum_{i=1}^{\ell} (\partial f/\partial x_i) \langle \theta, dx_i \rangle = \sum_{i=1}^{\ell} (\partial f/\partial x_i) \theta(x_i) = \theta(f).
\end{aligned}$$

This completes the proof. $\square$

Proposition 4.74 *For any arrangement $\mathcal{A}$, the interior product induces maps*

$$\langle\,,\,\rangle : D(\mathcal{A}) \times \Omega^p(\mathcal{A}) \longrightarrow \Omega^{p-1}(\mathcal{A}).$$

Proof. Let $\theta \in D(\mathcal{A})$ and $\omega \in \Omega^p(\mathcal{A})$. It follows from Proposition 4.69 that $dQ \wedge \omega \in \Omega^{p+1}[V]$. Since $\theta(Q) \in QS$, the calculation

$$\begin{aligned}
\langle \theta, dQ \wedge \omega \rangle &= \langle \theta, dQ \rangle \omega - dQ \wedge \langle \theta, \omega \rangle \\
&= \theta(Q)\omega - dQ \wedge \langle \theta, \omega \rangle
\end{aligned}$$

shows that $dQ \wedge \langle \theta, \omega \rangle \in \Omega^p[V]$. $\square$

It follows from Proposition 4.68 that $\Omega^0(\mathcal{A}) = S$. Thus, setting $p = 1$ in Proposition 4.74 gives an S–bilinear pairing

$$\langle\,,\,\rangle : D(\mathcal{A}) \times \Omega^1(\mathcal{A}) \longrightarrow S.$$

128   4. Free Arrangements

Theorem 4.75 *The S-modules $D(\mathcal{A})$ and $\Omega^1(\mathcal{A})$ are duals of each other: $D(\mathcal{A})^* \simeq \Omega^1(\mathcal{A})$ and $\Omega^1(\mathcal{A})^* \simeq D(\mathcal{A})$.*

Proof. The interior product pairing provides natural S–linear maps

$$\alpha : D(\mathcal{A}) \longrightarrow \Omega^1(\mathcal{A})^*,$$

$$\beta : \Omega^1(\mathcal{A}) \longrightarrow D(\mathcal{A})^*.$$

We will prove that these maps are isomorphisms.

α *is injective:* Let $\theta \in D(\mathcal{A})$. Suppose $\alpha(\theta) = 0$. This implies $0 = [\alpha(\theta)](\omega) = \langle \theta, \omega \rangle$ for all $\omega \in \Omega^1(\mathcal{A})$. Thus $\theta(f) = \langle \theta, df \rangle = 0$ for all $f \in S$. This shows that $\theta = 0$.

α *is surjective:* Let $\eta \in \Omega^1(\mathcal{A})^*$. Define a $\mathbb{K}$–linear map $\tilde{\eta} : S \to S$ by $\tilde{\eta}(f) = \eta(df)$ for $f \in S$. It is easy to check that $\tilde{\eta}$ is a derivation. We have

$$\tilde{\eta}(Q) = \eta(dQ) = Q\eta(dQ/Q) \in QS$$

because $dQ/Q \in \Omega^1(\mathcal{A})$. This implies that $\tilde{\eta} \in D(\mathcal{A})$. Moreover, we have

$$[\alpha(\tilde{\eta})](dx_i) = \langle \tilde{\eta}, dx_i \rangle = \tilde{\eta}(x_i) = \eta(dx_i)$$

for $1 \leq i \leq \ell$. Therefore $\alpha(\tilde{\eta}) = \eta$. Thus α is a bijection.

β *is injective:* Let $\omega \in \Omega^1(\mathcal{A})$. Suppose $\beta(\omega) = 0$. This implies $0 = [\beta(\omega)](\theta) = \langle \theta, \omega \rangle$ for all $\theta \in D(\mathcal{A})$. Write

$$\omega = (1/Q) \sum_{i=1}^{\ell} a_i dx_i,$$

using a basis for V^*. Then $0 = \langle QD_i, \omega \rangle = a_i$ for $1 \leq i \leq \ell$. This shows that $\omega = 0$.

β *is surjective:* Let $\omega \in D(\mathcal{A})^*$. Define a rational 1-form $\tilde{\omega}$ by

$$\tilde{\omega} = (1/Q) \sum_{i=1}^{\ell} \omega(QD_i) dx_i.$$

The coefficient of $dx_i \wedge dx_j$ ($i < j$) in $dQ \wedge \tilde{\omega}$ is equal to

$$(1/Q)[(\partial Q/\partial x_i)\omega(QD_j) - (\partial Q/\partial x_j)\omega(QD_i)] =$$
$$\omega((\partial Q/\partial x_i)D_j - (\partial Q/\partial x_j)D_i).$$

Since $(\partial Q/\partial x_i)D_j - (\partial Q/\partial x_j)D_i \in D(\mathcal{A})$, we know that each coefficient of $dQ \wedge \tilde{\omega}$ belongs to S. It follows that $dQ \wedge \tilde{\omega} \in \Omega^2[V]$. This implies that $\tilde{\omega} \in \Omega^1(\mathcal{A})$ by Proposition 4.69. Moreover we have

$$\begin{aligned}[][\beta(\tilde{\omega})](\theta) &= \langle \theta, \tilde{\omega} \rangle \\ &= \left\langle \theta, (1/Q) \sum_{i=1}^{\ell} \omega(QD_i) dx_i \right\rangle\end{aligned}$$

$$= (1/Q)\sum_{i=1}^{\ell} \omega(QD_i)\theta(x_i)$$
$$= (1/Q)\omega(Q\sum_{i=1}^{\ell} \theta(x_i)D_i)$$
$$= (1/Q)\omega(Q\theta)$$
$$= \omega(\theta).$$

Therefore $\beta(\widetilde{\omega}) = \omega$. Thus β is a bijection. □

Since $\Omega^1(\mathcal{A})$ and $D(\mathcal{A})$ are S–duals of each other, we have the following consequences.

Corollary 4.76 *The arrangement $\mathcal{A}$ is free if and only if $\Omega^1(\mathcal{A})$ is a free S-module.*

Corollary 4.77 *Assume that the arrangement $\mathcal{A}$ is free with $\exp \mathcal{A} = \{b_1, \ldots, b_\ell\}$. Then $\Omega^1(\mathcal{A})$ has a homogeneous basis $\omega_1, \omega_2, \ldots, \omega_\ell$ with $\mathrm{pdeg}(\omega_i) = -b_i$ for $1 \leq i \leq \ell$.*

Next we give more characterizations of logarithmic differential forms.

Proposition 4.78 *The following statements are equivalent:*
(1) $\omega \in \Omega^p(\mathcal{A})$,
(2) $Q\omega \in \Omega^p[V]$ and $dQ \wedge \omega \in \Omega^{p+1}[V]$,
(3) $Q\omega \in \Omega^p[V]$ and $Q(d\alpha/\alpha) \wedge \omega \in \Omega^{p+1}[V]$ for all $\alpha \in V^*$ with $\ker(\alpha) \in \mathcal{A}$,
(4) there exist $\xi_j \in \Omega^{p-1}[V]$ and $\eta_j \in \Omega^p[V]$ for $1 \leq j \leq \ell$ so that

$$(\partial Q/\partial x_j)\omega = (dQ/Q) \wedge \xi_j + \eta_j,$$

(5) there exist polynomials $f_1, \ldots, f_k \in S$ and forms $\xi_j \in \Omega^{p-1}[V]$ and $\eta_j \in \Omega^p[V]$ so that $f_1, \ldots, f_k$ and Q have no common factor of positive degree and $f_j\omega = (dQ/Q) \wedge \xi_j + \eta_j$ for $1 \leq j \leq k$.

Proof. The equivalence of (1), (2), and (3) was proved in Proposition 4.69.

(2) ⇒ (4) Let $\theta = D_j$, $\omega_1 = dQ$, and $\omega_2 = \omega$ in Lemma 4.73.1. Then we have

$$\langle D_j, dQ \wedge \omega \rangle = \langle D_j, dQ \rangle \omega - dQ \wedge \langle D_j, \omega \rangle$$
$$= (\partial Q/\partial x_j)\omega - (dQ/Q) \wedge \langle D_j, Q\omega \rangle.$$

Define $\xi_j = \langle D_j, Q\omega \rangle$ and $\eta_j = \langle D_j, dQ \wedge \omega \rangle$ for $1 \leq j \leq \ell$.

(4) ⇒ (5) This follows from Lemma 4.67, since the polynomials $\partial Q/\partial x_1, \ldots, \partial Q/\partial x_\ell$ and Q have no common factor of positive degree.

(5) ⇒ (2) We have

$$f_j(Q\omega) = dQ \wedge \xi_j + Q\eta_j \in \Omega^p[V],$$

and
$$f_j(dQ \wedge \omega) = dQ \wedge \eta_j \in \Omega^{p+1}[V].$$
Since the polynomials $f_1, \ldots, f_k$ and Q have no common factor of positive degree, $Q\omega \in \Omega^p[V]$ and $dQ \wedge \omega \in \Omega^{p+1}[V]$. □

Proposition 4.79 Let $\Omega^{\cdot}(\mathcal{A}) = \bigoplus_{p \geq 0} \Omega^p(\mathcal{A})$. The S-module $\Omega^{\cdot}(\mathcal{A})$ is closed under exterior product:
$$\Omega^p(\mathcal{A}) \times \Omega^q(\mathcal{A}) \longrightarrow \Omega^{p+q}(\mathcal{A}).$$

Proof. Let $\omega_1 \in \Omega^p(\mathcal{A})$ and $\omega_2 \in \Omega^q(\mathcal{A})$. By Proposition 4.78, there exist $\xi_j^{(i)} \in \Omega^{p-1}[V]$ and $\eta_j^{(i)} \in \Omega^p[V]$ with $i = 1, 2$ and $1 \leq j \leq \ell$ so that
$$(\partial Q/\partial x_j)\omega_i = (dQ/Q) \wedge \xi_j^{(i)} + \eta_j^{(i)} \quad (i = 1, 2).$$
Thus we have
$$(\partial Q/\partial x_j)^2 \omega_1 \wedge \omega_2 = (dQ/Q) \wedge (\xi_j^{(1)} \wedge \eta_j^{(2)} + \xi_j^{(2)} \wedge \eta_j^{(1)}) + (\eta_j^{(1)} \wedge \eta_j^{(2)}).$$
Since $(\partial Q/\partial x_1)^2, \ldots, (\partial Q/\partial x_\ell)^2$ and Q have no common factor of positive degree, we conclude from Proposition 4.78 that $\omega_1 \wedge \omega_2 \in \Omega^{p+q}(\mathcal{A})$. □

Recall the $\mathbb{K}$-algebra $R(\mathcal{A})$ from Definition 3.116. We noted in Proposition 4.66 that its generators are in $\Omega^1(\mathcal{A})$. It follows from Proposition 4.79 that $R(\mathcal{A}) \subseteq \Omega^{\cdot}(\mathcal{A})$. Since elements of $R(\mathcal{A})$ have total degree 0, we get
$$R(\mathcal{A}) \subseteq \bigoplus_{p+q=0} \Omega_q^p(\mathcal{A}).$$
Next we prove the analog of Saito's criterion 4.19 for differential forms.

Proposition 4.80 Given $\omega_1, \ldots, \omega_\ell$, the following two conditions are equivalent:
(1) $\omega_1, \ldots, \omega_\ell$ form a basis for $\Omega^1(\mathcal{A})$ over S,
(2) $\omega_1 \wedge \cdots \wedge \omega_\ell \doteq Q^{-1} dx_1 \wedge \cdots \wedge dx_\ell$.

Proof. Write
$$\omega_j = \sum_{k=1}^{\ell} a_{j,k} dx_k,$$
and let $\mathsf{N} = [a_{j,k}]_{1 \leq j,k \leq \ell}$. Then we have
$$\omega_1 \wedge \cdots \wedge \omega_\ell = (\det \mathsf{N})(dx_1 \wedge \cdots \wedge dx_\ell).$$

(1) $\Rightarrow$ (2) Since $\Omega^1(\mathcal{A})^* \simeq D(\mathcal{A})$ by Theorem 4.75, $D(\mathcal{A})$ has the dual basis $\theta_1, \ldots, \theta_\ell$ defined by $\langle \theta_i, \omega_j \rangle = \delta_{i,j}$. Recall the coefficient matrix
$$\mathsf{M} = \mathsf{M}(\theta_1, \ldots, \theta_\ell) = [\theta_j(x_i)]_{1 \leq i,j \leq \ell}$$

from Definition 4.11. Since

$$\delta_{i,j} = \langle \theta_i, \omega_j \rangle = \sum_{k=1}^{\ell} \theta_i(x_k) a_{j,k},$$

we have $\mathsf{NM} = \mathsf{I}$. By Saito's criterion 4.19, $\det \mathsf{M} \doteq Q$. Thus $\det \mathsf{N} \doteq Q^{-1}$.
(2) $\Rightarrow$ (1) We may assume that

$$\omega_1 \wedge \cdots \wedge \omega_\ell = Q^{-1} dx_1 \wedge \cdots \wedge dx_\ell.$$

Thus $\det \mathsf{N} = Q^{-1}$. For $1 \leq i \leq \ell$, define $\theta_i \in \Omega^1(\mathcal{A})^*$ by

$$\theta_i(\omega) Q^{-1} dx_1 \wedge \cdots \wedge dx_\ell = \omega_1 \wedge \cdots \wedge \omega_{i-1} \wedge \omega \wedge \omega_{i+1} \wedge \cdots \wedge \omega_\ell.$$

By identifying $\Omega^1(\mathcal{A})^*$ with $D(\mathcal{A})$, we may assume that $\theta_i \in D(\mathcal{A})$. Then $\langle \theta_i, \omega_j \rangle = \delta_{i,j}$ and hence

$$\mathsf{NM}(\theta_1, \ldots, \theta_\ell) = \mathsf{I}.$$

Thus $\det \mathsf{M}(\theta_1, \ldots, \theta_\ell) = Q$. Apply Saito's criterion 4.19 again to conclude that $D(\mathcal{A})$ is free. It follows that $\Omega^1(\mathcal{A}) \simeq D(\mathcal{A})^*$ is also free. □

Proposition 4.81 *If $\Omega^1(\mathcal{A})$ is a free S-module with basis $\omega_1, \ldots, \omega_\ell$, then for $1 \leq p \leq \ell$ the S-module $\Omega^p(\mathcal{A})$ is free with basis $\{\omega_{i_1} \wedge \ldots \wedge \omega_{i_p} \mid 1 \leq i_1 < \cdots < i_p \leq \ell\}$.*

Proof. By Proposition 4.80, we may assume that

$$\omega_1 \wedge \cdots \wedge \omega_\ell = Q^{-1} dx_1 \wedge \cdots \wedge dx_\ell.$$

Given a multiindex $K = (k_1, \ldots, k_s)$, define

$$\begin{aligned} \omega_K &= \omega_{k_1} \wedge \cdots \wedge \omega_{k_s}, \\ dx_K &= dx_{k_1} \wedge \cdots \wedge dx_{k_s}. \end{aligned}$$

Let $I = (i_1, \ldots, i_p)$ be an ordered multiindex of length p, $1 \leq i_1 < \cdots < i_p \leq \ell$. Let $I^c = (j_1, \ldots, j_{\ell-p})$ be the ordered complement of I, $1 \leq j_1 < \cdots < j_{\ell-p} \leq \ell$. Thus

$$\{i_1, \ldots, i_p\} \cup \{j_1, \ldots, j_{\ell-p}\} = \{1, \ldots, \ell\}.$$

Let $\sigma(I)$ be the sign of the permutation $(i_1, \ldots, i_p, j_1, \ldots, j_{\ell-p})$. Let $\mathcal{I}$ be the set of ordered multiindices of length p.

Let $\omega \in \Omega^p(\mathcal{A})$ and let $I \in \mathcal{I}$. Since $\Omega^{\cdot}(\mathcal{A})$ is closed under exterior product by Proposition 4.79, we have

$$\omega \wedge \omega_{I^c} \in \Omega^\ell(\mathcal{A}) = (1/Q) \Omega^\ell[V].$$

Define $f_I \in S$ by

$$\omega \wedge \omega_{I^c} = (f_I/Q)(dx_1 \wedge \cdots \wedge dx_\ell).$$

Let $\eta = \omega - \sum_{I \in \mathcal{I}} \sigma(I) f_I \omega_I$. Then $\eta \wedge \omega_{I^c} = 0$ for all $I \in \mathcal{I}$. Since $dx_1, \ldots, dx_\ell \in \Omega^1(\mathcal{A})$, each dx_i is a linear combination of $\omega_1, \ldots, \omega_\ell$ over S. Thus for any $K = (k_1, \ldots, k_{\ell-p})$, dx_K is a linear combination of $\{\omega_{I^c} \mid I \in \mathcal{I}\}$. This implies that $\eta \wedge dx_K = 0$. Therefore $\eta = 0$ and
$$\omega = \sum_{I \in \mathcal{I}} \sigma(I) f_I \omega_I.$$
This shows that $\{\omega_{i_1} \wedge \ldots \wedge \omega_{i_p} \mid 1 \leq i_1 < \cdots < i_p \leq \ell\}$ spans $\Omega^p(\mathcal{A})$. To show that this set is S-independent, assume $\sum_{I \in \mathcal{I}} f_I \omega_I = 0$. By taking exterior product with ω_{I^c}, we get $f_I = 0$. □

Example 4.82 Let $\mathcal{A}$ be a nonempty 2-arrangement defined by $Q = Q(\mathcal{A})$. We showed in Example 4.20 that $\mathcal{A}$ is free. It follows from Corollary 4.76 and Proposition 4.81 that $\Omega^{\cdot}(\mathcal{A})$ is free. Let $\alpha \in V^*$ with $\ker(\alpha) \in \mathcal{A}$. Then $\{d\alpha/\alpha, (y dx - x dy)/Q\}$ is a basis for $\Omega^1(\mathcal{A})$ with pdegrees $-1, -(n-1)$. The module $\Omega^2(\mathcal{A})$ has basis $(dx \wedge dy)/Q$.

Next we prove some facts about product arrangements. Let $(\mathcal{A}_1, V_1)$ and $(\mathcal{A}_2, V_2)$ be arrangements. Recall their product $\mathcal{A}_1 \times \mathcal{A}_2$ from Definition 2.13. Let $V = V_1 \oplus V_2$, $S = S(V^*)$, and for $i = 1, 2$ let $S_i = S(V_i^*)$. We may identify $S = S_1 \otimes_{\mathbb{K}} S_2$. For $i = 1, 2$ let $Q_i \in S_i$ be a defining polynomial for $\mathcal{A}_i$.

Lemma 4.83 *For $1 \leq i \leq m$ let $\eta_i \in (1/Q_1)\Omega^p[V_1]$ and let $\xi_i \in (1/Q_2)\Omega^q[V_2]$. If ξ_i are linearly independent over $\mathbb{K}$ and $\sum_{i=1}^m \eta_i \wedge \xi_i \in \Omega^n(\mathcal{A}_1 \times \mathcal{A}_2)$, then $\eta_i \in \Omega^p(\mathcal{A}_1)$ for $1 \leq i \leq m$.*

Proof. Note that
$$Q(dQ) \wedge (\sum_{i=1}^m \eta_i \wedge \xi_i)$$
$$= \sum_{i=1}^m (dQ_1 \wedge Q_1 \eta_i) \wedge Q_2(Q_2 \xi_i) + (-1)^p \sum_{i=1}^m Q_1(Q_1 \eta_i) \wedge (dQ_2 \wedge Q_2 \xi_i).$$
Since $\sum_{i=1}^m \eta_i \wedge \xi_i \in \Omega^n(\mathcal{A}_1 \times \mathcal{A}_2)$, we have $Q(dQ) \wedge (\sum_{i=1}^m \eta_i \wedge \xi_i) \in Q\Omega^{n+1}[V]$. Since $Q_2^2 \xi_i$ are linearly independent over $\mathbb{K}$, they are linearly independent over S_1. Thus for $1 \leq i \leq m$, we have $dQ_1 \wedge Q_1 \eta_i \in Q_1 \Omega^{p+1}[V_1]$, so $\eta_i \in \Omega^p(\mathcal{A}_1)$. □

Proposition 4.84 $\Omega^n(\mathcal{A}_1 \times \mathcal{A}_2) \simeq \bigoplus_{p+q=n} \Omega^p(\mathcal{A}_1) \otimes_{\mathbb{K}} \Omega^q(\mathcal{A}_2)$.

Proof. Let $\omega \in \Omega^n(\mathcal{A}_1 \times \mathcal{A}_2)$. To show that $\omega \in \bigoplus_{p+q=n} \Omega^p(\mathcal{A}_1) \otimes_{\mathbb{K}} \Omega^q(\mathcal{A}_2)$, we may assume that $\omega \in (1/Q_1)\Omega^p[V_1] \otimes (1/Q_2)\Omega^q[V_2]$. Write $\omega = \sum_{i=1}^m \eta_i \wedge \xi_i$, where for $1 \leq i \leq m$ we have $\eta_i \in (1/Q_1)\Omega^p[V_1]$, $\xi_i \in (1/Q_2)\Omega^q[V_2]$, and the ξ_i are linearly independent over $\mathbb{K}$. It follows from Lemma 4.83 that $\eta_i \in \Omega^p(\mathcal{A}_1)$ for $1 \leq i \leq m$. Thus $\omega \in \Omega^p(\mathcal{A}_1) \otimes_{\mathbb{K}} (1/Q_2)\Omega^q[V_2]$. Write $\omega = \sum_{i=1}^k \sigma_i \otimes \tau_i$, where for $1 \leq i \leq k$ we have $\sigma_i \in \Omega^p(\mathcal{A}_1)$, $\tau_i \in (1/Q_2)\Omega^q[V_2]$, and the σ_i are linearly independent over $\mathbb{K}$. It follows from Lemma 4.83 that $\tau_i \in \Omega^q(\mathcal{A}_2)$ for $1 \leq i \leq k$. Thus $\omega \in \Omega^p(\mathcal{A}_1) \otimes_{\mathbb{K}} \Omega^q(\mathcal{A}_2)$. □

The Acyclic Complex $(\Omega^{\cdot}(\mathcal{A}), \partial)$

Assume that $\mathcal{A}$ is a nonempty arrangement. Fix a hyperplane $H \in \mathcal{A}$. Let $\alpha = \alpha_H \in V^*$ with $H = \ker(\alpha)$. Recall that $d\alpha/\alpha \in \Omega^1(\mathcal{A})$. We showed in Proposition 4.79 that $\Omega^{\cdot}(\mathcal{A})$ is closed under exterior product.

Definition 4.85 *Define* $\partial : \Omega^p \to \Omega^{p+1}(\mathcal{A})$ *by* $\partial(\omega) = (d\alpha/\alpha) \wedge \omega$. *This map is homogeneous of pdegree* -1. *Since* $\partial \cdot \partial = 0$, *we have a complex*

$$0 \to \Omega^0(\mathcal{A}) \xrightarrow{\partial} \Omega^1(\mathcal{A}) \xrightarrow{\partial} \cdots \xrightarrow{\partial} \Omega^{\ell}(\mathcal{A}) \to 0.$$

Proposition 4.86 *The complex* $(\Omega^{\cdot}(\mathcal{A}), \partial)$ *is acyclic.*

Proof. Recall the Euler derivation $\theta_E = \sum_{i=1}^{\ell} x_i D_i$ from Definition 4.7. Let $\omega \in \Omega^p(\mathcal{A})$ be a cocycle, $\partial(\omega) = (d\alpha/\alpha) \wedge \omega = 0$. It follows from Proposition 4.74 that $\langle \theta_E, \omega \rangle \in \Omega^{p-1}(\mathcal{A})$. By Lemma 4.73.1, we have

$$\begin{aligned} \partial(\langle \theta_E, \omega \rangle) &= (d\alpha/\alpha) \wedge \langle \theta_E, \omega \rangle \\ &= -\langle \theta_E, (d\alpha/\alpha) \wedge \omega \rangle + \langle \theta_E, (d\alpha/\alpha) \rangle \wedge \omega \\ &= (\theta_E(d\alpha)/\alpha) \wedge \omega \\ &= \omega \end{aligned}$$

as required. □

The η-Complex $(\Omega^{\cdot}(\mathcal{A}), \partial_h)$

We will define other boundary maps in $\Omega^{\cdot}(\mathcal{A})$. Let $\mathcal{A}$ be a possibly empty arrangement. Let d be a positive integer. Let $\eta \in \Omega^1_d[V]$ be a homogeneous regular differential 1-form of pdegree d on V. Note that $\eta \in \Omega^1(\mathcal{A})$ because $\Omega^1[V] \subseteq \Omega^1(\mathcal{A})$.

Definition 4.87 *Define* $\partial_\eta : \Omega^p(\mathcal{A}) \to \Omega^{p+1}(\mathcal{A})$ *by* $\partial_\eta(\omega) = \eta \wedge \omega$. *This map is homogeneous of pdegree* d. *Since* $\partial_\eta \cdot \partial_\eta = 0$, *we have a complex*

$$0 \to \Omega^0(\mathcal{A}) \xrightarrow{\partial_\eta} \Omega^1(\mathcal{A}) \xrightarrow{\partial_\eta} \cdots \xrightarrow{\partial_\eta} \Omega^{\ell}(\mathcal{A}) \to 0.$$

We call it the η-*complex* $(\Omega^{\cdot}(\mathcal{A}), \partial_\eta)$.

In the rest of this section we assume that the field $\mathbb{K}$ is algebraically closed. Our aim is to prove that the cohomology groups of the η-complex are finite dimensional over $\mathbb{K}$ for a **generic** $\eta \in \Omega^1_d[V]$. In order to define the term generic, note that we can introduce the Zariski topology in S_d and $\Omega^1_d[V]$ because they can be identified with affine spaces:

$$\begin{aligned} S_d &\simeq \mathbb{K}^{\dim S_d}, \\ \Omega^1_d[V] &\simeq S_d^{\ell} \simeq \mathbb{K}^{\ell \dim S_d}. \end{aligned}$$

4. Free Arrangements

Let Y be a topological space. We say that property P is true for a **generic** $\omega \in Y$ if Y has an open dense subset $Z \subseteq Y$ such that property P is true for all $\omega \in Z$.

Lemma 4.88 *Let W be a k-dimensional vector space over $\mathbb{K}$. A generic $\omega \in \Omega_d^1[W]$ vanishes only at the origin.*

Proof. Let $x_1, \ldots, x_k$ be a basis for the dual space W^*. Write
$$\omega = f_1 dx_1 + \cdots + f_k dx_k$$
for $f_i \in \mathbb{K}[x_1, \ldots, x_k]$ with $\deg f_i = d$ for $1 \leq i \leq k$. It is well-known that there is an open dense set of polynomials $(f_1, \ldots, f_k) \in S_d^k$ such that the system of equations
$$f_1 = 0, f_2 = 0, \ldots, f_k = 0$$
has only the trivial solution. □

Suppose $\mathcal{A}$ is a nonempty arrangement. Consider $X \in L(\mathcal{A})$ with $\dim X > 0$. Define
$$\Omega_d^1[X]^\circ = \{\omega \in \Omega_d^1[X] \mid \omega \text{ vanishes only at the origin }\}.$$

It follows from Lemma 4.88 that $\Omega_d^1[X]^\circ$ is an open dense set in $\Omega_d^1[X]$. Let S^X be the symmetric algebra of the dual space X^* of X. The natural restriction map $S_d \to S_d^X$ is continuous with respect to the Zariski topology. Denote the image of $f \in S_d$ under this restriction map by $\bar{f} \in S_d^X$. We also have the restriction map $r_{V,X} : \Omega_d^1[V] \to \Omega_d^1[X]$. If we choose a basis $x_1, \ldots, x_\ell$ for V^* so X is defined by $x_{k+1} = \cdots = x_\ell = 0$, then $r_{V,X}$ is given by
$$r_{V,X}(f_1 dx_1 + \cdots + f_\ell dx_\ell) = \bar{f}_1 d\bar{x}_1 + \cdots + \bar{f}_k d\bar{x}_k.$$

The next result is immediate from the definitions.

Lemma 4.89 *For $X \in L(\mathcal{A})$, the restriction map $r_{V,X} : \Omega_d^1[V] \to \Omega_d^1[X]$ is continuous with respect to the Zariski topology.*

Let $N_d^X = r_{V,X}^{-1}(\Omega_d^1[X]^\circ)$. It follows from Lemmas 4.88 and 4.89 that N_d^X is an open dense set in $\Omega_d^1[V]$. Let
$$N_d = \bigcap_{\substack{X \in L(\mathcal{A}) \\ \dim X > 0}} N_d^X.$$

Since $L(\mathcal{A})$ is a finite set, N_d is an open dense set in $\Omega_d^1[V]$. In particular, N_d is nonempty.

Lemma 4.90 *Let $\eta \in N_d$. Then the radical of the ideal*

$$I(\eta) = \{\langle\theta,\eta\rangle \mid \theta \in D(\mathcal{A})\}$$

contains the maximal ideal $S_+ = \bigoplus_{p>0} S_p$.

Proof. By Hilbert's Nullstellensatz, it suffices to show that the zero locus $V(I(\eta))$ of $I(\eta)$ is contained in $\{0\}$. Let $v \in V \setminus \{0\}$ and let $X = \bigcap_{v \in H \in \mathcal{A}} H$. Thus $v \in X$ and $v \notin Y$ for any $Y \in L(\mathcal{A})$ with $Y \subset X$. Choose a basis $x_1, \ldots, x_\ell$ for V^* so X is defined by $x_{k+1} = \cdots = x_\ell = 0$. Let $\mathcal{A}_1 = \{H \in \mathcal{A} \mid H \not\supseteq X\}$. Let Q_1 be a defining polynomial for $\mathcal{A}_1$. Note that $Q_1(\partial/\partial x_i) \in D(\mathcal{A})$ for $1 \leq i \leq k$. Write $\eta = f_1 dx_1 + \cdots + f_\ell dx_\ell$ with $f_i \in S$ and $1 \leq i \leq \ell$. Then $\langle Q_1(\partial/\partial x_i), \eta \rangle = Q_1 f_i$. Since $\eta \in N_d^X$,

$$r_{V,X}(\eta) = \bar{f}_1 d\bar{x}_1 + \cdots \bar{f}_k d\bar{x}_k$$

vanishes only at the origin. Thus there exists i with $1 \leq i \leq k$ so $f_i(v) = \bar{f}_i(v) \neq 0$. Since $Q_1(v) \neq 0$, we have $v \notin V(I(\eta))$. □

Proposition 4.91 *If $\eta \in \Omega_d^1[V]$ is generic, then the cohomology groups of the η-complex are finite dimensional over $\mathbb{K}$.*

Proof. Let $\eta \in N_d$. Let H^p denote the p-th cohomology group of the complex in Definition 4.87. Note that H^p is an S–module. Since $\Omega^p(\mathcal{A})$ is a finitely generated S–module, so is H^p. By Lemma 4.90 and Proposition A.25, it suffices to prove that $I(\eta)$ annihilates H^p. Let $\omega \in \Omega^p(\mathcal{A})$ be a cocycle and let $\theta \in D(\mathcal{A})$. We have

$$0 = \langle\theta, \eta \wedge \omega\rangle = \langle\theta,\eta\rangle\omega - \eta \wedge \langle\theta,\omega\rangle = \langle\theta,\eta\rangle\omega - \partial_\eta(\langle\theta,\omega\rangle).$$

Thus $\langle\theta,\eta\rangle\omega = \partial_\eta(\langle\theta,\omega\rangle)$ is a coboundary, since $\langle\theta,\omega\rangle \in \Omega^{p-1}(\mathcal{A})$ by Proposition 4.74. □

4.5 Lattice Homology

In this section we associate to the lattice $L(\mathcal{A})$ a simplicial complex $\mathsf{F}(\mathcal{A})$ first studied by Folkman [85] and Rota [195], compute its homology groups, and determine its homotopy type. There is an active area of research concerned with the topological properties of complexes obtained from partially ordered sets, such as the poset of all subgroups of a group. Since $L(\mathcal{A})$ is a geometric lattice, we need not be concerned with the deeper aspects of that theory. We shall present only as much as we need for arrangements. We use books by Dold [67] and Spanier [215] as general references in topology and papers by Rota [195], Folkman [85], Björner [29], and Quillen [186] as general references for combinatorial topology. The first part of this exposition follows [74].

The Order Complex

Definition 4.92 *Let P be a partially ordered set. Let $K = K(P)$ be the simplicial complex associated to P as follows:*
(1) the vertices of K are the elements of P,
(2) a set of vertices $\{X_0, \ldots, X_q\}$ spans a q-simplex if and only if it is a linearly ordered subset of P; after relabeling,

$$X_0 < \cdots < X_q.$$

Definition 4.93 *Given a poset P and the associated simplicial complex $K(P)$, let $\mathsf{K}(P)$ be the corresponding geometric complex called the **order complex**.*

If P_1 and P_2 are posets, then there is a natural partial order on the set $P_1 \times P_2$ given in Proposition 2.14:

$$(X_1, X_2) \leq (Y_1, Y_2) \iff X_1 \leq Y_1 \text{ and } X_2 \leq Y_2.$$

Proposition 4.94 *Let P be any poset and let $Q = \{0 < 1\}$ be a poset with two elements. Then $\mathsf{K}(P \times Q)$ is a subdivision of $\mathsf{K}(P) \times I$.*

Proof. The space $\mathsf{K}(Q) = I$ is the unit interval. It is sufficient to prove the special case when $P = \{X_0 < \cdots < X_p\}$ is a linearly ordered set, so $\mathsf{K}(P) = \Delta^p$ is a simplex. Consider Δ^p as a simplex of the standard simplicial subdivision of the p-cube I^p; see [67, p.118]. The complex $\mathsf{K}(P \times Q)$ is a simplicial subdivision of the subspace $\Delta^p \times I$ in $I^p \times I = I^{p+1}$. Figure 4.2 illustrates the case $p = 2$. □

Corollary 4.95 *Let P be a poset and let $f : P \to P$ be an order preserving map with the property that $f(X) \leq X$ for all $X \in P$. Then the induced map of topological spaces $f : \mathsf{K}(P) \to \mathsf{K}(P)$ is homotopic to the identity. If $P_0 \subseteq P$ is a subset such that $f|_{P_0} = id_{P_0}$, then the homotopy is relative to $\mathsf{K}(P_0)$.*

Proof. Let $Q = \{0 < 1\}$. Define $F : P \times Q \to P$ by $F(X, 0) = f(X)$ and $F(X, 1) = X$. Since $f(X) \leq X$, F is order preserving. It induces a map

$$F : \mathsf{K}(P \times Q) \to \mathsf{K}(P).$$

By Proposition 4.94, we may view F as a homotopy between f and the identity:

$$F : \mathsf{K}(P) \times I \to \mathsf{K}(P).$$

If $f|_{P_0} = id_{P_0}$, then F is the identity on $\mathsf{K}(P_0)$. □

Lemma 4.96 *(1) Suppose P has a unique minimal element V. Then $\mathsf{K}(P)$ is a cone with base $\mathsf{K}(P \setminus \{V\})$. Thus $\mathsf{K}(P)$ is a contractible space.*

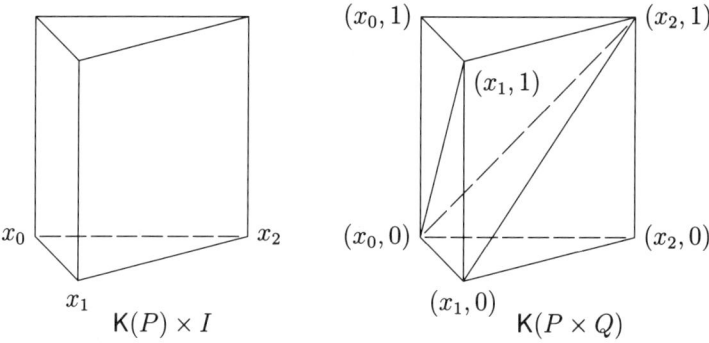

Fig. 4.2. Subdivision of $\Delta^2 \times I$

(2) Suppose P has a unique maximal element T. Then $\mathsf{K}(P)$ is a cone with base $\mathsf{K}(P \setminus \{T\})$. Thus $\mathsf{K}(P)$ is a contractible space.

Proof. We prove (1). If $\sigma^p = [X_0, \ldots, X_p] \in K(P \setminus \{V\})$, then $\tau^{p+1} = [V, X_0, \ldots, X_p] \in K(P)$. Moreover, every simplex of $K(P) \setminus K(P \setminus \{V\})$ has the form $V\sigma$ for some $\sigma \in K(P \setminus \{V\})$. The argument is similar for (2). □

The Folkman Complex

Definition 4.97 *Let $\mathcal{A}$ be an arrangement and let $L = L(\mathcal{A})$. Suppose $r(\mathcal{A}) \geq 2$. Let $K(L \setminus \{V, T\})$ be the simplicial complex associated to the poset obtained from L by deleting its minimal and its maximal elements. Let the **Folkman complex** $\mathsf{F}(\mathcal{A}) = \mathsf{K}(L \setminus \{V, T\})$ be the corresponding geometric complex.*

Note that $\dim \mathsf{F}(\mathcal{A}) = r(\mathcal{A}) - 2$. If $r(\mathcal{A}) = 2$, then $\mathsf{F}(\mathcal{A})$ consists of $|\mathcal{A}|$ points.

Example 4.98 *Let $\mathcal{B}(\ell + 1)$ denote the Boolean arrangement defined by $Q = x_0 x_1 \cdots x_\ell$. Let $H_i = \ker(x_i)$. Then F is the $(\ell - 1)$-complex consisting of the barycentric subdivision of the boundary of an ℓ-simplex with vertices $H_0, \ldots, H_\ell$. Thus F is homeomorphic to $S^{\ell-1}$.*

Definition 4.99 *Let $(\mathcal{A}, \mathcal{A}', \mathcal{A}'')$ be a triple with respect to $H_0 \in \mathcal{A}$. Let $L' = L(\mathcal{A}')$, $L'' = L(\mathcal{A}'')$, and $T = T(\mathcal{A})$. Define*

$$\mathsf{F}'' = \mathsf{F}''(\mathcal{A}) = \mathsf{F}(\mathcal{A}'')$$

and

$$\mathsf{F}' = \mathsf{F}'(\mathcal{A}) = \begin{cases} |K(L' \setminus \{V, T\})| & \text{if } T \in L', \\ |K(L' \setminus \{V\})| & \text{if } T \notin L'. \end{cases}$$

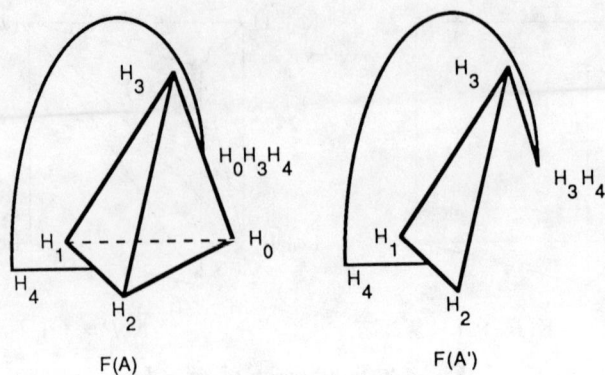

Fig. 4.3. Folkman complexes for $Q = xyz(x+y)(x+y-z)$

This case distinction is essential in several proofs. Recall from Definition 2.58 that H_0 is a separator if $T \notin L(\mathcal{A}')$. The poset $L' \setminus \{V\}$ has a unique maximal element $T' = T(\mathcal{A}')$. If H_0 is a separator, then Lemma 4.96 implies that the space $\mathsf{F}'(\mathcal{A})$ is contractible. If H_0 is not a separator, then $\mathsf{F}'(\mathcal{A}) = \mathsf{F}(\mathcal{A}')$.

Example 4.100 Let $\mathcal{A}$ be the 3–arrangement in Example 3.37 defined by

$$Q(\mathcal{A}) = xyz(x+y)(x+y-z).$$

Recall the notation of Example 3.37: $H_0 = \ker(x+y-z)$, $H_1 = \ker(x)$, $H_2 = \ker(y)$, $H_3 = \ker(z)$, $H_4 = \ker(x+y)$. The 1–complex $\mathsf{F}(\mathcal{A})$ is illustrated in Figure 4.3. The 0–complex $\mathsf{F}''(\mathcal{A})$ consists of the three points $H_0 \cap H_1$, $H_0 \cap H_2$, $H_0 \cap H_3 \cap H_4$. Here $T \in L'$, so H_0 is not a separator. The 1–complex F' is also illustrated in Figure 4.3.

Example 4.101 In $\mathcal{B}(\ell+1)$, the complex F'' is homeomorphic to $S^{\ell-2}$. Note here that $T \notin L'$, so H_0 is a separator. The complex F' is the $(\ell-1)$–simplex opposite the vertex H_0. These complexes are illustrated for $\ell = 3$ in Figure 4.4.

Lemma 4.102 *If $\mathcal{A}$ is an arrangement of rank 2, then $\mathsf{F}(\mathcal{A})$ consists of $\mu(\mathcal{A})+1$ points.*

Proof. We observed that F consists of $|\mathcal{A}|$ points and the Möbius function gives $1 - |\mathcal{A}| + \mu(\mathcal{A}) = 0$. □

Lemma 4.103 *If $\mathcal{A}$ is an arrangement with $r(\mathcal{A}) \geq 3$, then $\mathsf{F}(\mathcal{A})$ is path connected.*

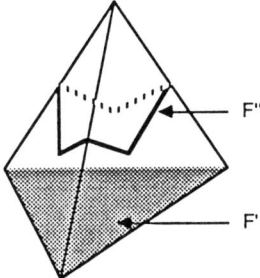

Fig. 4.4. Complexes for the Boolean arrangement

Proof. Suppose X is a vertex of $\mathsf{F}(\mathcal{A})$. There exists $H \in \mathcal{A}$ such that $X \geq H$ and $H \in \mathsf{F}(\mathcal{A})$. If $X \neq H$, then the 1–simplex $[H, X] \subseteq \mathsf{F}(\mathcal{A})$. Thus every vertex and hence every point of $\mathsf{F}(\mathcal{A})$ is connected by a path to some vertex of $\mathsf{F}(\mathcal{A})$ which corresponds to a hyperplane. It remains to show that vertices corresponding to distinct hyperplanes $H_1, H_2 \in \mathcal{A}$ are connected. Since $r(H_1 \cap H_2) = 2 < r(\mathcal{A})$, we have $H_1 \cap H_2 \in \mathsf{F}(\mathcal{A})$. Thus the 1–simplexes $[H_1, H_1 \cap H_2]$ and $[H_2, H_1 \cap H_2]$ are in $\mathsf{F}(\mathcal{A})$. □

Let K be a simplicial complex and let $|K|$ be its geometric complex. Let v be a vertex of K. Recall that the **star** of v is a subset of $|K|$ consisting of all open simplexes whose closure contains v:

$$st(v) = \{ | \overset{\circ}{\sigma} | \ \ | \ v \in \sigma \}.$$

Note that its closure, $\overline{st(v)}$, is a cone with cone point v.

Proposition 4.104 *Let $(\mathcal{A}, \mathcal{A}', \mathcal{A}'')$ be a triple with respect to H_0 and let $\mathsf{F} = \mathsf{F}(\mathcal{A}), \mathsf{F}' = \mathsf{F}'(\mathcal{A})$. There is a strong deformation retraction*

$$\rho : \mathsf{F} \setminus st(H_0) \to \mathsf{F}'.$$

Proof. Note first that $\mathsf{F} \setminus st(H_0) = |K(L \setminus \{V, H_0, T\})|$. Define a poset map

$$\rho : L \setminus \{V, H_0, T\} \to L \setminus \{V, H_0, T\}$$

by

$$\rho(X) = \bigcap_{H \in \mathcal{A}_X \setminus \{H_0\}} H.$$

If H_0 is a separator, then $\mathrm{im}(\rho) \subseteq L' \setminus \{V\}$. If H_0 is not a separator, then $\mathrm{im}(\rho) \subseteq L' \setminus \{V, T\}$. Extend ρ linearly to $\mathsf{F} \setminus st(H_0)$ and call the resulting map again ρ. It follows that $\mathrm{im}(\rho) \subseteq \mathsf{F}'$. Clearly $r(\rho(X)) \leq r(X)$ and $\rho|_{\mathsf{F}'} = id_{\mathsf{F}'}$. It follows from Corollary 4.95 that ρ is a strong deformation retraction. □

Theorem 4.105 *If $\mathcal{A}$ is an arrangement with $r(\mathcal{A}) \geq 4$, then $\mathsf{F}(\mathcal{A})$ is simply connected.*

Proof. We use induction on $|\mathcal{A}|$. Since $|\mathcal{A}| \geq r(\mathcal{A})$, the induction starts with $|\mathcal{A}| = r(\mathcal{A})$. In this case, $\mathcal{A}$ is isomorphic to the Boolean arrangement $\mathcal{B}(q)$ for $q = r(\mathcal{A})$. Example 4.98 showed that $\mathsf{F}(\mathcal{B}(q))$ is homeomorphic to S^{q-2}. Since $q = r(\mathcal{A}) \geq 4$, the assertion holds for $|\mathcal{A}| = r(\mathcal{A})$. For the induction step, choose $H_0 \in \mathcal{A}$ and consider the associated spaces $\mathsf{F}, \mathsf{F}', \mathsf{F}''$. We have

(1) $$\mathsf{F} = \overline{st(H_0)} \cup (\mathsf{F} \setminus st(H_0))$$

and

(2) $$\mathsf{F}'' = \overline{st(H_0)} \cap (\mathsf{F} \setminus st(H_0)).$$

Now $\overline{st(H_0)}$ is a cone over F'' with cone point H_0. In particular, it is simply connected. We showed in Proposition 4.104 that $\mathsf{F} \setminus st(H_0)$ has the homotopy type of F'. If H_0 is a separator, then F' is contractible. If H_0 is not a separator, then $\mathsf{F}' = \mathsf{F}(\mathcal{A}')$ and $r(\mathcal{A}') = r(\mathcal{A})$. Since $|\mathcal{A}'| < |\mathcal{A}|$, the induction hypothesis implies that F' is simply connected. Finally, $r(\mathcal{A}'') = r(\mathcal{A}) - 1 \geq 3$, so it follows from Lemma 4.103 that F'' is path connected. Thus van Kampen's theorem implies that F is simply connected. □

The Homology Groups

Next we want to compute the homology groups of $\mathsf{F}(\mathcal{A})$. Integer coefficients are understood unless otherwise indicated, and $\tilde{H}$ denotes reduced homology. Consider the Mayer–Vietoris sequence of reduced homology for the excisive couple $\{\mathsf{F} \setminus st(H_0), \overline{st(H_0)}\}$. Using (1) and (2), we get the long exact sequence

$$\cdots \to \tilde{H}_p(\mathsf{F} \setminus st(H_0)) \oplus \tilde{H}_p(\overline{st(H_0)}) \xrightarrow{(i_1,i_2)_*} \tilde{H}_p(\mathsf{F}) \xrightarrow{\partial_*}$$

$$\tilde{H}_{p-1}(\mathsf{F}'') \xrightarrow{(j_1,-j_2)_*} \tilde{H}_{p-1}(\mathsf{F} \setminus st(H_0)) \oplus \tilde{H}_{p-1}(\overline{st(H_0)}) \to \cdots$$

The fact that $\overline{st(H_0)}$ is contractible and Proposition 4.104 give

(3) $$\cdots \to \tilde{H}_p(\mathsf{F}') \xrightarrow{i_{1*}} \tilde{H}_p(\mathsf{F}) \xrightarrow{\partial_*} \tilde{H}_{p-1}(\mathsf{F}'') \xrightarrow{j_{1*}} \tilde{H}_{p-1}(\mathsf{F}') \to \cdots$$

The next result is due to Folkman [85].

Theorem 4.106 *Let $\mathcal{A}$ be an arrangement with Folkman complex $\mathsf{F} = \mathsf{F}(\mathcal{A})$. Then*
$$\tilde{H}_i(\mathsf{F}) = \begin{cases} 0 & \text{if } i \neq r(\mathcal{A}) - 2, \\ \text{free of rank } |\mu(\mathcal{A})| & \text{if } i = r(\mathcal{A}) - 2. \end{cases}$$

Proof. We use induction on $r(\mathcal{A})$, and for fixed $r(\mathcal{A})$ on $|\mathcal{A}|$. The assertion is correct for $r(\mathcal{A}) = 2$ and arbitrary $|\mathcal{A}|$ by Lemma 4.102. The assertion is also correct for arbitrary $r(\mathcal{A})$ when $|\mathcal{A}| = r(\mathcal{A})$, since in that case $\mathcal{A}$ is the Boolean arrangement and we noted in Example 4.98 that $\mathsf{F}(\mathcal{A})$ is an $(r(\mathcal{A}) - 2)$-sphere, while it follows from Proposition 2.44 that $|\mu(\mathcal{A})| = 1$. For the induction step we assume that the result holds for all arrangements $\mathcal{B}$ with $r(\mathcal{B}) < r(\mathcal{A})$ and for all arrangements $\mathcal{B}$ with $r(\mathcal{A}) = r(\mathcal{B})$ and $|\mathcal{B}| < |\mathcal{A}|$. Fix $H_0 \in \mathcal{A}$. Consider the exact sequence (3). For $p \neq r(\mathcal{A}) - 2$, the induction hypothesis implies that $\tilde{H}_p(\mathsf{F}') = \tilde{H}_{p-1}(\mathsf{F}'') = 0$ and hence $\tilde{H}_p(\mathsf{F}) = 0$. For $p = r(\mathcal{A}) - 2$, the induction hypothesis implies that $\tilde{H}_{p-1}(\mathsf{F}'')$ is free of rank $|\mu(\mathcal{A}'')|$. If H_0 is not a separator, then $\mathsf{F}' = \mathsf{F}(\mathcal{A}')$, so the induction hypothesis implies that $\tilde{H}_p(\mathsf{F}')$ is free of rank $|\mu(\mathcal{A}')|$. Thus $\tilde{H}_p(\mathsf{F})$ is free of rank $|\mu(\mathcal{A}')| + |\mu(\mathcal{A}'')|$. If H_0 is a separator, then F' is contractible, so $\tilde{H}_p(\mathsf{F})$ is free of rank $|\mu(\mathcal{A}'')|$. The conclusion follows from Corollary 2.59. □

Corollary 4.107 *Let $\mathcal{A}$ be an arrangement with Folkman complex $\mathsf{F} = \mathsf{F}(\mathcal{A})$. Let $\mathcal{K}$ be a commutative ring. Then*

$$\tilde{H}_i(\mathsf{F}; \mathcal{K}) = \begin{cases} 0 & \text{if } i \neq r(\mathcal{A}) - 2, \\ \text{free } \mathcal{K}\text{-module of rank } |\mu(\mathcal{A})| & \text{if } i = r(\mathcal{A}) - 2. \end{cases}$$

The Homotopy Type

Definition 4.108 *Let $(S_1^k, P_1), \ldots, (S_m^k, P_m)$ be m disjoint k-spheres with base points P_j. Their **wedge** is the based space $(\vee_m S^k, P)$ obtained by identifying the base points $P_1 = \ldots = P_m = P$.*

It is clear that $(\vee_m S^k, P)$ is a cell complex. We may write $\vee_m S^k$ for brevity. We have

$$\pi_i(\vee_m S^k) = \begin{cases} 0 & \text{for } i < k \\ \text{free of rank } m & \text{for } i = k, \end{cases}$$

and

$$\tilde{H}_i(\vee_m S^k; \mathbb{Z}) = \begin{cases} 0 & \text{for } i \neq k \\ \text{free of rank } m & \text{for } i = k. \end{cases}$$

The next result was stated by Quillen [186] without proof. It is reasonable to assume that he had in mind the argument below. Björner and Walker [31] proved the result without appeal to facts in homotopy theory.

Theorem 4.109 *Let $\mathcal{A}$ be an arrangement with $r(\mathcal{A}) \geq 2$. Then its Folkman complex $\mathsf{F} = \mathsf{F}(\mathcal{A})$ has the homotopy type of $\vee_m S^k$ with $k = r(\mathcal{A}) - 2$ and $m = |\mu(\mathcal{A})|$.*

Proof. For $r(\mathcal{A}) = 2$, this follows from Lemma 4.102. For $r(\mathcal{A}) = 3$, the complex F is 1–dimensional and hence it has the homotopy type of a wedge of circles whose number equals the rank of $H_1(\mathsf{F})$. We showed in Theorem 4.106 that this rank is m. For $r(\mathcal{A}) \geq 4$, the complex F is simply connected by Theorem 4.105.

It follows from Theorem 4.106 and the Hurewicz isomorphism theorem [215, p.398] that $\pi_i(\mathsf{F}) = 0$ for $1 \leq i < k$ and $\pi_k(\mathsf{F}) \simeq H_k(\mathsf{F}; \mathbb{Z})$. The last group is free of rank m by Theorem 4.106. For $1 \leq i \leq m$, let $\rho_i : S_i^k \to \mathsf{F}$ be generators of $\pi_k(\mathsf{F})$. Let $\rho : \vee_m S^k \to \mathsf{F}$ be the sum of ρ_i. Then ρ induces an isomorphism in homology by construction. Since the spaces are simply connected cell complexes, it follows from standard results in homotopy theory [215, pp.405-406] that ρ is a homotopy equivalence. □

Whitney Homology

Next we associate to $L(\mathcal{A})$ another chain complex, studied by Deheuvels [60] and Baclawski [16], compute its homology, and relate it to the homology of $\mathsf{F}(\mathcal{A})$ and to the algebra $B(\mathcal{A})$. Let $\mathcal{K}$ be a commutative ring. Recall the spaces $\mathcal{T}_p$ from Definition 3.93 and the set of chains chL from Definition 2.36.

Definition 4.110 *Let $\mathcal{A}$ be an arrangement with lattice $L = L(\mathcal{A})$. Define a chain complex $(\mathcal{C}, \delta)$ as follows. Let $\mathcal{C}_0 = \mathcal{K}$ and for $p > 0$ let $\mathcal{C}_p \subseteq \mathcal{T}_p$ have a $\mathcal{K}$-basis consisting of all p-chains $(X_1, \ldots, X_p) \in ch(L \setminus \{V\})$. Let $\mathcal{C} = \oplus_{p=0}^{\ell} \mathcal{C}_p$. Define a $\mathcal{K}$-linear map $\delta : \mathcal{C} \to \mathcal{C}$ by $\delta(1) = 0$, $\delta(X) = 0$ for $X \in L - \{V\}$ and for $p \geq 2$*

$$\delta(X_1, \ldots, X_p) = \sum_{k=1}^{p-1} (-1)^{k-1}(X_1, \ldots, \hat{X}_k, \ldots, X_p).$$

The map δ differs from the usual boundary operator in that X_p is never deleted. We still have $\delta^2 = 0$, so $(\mathcal{C}, \delta)$ is a chain complex, which we call the **Whitney complex** of $\mathcal{A}$. The Poincaré polynomial of its homology was first computed by Baclawski [16]. We give a generalization of his result.

For $2 \leq k \leq p$, define $\mathcal{K}$-linear maps $\delta_k : \mathcal{C}_p \to \mathcal{C}_{p-1}$ by

$$\delta_k(X_1, \ldots, X_p) = (-1)^{k-1}(X_1, \ldots, \hat{X}_k, \ldots, X_p).$$

We agree that $\delta_1 : \mathcal{C}_1 \to \mathcal{C}_0$ is the zero map. Recall the map $\tau : \mathcal{T} \to \mathcal{T}$ of Definition 3.105. If $x \in \mathcal{C}_p$, then $\tau(x) = \delta_p(x)$ and $\delta(x) = \delta_1(x) + \cdots + \delta_{p-1}(x)$. Recall the $\mathcal{K}$-algebra $B = B(\mathcal{A})$ of Section 3.4 and note that $B \subseteq \mathcal{C}$.

Lemma 4.111 *(1) The elements of B are cycles of $\mathcal{C}$, so $\delta B = 0$.*
(2) Let $r = r(\mathcal{A})$. If $x \in \mathcal{C}_r$ is a cycle, then $\tau \tau x = \delta \tau x = 0$.

Proof. (1) It suffices to show that $\delta b_S = 0$ for all $S \in \mathbf{S}$. Let $S = (H_1, \ldots, H_p)$. In $\delta b_{(H_1, \ldots, H_p)}$, each term

$$(\operatorname{sign}\pi)(-1)^{k-1}(H_{\pi 1}, \ldots, \widehat{H_{\pi 1} \cap \cdots \cap H_{\pi k}}, \ldots, H_{\pi 1} \cap \cdots \cap H_{\pi p})$$

is cancelled against the term in which πk and $\pi(k+1)$ are transposed. For (2) note that x is a linear combination of chains consisting of elements of ranks

$1, 2, \ldots, r$. So $\delta_i x$ is a linear combination of chains consisting of elements of ranks $1, 2, \ldots, i-1, i+1, \ldots, r$. Thus $0 = \delta x = \delta_1 x + \cdots + \delta_{r-1} x$ implies $\delta_i x = 0$ for $1 \leq i \leq r - 1$. We have

$$\tau \tau x = \delta_{r-1} \delta_r x = \delta_{r-1} \delta_{r-1} x = 0$$

and

$$\delta \tau x = (\delta_1 + \cdots + \delta_{r-2}) \delta_r x = \delta_{r-1} (\delta_1 + \cdots + \delta_{r-2}) x = 0.$$

This completes the proof. □

For $X \in L \setminus \{V\}$, let $\mathcal{C}_X$ be the subspace of $\mathcal{C}$ spanned by all $(X_1, \ldots, X_p)$ with $X_p = X$ and let $\mathcal{C}_V = \mathcal{K}$. Then $\mathcal{C} = \oplus_{X \in L} \mathcal{C}_X$ and $\delta : \mathcal{C}_X \to \mathcal{C}_X$. Thus $(\mathcal{C}_X, \delta)$ is a subcomplex. Let $\mathcal{H}_X$ be its homology. Then

(1) $$\mathcal{H} = \bigoplus_{X \in L} \mathcal{H}_X.$$

The natural identification $\mathcal{C}_X(\mathcal{A}_X) \simeq \mathcal{C}_X(\mathcal{A})$ implies

(2) $$\mathcal{H}_X(\mathcal{A}_X) \simeq \mathcal{H}_X(\mathcal{A}).$$

Theorem 4.112 *The map $B(\mathcal{A}) \to \mathcal{H}(\mathcal{A}; \mathcal{K})$ which sends b_S to its homology class $[b_S]$ is an isomorphism of free $\mathcal{K}$-modules.*

Proof. By Lemma 3.108, (1), and (2), it suffices to show that the natural map $B_T(\mathcal{A}) \to \mathcal{H}_T(\mathcal{A}; \mathcal{K})$ is an isomorphism. Let $\ell = r(T)$. Since the map is induced by inclusion, it is injective because $\mathcal{C}_{\ell+1} = 0$. We will show surjectivity by induction on $r(\mathcal{A}) = \ell$. The assertion is clear for $\ell = 1$. Denote the natural projection from $\mathcal{C} = \oplus_{X \in L} \mathcal{C}_X$ onto $\mathcal{C}_X$ by π_X. Let $x \in \mathcal{C}_\ell$ with $\delta x = 0$ and write

$$\tau x = \sum_{X \in L_{\ell-1}} \pi_X(\tau x).$$

Lemma 4.111.2 implies that

$$0 = \delta \tau x = \sum_{X \in L_{\ell-1}} \delta \pi_X(\tau x) = \sum_{X \in L_{\ell-1}} \pi_X(\delta \tau x).$$

Thus $0 = \pi_X(\delta \tau x) = \delta(\pi_X \tau x)$, and therefore $\pi_X(\tau x) \in \mathcal{H}_X(\mathcal{A}_X)$ for all $X \in L_{\ell-1}$. By the induction assumption, $\mathcal{H}_X(\mathcal{A}_X) = B_X(\mathcal{A}_X)$ for $X < T$. Thus $\tau x \in B_{\ell-1}(\mathcal{A})$. Recall that $\tau \tau x = 0$ by Lemma 4.111.2. Since the complex (B, τ) is acyclic by Lemma 3.109.3, we have $\tau x \in \tau B_\ell = \tau B_T$. Since $\tau : \mathcal{C}_T \to \mathcal{C}$ is injective, we conclude that $x \in B_T$. □

The following results are consequences of Corollaries 3.111, 3.112 and 3.113, respectively.

Corollary 4.113 *The module $\mathcal{H}(\mathcal{A}; \mathcal{K}) = \oplus_p \mathcal{H}_p(\mathcal{A}; \mathcal{K})$ is a free graded $\mathcal{K}$-module.*

Corollary 4.114 *The Poincaré polynomial of* $\mathcal{H}(\mathcal{A};\mathcal{K})$ *is*

$$\mathrm{Poin}(\mathcal{H}(\mathcal{A};\mathcal{K}),t) = \pi(\mathcal{A},t).$$

Corollary 4.115 *If* $X \in L(\mathcal{A})$, *then* $\mathrm{rank}\mathcal{H}_X(\mathcal{A};\mathcal{K}) = (-1)^{r(X)}\mu(X)$.

Connection with the Folkman Complex

These constructions are closely related. In fact, the homology groups of the Whitney complex equal direct sums of the homology groups of all Folkman subcomplexes of $\mathcal{A}$ in the following sense.

Theorem 4.116 *Let* $\mathcal{A}$ *be an arrangement. Then*

$$\mathcal{H}_0(\mathcal{A};\mathcal{K}) = \mathcal{K}, \quad \mathcal{H}_1(\mathcal{A};\mathcal{K}) = \bigoplus_{H \in \mathcal{A}} \mathcal{K}(H),$$

and for $p \geq 2$ *the map* $\tau : \mathcal{T} \to \mathcal{T}$ *of Definition 3.105 induces isomorphisms*

$$\mathcal{H}_p(\mathcal{A};\mathcal{K}) \simeq \bigoplus_{X \in L_p(\mathcal{A})} \tilde{H}_{p-2}(\mathsf{F}(\mathcal{A}_X);\mathcal{K}).$$

Proof. Recall that we identified B_p with the group of p–cycles of $\mathcal{C}$. Since $B_p = \oplus_{X \in L_p} B_X$ and $B_X(\mathcal{A}) = B_X(\mathcal{A}_X)$, it suffices to prove the assertion for $X = T(\mathcal{A})$ and $r(\mathcal{A}) \geq 2$. Write $T = T(\mathcal{A})$ and $r(\mathcal{A}) = \ell$. If $\ell = 2$, then

$$B_T \simeq (\bigoplus_{H \in \mathcal{A}} \mathcal{K}(H))/\mathcal{K}(\sum_{H \in \mathcal{A}} H) \simeq \tilde{H}_0(\mathsf{F}(\mathcal{A});\mathcal{K}).$$

Assume $\ell \geq 3$. Since there are no $(\ell-2)$–boundaries, $\tilde{H}_{\ell-2}(\mathsf{F};\mathcal{K}) = H_{\ell-2}(\mathsf{F};\mathcal{K}) = Z_{\ell-2}(\mathsf{F};\mathcal{K})$. We identify the cycle group $Z_{\ell-2}$ with a subspace of $\mathcal{T}_{\ell-1}$. Note that $B_T \subseteq \mathcal{T}_\ell$. We show that $\tau B_T \subseteq Z_{\ell-2}(\mathsf{F};\mathcal{K})$. Let $S = (H_1, \ldots, H_\ell) \in \mathbf{S}_\ell$. If S is dependent, then $\tau b_S = \tau 0 = 0$. Suppose S is independent. Let $\mathcal{B}$ denote the subarrangement of $\mathcal{A}$ whose elements are the hyperplanes of S. This is a fine but useful distinction: $\mathcal{B}$ is a set, S is an ordered set with the same elements. Then $L(\mathcal{B})$ is a Boolean lattice. We showed that $H_{\ell-2}(\mathsf{F}(\mathcal{B});\mathcal{K})$ is one dimensional, generated by the cycle

$$z_S = \sum_{\pi \in Sym(\ell)} (-1)^{\ell-1}(\mathrm{sign}\pi)(H_{\pi 1}, H_{\pi 1} \cap H_{\pi 2}, \ldots, H_{\pi 1} \cap \cdots \cap H_{\pi(\ell-1)}).$$

Since $\tau b_S = z_S$, $\tau B_T \subseteq Z_{\ell-2}(\mathsf{F};\mathcal{K})$. But $\tau : \mathcal{C}_T \to \mathcal{C}$ is a monomorphism and thus $\tau : B_T \to Z_{\ell-2}(\mathsf{F};\mathcal{K})$ is a monomorphism. Let $x \in Z_{\ell-2}(\mathsf{F};\mathcal{K})$. Then $x \in \mathcal{C}_{\ell-1}$. Define $y \in \mathcal{C}_\ell$ by adding T at the end of each chain appearing in x. Then $\tau y = x$ and $\delta y = 0$. Thus $y \in \mathcal{H}_\ell$. Since $B_\ell = \mathcal{H}_\ell$ by Theorem 4.112, we have $y \in B_T$. This proves that $\tau : B_T \to Z_{\ell-2}(\mathsf{F};\mathcal{K})$ is surjective. □

4.6 The Characteristic Polynomial

Recall the characteristic polynomial of an arrangement from Definition 2.52 and the graded S–module $\Omega^p(\mathcal{A})$ from Proposition 4.72. The central object of this section is the Poincaré series of $\Omega^p(\mathcal{A})$.

Definition 4.117 *Suppose M is a finitely generated graded S–module and each M_p is finite dimensional over $\mathbb{K}$. The **Poincaré series** $\mathrm{Poin}(M, x) \in \mathbb{Z}[x^{-1}][[x]]$ of the graded S–module M is*

$$\mathrm{Poin}(M, x) = \sum_{p \in \mathbb{Z}} (\dim_{\mathbb{K}} M_p) x^p.$$

First we study a generalization of the Folkman complex and Whitney complex, called the order complex with functors. It could be stated in terms of sheaf theory on ordered sets, but we choose not to use any sheaf theory here. It is used to prove Theorem 4.136, which amounts to the following formula for the characteristic polynomial:

$$\chi(\mathcal{A}, t) = \lim_{x \to 1} \sum_{p=0}^{\ell} \mathrm{Poin}(\Omega^p(\mathcal{A}), x)(t(1-x) - 1)^p.$$

When this formula is applied to a free arrangement, we obtain the Factorization Theorem 4.137, which asserts that if $\mathcal{A}$ is a free arrangement with $\exp \mathcal{A} = \{b_1, \ldots, b_\ell\}$, then

$$\pi(\mathcal{A}, t) = \prod_{i=1}^{\ell} (1 + b_i t).$$

In Theorem 4.61 we proved this factorization for recursively free arrangements.

The Order Complex with Functors

A poset $\mathcal{P}$ may be regarded as a category. Its objects are the elements of $\mathcal{P}$. Its morphisms are induced by the partial order

$$\mathrm{Hom}(X, Y) = \begin{cases} \{X \leq Y\} & \text{if } X \leq Y, \\ \emptyset & \text{otherwise.} \end{cases}$$

The composition of two morphisms $X \leq Y$ and $Y \leq Z$ is $X \leq Z$.

Let R be a commutative ring and let $(R\text{-Mod})$ be the category of R–modules. Suppose F is a covariant functor from $\mathcal{P}$ to $(R\text{-Mod})$. When $X, Y \in \mathcal{P}$ with $X \leq Y$, the induced morphism from $F(X)$ to $F(Y)$ is denoted $\nu_{X,Y} : F(X) \to F(Y)$. When we are working with a covariant functor F, we assume that $\mathcal{P}$ has a unique maximal element T and set $P = \mathcal{P} \setminus \{T\}$. Suppose F is a contravariant functor from $\mathcal{P}$ to $(R\text{-Mod})$. Then we have the induced morphism $\nu_{X,Y} : F(Y) \to F(X)$. When we are working with a contravariant functor F, we assume that $\mathcal{P}$ has a unique minimal element V and set $P = \mathcal{P} \setminus \{V\}$. Let p be a positive integer. Recall from Definition 2.36 that a p-chain in P is a p-tuple $c = (X_1, \ldots, X_p)$ of elements $X_i \in P$ satisfying $X_1 < \cdots < X_p$. Let $\mathcal{C}_p(P)$ be the free R-module with basis the set of all p-chains in P.

Definition 4.118 *Suppose $F : \mathcal{P} \to (R\text{-Mod})$ is a covariant functor. Recall that $P = \mathcal{P} \setminus \{T\}$. We extend Definition 4.110 to the* **order complex with F coefficients**, *$(\mathcal{C}_\cdot(\mathcal{P}, F), \partial)$, as follows. Let $\mathcal{C}_0(\mathcal{P}, F) = F(T)$ and for $p \geq 1$ let $\mathcal{C}_p(\mathcal{P}, F)$ be the R-module spanned by*

$$y \otimes (X_1, \ldots, X_p) \in (\bigoplus_{X \in P} F(X)) \otimes_R \mathcal{C}_p(P)$$

where $y \in F(X_1)$ and $X_1 < \ldots < X_p < T$. The boundary operator

$$\partial : \mathcal{C}_p(\mathcal{P}, F) \to \mathcal{C}_{p-1}(\mathcal{P}, F)$$

is given by $\partial(y \otimes (X_1)) = \nu_{X_1, T}(y)$ and for $p > 1$

$$\partial(y \otimes (X_1, \ldots, X_p)) =$$

$$\nu_{X_1, X_2}(y) \otimes (X_2, \ldots, X_p) + \sum_{k=2}^{p} (-1)^{k-1} y \otimes (X_1, \ldots, \widehat{X_k}, \ldots, X_p).$$

When $F : \mathcal{P} \to (R\text{-Mod})$ is a contravariant functor, the following modifications are necessary. Here $P = L \setminus \{V\}$. Let $\mathcal{C}_0(\mathcal{P}, F) = F(V)$ and for $p \geq 1$ let $\mathcal{C}_p(\mathcal{P}, F)$ be the R-module spanned by

$$y \otimes (X_1, \ldots, X_p) \in (\bigoplus_{X \in P} F(X)) \otimes_R \mathcal{C}_p(P)$$

where $y \in F(X_p)$ and $V < X_1 < \ldots < X_p$. The boundary operator

$$\partial : \mathcal{C}_p(\mathcal{P}, F) \to \mathcal{C}_{p-1}(\mathcal{P}, F)$$

is given by $\partial(y \otimes (X_1)) = \nu_{V, X_1}(y)$ and for $p > 1$

$$\partial(y \otimes (X_1, \ldots, X_p)) =$$

$$\sum_{k=1}^{p-1} (-1)^{k-1} y \otimes (X_1, \ldots, \widehat{X_k}, \ldots, X_p) + (-1)^{p-1} \nu_{X_{p-1}, X_p}(y) \otimes (X_1, \ldots, X_{p-1}).$$

The following two examples were studied in the last section.

Example 4.119 Let $\mathcal{A}$ be an arrangement in V. Let $\mathcal{P} = L(\mathcal{A}) \setminus \{T(\mathcal{A})\}$. Define the contravariant functor F by $F(X) = \mathbb{Z}$ for $X \in \mathcal{P}$, and let $\nu_{X,Y}$ be the identity map for all X, Y with $X \leq Y$. In this case the chain complex $(\mathcal{C}_\cdot(\mathcal{P}, F), \partial)$ is equal to the simplicial chain complex associated with the Folkman complex.

Example 4.120 Let $\mathcal{A}$ be an arrangement in V and let $\mathcal{P} = L(\mathcal{A})$. Let $R = \mathcal{K}$ be a commutative ring. Define the contravariant functor F by $F(X) = \mathcal{K}$. Here $\nu_{X,X}$ is the identity map and $\nu_{X,Y} = 0$ for $X < Y$. In this case the chain complex $(\mathcal{C}_\cdot(L(\mathcal{A}), F), \partial)$ is equal to the Whitney complex.

Local Functors

Let $\mathcal{A}$ be an arrangement in V. Write $L = L(\mathcal{A})$ and recall that $S = S(V^*)$ is the symmetric algebra of the dual space V^* of V. Given a prime ideal $\wp \in \mathrm{Spec} S$ and an element $X \in L$, define $X(\wp) \in L$ by

$$X(\wp) = \bigcap_{\substack{H \in \mathcal{A}_X \\ \alpha_H \in \wp}} H.$$

It follows from the definition that $X(\wp) \supseteq X$, so $X(\wp) \leq X$.

Definition 4.121 *A covariant functor $F : L(\mathcal{A}) \to (S\text{-Mod})$ is called* **local** *if the localization at $\wp$ of the map*

$$\nu_{X(\wp),X} : F(X(\wp)) \to F(X)$$

is an isomorphism for all $\wp \in \mathrm{Spec} S$ and for all $X \in L$.

If F is a contravariant functor, then F is local if the localization at $\wp$ of the map $\nu_{X(\wp),X} : F(X) \to F(X(\wp))$ is an isomorphism for all $\wp \in \mathrm{Spec} S$ and for all $X \in L$.

Example 4.122 Recall the S–modules $\Omega^q(\mathcal{A})$ from Definition 4.64. Let q be a nonnegative integer. Define the covariant functor $F : L \to (S\text{-Mod})$ by $F(X) = \Omega^q(\mathcal{A}_X)$. If $X \leq Y$, then $\mathcal{A}_X \subseteq \mathcal{A}_Y$. It follows from Proposition 4.70 that there are inclusions

$$\nu_{X,Y} : \Omega^q(\mathcal{A}_X) \hookrightarrow \Omega^q(\mathcal{A}_Y).$$

To show that F is a local functor, let $\wp \in \mathrm{Spec} S$ and let $X \in L$. The localization at $\wp$ of the inclusion

$$\nu_{X(\wp),X} : \Omega^q(\mathcal{A}_{X(\wp)}) \hookrightarrow \Omega^q(\mathcal{A}_X)$$

is injective because localization is an exact functor; see Theorem A.9. Let $\omega \in \Omega^q(\mathcal{A}_X)$. It follows from Proposition 4.70 that if f is a defining polynomial for the arrangement $\mathcal{A}_X \setminus \mathcal{A}_{X(\wp)}$, then $f\omega \in \Omega^q(\mathcal{A}_{X(\wp)})$. Since $f \notin \wp$, we have $\omega = (f\omega)/f \in \Omega^q(\mathcal{A}_{X(\wp)})_\wp$. This implies that the localization of $\nu_{X(\wp),X}$ is bijective.

Example 4.123 Recall the S–modules $D(\mathcal{A}_X)$ from Definition 4.12. Let q be a nonnegative integer. Define the contravariant functor $F : L \to (S\text{-Mod})$ by $F(X) = D(\mathcal{A}_X)$. If $X \leq Y$, then $\mathcal{A}_X \subseteq \mathcal{A}_Y$. It follows from Proposition 4.9 that there are inclusions

$$\nu_{X,Y} : D(\mathcal{A}_Y) \hookrightarrow D(\mathcal{A}_X).$$

The functor F is local. The proof is similar to the argument above.

The Homology $H_p(\mathcal{A}, F)$

Let $\mathcal{A}$ be an ℓ-arrangement in V of rank $r(\mathcal{A})$. Let $\mathcal{P} = L(\mathcal{A})$ and let F be a local covariant functor from $L(\mathcal{A})$ to $(S\text{-Mod})$. Recall that $P = L(\mathcal{A}) \setminus \{T\}$. Let $(C_\cdot(\mathcal{A}, F), \partial) = (C_\cdot(L(\mathcal{A}), F), \partial)$ denote the order complex with F coefficients. Let $q \geq 0$ and let $H_q(\mathcal{A}, F)$ denote the q-th homology of $(C_\cdot(\mathcal{A}, F), \partial)$. It is an S-module. Suppose that $\wp \in \operatorname{Spec} S$ satisfies $\operatorname{ht} \wp < r(\mathcal{A})$. We want to show that $H_q(\mathcal{A}, F)_\wp = 0$. In order to prove this result, we modify the chain complex $(C_\cdot(\mathcal{A}, F), \partial)$ to obtain a new chain complex $(\tilde{C}_\cdot(\mathcal{A}, F), \tilde{\partial})$ by allowing chains with repetitions. Let $\tilde{C}_0(\mathcal{A}, F) = C_0(\mathcal{A}, F) = F(T)$. For $q > 0$ the free S-module $\tilde{C}_q(\mathcal{A}, F)$ is spanned by

$$\{y \otimes (X_1, \ldots, X_q) \mid y \in F(X_1),\ X_1 \leq \cdots \leq X_q < T\}.$$

Clearly, $C_q(\mathcal{A}, F) \subseteq \tilde{C}_q(\mathcal{A}, F)$. The boundary maps

$$\tilde{\partial} : \tilde{C}_q(\mathcal{A}, F) \to \tilde{C}_{q-1}(\mathcal{A}, F)$$

are defined by the same formulas as the boundary maps in $(C_\cdot(\mathcal{A}, F), \partial)$. Define the S-linear map $\pi : \tilde{C}_q(\mathcal{A}, F) \to C_q(\mathcal{A}, F)$ by

$$\pi(y \otimes (X_1, \ldots, X_q)) = \begin{cases} y \otimes (X_1, \ldots, X_q) & \text{if } (X_1, \ldots, X_q) \text{ has no repetition,} \\ 0 & \text{otherwise.} \end{cases}$$

Lemma 4.124 *The map π is a chain map: $\pi \tilde{\partial} = \partial \pi$.*

Proof. If $(X_1, \ldots, X_q)$ has no repetition, then the terms of

$$\sum_{k=1}^{q} (-1)^{k-1} (X_1, \ldots, \widehat{X_k}, \ldots, X_q)$$

have no repetition. Thus

$$\pi \tilde{\partial}(y \otimes (X_1, \ldots, X_q)) = \partial \pi(y \otimes (X_1, \ldots, X_q))$$

for $y \otimes (X_1, \ldots, X_q) \in \tilde{C}_q(\mathcal{A}, F)$. Suppose that $(X_1, \ldots, X_q)$ has a repetition $X_i = X_{i+1}$ for $1 \leq i \leq q-1$. Then

$$\sum_{k=1}^{q}(-1)^{k-1}(X_1,\ldots,\widehat{X_k},\ldots,X_q) = \sum_{k=1}^{i-1}(-1)^{k-1}(X_1,\ldots,\widehat{X_k},\ldots,X_q)$$
$$+ \sum_{k=i+2}^{q}(-1)^{k-1}(X_1,\ldots,\widehat{X_k},\ldots,X_q).$$

Since $(X_1, \ldots, \widehat{X_k}, \ldots, X_q)$ with $i \neq k \neq i+1$ has the repetition $X_i = X_{i+1}$, we have $\pi \tilde{\partial}(y \otimes (X_1, \ldots, X_q)) = 0 = \partial \pi(y \otimes (X_1, \ldots, X_q))$. □

Proposition 4.125 *Let $\mathcal{A}$ be an arrangement of rank r and let $F : L(\mathcal{A}) \to (S\text{-Mod})$ be a covariant local functor. Suppose that $\wp \in \operatorname{Spec} S$ satisfies $\operatorname{ht} \wp < r$. Then $H_q(\mathcal{A}, F)_\wp = 0$.*

4.6 The Characteristic Polynomial

Proof. Let $\wp \in \text{Spec}\, S$. Recall that $T = T(\mathcal{A})$ and $r = r(T)$. By assumption, $\text{ht}\,\wp < r(T)$. Note that $r(T(\wp)) \leq \text{ht}\,\wp < r(T)$. Thus $T(\wp) \in P = L \setminus \{T\}$. Identify $F(X)_\wp$ with $F(X(\wp))_\wp$ for all $X \in L$. Define the S-linear map $h : \mathcal{C}_q(\mathcal{A}, F)_\wp \to \mathcal{C}_{q+1}(\mathcal{A}, F)_\wp$ as follows. For $q = 0$, let $h(y) = y \otimes (T(\wp))$ where $y \in F(T)_\wp = F(T(\wp))_\wp$. For $q > 0$, let

$$h(y \otimes (X_1, \ldots, X_q)) = \\ (-1)^q y \otimes (X_1(\wp), \ldots, X_q(\wp), T(\wp)) \\ + \sum_{j=1}^{q} (-1)^{j-1} y \otimes (X_1(\wp), \ldots, X_j(\wp), X_j, \ldots, X_q))$$

where $y \in F(X_1)_\wp = F(X_1(\wp))_\wp$. A lengthy computation, which we omit, yields $\tilde{\partial} h + h\partial = \text{id}$ and hence $\pi\tilde{\partial} h + \pi h \partial = \text{id}$. Using Lemma 4.124, we have $\partial(\pi h) + (\pi h)\partial = \text{id}$. This shows that πh is a chain homotopy between the identity map and the zero map. Thus $(\mathcal{C}_\cdot(\mathcal{A}, F)_\wp, \partial)$ is an acyclic complex. It follows from Theorem A.9 that localization is an exact functor. Thus

$$H_q(\mathcal{A}, F)_\wp = H_q(\mathcal{C}_\cdot(\mathcal{A}, F), \partial)_\wp = H_q(\mathcal{C}_\cdot(\mathcal{A}, F)_\wp, \partial) = 0.$$

This completes the proof. $\square$

Theorem 4.126 *Let $q \geq 0$. If $F : L(\mathcal{A}) \to (S\text{-Mod})$ is a local covariant functor, then*

$$\dim_S H_q(\mathcal{A}, F) \leq \dim T.$$

Proof. Let $M = H_q(\mathcal{A}, F)$. By Theorem A.18, we have

$$\dim_S M = \ell - \min_{\wp \in \text{Supp}(M)} \text{ht}\,\wp.$$

Proposition 4.125 asserts that $\wp \notin \text{Supp}(M)$ if $\text{ht}\,\wp < r(T)$. Consequently, if $\wp \in \text{Supp}(M)$, then $\text{ht}\,\wp \geq r(T)$. Thus $\dim_S M \leq \ell - r(T) = \dim T$. $\square$

The next result follows from Theorems 4.126 and A.22.

Corollary 4.127 *Let $q \geq 0$. If $F : L(\mathcal{A}) \to (S\text{-Mod})$ is a local covariant functor, then $\text{Poin}(H_q(\mathcal{A}, F), x)$ has a pole at $x = 1$ of order at most $\dim T$.*

Let $ch_q(P)$ be the set of all q-chains in P:

$$ch_q(P) = \{(X_1, \ldots, X_q) \mid X_i \in P,\ X_1 < \cdots < X_q\}.$$

Then $ch(P) = \bigcup_q ch_q(P)$. Recall from Definition 2.36 that for $c \in ch(P)$, the first element of c is $\underline{c}$, the last element of c is $\overline{c}$, and the cardinality of c is $|c|$.

Theorem 4.128 *Let $X \in L$. If $F : L(\mathcal{A}) \to (S\text{-Mod})$ is a local covariant functor, then*

$$\sum_{Y \in L_X} \mu(Y, X) \mathrm{Poin}(F(Y), x)$$

has a pole at $x = 1$ of order at most $\dim X$.

Proof. Since the sum involves elements $Y \leq X$, we may assume that $X = T$. Let $P = L \setminus \{T\}$. Note that $\mathcal{C}_0(\mathcal{A}, F) = F(T)$ and $\mathcal{C}_q(\mathcal{A}, F) \simeq \bigoplus_{c \in ch_q(P)} F(\underline{c})$. Therefore we have

$$\sum_{q=0}^{\ell} (-1)^q \mathrm{Poin}(H_q(\mathcal{A}, F), x)$$

$$= \sum_{q=0}^{\ell} (-1)^q \mathrm{Poin}(\mathcal{C}_q(\mathcal{A}, F), x)$$

$$= \mathrm{Poin}(F(T), x) + \sum_{q=0}^{\ell} (-1)^q \sum_{c \in ch_q(P)} \mathrm{Poin}(F(\underline{c}), x)$$

$$= \mathrm{Poin}(F(T), x) + \sum_{c \in ch(P)} (-1)^{|c|} \mathrm{Poin}(F(\underline{c}), x)$$

$$= \mathrm{Poin}(F(T), x) + \sum_{Y \in P} \mathrm{Poin}(F(Y), x) \sum_{\substack{c \in ch(P) \\ \underline{c} = Y}} (-1)^{|c|}.$$

Let $ch(L)$ be the set of all chains in L. It follows from Proposition 2.37 that for all $Y \in P$

$$\sum_{\substack{c \in ch(P) \\ \underline{c} = Y}} (-1)^{|c|} = \sum_{c \in ch[Y,T]} (-1)^{|c|-1} = \mu(Y, T).$$

Thus we have

$$\sum_{q=0}^{\ell} (-1)^q \mathrm{Poin}(H_q(\mathcal{A}, F), x) = \sum_{Y \in L} \mu(Y, T) \mathrm{Poin}(F(Y), x),$$

which has a pole at $x = 1$ of order at most $\dim T$ by Corollary 4.127. □

The following theorem is an immediate consequence of Theorem 4.128 and Example 4.122.

Theorem 4.129 *Let $X \in L$ and let p be a nonnegative integer. Then*

$$\sum_{Y \in L_X} \mu(Y, X) \mathrm{Poin}(\Omega^p(\mathcal{A}_Y), x)$$

has a pole at $x = 1$ of order at most $\dim X$.

The Polynomial $\Psi(\mathcal{A}, x, t)$

Next we prove the main result of this section, the following formula for the characteristic polynomial:

(1) $$\chi(\mathcal{A},t) = \lim_{x \to 1} \sum_{p=0}^{\ell} \text{Poin}(\Omega^p(\mathcal{A}),x)(t(1-x)-1)^p.$$

Note first that in the proof of (1) we may assume that the field $\mathbb{K}$ is algebraically closed. Let $\mathbb{L}$ be a field extension of $\mathbb{K}$. Let $V_{\mathbb{L}} = \mathbb{L} \oplus_{\mathbb{K}} V$ and let $\mathcal{A}_{\mathbb{L}}$ be the corresponding arrangement in $V_{\mathbb{L}}$. Clearly $\chi(\mathcal{A}_{\mathbb{L}},t) = \chi(\mathcal{A},t)$. Moreover, $\mathbb{L} \oplus_{\mathbb{K}} \Omega^p(\mathcal{A})$ is isomorphic to $\Omega^p(\mathcal{A}_{\mathbb{L}})$. Thus the right-hand side of (1) is also independent of field extension.

Definition 4.130 *Let*

$$\Psi(\mathcal{A};x,t) = \sum_{p=0}^{\ell} \text{Poin}(\Omega^p(\mathcal{A}),x)(t(1-x)-1)^p.$$

Proposition 4.131 *If $\mathcal{A}$ is free with $\exp \mathcal{A} = \{b_1,\ldots,b_\ell\}$, then*

$$\Psi(\mathcal{A};x,t) = \prod_{i=1}^{\ell}(tx^{-b_i} - (x^{-1} + x^{-2} + \cdots + x^{-b_i})).$$

In particular, $\Psi(\Phi_\ell;x,t) = t^\ell$ for the empty ℓ-arrangement.

Proof. It follows from Corollary 4.77 that

$$\text{Poin}(\Omega^p(\mathcal{A}),x) = \sum x^{-b_{i_1}-\cdots-b_{i_p}}/(1-x)^\ell$$

where the sum is over the set $\{(i_1,\ldots,i_p) \mid 1 \leq i_1 < \cdots < i_p \leq \ell\}$. Let y be an indeterminate. Then

$$\sum_{p=0}^{\ell} \text{Poin}(\Omega^p(\mathcal{A}),x)y^p = \prod_{i=1}^{\ell}((1+x^{-b_i}y)/(1-x)).$$

Now set $y = t(1-x) - 1$ and divide by $1-x$ in each factor. □

A priori $\Psi(\mathcal{A};x,t)$ may have a pole at $x = 1$ because each $\Omega^p(\mathcal{A})$ is a finite S-module. The order of the pole is at most ℓ by Proposition A.22. We will prove in Proposition 4.133 that $\Psi(\mathcal{A};x,t)$ has no pole at $x = 1$ and that $\Psi(\mathcal{A};x,t) \in \mathbb{Z}[x,x^{-1},t]$. Thus we will be able to rewrite (1) as

(2) $$\chi(\mathcal{A},t) = \Psi(\mathcal{A};1,t).$$

Proposition 4.132 *If $\mathcal{A}$ is nonempty, then $\Psi(\mathcal{A};x,1) = 0$.*

Proof. In Definition 4.85 we constructed for a nonempty arrangement $\mathcal{A}$ the complex $(\Omega^{\cdot}(\mathcal{A}),\partial)$. Its boundary operator $\partial\omega = (d\alpha/\alpha) \wedge \omega$ with $\ker(\alpha) \in \mathcal{A}$ has pdegree -1. In Proposition 4.86 we showed that this complex is acyclic. Thus

$$\Psi(\mathcal{A}; x, 1) = \sum_{p=0}^{\ell} \text{Poin}(\Omega^p(\mathcal{A}), x)(-x)^p = 0$$

as required. □

Proposition 4.133 *For any arrangement, $\Psi(\mathcal{A}; x, t) \in \mathbb{Z}[x, x^{-1}, t]$. In particular, $\Psi(\mathcal{A}; x, t)$ has no pole at $x = 1$.*

Proof. Let m and n be minimal nonnegative integers so that

$$P(x, t) = x^n (1 - x)^m \Psi(x, t)$$

is a polynomial in x and t. Note that $m \leq \ell$ because the order of the pole at $x = 1$ of $\text{Poin}(\Omega^p(\mathcal{A}), x)$ is at most ℓ by Proposition A.22. Thus for $d > 0$

$$\begin{aligned}
P(x, (1 - x^{-d})/(1 - x)) &= x^n (1 - x)^m \Psi(x, (1 - x^{-d})/(1 - x)) \\
&= x^n (1 - x)^m \sum_{p=0}^{\ell} \text{Poin}(\Omega^p(\mathcal{A}), x)(-x^{-d})^p.
\end{aligned}$$

Choose a generic $\eta \in \Omega_d^1[V]$ as in Proposition 4.91. Recall the η-complex from Definition 4.87

$$0 \to \Omega^0(\mathcal{A}) \xrightarrow{\partial_\eta} \Omega^1(\mathcal{A}) \xrightarrow{\partial_\eta} \cdots \xrightarrow{\partial_\eta} \Omega^\ell(\mathcal{A}) \to 0$$

Here $\partial_\eta \omega = \eta \wedge \omega$. Proposition 4.91 asserts that its cohomology groups are finite dimensional over $\mathbb{K}$. Since each boundary map ∂_η is of pdegree d,

$$\sum_{p=0}^{\ell} \text{Poin}(\Omega^p(\mathcal{A}), x)(-x^{-d})^p = \sum_{p=0}^{\ell} \text{Poin}(H^p(\Omega^\cdot(\mathcal{A})), x)(-x^{-d})^p$$

has finite value at $x = 1$. If $m > 0$, then

$$P(x, (1 - x^{-d})/(1 - x)) = x^n (1 - x)^m \sum_{p=0}^{\ell} \text{Poin}(\Omega^p(\mathcal{A}), x)(-x^{-d})^p$$

vanishes at $x = 1$. Thus

$$P(1, -d) = P(x, (1 - x^{-d})/(1 - x))|_{x=1} = 0.$$

This is true for infinitely many positive integers d. It follows that $P(1, t)$ is identically 0. This contradicts the minimality of m. Therefore $m = 0$. □

Proposition 4.134 *If $X \in L(\mathcal{A})$, then $t^{\dim X}$ divides $\Psi(\mathcal{A}_X; x, t)$.*

Proof. Let $d = \dim X$. Note that $\mathcal{A}_X = \mathcal{A}_1 \times \Phi_d$ for some $(\ell - d)$-arrangement $\mathcal{A}_1$ and the empty arrangement Φ_d in X. By Proposition 4.84, we have

4.6 The Characteristic Polynomial

$$\Psi(\mathcal{A}_X; x, t) = \sum_{n=0}^{\ell} \text{Poin}(\Omega^n(\mathcal{A}_X), x)(t(1-x) - 1)^n$$

$$= (\sum_{p=0}^{\ell-d} \text{Poin}(\Omega^p(\mathcal{A}_1), x)(t(1-x) - 1)^p)$$

$$\cdot (\sum_{q=0}^{d} \text{Poin}(\Omega^q(\Phi_d), x)(t(1-x) - 1)^q)$$

$$= \Psi(\mathcal{A}_1; x, t)\Psi(\Phi_d; x, t).$$

By Proposition 4.131, we have $\Psi(\Phi_d; x, t) = t^d$. □

We need the following characterization of $\chi(\mathcal{A}, t)$.

Proposition 4.135 *Let $L = L(\mathcal{A})$. Suppose that a map $G : L \to \mathbb{Z}[t]$ satisfies the following four conditions:*
(1) $G(V) = t^\ell$,
(2) $G(X)|_{t=1} = 0$ for $X \neq V$,
(3) $t^{\dim X}$ divides $G(X)$ for all $X \in L$,
(4) the degree of t in $\sum_{Y \in L_X} \mu(Y, X)G(Y)$ does not exceed $\dim X$ for any $X \in L$.
Then $G(X) = \chi(\mathcal{A}_X, t)$ for all $X \in L$.

Proof. Let
$$G'(X) = \sum_{Y \in L_X} \mu(Y, X)G(Y).$$

If $Y \leq X$, then $\dim Y \geq \dim X$. By (3) $t^{\dim Y}$ divides $G(Y)$, so $t^{\dim X}$ divides $G'(X)$. On the other hand, it follows from (4) that $\deg G'(X) \leq \dim X$. Therefore we can write $G'(X) = g(X)t^{\dim X}$ for some map $g : L \to \mathbb{Z}$. We get

$$g(X) = G'(X)|_{t=1} = \sum_{Y \in L_X} \mu(Y, X)G(Y)|_{t=1} = \mu(V, X)$$

by using (1) and (2). Thus $G'(X) = \mu(V, X)t^{\dim X}$. It follows from the Möbius inversion formulas of Proposition 2.39 and from Definition 2.52 that

$$G(X) = \sum_{Y \in L_X} G'(Y) = \sum_{Y \in L_X} \mu(V, Y)t^{\dim Y} = \chi(\mathcal{A}_X, t).$$

This completes the argument. □

Theorem 4.136 *The characteristic polynomial of an ℓ-arrangement $\mathcal{A}$ is given by*
$$\chi(\mathcal{A}, t) = \Psi(\mathcal{A}; 1, t).$$

Proof. We verify conditions (1)–(4) of Proposition 4.135 for $G(X) = \Psi(\mathcal{A}_X; 1, t)$.
(1) It follows from Proposition 4.131 that $G(V) = \Psi(\Phi_\ell; 1, t) = t^\ell$.

(2) It follows from Proposition 4.132 that for $X \neq V$
$$G(X)|_{t=1} = \Psi(\mathcal{A}_X; 1, 1) = \Psi(\mathcal{A}_X; x, 1)|_{x=1} = 0.$$

(3) This follows from Proposition 4.134 with $x = 1$.

(4) Fix $X \in L$. We compute

$$\sum_{Y \in L_X} \mu(Y, X) G(Y)$$
$$= \sum_{Y \in L_X} \mu(Y, X) \Psi(\mathcal{A}_Y; 1, t)$$
$$= \sum_{p=0}^{\ell} \sum_{Y \in L_X} \mu(Y, X) \text{Poin}(\Omega^p(\mathcal{A}_Y), x)(t(1-x)-1)^p|_{x=1}$$
$$= \sum_{p=0}^{\ell} M_p(x)(t(1-x)-1)^p|_{x=1},$$

where
$$M_p(x) = \sum_{Y \in L_X} \mu(Y, X) \text{Poin}(\Omega^p(\mathcal{A}_Y), x).$$

By Theorem 4.129, $(1-x)^{\dim X} M_p(x)$ has no pole at $x = 1$. Thus the coefficient of t^n in $M_p(x)(t(1-x)-1)^p$, $(-1)^{p-n} \binom{p}{n} M_p(x)(1-x)^n$, lies in $(1-x)\mathbb{Z}[x, x^{-1}]$ if $n > \dim X$. Thus for each p, the degree of t in $M_p(x)(t(1-x)-1)^p|_{x=1}$ does not exceed $\dim X$. Therefore $\sum_{Y \in L_X} \mu(Y, X) G(Y)$ has the same property. □

The Factorization Theorem

Theorem 4.137 (Factorization) *If $\mathcal{A}$ is a free arrangement with $\exp \mathcal{A} = \{b_1, \ldots, b_\ell\}$, then*

$$\pi(\mathcal{A}, t) = \prod_{i=1}^{\ell} (1 + b_i t).$$

Proof. We computed $\Psi(\mathcal{A}; x, t)$ in Proposition 4.131. Set $x = 1$ and use Theorem 4.136 to obtain the equivalent statement, $\chi(\mathcal{A}, t) = \prod_{i=1}^{\ell} (t - b_i)$. □

The Factorization Theorem 4.137 shows the exponents of a free arrangement are determined by $L(\mathcal{A})$.

Conjecture 4.138 (Terao) *For fixed $\mathbb{K}$, the property that $\mathcal{A}$ is free depends only on $L(\mathcal{A})$.*

We say that $\pi(\mathcal{A}, t)$ **factors** if $\pi(\mathcal{A}, t) = \prod_{i=1}^{\ell}(1 + b_i t)$ where the $b_i \in \mathbb{Z}$. The next example is due to Stanley [217] in the setting of matroids and independently to Falk and Randell [81]. This example shows that the implication in Theorem 4.137 cannot be reversed.

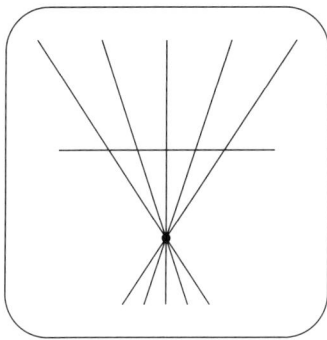

Fig. 4.5. $\pi(\mathcal{A}, t)$ factors, but $\mathcal{A}$ is not free

Example 4.139 Let $\mathcal{A}$ be the 3–arrangement in Figure 4.5. Then $\pi(\mathcal{A}, t)$ factors, but $\mathcal{A}$ is not free. Direct computation gives $\pi(\mathcal{A}, t) = (1+t)(1+3t)(1+3t)$. Remove the horizontal line to obtain $\mathcal{A}'$. Both $\mathcal{A}'$ and $\mathcal{A}''$ are free with $\exp \mathcal{A}' = \{1, 1, 4\}$ and $\exp \mathcal{A}'' = \{1, 5\}$. If we assume that $\mathcal{A}$ is free, then we contradict Theorem 4.46.

We conclude this section with a discussion of restrictions.

Definition 4.140 Call $\mathcal{A}$ **k–free** if $\mathcal{A}^X$ is free for all $X \in L(\mathcal{A})$ with $r(X) \leq k$. Call $\mathcal{A}$ **hereditarily free** if $\mathcal{A}$ is k–free for all k.

Since $\mathcal{A} = \mathcal{A}^V$, a 0–free arrangement is free. The terminology is motivated by the fact that in a hereditarily free arrangement all restrictions are again hereditarily free. Orlik conjectured in 1981 that every free arrangement is hereditarily free. The following counterexample was constructed by Edelman and Reiner [70] in 1991.

Example 4.141 (Edelman–Reiner) Define a 5–arrangement $\mathcal{A}$ of 21 hyperplanes by the kernels of the linear forms

$$x_i \text{ where } 1 \leq i \leq 5,$$
$$x_1 + a_2 x_2 + a_3 x_3 + a_4 x_4 + a_5 x_5 \text{ where } a_i = \pm 1 \text{ for all } i.$$

The arrangement $\mathcal{A}$ is free. Let $H = \ker(x_1 - x_2 - x_3 - x_4 - x_5) \in \mathcal{A}$. The restriction $\mathcal{A}^H$ is not free. Edelman and Reiner checked that $\mathcal{A}$ is free with $\exp \mathcal{A} = \{1, 5, 5, 5, 5\}$ using the computer program Macaulay, which provided an explicit basis for $D(\mathcal{A})$. They used the computer program Mathematica to show that this basis satisfies Saito's criterion 4.19.

The restriction $\mathcal{A}^H$ is a 4–arrangement of 15 hyperplanes defined by the kernels of the linear forms

$$b_1 x_1 + b_2 x_2 + b_3 x_3 + b_4 x_4 \text{ where } b_i = 0, 1.$$

4. Free Arrangements

Direct computation gives

$$\pi(\mathcal{A}^H, t) = (1+t)(1+4t)(1+10t+26t^2).$$

It follows from the Factorization Theorem 4.137 that $\mathcal{A}^H$ is not free.

Note that $\mathcal{A}$ is not inductively free. Restriction to one of the coordinate hyperplanes, say $K = \ker(x_5)$, results in a free arrangement $\mathcal{A}^K$ with $\exp \mathcal{A}^K = \{1, 3, 3, 5\}$. It follows from the Addition–Deletion Theorem 4.51 that $\mathcal{A}$ is not inductively free.

It would be interesting to know whether there are k–free arrangements which are not $(k+1)$–free for all k.

5. Topology

In this chapter we return to the convention that an arrangement is not necessarily central. The subject of this chapter is the topology of the complement of a complex arrangement, $M(\mathcal{A})$. Call the complex arrangements $\mathcal{A} = (\mathcal{A}, V)$ and $\mathcal{B} = (\mathcal{B}, V)$ **diffeomorphic**, **homeomorphic**, or **homotopy equivalent** if $M(\mathcal{A})$ and $M(\mathcal{B})$ are diffeomorphic, homeomorphic, or homotopy equivalent. It is natural to ask how these topological equivalence classes relate to the combinatorial equivalence classes defined earlier. For example, we will show in Section 5.4 that $M(\mathcal{A})$ and $M(\mathcal{B})$ have the same Betti numbers if and only if $\mathcal{A}$ and $\mathcal{B}$ are π–equivalent, and that $M(\mathcal{A})$ and $M(\mathcal{B})$ have isomorphic cohomology rings if and only if $\mathcal{A}$ and $\mathcal{B}$ are A–equivalent.

In Section 5.1 we prove some elementary facts about $M = M(\mathcal{A})$ and discuss a few examples. We also review fundamental work of Arnold, Brieskorn, Deligne and Hattori. The rest of the chapter does not follow the chronology of discovery. In Section 5.2 we construct a finite simplicial complex M of the homotopy type of M. The construction uses an embedding in V of the order complex of the face poset of a real arrangement. In the special case of a complexified real arrangement, Salvetti [203] constructed a smaller complex W of the homotopy type of M and Arvola [15] constructed a simplicial map M $\to$ W, which is a homotopy equivalence. In principle, M contains all information about the homotopy type of M. In practice, M and W are very large and unsuited for explicit calculations. It is therefore desirable to find simple algorithms to compute topological invariants of M.

Arvola's presentation of the fundamental group of M is in Section 5.3. It generalizes Randell's presentation of the fundamental group of the complexification of a real arrangement. In Section 5.4 we consider the cohomology groups of $M(\mathcal{A})$. We use our results on $R(\mathcal{A})$ from Section 3.5 to prove that given a triple $(\mathcal{A}, \mathcal{A}', \mathcal{A}'')$, there are split short exact sequences for all $k \geq 0$

$$0 \to H^{k+1}(M(\mathcal{A}')) \to H^{k+1}(M(\mathcal{A})) \to H^k(M(\mathcal{A}'')) \to 0.$$

Thus the map $R(\mathcal{A}) \to H^*(M(\mathcal{A}))$ induced by $\omega_H \mapsto [(1/2\pi i)\omega_H]$ is an algebra isomorphism. Together with the algebra isomorphism $R(\mathcal{A}) \simeq A(\mathcal{A})$ established in Section 3.5, this provides a presentation of the cohomology algebra in terms of generators and relations. This is the topological interpretation of $A(\mathcal{A})$. Thus the cohomology algebra of $M(\mathcal{A})$ depends only on $L(\mathcal{A})$.

We give an elementary proof of Brieskorn's Lemma 5.91. It also follows that the Poincaré polynomial of the complement is

$$\text{Poin}(M(\mathcal{A}), t) = \pi(\mathcal{A}, t).$$

Thus the coefficients of $\pi(\mathcal{A}, t)$ are also the Betti numbers of the complement. It also shows that if $\mathcal{A}$ is a real arrangement, then $M(\mathcal{A})$ has the M–property. In Section 5.5 we prove that the complement of a supersolvable arrangement admits a strictly linear fibration. In Section 5.6 we describe some related recent results: work of Falk and Kohno on minimal models, Manin and Schechtman's work on discriminantal arrangements, Falk's geometric linking, the cohomology of the Milnor fiber of a generic arrangement, and the results of Goresky and MacPherson on arrangements of subspaces of arbitrary codimension.

We have been using the basis $x_1, \ldots, x_\ell$ for V^*. We continue to use it, except when we want to introduce real variables in complex space. Then we use $z_1, \ldots, z_\ell$ as a basis for V^* and write $z_k = x_k + iy_k$. Note also that, depending on the application, we may write these real variables ordered either $x_1, \ldots, x_\ell, y_1, \ldots, y_\ell$, or $x_1, y_1, \ldots, x_\ell, y_\ell$.

5.1 The Complement $M(\mathcal{A})$

In this section we prove some elementary facts about the topology of the complement of an arrangement over the complex numbers. In addition, we outline some of the work which started the recent activity in the area by Arnold [6], Brieskorn [41], Deligne [61], and Hattori [108]. The coning construction of Definition 1.15 has a topological interpretation. Recall the Hopf bundle $p : \mathbb{C}^{\ell+1} \setminus \{0\} \to P^\ell_\mathbb{C}$ with fiber $\mathbb{C}^*$, which identifies $z \in \mathbb{C}^{\ell+1}$ with λz for $\lambda \in \mathbb{C}^*$.

Proposition 5.1 *Let $\mathcal{A}$ be an affine arrangement and let $\mathbf{c}\mathcal{A}$ be the cone over $\mathcal{A}$. The restriction of the Hopf bundle $p : M(\mathbf{c}\mathcal{A}) \to M(\mathcal{A})$ is a trivial bundle, so*

$$M(\mathbf{c}\mathcal{A}) \approx M(\mathcal{A}) \times \mathbb{C}^*.$$

Similarly, if $\mathcal{A}$ is a central arrangement, then $M(\mathcal{A}) \approx M(\mathbf{d}\mathcal{A}) \times \mathbb{C}^*$.

Proof. The identification $p(M(\mathbf{c}\mathcal{A})) = M(\mathcal{A})$ is immediate from Definition 1.15. Let $K_0 \in \mathbf{c}\mathcal{A}$. The restriction of p to $M_{K_0} = \mathbf{c}V \setminus \{K_0\}$ has base space $P^\ell_\mathbb{C} \setminus P^{\ell-1}_\mathbb{C} \approx \mathbb{C}^\ell$. Thus $p : M_{K_0} \to \mathbb{C}^\ell$ is a trivial bundle and $p : M(\mathbf{c}\mathcal{A}) \to M(\mathcal{A})$ is a restriction. □

Proposition 5.2 *Let $\mathcal{A}$ be a complex central arrangement defined by $Q = Q(\mathcal{A})$. The map $Q : M \to \mathbb{C}^*$ is the projection of a smooth fiber bundle, called the **Milnor fibration**. The typical fiber $F = Q^{-1}(1)$ is called the **Milnor fiber**.*

Proof. It follows from work of Milnor [155] that the restriction of Q to a suitable neighborhood of the origin is a fibration. Since Q is homogeneous, we may take this neighborhood to be all of M. □

5.1 The Complement $M(\mathcal{A})$

Proposition 5.3 *The complement $M = M(\mathcal{A})$ is an open smooth parallelizable manifold of real dimension 2ℓ which has the homotopy type of a finite cell complex.*

Proof. The vector space V is an open smooth parallelizable manifold of real dimension 2ℓ and M is open in V. By Proposition 5.1, it suffices to prove the last assertion for central arrangements. It is clear if $\mathcal{A}$ is empty. Otherwise, we use the Milnor fibration 5.2. Milnor [155] proved that the fiber $F = Q^{-1}(1)$ has the homotopy type of a finite cell complex. It follows that the same holds for M. □

Suppose $\mathcal{A}$ is a central ℓ-arrangement. Its Milnor fiber admits a free action by the cyclic group of order $n = |\mathcal{A}|$. The quotient space is naturally identified with $M(\mathbf{d}\mathcal{A})$. The map $p : M(\mathcal{A}) \to M(\mathbf{d}\mathcal{A})$ is the orbit map of the standard $\mathbb{C}^*$-action

$$t(x_1, \ldots, x_\ell) = (tx_1, \ldots, tx_\ell).$$

Let $G(n)$ denote the cyclic subgroup of $\mathbb{C}^*$ of order n. Since Q is homogeneous of degree n, $G(n)F \subseteq F$. It follows that $M(\mathbf{d}\mathcal{A}) = M(\mathcal{A})/\mathbb{C}^* = F/G(n)$. Let $\zeta = e^{2\pi i/n}$ be a generator of $G(n)$. The map $F \to F$ induced by multiplication by ζ is called the **monodromy** of the Milnor fiber. Let $\pi : F \to F/G(n)$ and let $\pi' : \mathbb{C}^* \to \mathbb{C}^*/G(n)$. The commutative diagram below connects the two fibrations. Here id denotes the identity map.

$$\begin{array}{ccc} & F & \stackrel{\pi}{\to} F/G(n) \\ & \downarrow i & \downarrow id \\ \mathbb{C}^* & \to M(\mathcal{A}) \stackrel{p}{\to} & M(\mathbf{d}\mathcal{A}) \\ \downarrow \pi' & \downarrow Q & \\ \mathbb{C}^*/G(n) & \stackrel{id}{\to} \mathbb{C}^* & \end{array}$$

We return in Section 5.6 to the calculation of the cohomology of the Milnor fiber for certain arrangements.

$K(\pi, 1)$-Arrangements

Example 5.4 Let $\mathcal{A}$ be the arrangement of Example 1.5 defined by $Q(\mathcal{A}) = xy(x + y)$. The complement $M(\mathcal{A})$ has the homotopy type of $(S^1 \vee S^1) \times S^1$. We use Proposition 5.1. If we let $K_0 = \ker(y)$ go to the line at infinity, then the corresponding 1-arrangement $\mathbf{d}\mathcal{A}$ is defined by $Q(\mathbf{d}\mathcal{A}) = x(x + 1)$. Thus $M(\mathbf{d}\mathcal{A})$ is the complement of the two points $x = 0$ and $x = -1$ in $\mathbb{C}$. It follows that $M(\mathbf{d}\mathcal{A})$ has the homotopy type of $S^1 \vee S^1$. Since $\mathbb{C}^*$ has the homotopy type of S^1, the conclusion follows from Proposition 5.1. Note that $\pi_i(M(\mathcal{A})) = 0$ for $i \geq 2$, so $M(\mathcal{A})$ is a $K(\pi, 1)$ space.

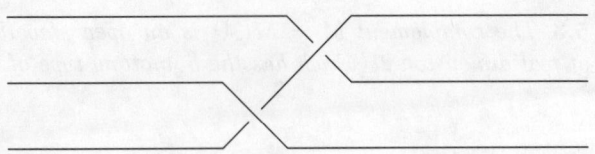

Fig. 5.1. A braid on three strands

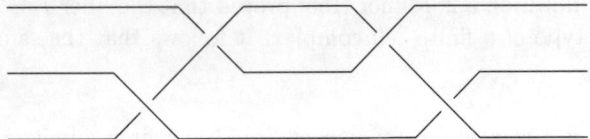

Fig. 5.2. A pure braid on three strands

Definition 5.5 *We call $\mathcal{A}$ a $K(\pi,1)$-**arrangement** if $M(\mathcal{A})$ is a $K(\pi,1)$ space.*

Proposition 5.6 *Every central 2–arrangement is $K(\pi,1)$.*

Proof. The argument in Example 5.4 shows that if $\mathcal{A}$ is a central 2–arrangement with $|\mathcal{A}| = n$, then $M(\mathcal{A})$ has the homotopy type of $(\vee_{n-1} S^1) \times S^1$. □

Proposition 5.7 *The Boolean arrangement is $K(\pi,1)$.*

Proof. Since M is the complement of the coordinate hyperplanes $M = (\mathbb{C}^*)^\ell$, it has the homotopy type of an ℓ–torus. □

Example 5.8 Before we show that the braid arrangement is also $K(\pi,1)$, it is appropriate to justify its name and describe some of its history. Braids and the braid group were defined by Artin [12]. Figure 5.1 shows a braid on 3 strands. Braids with ℓ strands may be composed by juxtaposition. There is a suitable notion of isotopy of braids which makes this an associative multiplication. The inverse of a braid is the braid which untangles it. Isotopy classes of braids on ℓ strands form a group called the **braid group**, $B(\ell)$. See Birman's book [27] for details. There is a natural surjection $B(\ell) \to Sym(\ell)$ which sends each braid to the permutation of its ends. The image of the braid in Figure 5.1 is the 3–cycle (1,2,3). The kernel of this map is called the **pure braid group** $PB(\ell)$. The corresponding pure braids have the property that each strand returns to its point of origin. Figure 5.2 shows a pure braid on 3 strands. The braid group is generated by the braids a_i for $1 \le i < \ell - 1$ indicated in Figure 5.3. The following relations are sufficient to give a presentation [27, p.11]:

$$a_i a_j = a_j a_i \qquad \text{if } |i-j| \ge 2,$$
$$a_i a_{i+1} a_i = a_{i+1} a_i a_{i+1} \qquad 1 \le i \le \ell - 2.$$

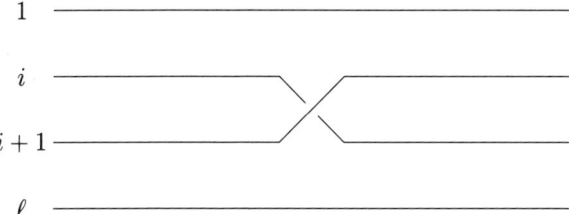

Fig. 5.3. The generator a_i

The fact that the pure braid group is the fundamental group of the pure braid space as we described it in Section 1.2 first appeared in a paper by Fox and Neuwirth [86]. To make this statement precise, let $\mathcal{A}_\ell$ denote the complexified braid arrangement of Definition 1.9. Let $M_\ell = M(\mathcal{A}_\ell)$. We described in Section 1.2 how a pure braid gives rise to a map of the circle into M_ℓ and hence to an element of its fundamental group. For the converse, choose a base point $x \in M_\ell$. An element of $\pi_1(M_\ell, x)$ is represented by a map $f : (I, \{0, 1\}) \to (M_\ell, x)$, which we may assume to be a smooth embedding. The coordinate functions of f are the strands of the pure braid. Thus $\pi_1(M_\ell) = PB(\ell)$. Note that M_ℓ admits a free action of $Sym(\ell)$ by permuting the coordinates. Let $B_\ell = M_\ell/Sym(\ell)$ be the orbit space and let $p : M_\ell \to B_\ell$ be the projection of this covering. A similar argument shows that $\pi_1(B_\ell) = B(\ell)$.

The covering $p : M_\ell \to B_\ell$ is a fibration with discrete fiber F of cardinality $|Sym(\ell)|$. The homotopy long exact sequence of this fibration gives the short exact sequence

$$1 \to \pi_1(M_\ell) \to \pi_1(B_\ell) \to \pi_0(F) \to 1,$$

which we may identify with

$$1 \to PB(\ell) \to B(\ell) \to Sym(\ell) \to 1.$$

For $k \geq 2$, the homotopy long exact sequence gives isomorphisms $\pi_k(M_\ell) \simeq \pi_k(B_\ell)$. Fadell and Neuwirth [73] showed that M_ℓ is a $K(\pi, 1)$ space.

Theorem 5.9 *The braid arrangement $\mathcal{A}_\ell$ is $K(\pi, 1)$.*

Proof. The projection map $\mathbb{C}^\ell \to \mathbb{C}^{\ell-1}$ defined by $(x_1, \ldots, x_\ell) \to (x_1, \ldots, x_{\ell-1})$ induces a locally trivial fibration $M_\ell \to M_{\ell-1}$. The fiber over $(\xi_1, \ldots, \xi_{\ell-1})$ is $\mathbb{C} \setminus \{\xi_1, \ldots, \xi_{\ell-1}\}$. The fiber retracts onto a wedge of $(\ell-1)$ circles, so it is a $K(\pi, 1)$–space. Since $M_2 = \{(x_1, x_2) \mid x_1 \neq x_2\} = \mathbb{C} \times \mathbb{C}^*$, we are done by induction. □

The representation of M as the total space of a sequence of fibrations is an important tool. The next two definitions and the following results are due to Falk and Randell [80]. In Section 5.5 the existence of a fibration is proved from the existence of modular elements in $L(\mathcal{A})$.

Definition 5.10 *Let $\mathcal{A}$ be an ℓ-arrangement. Call $\mathcal{A}$ **strictly linearly fibered** if, after a suitable linear change of coordinates, the restriction of the projection of $M(\mathcal{A})$ to the first $(\ell - 1)$ coordinates is a fiber bundle projection whose base space B is the complement of an arrangement in $\mathbb{C}^{\ell-1}$, and whose fiber is the complex line $\mathbb{C}$ with finitely many points removed.*

Definition 5.11 *(1) The 1-arrangement $(\{0\}, \mathbb{C})$ is **fiber type**.*
*(2) For $\ell \geq 2$, the ℓ-arrangement $\mathcal{A}$ is **fiber type** if $\mathcal{A}$ is strictly linearly fibered with base $B = M(\mathcal{B})$ and $\mathcal{B}$ is an $(\ell - 1)$-arrangement of fiber type.*

Proposition 5.12 *If $\mathcal{A}$ is fiber type, then $\mathcal{A}$ is $K(\pi, 1)$.*

Proof. It follows from the definition that there exist k-arrangements $\mathcal{A}_k$ for $1 \leq k \leq \ell$ with $\mathcal{A} = \mathcal{A}_\ell$ and a tower of fibrations

$$M(\mathcal{A}_\ell) \stackrel{\pi_{\ell-1}}{\to} M(\mathcal{A}_{\ell-1}) \stackrel{\pi_{\ell-2}}{\to} \cdots \stackrel{\pi_2}{\to} M(\mathcal{A}_2) \stackrel{\pi_1}{\to} M(\mathcal{A}_1) = \mathbb{C}^*$$

with the fiber F_k of π_k homeomorphic to $\mathbb{C}$ with d_k points removed. The conclusion follows by repeated application of the homotopy exact sequence of a fibration. □

The arrangement $\mathcal{A}$ of Example 2.61 is strictly linearly fibered. In fact, it is easy to see from Figure 2.8 that $M(\mathcal{A})$ is homeomorphic to $C_3 \times C_3 \times C_1$, where C_k denotes the complex line with k points removed. Falk and Randell [80] proved that in fiber type arrangements each projection map admits a section and in each fibration the fundamental group of the base acts trivially in the cohomology of the fiber. This is sufficient to show the following.

Theorem 5.13 *If $\mathcal{A}$ is a fiber type ℓ-arrangement, then there is an isomorphism of graded $\mathbb{Z}$-modules*

$$H^*(M(\mathcal{A})) \simeq H^*(F_1) \otimes \cdots \otimes H^*(F_\ell).$$

In a 1971 Bourbaki Seminar talk Brieskorn [41] generalized Arnold's results. He replaced the symmetric group and the braid arrangement by a Coxeter group W acting in an ℓ-dimensional real vector space $V_{\mathbb{R}}$. Let V be the complexification of $V_{\mathbb{R}}$. Then W acts as a reflection group in V. Let $\mathcal{A} = \mathcal{A}(W)$ be its reflection arrangement. Brieskorn conjectured that $\mathcal{A}(W)$ is a $K(\pi, 1)$-arrangement for all Coxeter groups W. He proved this for some of the groups by representing M as the total space of a sequence of fibrations, some of which are not strictly linear. Deligne [61] settled the question by proving the much stronger result stated below.

Definition 5.14 *Let $(\mathcal{A}_{\mathbb{R}}, V_{\mathbb{R}})$ be a real arrangement. Call $\mathcal{A}_{\mathbb{R}}$ a **simplicial arrangement** if every component of $M(\mathcal{A}_{\mathbb{R}})$ is an open simplicial cone.*

5.1 The Complement $M(\mathcal{A})$

Theorem 5.15 (Deligne) *Let $(\mathcal{A}_\mathbb{R}, V_\mathbb{R})$ be a simplicial arrangement. Then its complexification $(\mathcal{A}, V)$ is $K(\pi, 1)$.*

This result proves Brieskorn's conjecture because the arrangement of a Coxeter group is simplicial [38]. There exist complex reflection groups [210] which are not Coxeter groups. It is natural to ask if their reflection arrangements are $K(\pi, 1)$. For a subclass of complex reflection groups called Shephard groups this was proved in [178]. We give an outline of the argument in Section 6.6. The conjecture is still open for some remaining complex reflection groups. The arrangement $\mathcal{A}$ of Example 2.61 shows that Theorem 5.12 is not a consequence of Theorem 5.15.

Free Arrangements

The module $D(\mathcal{A})$ has topological interpretation in case $\mathbb{K} = \mathbb{C}$. Since $N = \cup_{H \in \mathcal{A}} H$ is a singular variety, it has no tangent space. But V has a stratification induced by N and each stratum has a tangent space. We show next that at every point of V, the evaluation of $D(\mathcal{A})$ at the point spans the tangent space there.

Definition 5.16 For $X \in L(\mathcal{A})$, let $M^X = M(\mathcal{A}^X)$. Then M^X is an open submanifold of X and we have a stratification of V:

$$V = \bigcup_{X \in L} M^X.$$

For $v \in V$, let TV_v denote the tangent space of V at v. Then TV_v is a $\mathbb{C}$-vector space with basis D_i. Thus we can define an evaluation map $\rho_v : D(\mathcal{A}) \to TV_v$ as follows. Given $\theta \in D(\mathcal{A})$, write $\theta = \sum h_i D_i$ and let $\rho_v(\theta) = \sum h_i(v) D_i$. Write $D(\mathcal{A})_v = \rho_v(D(\mathcal{A}))$.

Proposition 5.17 *If $v \in M^X$, then $D(\mathcal{A})_v = TM_v^X$.*

Proof. Suppose $r(X) = p$. Choose coordinates so $X = H_1 \cap \cdots \cap H_p$ where $H_j = \ker x_j$ for $1 \leq j \leq p$. Note that this makes $Q(\mathcal{A})$ divisible by $x_1 \cdots x_p$. We may choose D_i for $p+1 \leq i \leq \ell$ as a basis for $TX_v = TM_v^X$. In the rest of this paragraph, let $1 \leq j \leq p$. If $v \in M^X$, then $x_j(v) = 0$. If $\theta \in D(\mathcal{A})$, then by Lemma 4.8 $\theta \in D(x_j)$ and hence $\theta(x_j) \in x_j S$. Write $\theta = \sum_{i=1}^\ell h_i D_i$. It follows that h_j is divisible by x_j and hence $h_j(v) = 0$. Thus $\rho_v(\theta) = \sum_{i=p+1}^\ell h_i(v) D_i \in TM_v^X$.

For the converse, let $Q = Q_1 Q_2$ where $Q_1 = Q(\mathcal{A}_X)$ and $Q_2 = Q(\mathcal{A} \setminus \mathcal{A}_X)$. By our choice of coordinates above, $Q_1 \in \mathbb{C}[x_1, \ldots, x_p]$ and if $v \in M^X \subset X$, then $Q_2(v) = c \neq 0$. It is easy to check that for $p+1 \leq i \leq \ell$ we have $\theta_i = Q_2 D_i \in D(\mathcal{A})$. Since $\rho_v \theta_i = c D_i$, it follows that $TM_v^X \subset D(\mathcal{A})_v$. □

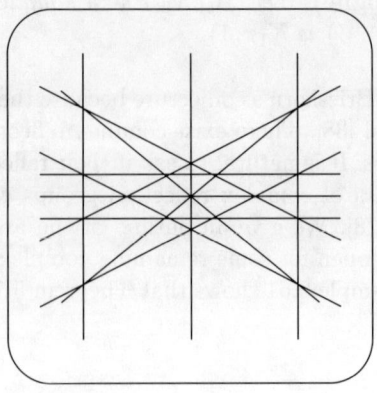

Fig. 5.4. $K(\pi, 1)$, but not free

It is not known what additional topological properties may be implied if $\mathcal{A}$ is free. Grünbaum [100] gives a list of simplicial 3–arrangements. It follows from Theorem 5.15 that these are $K(\pi, 1)$. This list was studied in [226] and several of these arrangements are not free. The smallest example is shown in Figure 5.4. It is labeled $A_4(13)$ by Grünbaum [100]. To see that it is not free we compute $\pi(\mathcal{A}, t) = (1+t)(1+12t+39t^2)$ and use the Factorization Theorem 4.137. Saito's conjecture concerns the reverse implication.

Conjecture 5.18 (Saito) If $\mathcal{A}$ is free, then $\mathcal{A}$ is $K(\pi, 1)$.

Generic Arrangements

So far we have listed a number of ways for an arrangement to be $K(\pi, 1)$. However, this is not generic behavior. In fact, it follows from a theorem of Hattori [108], which we state below, that most arrangements are not $K(\pi, 1)$. First we need two definitions.

Definition 5.19 An ℓ-arrangement $\mathcal{A}$ is called a **general position** arrangement if for every subset $\{H_1, \ldots, H_p\} \subseteq \mathcal{A}$ with $p \leq \ell$,

$$r(H_1 \cap \cdots \cap H_p) = p,$$

and when $p > \ell$

$$H_1 \cap \cdots \cap H_p = \emptyset.$$

Note that if $\mathcal{A}$ is a central general position ℓ-arrangement, then $|\mathcal{A}| \leq \ell$. Thus the only interesting general position arrangements are centerless.

5.1 The Complement $M(\mathcal{A})$ 165

Definition 5.20 Let $\mathbf{n} = \{1,\ldots,n\}$. If $I \subseteq \mathbf{n}$, let $|I|$ be its cardinality. Define the subtorus T_I of T^n by

$$T_I = \{(z_1,\ldots,z_n) \in T^n \mid z_j = 1 \text{ for } j \notin I\}.$$

Theorem 5.21 (Hattori) Let $\mathcal{A}$ be an ℓ-arrangement in general position and assume that $n = |\mathcal{A}| \geq \ell + 1$. Then $M = M(\mathcal{A})$ has the homotopy type of

$$M_0 = \bigcup_{|I|=\ell} T_I.$$

Hattori [108] also proved that $\pi_1(M)$ is free abelian of rank n, and that the universal covering space $\tilde{M}$ of M has trivial homology in dimensions $\neq 0, \ell$. He also gave a free $\mathbb{Z}[\pi_1(M)]$ resolution of $H_\ell(\tilde{M}; \mathbb{Z})$. In particular, if $n = \ell + 1$, then $H_\ell(\tilde{M}; \mathbb{Z})$ is a free $\mathbb{Z}[\pi_1(M)]$-module of rank 1.

Definition 5.22 Let $\mathcal{A}$ be a central ℓ-arrangement with $\ell \geq 2$. Call $\mathcal{A}$ a **generic** arrangement if the hyperplanes of every subarrangement $\mathcal{B} \subseteq \mathcal{A}$ with $|\mathcal{B}| = \ell$ are linearly independent.

Thus a central arrangement is generic if and only if it is the cone over a general position arrangement.

Corollary 5.23 Generic arrangements are not $K(\pi, 1)$.

Proof. We may assume that the given arrangement is $\mathbf{c}\mathcal{A}$, the cone over a general position affine arrangement $\mathcal{A}$. Recall the Hopf bundle $p : M(\mathbf{c}\mathcal{A}) \to M(\mathcal{A})$ from Proposition 5.1. Since the universal cover $\tilde{M}(\mathcal{A})$ is simply connected, it follows from Hattori's results and the Hurewicz isomorphism theorem that $\pi_i(\tilde{M}(\mathcal{A})) = H_i(\tilde{M}(\mathcal{A}); \mathbb{Z}) = 0$ for $1 \leq i < \ell - 1$ and $\pi_{\ell-1}(\tilde{M}(\mathcal{A})) = H_{\ell-1}(\tilde{M}(\mathcal{A}); \mathbb{Z})$. By Hattori's theorem, $\pi_{\ell-1}(M(\mathcal{A})) \neq 0$ and $M(\mathcal{A})$ is not a $K(\pi,1)$-space. The conclusion follows from the fact that $\pi_i(M(\mathbf{c}\mathcal{A})) = \pi_i(M(\mathcal{A}))$ for $i \geq 2$. □

Example 5.24 Define the generic arrangement $\mathbf{c}\mathcal{A}$ by $Q = xyz(x+y-z)$. Then $\mathbf{c}\mathcal{A}$ is not a $K(\pi,1)$-arrangement. Setting $K_0 = \ker(z)$, the 2-arrangement $\mathcal{A}$ is defined by $Q(\mathcal{A}) = xy(x+y-1)$. This is the arrangement in Example 1.6. By Proposition 5.1, we have $M(\mathbf{c}\mathcal{A}) = M(\mathcal{A}) \times \mathbb{C}^*$. Clearly $\mathcal{A}$ is a general position arrangement, so we may use Theorem 5.21. Here $n = 3$ and $\ell = 2$. Thus

$$M(\mathcal{A})_0 = S^1 \times S^1 \times 1 \cup S^1 \times 1 \times S^1 \cup 1 \times S^1 \times S^1 \subset S^1 \times S^1 \times S^1.$$

We can visualize $M(\mathcal{A})_0$ as the identification space obtained from the boundary of a cube by identifying opposite faces and some of their edges. Figure 5.5 shows the edge identifications on the front three faces of the cube. Each face becomes a torus. Since $n = 3 = \ell + 1$, it follows that $\pi_2(M(\mathcal{A})) = \mathbb{Z}$. A nontrivial element

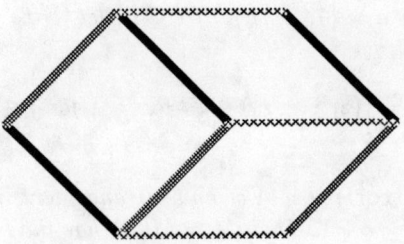

Fig. 5.5. Three lines in general position

of $\pi_2(M(\mathcal{A})_0) = \pi_2(M(\mathbf{c}\mathcal{A}))$ is obtained by any map which sends S^2 onto the boundary of the cube, followed by the identifications.

Deformation

Part of Hattori's argument consists of showing that every general position arrangement may be deformed through general position arrangements into the complexification of a real general position arrangement. The idea of such a deformation was generalized by Randell [190].

Definition 5.25 *Let $a_j : \mathbb{R} \to \mathbb{C}$ be smooth functions, and let $\alpha_t(x) = \sum_{j=1}^{\ell} a_j(t)x_j$. A smooth 1–parameter family of hyperplanes is defined by $H_t = \ker \alpha_t$. A smooth 1–parameter family of arrangements $\mathcal{A}_t$ in V is a finite collection of smooth 1–parameter families of hyperplanes.*

Definition 5.26 *Let $\mathcal{A} = \{H_1, \ldots, H_n\}$, $\mathcal{A}' = \{H'_1, \ldots, H'_n\}$ be arrangements in V of the same cardinality. We say that they have the **same lattice** if for all $I \subseteq \{1, \ldots, n\}$,*

$$\dim \cap_{i \in I} H_i = \dim \cap_{i \in I} H'_i.$$

An L–equivalence asserts that there is a linear order of the hyperplanes which gives rise to an isomorphism of the lattices. The present notion includes a fixed linear order of the hyperplanes, thus it may be viewed as an explicit L–equivalence.

Definition 5.27 *A 1–parameter family $\mathcal{A}_t$ is a **lattice isotopy** in V if for any p, q the arrangements $\mathcal{A}_p$ and $\mathcal{A}_q$ have the same lattice.*

Theorem 5.28 *Let $\mathcal{A}_t$ be a lattice isotopy in V. Then $M(\mathcal{A}_0)$ is diffeomorphic to $M(\mathcal{A}_1)$, and the pair $(V, N(\mathcal{A}_0))$ is homeomorphic to the pair $(V, N(\mathcal{A}_1))$.*

5.1 The Complement $M(\mathcal{A})$

Example 5.29 The two arrangements in Figure 2.6 have different face posets, but they fit in the 1-parameter family $\mathcal{A}_t$ defined by

$$Q(\mathcal{A}_t) = (x-z)(x+z)(y-z)(y+z)(y-x+(2+e^{i\pi t})z).$$

It is not hard to check that this is a lattice isotopy. Thus the complements of their complexifications are diffeomorphic.

J. Keaty constructed two real 3-arrangements which are L-equivalent but not lattice isotopic. It is not known whether the complements of their complexifications are homotopy equivalent.

Arnold's Conjectures

The main result of Arnold's paper [6] was the calculation of the Poincaré polynomial of the pure braid space M_ℓ and the cohomology ring structure of $H^*(M_\ell)$. Arnold showed that

$$\text{Poin}(M_\ell, t) = (1+t)(1+2t)\cdots(1+(\ell-1)t).$$

The reader should compare this formula with the Poincaré polynomial of the braid arrangement computed in Proposition 2.54. Arnold also showed that $H^*(M_\ell)$ is generated by the 1-dimensional elements

$$\omega_{p,q} = \frac{1}{2\pi i}\cdot\frac{dz_p - dz_q}{z_p - z_q},$$

and that all relations among these generators are consequences of the relations:

$$\omega_{p,q}\omega_{q,r} + \omega_{q,r}\omega_{r,p} + \omega_{r,p}\omega_{p,q} = 0.$$

Arnold stated two conjectures for an arbitrary arrangement $\mathcal{A}$. The first said that $H^*(M(\mathcal{A});\mathbb{Z})$ is torsion free. The second may be stated as follows. Define holomorphic differential forms $\omega_H = d\alpha_H/\alpha_H$ for $H \in \mathcal{A}$ and let $[\omega_H]$ denote the corresponding cohomology class. Let $R(\mathcal{A}) = \oplus_{p=0}^\ell R_p$ be the graded $\mathbb{C}$-algebra of holomorphic differential forms on $M(\mathcal{A})$ generated by the ω_H and 1. Arnold conjectured that the natural map $\omega \to [(1/2\pi i)\omega_H]$ of $R(\mathcal{A}) \to H^*(M(\mathcal{A});\mathbb{Z})$ is an isomorphism of graded algebras. This was proved by Brieskorn [41], who showed in fact that the $\mathbb{Z}$-subalgebra of $R(\mathcal{A})$ generated by the forms $(1/2\pi i)\omega_H$ and 1 is isomorphic to the singular cohomology $H^*(M(\mathcal{A});\mathbb{Z})$. In [171] it was shown that for an arbitrary arrangement $\mathcal{A}$, the Poincaré polynomial of $M(\mathcal{A})$ equals the Poincaré polynomial of $\mathcal{A}$. We prove this in Section 5.4. The structure of the algebra $R(\mathcal{A})$ was also obtained in [171]. We presented this work in Section 3.5.

5.2 The Homotopy Type of $M(\mathcal{A})$

We showed in Proposition 5.3 that $M(\mathcal{A})$ has the homotopy type of a finite cell complex. In this section we construct a finite simplicial complex $\mathsf{M}(\mathcal{A})$ in the complement $M(\mathcal{A})$ and prove that $\mathsf{M}(\mathcal{A})$ is a strong deformation retract of $M(\mathcal{A})$. The construction applies equally well to all subspace arrangements; see [169]. It combines two ideas. First, the use of the face poset for real hyperplane arrangements. On the combinatorial side, this approach was pioneered by Zaslavsky [256] and continued in the wider setting of oriented matroids by Folkman, Lawrence, and others; see [30]. On the topological side, it appears in the work of Salvetti [203]. The second, due to Goresky and MacPherson [91], shows that from the point of view of singularity theory, the natural objects of study are not just hyperplane arrangements, but arrangements of arbitrary affine subspaces. In this setting, complex arrangements are just special cases of real arrangements. See also related work of Gelfand and Rybnikov [88], and Björner and Ziegler [34].

For central complex arrangements it is interesting to note that although we prove here that the homotopy type of the total space of the Milnor fibration is determined by combinatorial data, the Milnor fiber is a more subtle analytic invariant. When $\mathcal{A}$ is the complexification of a real arrangement, Salvetti [203] constructed a simplicial complex $\mathsf{W}(\mathcal{A})$ which has the homotopy type of $M(\mathcal{A})$. The complexes $\mathsf{M}(\mathcal{A})$ and $\mathsf{W}(\mathcal{A})$ are different. Arvola [15] has found a simplicial map between these complexes which is a homotopy equivalence.

Real Arrangements

Assume $\mathcal{H}$ is a nonempty real arrangement.

Lemma 5.30 *Maximal elements of $L(\mathcal{H})$ are parallel subspaces.*

Proof. We showed in Lemma 2.4 that maximal elements have the same dimension. If $T \in L(\mathcal{H})$ is a maximal element and $H \in \mathcal{H}$ does not contain T, then H is parallel to T. The same holds for the maximal element T'. Thus every hyperplane which contains T' either contains T or is parallel to T. It follows that T and T' are parallel. □

Thus we may assume that $\mathcal{H}$ is essential. Recall the face poset $\mathcal{L} = \mathcal{L}(\mathcal{H})$ associated to $\mathcal{H}$ in Definition 2.18 and the order complex $\mathsf{K}(P)$ associated to any poset P in Definition 4.93. We follow Salvetti [203] to show that there is a natural embedding of the order complex, $\mathsf{K}(\mathcal{L}(\mathcal{H}))$ in V. Given two points $u_0, u_1 \in V$, their **join** $u_0 * u_1$ is the line segment between u_0 and u_1:

$$u_0 * u_1 = \{(1-\lambda)u_0 + \lambda u_1, \ \lambda \in [0,1]\}.$$

This may be iterated for the affine independent points $u_0, \ldots, u_k$, $k \leq \ell$, to obtain their convex hull, a k-simplex denoted $u_0 * \cdots * u_k$.

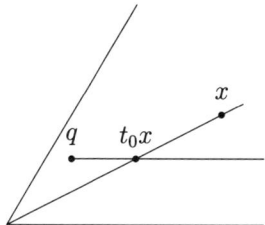

Fig. 5.6. The critical half-line

Definition 5.31 *Let $\mathcal{H}$ be an essential real hyperplane arrangement. Define a map $\phi : \mathcal{L} \to V$ which sends each face Q to a point $\mathsf{v}(Q) \in Q$. Let $\mathsf{V} = \phi(\mathcal{L})$. Then $\phi : \mathcal{L} \to \mathsf{V}$ is a bijection. Extend it to a map $\phi : \mathsf{K}(\mathcal{L}) \to V$ as follows. Given a k-simplex $[Q_0, \ldots, Q_k]$ of $\mathsf{K}(\mathcal{L})$, let*

$$\phi([Q_0, \ldots, Q_k]) = \mathsf{v}(Q_0) * \cdots * \mathsf{v}(Q_k).$$

In order to prove that ϕ is an embedding with very good properties, we need to consider central arrangements first.

Lemma 5.32 *Let $\mathcal{H}$ be a central arrangement. Let $Q \neq \{0\}$ be a face and let $q \in Q$. For every $x \in Q$ there exists a number $t_0 > 0$ such that the ray from q through tx, $t \geq 0$, meets $\partial \bar{Q}$ if $t < t_0$, and does not meet any hyperplane of $\mathcal{H} \setminus \mathcal{H}_{|Q|}$ if $t \geq t_0$.*

Proof. Let $\mathcal{B} = \mathcal{H} \setminus \mathcal{H}_{|Q|}$. By assumption $\{0\} \in \mathcal{L}$ and $Q \neq \{0\}$. Thus $\mathcal{B} \neq \emptyset$. Let $H \in \mathcal{B}$. There exists $t(H) \in \mathbb{R}$, $0 < t(H) < \infty$, such that the ray from q to tx intersects H if $t < t(H)$, is parallel to H if $t = t(H)$, and is disjoint from H if $t > t(H)$. In the special case when x is in the ray $\{tq \mid t > 0\}$, the ray from q to tx degenerates to a point when $t = t(H)$. Let $t_0 = \max\{t(H) \mid H \in \mathcal{B}\}$, and let $t_0 = t(H_0)$. Then the segment $[0, t_0 x]$ projects from q onto a ray contained in $H_0 \cap \partial \bar{Q}$. □

Lemma 5.33 *Let $\mathcal{H}$ be a central arrangement. Suppose $x \in V \setminus \{0\}$. Then*
  (1) there exists a unique linearly ordered subset $[Q_0, \ldots, Q_k]$ of $\mathcal{L}$ such that the ray $s(x) = \{tx \mid t > 0\}$ intersects the simplex $\mathsf{v}(Q_0) * \cdots * \mathsf{v}(Q_k)$, and
  (2) the intersection is a single point x', which may be expressed in barycentric coordinates as $x' = \sum \lambda_i \mathsf{v}(Q_i)$ where $\sum \lambda_i = 1$ and $\lambda_i \in (0, 1]$ for $0 \leq i \leq k$.

Proof. The interior of the ray $s(x)$ is in some face, say Q_0. If $\mathsf{v}(Q_0) \in s(x)$, then the linearly ordered subset is $[Q_0]$, $x' = \mathsf{v}(Q_0)$, and both assertions hold. Otherwise use Lemma 5.32 with $q = \mathsf{v}(Q_0)$ to project $s(x)$ onto a ray $s_1(x) \subseteq \partial \bar{Q}_0$. The interior of $s_1(x)$ is in some face, say Q_1. Since $Q_1 \subseteq \partial \bar{Q}_0$, we have $Q_0 < Q_1$. The argument is repeated until $\mathsf{v}(Q_k) \in s_k(x)$. Since $\dim Q_{i+1} < \dim Q_i$, the process is finite. □

Theorem 5.34 *Let $\mathcal{H}$ be an essential real hyperplane arrangement.*

(1) The map ϕ is an embedding. Let $\mathsf{K}(\mathcal{H}) = \mathrm{im}\,\phi$. Then $\mathsf{K}(\mathcal{H})$ is a simplicial triangulation of a disk $D(\mathcal{H})$.

(2) If $X \in L(\mathcal{H})$, then $\mathsf{K}(\mathcal{H}) \cap X = \mathsf{K}(\mathcal{H}^X)$ is a full subcomplex which is a simplicial triangulation of the disk $D(\mathcal{H}) \cap X$.

(3) There is a strong deformation retraction of V onto $D(\mathcal{H})$ which respects the stratification: each $X \in L(\mathcal{H})$ is retracted in X to $D(\mathcal{H}) \cap X$.

Proof. Assume first that $\mathcal{H}$ is a central arrangement. The map $\phi : \mathsf{K}(\mathcal{L}) \to \mathsf{K}(\mathcal{H})$ is PL and surjective by definition. It follows from Lemma 5.33 that $\mathsf{K}(\mathcal{H})$ is a simplicial complex and ϕ is injective. The bijection $\phi : \mathcal{L} \to \mathsf{V}$ induces a partial order on V which makes ϕ order preserving. It is natural to view $\mathsf{K}(\mathcal{H}) = \mathsf{K}(\mathsf{V})$. Since $\mathcal{H}$ is central, the origin $\{0\}$ is the unique maximal element of V. It follows from Lemma 4.96 that $\mathsf{K}(\mathcal{H})$ is the cone over $\mathsf{K}(\mathsf{V} \setminus \{0\})$. The map $x' \mapsto x'/|x'|$ applied to the points x' constructed in Lemma 5.33 shows that radial projection of $\mathsf{K}(\mathsf{V}\setminus\{0\})$ is a simplicial triangulation of a sphere $S^{\ell-1}$. Thus radial projection of $\mathsf{K}(\mathcal{H})$ is a simplicial triangulation of a disk $D(\mathcal{H})$. Part (2) is immediate from the construction. The strong deformation retraction of (3) follows the rays considered in Lemma 5.33.

Now suppose $\mathcal{H}$ is centerless. Let $\mathcal{B} = \mathbf{c}\mathcal{H}$ be the cone over $\mathcal{H}$; see Definition 1.15. Let $K_0 = \ker(x_0)$. Since $\mathcal{B}$ is central, we may assume that $\mathsf{K}(\mathcal{B})$ is a simplicial trangulation of a disk. Its boundary S is a simplicial triangulation of S^ℓ. Note that $\mathsf{E} = \mathsf{S} \cap \{x_0 \leq 0\}$ is a subcomplex which is again a disk. Identify the total space of $\mathcal{H}$ with $\ker(x_0 - 1)$ and by radial projection with $U = \mathsf{S} \setminus \mathsf{E}$. Then $\mathsf{K}(\mathcal{H}) \subset U$, and the closure of $\mathsf{S} \setminus \mathsf{K}(\mathcal{H})$ is a regular neighborhood of E. It follows that $\mathsf{K}(\mathcal{H})$ is also a simplicial disk. An alternate proof may be given using the fact that the union of two balls which intersect in a codimension 1 face is a ball. Properties (2) and (3) are immediate as before. □

Lemma 5.35 *Let $P_k \in \mathcal{L}$ be a face of codimension k. Let S be the set of all saturated chains $Q_0 < \cdots < Q_{k-1} < P_k$. Then*

$$D_P^k = \bigcup_S \mathsf{v}(Q_0) * \cdots * \mathsf{v}(Q_{k-1}) * \mathsf{v}(P_k)$$

is a triangulated k-cell in V whose boundary is

$$S_P^{k-1} = \bigcup_S \mathsf{v}(Q_0) * \cdots * \mathsf{v}(Q_{k-1}).$$

Proof. Given $P \in \mathcal{L}$, let $\mathrm{pr}_P : V \to V/|P|$ be the affine projection which sends the affine hyperplanes containing P into hyperplanes of a central arrangement $(\mathcal{H}/P, V/|P|)$. Let $\mathcal{L}/P$ be the set of faces of $\mathcal{H}/P$. Let $\pi_P : \mathcal{L} \to \mathcal{L}/P$ be the map which sends $Q \in \mathcal{L}$ to the smallest face of $\mathcal{L}/P$ containing $\mathrm{pr}_P(Q)$. If $Q \in \mathcal{L}$, then $\mathrm{codim}_V(Q) \geq \mathrm{codim}_{V/|P|}(\pi_P(Q))$. Let $\mathcal{L}_P \subseteq \mathcal{L}$ be the set of faces which contain P. Then the restriction $\pi_P : \mathcal{L}_P \to \mathcal{L}/P$ is a bijection

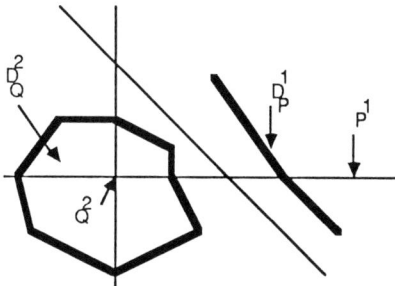

Fig. 5.7. Dual cells

which preserves codimension. The assertion follows from the corresponding fact proved for a central arrangement in Theorem 5.34. □

Note that we assigned to the face P_k of codimension k a cell D_P^k of dimension k. This is quite natural. We may view P_k as an $(\ell-k)$–cell of the cell complex $\mathcal{L}$ whose union is V. The cell dual to P_k is D_P^k and $\cup_{P_k \in \mathcal{L}} D_P^k$ is a cell decomposition of $\mathsf{K}(\mathcal{H})$. Figure 5.7 indicates some of these dual cells.

The Homotopy Type

Definition 5.36 *Let $\mathcal{A}$ be an arrangement in V and let $\mathcal{H}$ be a real hyperplane arrangement in V. We say that $\mathcal{A}$ is **embedded** in $\mathcal{H}$ if $\mathcal{A} \subseteq L(\mathcal{H})$, and write $\mathcal{A} \sqsubset \mathcal{H}$.*

Proposition 5.37 *Every arrangement $\mathcal{A}$ in V is embedded in some real hyperplane arrangement $\mathcal{H}$ in V.*

Proof. Define the real hyperplane arrangement $\mathcal{H}$ by the kernels of the real and imaginary parts of the α_H for $H \in \mathcal{A}$. □

Definition 5.38 *If $\mathcal{H}$ is a real hyperplane arrangement, define for $X \in L(\mathcal{H})$*

$$\mathcal{N}_{\mathcal{H}}(X) = \bigcup \{Q \in \mathcal{L}(\mathcal{H}) \mid X \leq |Q|\}.$$

If $\mathcal{A}$ is a complex arrangement in V, then embed $\mathcal{A}$ in a real hyperplane arrangement $\mathcal{H}$ in V. Since $\mathcal{A} \subseteq L(\mathcal{H})$, we may define $\mathcal{N}_{\mathcal{H}}(\mathcal{A}) \subseteq \mathcal{L}(\mathcal{H})$ and $\mathcal{M}_{\mathcal{H}}(\mathcal{A}) \subseteq \mathcal{L}(\mathcal{H})$ by

$$\mathcal{N}_{\mathcal{H}}(\mathcal{A}) = \bigcup_{X \in \mathcal{A}} \mathcal{N}_{\mathcal{H}}(X), \qquad \mathcal{M}_{\mathcal{H}}(\mathcal{A}) = \mathcal{L}(\mathcal{H}) \setminus \mathcal{N}_{\mathcal{H}}(\mathcal{A}).$$

Note that $\mathcal{N}_{\mathcal{H}}(X)$ is the set of faces whose union is X. Thus $\mathcal{N}_{\mathcal{H}}(\mathcal{A})$ is the set of faces whose union is $N(\mathcal{A})$ and $\mathcal{M}_{\mathcal{H}}(\mathcal{A})$ is the set of faces whose union is $M(\mathcal{A})$.

Proposition 5.39 *Let $D = D(\mathcal{H})$ be the disk in V with triangulation $\mathsf{K}(\mathcal{H})$ constructed in Theorem 5.34. Let $\mathsf{N}_{\mathcal{H}}(\mathcal{A}) = D \cap N(\mathcal{A})$. Then $\mathsf{N}_{\mathcal{H}}(\mathcal{A})$ is a full subcomplex of $\mathsf{K}(\mathcal{H})$, which is a geometric realization of $\mathsf{K}(\mathcal{N}_{\mathcal{H}}(\mathcal{A}))$.*

Proof. In general the union of full subcomplexes may not be a full subcomplex. In our case, minimal elements are incomparable and we are taking saturated intervals above them. $\square$

Theorem 5.40 *Let $\mathsf{M}_{\mathcal{H}}(\mathcal{A})$ be the largest subcomplex of $\mathsf{K}(\mathcal{H})$ disjoint from $\mathsf{N}_{\mathcal{H}}(\mathcal{A})$. Then $\mathsf{M}_{\mathcal{H}}(\mathcal{A})$ is a finite simplicial complex with the following properties:*
(1) $\mathsf{M}_{\mathcal{H}}(\mathcal{A})$ is a geometric realization of $\mathsf{K}(\mathcal{M}_{\mathcal{H}}(\mathcal{A}))$,
(2) $\mathsf{M}_{\mathcal{H}}(\mathcal{A})$ is a strong deformation retract of the complement $M(\mathcal{A})$,
(3) the homotopy type of $\mathsf{M}_{\mathcal{H}}(\mathcal{A})$ is independent of $\mathcal{H}$.

Proof. Part (1) follows from Proposition 5.39. The deformation retraction of $M(\mathcal{A})$ onto $\mathsf{M}_{\mathcal{H}}(\mathcal{A})$ is done in two stages. First, use Theorem 5.34.3 to get a strong deformation retraction of the pair $(V, N(\mathcal{A}))$ onto $(D, D \cap N(\mathcal{A}))$. This shows that $M(\mathcal{A})$ has the same homotopy type as $D \setminus D \cap N(\mathcal{A})$. It follows from Proposition 5.39 that $(D, D \cap N(\mathcal{A}))$ is a polyhedral pair with triangulation $(\mathsf{K}(\mathcal{H}), \mathsf{N}_{\mathcal{H}}(\mathcal{A}))$. Since $\mathsf{N}_{\mathcal{H}}(\mathcal{A})$ is a full subcomplex, it follows from a standard result [161, Lemma 70.1] that $\mathsf{M}_{\mathcal{H}}(\mathcal{A})$ is a strong deformation retract of $D \setminus D \cap N(\mathcal{A})$. Part (3) follows from (2). $\square$

Although the homotopy type of $\mathsf{M}_{\mathcal{H}}(\mathcal{A})$ is independent of the choice of $\mathcal{H}$, we need the embedding $\mathcal{A} \sqsubset \mathcal{H}$ for the construction. Thus it is not clear how much combinatorial information about $\mathcal{A}$ determines the homotopy type of the complement. In particular, it is not known whether $L(\mathcal{A})$ determines the homotopy type of $M(\mathcal{A})$.

Definition 5.41 *For $P \in \mathcal{L}$, let $\mathcal{L}_P = \{Q \in \mathcal{L} \mid Q \leq P\}$ be the segment below P. It consists of all faces whose closure contains P. Let $\mathsf{E}(P) = \mathsf{K}(\mathcal{L}_P)$. It follows from Lemma 5.35 that $\mathsf{E}(P)$ is a triangulated disk with center $\mathsf{v}(P)$, whose dimension equals $\mathrm{codim} P$.*

Proposition 5.42 *The space $\mathsf{M}_{\mathcal{H}}(\mathcal{A})$ is the total space of a regular cell complex*

$$\mathsf{M}_{\mathcal{H}}(\mathcal{A}) = \bigcup_{P \in \mathcal{M}_{\mathcal{H}}(\mathcal{A})} \mathsf{E}(P)$$

with $\partial \mathsf{E}(P) = \cup_{Q < P} \mathsf{E}(Q)$.

5.2 The Homotopy Type of $M(\mathcal{A})$

Although the description of the complex $M_\mathcal{H}(\mathcal{A})$ is easy, even the simplest examples contain far too many cells to be of much use for explicit calculations; see Examples 5.46 and 5.47 below.

Complexified Real Arrangements

We turn to the special case of the complexification of a real arrangement. We show first that it suffices to consider central arrangements. Suppose $\mathcal{A}$ is a complex hyperplane arrangement. If $\mathcal{A}$ is not central, then consider $\mathbf{c}\mathcal{A}$ and recall that $M(\mathbf{c}\mathcal{A}) = M(\mathcal{A}) \times \mathbb{C}^*$. Thus for the topology of the complement of a complex arrangement it suffices to study central arrangements. Now suppose $\mathcal{A}$ is a real arrangement and $\mathcal{A}_\mathbb{C}$ is the complexification of $\mathcal{A}$. Then $\mathcal{A}_\mathbb{C}$ is central if and only if $\mathcal{A}$ is central. Thus we may assume that $\mathcal{A}$ is central.

Proposition 5.43 *Let $\mathcal{A}$ be a central real arrangement in V. There is a natural embedding of its complexification $\mathcal{A}_\mathbb{C}$ in $\mathcal{H} = \mathcal{A} \times \mathcal{A}$.*

Proof. View $V \otimes \mathbb{C} = V \oplus iV$. Let $H \in \mathcal{A}$ be the kernel of the real linear form α_H. Since $\alpha_H(x+iy) = \alpha_H(x) + i\alpha_H(y)$, it follows that $x+iy \in \ker \alpha_H$ if and only if $x \in \ker \alpha_H$ and $y \in \ker \alpha_H$. □

Proposition 5.44 *Let $\mathcal{A}$ be a central real arrangement in V. Consider the natural embedding of its complexification $\mathcal{A}_\mathbb{C}$ in $\mathcal{H} = \mathcal{A} \times \mathcal{A}$. Identify $L(\mathcal{H})$ with $L(\mathcal{A}) \times L(\mathcal{A})$. Then $M_\mathcal{H}(\mathcal{A}_\mathbb{C})$ consists of elements $(P, P') \in L(\mathcal{H})$ with $\zeta(P) \cap \zeta(P') = \emptyset$.*

Proof. Recall from Definition 2.20 that $\zeta(P)$ determines the hyperplanes which contain the face P. Thus $k \in \zeta(P) \cap \zeta(P')$ if and only if $|P| \times |P'| \subset H_k \times H_k$. Since $H_k \times H_k \in L(\mathcal{A}_\mathbb{C})$, it follows that $(P, P') \in \mathcal{N}_\mathcal{H}(\mathcal{A}_\mathbb{C})$. The converse is clear. □

It follows that if $L(\mathcal{A}) \simeq L(\mathcal{B})$, then $M(\mathcal{A}_\mathbb{C})$ and $M(\mathcal{B}_\mathbb{C})$ are homotopy equivalent. This was first established by Salvetti [203]. Example 5.29 shows that the converse is false. Given a central real arrangement $\mathcal{A}$ with canonical embedding of $\mathcal{A}_\mathbb{C} \sqsubset \mathcal{A} \times \mathcal{A}$, write $\mathcal{M}(\mathcal{A}) = \mathcal{M}_{\mathcal{A}\times\mathcal{A}}(\mathcal{A}_\mathbb{C})$, $\mathsf{M}(\mathcal{A}) = \mathsf{M}_{\mathcal{A}\times\mathcal{A}}(\mathcal{A}_\mathbb{C})$, etc. When $\mathcal{A}$ is fixed, we write $\mathcal{M}$, M, etc. For a complexified real arrangement, if $(P,Q) \in \mathcal{M}$, then $\mathcal{L}_P \times \mathcal{L}_Q \subset \mathcal{M}$. Let $\mathsf{E}(P,Q) = \mathsf{K}(\mathcal{L}_P \times \mathcal{L}_Q)$. Thus $\mathsf{E}(P,Q)$ is a triangulated disk with center $\mathsf{v}(P,Q)$, whose dimension equals codim$P$ + codim$Q$. It follows from Lemma 5.35 that $\partial \mathsf{E}(P,Q) = \cup_{(R,S)<(P,Q)} \mathsf{E}(R,S)$.

Proposition 5.45 *The complex M has the structure of a regular cell complex:*

$$\mathsf{M} = \bigcup_{(P,Q)\in\mathcal{M}} \mathsf{E}(P,Q).$$

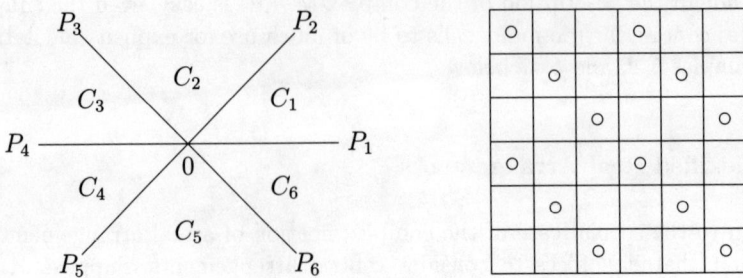

Fig. 5.8. Three concurrent lines in $\mathbb{R}^2$ and in $\mathbb{C}^2$

Example 5.46 Let $V = \mathbb{R}^2$ and let $\mathcal{A}$ consist of the two coordinate lines. Then $\mathcal{A}_\mathbb{C}$ consists of the two complex coordinate lines in $V \otimes \mathbb{C} = \mathbb{C}^2$. Note that $\mathsf{K}(\mathcal{A})$ is the barycentric triangulation of a square. It is a geometric realization of $\mathsf{K}(J^2)$. Similarly, $\mathsf{K}(\mathcal{H})$ is the barycentric triangulation of a four-dimensional cube. It is a geometric realization of $\mathsf{K}(J^4)$. The complex $\mathcal{M}$ has 16 2–cells. There are eight cells $\mathsf{E}(0, C)$, $\mathsf{E}(C, 0)$ where C is a chamber and eight cells $\mathsf{E}(P, Q)$ where P, Q are faces of dimension 1 with $|P| \neq |Q|$. To check how these cells fit together, we must compute their boundaries. It is easy to see that they form a torus, which is known to have the homotopy type of the complement $M(\mathcal{A}_\mathbb{C})$.

Example 5.47 Let $V = \mathbb{R}^2$ and let $\mathcal{A}$ consist of three concurrent lines. Then $\mathcal{A}_\mathbb{C}$ consists of three concurrent complex lines in $V \otimes \mathbb{C} = \mathbb{C}^2$. Label the faces of $\mathcal{L}(\mathcal{A})$ as in Figure 5.8. The complex $\mathsf{K}(\mathcal{A})$ is the barycentric triangulation of a hexagon. Thus $\mathsf{K}(\mathcal{H})$ is the product triangulation of a 4-ball. The number of 2–cells in $\mathcal{M}$ is 36. There are 24 barycentrically triangulated squares $\mathsf{E}(P_i, P_j)$, $|P_i| \neq |P_j|$, and 12 barycentrically triangulated hexagons $\mathsf{E}(0, C_i)$, $\mathsf{E}(C_i, 0)$. These cells fit together as indicated in Figure 5.8. The sides of the 6×6 square are identified to form a torus T. Of the 36 squares in T, 24 are filled with $\mathsf{E}(P_i, P_j)$. The remaining 12 squares are missing. These are marked with circles. The 12 hexagons are attached to 12 circles in T, which appear as the six horizontal and six vertical lines in Figure 5.8. We showed in Example 5.4 that the complement has the homotopy type of $(S^1 \vee S^1) \times S^1$. This is not obvious from our description of $\mathcal{M}$.

Arvola [15] gave another description of $\mathcal{M}$. Call two faces $P, Q \in \mathcal{L}$ **noncoplanar** if no hyperplane in $\mathcal{A}$ contains both. It is immediate from Proposition 5.44 that $\mathcal{M}$ consists of noncoplanar pairs. Recall the vector product operation in $\mathcal{L}$ from Definition 2.21.

Proposition 5.48 *Two faces $P, Q \in \mathcal{L}$ are noncoplanar if and only if their vector product PQ is a chamber. Thus*

5.2 The Homotopy Type of $M(\mathcal{A})$

$$\mathcal{M} = \{(P,Q) \in \mathcal{L} \times \mathcal{L} \mid PQ \in \mathcal{C}\}.$$

Salvetti's Complex

Arvola [15] described Salvetti's complex, W as follows. For $P \in \mathcal{L}$ choose a point $\mathsf{w}(P) \in P$. For each pair $P \times C \in \mathcal{L} \times \mathcal{C}$ define a point $\mathsf{w}(P, C) \in V \oplus V$ by

$$\mathsf{w}(P, C) = \mathsf{w}(P) \oplus (\mathsf{w}(C) - \mathsf{w}(P))$$

where minus indicates vector subtraction. The vertex set of W is the collection of points

$$\{\mathsf{w}(P, C) \mid P \times C \in \mathcal{L} \times \mathcal{C},\ P \geq C\}.$$

We assign a simplex $\Delta = \Delta(P_1 \geq \cdots \geq P_k; C)$ to a chain of faces $P_1 \geq \cdots \geq P_k$ and a chamber C such that $P_1 \geq C$. The chamber C need not be comparable to any of the faces P_i other than P_1. Recall from Proposition 2.22 that $P_iC \in \mathcal{C}$ and $P_i \geq P_iC$. Then Δ is the convex hull in $V \oplus V$ of the vertices

$$\{\mathsf{w}(P_1, P_1C), \mathsf{w}(P_2, P_2C), \ldots, \mathsf{w}(P_k, P_kC)\}$$

and

$$\mathsf{W} = \{\Delta(P_1 \geq \cdots \geq P_k; C) \mid P_i \in \mathcal{L},\ C \in \mathcal{C},\ P_1 \geq C\}.$$

Salvetti's theorem [203] may be stated with this notation.

Theorem 5.49 *Let $\mathcal{A}$ be a central real arrangement. The collection $\mathsf{W}(\mathcal{A})$ of convex sets in $V \oplus V$ is the geometric realization of a simplicial complex. There is a strong deformation retraction of the complement $M(\mathcal{A}_\mathbb{C})$ onto $\mathsf{W}(\mathcal{A})$.*

The complex W is also a regular cell complex. Given $(P, C) \in \mathcal{L} \times \mathcal{C}$ with $P \geq C$, let

$$\mathcal{D}(P, C) = \{\mathsf{w}(Q, QC) \mid Q \leq P\}$$

viewed as a set of vertices of W. Introduce a partial order in $\mathcal{D}(P, C)$ by $\mathsf{w}(R, RC) \leq \mathsf{w}(S, SC)$ when $R \leq S$. Let $\mathsf{K}(\mathcal{D}(P, C))$ be its order complex. It follows from the definition of W that the inclusion of vertices $\mathsf{K}(\mathcal{D}(P, C)) \to \mathsf{W}$ extends to a simplicial map which is an embedding. Let $\mathsf{D}(P, C)$ denote its image. It follows from Lemma 5.35 that $\mathsf{D}(P, C)$ is a triangulated disk with center $\mathsf{w}(P, C)$ whose dimension is $\mathrm{codim} P$, and $\partial \mathsf{D}(P, C) = \cup_{Q < P} \mathsf{D}(Q, QC)$.

Proposition 5.50 *The complex W has the structure of a regular cell complex:*

$$\mathsf{W} = \bigcup_{P \geq C} \mathsf{D}(P, C).$$

We return to Example 5.46 of two lines in the plane. Here W has four 2–cells of the form $\mathsf{D}(0, C)$ where 0 is the origin and C is one of the four chambers of

$\mathcal{L}$. To check how these cells fit together, we must compute their boundaries. It is easy to see that they form a torus, which is known to have the homotopy type of the complement.

Example 5.47 of three concurrent lines in the plane is more interesting. Here W has six 2–cells $D(0, C_i)$, 12 1–cells $D(P_i, C_j)$ with $P_i > C_j$, and six 0–cells $D(C_i, C_i)$. The attaching maps are computed using the boundary operator: $\partial D(0, C_k)$ is the union of six 1–cells and six 0–cells. The 1–cells are in cyclic order $D(P_{k+1}, C_k)$, $D(P_{k+2}, C_{k+1})$, $D(P_{k+3}, C_{k+2})$, $D(P_{k+4}, C_{k+4})$, $D(P_{k+5}, C_{k+5})$, $D(P_k, C_k)$. Recall from Example 5.4 that the complement has the homotopy type of $(S^1 \vee S^1) \times S^1$. This is not obvious even from the complex W.

The Homotopy Equivalence

Arvola [15] constructed an explicit simplicial map $M \to W$ which is a homotopy equivalence.

Lemma 5.51 *For each noncoplanar pair $(P, Q) \in \mathcal{L} \times \mathcal{L}$ define $\phi(v(P,Q)) = w(P, PQ)$. Linear extension of ϕ defines a surjective simplicial map $\phi : M \to W$.*

Proof. Note that ϕ is defined for every vertex of M. It follows from Proposition 5.48 that its image is in the vertex set of W. To show that ϕ is simplicial, let

$$\delta = \{v(P_1, Q_1), \ldots, v(P_k, Q_k)\}$$

be a simplex in M. Here $P_1 \geq \cdots \geq P_k$, $Q_1 \geq \cdots \geq Q_k$, and (P_i, Q_i) are noncoplanar for $1 \leq i \leq k$. Then $\phi(\delta)$ has vertices

$$\{w(P_1, P_1Q_1), \ldots, w(P_k, P_kQ_k)\}.$$

Let $C = P_1Q_1$. We claim that $\phi(\delta) = \Delta(P_1 \geq \cdots \geq P_k; C)$. It suffices to prove that $P_iC = P_iQ_i$ for $1 \leq i \leq k$. Since $P_1 \geq P_i$, it follows that $\zeta(P_i) \subseteq \zeta(P_1)$ and hence $P_iP_1 = P_i$. Thus $P_iC = P_iP_1Q_1 = P_iQ_1$. It remains to show that $P_iQ_1 = P_iQ_i$. This is clear for $k \notin \zeta(P_i)$. If $k \in \zeta(P_i)$, then $k \notin \zeta(Q_i)$ because $\zeta(P_i) \cap \zeta(Q_i) = \emptyset$. Suppose $k \in \zeta(Q_1)$. Then $k \notin \zeta(P_1)$ because $\zeta(P_1) \cap \zeta(Q_1) = \emptyset$, contradicting $\zeta(P_i) \subseteq \zeta(P_1)$.

Given the simplex $\Delta = \Delta(P_1 \geq \cdots \geq P_k; C)$ in W with $P_1 \geq C$, let δ be the simplex of M with vertices $\{v(P_1, C), \ldots, v(P_k, C)\}$. Then $\phi(\delta) = \Delta$. □

In the next result we need the following theorem of M. Cohen [50].

Theorem 5.52 *A simplicial map of a finite complex onto another which has contractible fibers is a homotopy equivalence.*

Theorem 5.53 *The simplicial map $\phi : M \to W$ is a homotopy equivalence.*

Proof. By Theorem 5.52, it suffices to show that ϕ has contractible fibers. Fix a simplex $\Delta = \Delta(P_1 \geq \cdots \geq P_k; C)$ with $P_1 \geq C$ in W. Let $M_\Delta = \phi^{-1}(\Delta)$.

Let $\mathcal{L}_\Delta = \{P_1, \ldots, P_k\}$ be the corresponding linearly ordered subset of $\mathcal{L}$. The subarrangement $\mathcal{A}_{|P_1|}$ consists of those hyperplanes which contain P_1. There is a unique chamber $D \in \mathcal{C}(\mathcal{A}_{|P_1|})$ with $C \subseteq D$. Let $\mathcal{T}_\Delta = \{Q \in \mathcal{L} \mid Q \subseteq D\}$. It is important to note here that in general $D \notin \mathcal{L}$, so Q and D are related only as subsets of V. Recall that each face Q may be viewed as a map $Q : \mathcal{A} \to \{+, -, 0\}$. With this notation, $\mathcal{T}_\Delta = \{Q \in \mathcal{L} \mid Q(H) = C(H)$ for all $H \in \mathcal{A}_{|P_1|}\}$. Let $\mathcal{M}_\Delta = \mathcal{L}_\Delta \times \mathcal{T}_\Delta$. Note that $\mathcal{M}_\Delta \subset \mathcal{M}$. It suffices to show that $(P_i, Q) \in \mathcal{M}_\Delta$ are noncoplanar. Since Q is contained in a chamber of $\mathcal{A}_{|P_1|}$, $\zeta(P_1) \cap \zeta(Q) = \emptyset$. The conclusion follows from $\zeta(P_i) \subseteq \zeta(P_1)$. Thus $\mathsf{K}(\mathcal{M}_\Delta) \subseteq \mathsf{M}$.

We show next that $\mathsf{K}(\mathcal{M}_\Delta) = \mathcal{M}_\Delta$. Suppose $Q_1 \geq \cdots \geq Q_k$, the (P_i, Q_i) are noncoplanar for $1 \leq i \leq k$, and $P_1 Q_1 = C$. Let $\delta = \{\mathsf{v}(P_1, Q_1), \ldots, \mathsf{v}(P_k, Q_k)\}$. To show that $\phi(\delta) = \Delta$, it suffices to prove that for all $H \in \mathcal{A}_{|P_1|}$ we have $P_i Q_i(H) = P_i C(H)$ for $1 \leq i \leq k$ if and only if $Q_i(H) = C(H)$ for $1 \leq i \leq k$. Fix $H \in \mathcal{A}_{|P_1|}$. For those values of i with $P_i(H) = 0$, we have $Q_i(H) = P_i Q_i(H)$ and $P_i C(H) = C(H)$, so the equations are equivalent. For those values of i with $P_i(H) \neq 0$, both equations hold. We have $P_i Q_i(H) = P_i(H) = P_i C(H)$ from the definition. On the other hand, $P_1(H) = 0$. Since $C(H) = P_1 Q_1(H) = Q_1(H) \neq 0$ and $Q_1 \geq Q_i$, we have $Q_i(H) = Q_1(H) = C(H)$.

Set $\mathsf{T}_\Delta = \mathsf{K}(\mathcal{T}_\Delta)$. Our argument shows that there is a homeomorphism of underlying topological spaces: $|\mathsf{M}_\Delta| \simeq |\Delta| \times |\mathsf{T}_\Delta|$. Since $\mathcal{T}_\Delta$ is the face poset of an arrangement in the affine space D, the complex T_Δ is the triangulation of a disk by Lemma 5.34. In particular, it is contractible. □

Corollary 5.54 *If* $\mathsf{E}(P, Q)$ *is a cell in* M, *then* $\phi \mathsf{E}(P, Q) \subseteq \mathsf{D}(P, PQ)$. *Thus the map* $\phi : \mathsf{M} \to \mathsf{W}$ *is a cellular homotopy equivalence. Given* $C \in \mathcal{C}$ *and* $P \geq C$, *we have*

$$\phi^{-1}(\mathsf{D}(P, C)) = \{\mathsf{E}(P, Q) \mid Q \in \mathcal{T}_{\mathsf{w}(P, Q)}\}.$$

Recall that in the complexification of two lines in $\mathbb{R}^2$, the complex M has 16 2–cells and the complex W has four 2–cells. The map ϕ sends the four 2–cells $\mathsf{E}(0, C)$ onto the four 2–cells $\mathsf{D}(0, C)$ and collapses the 2–cells $\mathsf{E}(C, 0)$ onto the 0–cells $\mathsf{D}(C, C)$. The remaining eight 2–cells are of the form $\mathsf{E}(P, Q)$ where P, Q are noncoplanar faces of dimension 1. These are sent to the 1–cells $\mathsf{D}(P, PQ)$.

In the complexification of three concurrent lines in $\mathbb{R}^2$, the six 2–cells $\mathsf{E}(0, C_i)$ map onto the six 2–cells $\mathsf{D}(0, C_i)$. The six 2–cells $\mathsf{E}(C_i, 0)$ are collapsed to the 0–cells $\mathsf{D}(C_i, C_i)$. The 24 2–cells $\mathsf{E}(P_i, P_j)$ are collapsed to the 12 1–cells $\mathsf{D}(P_i, P_i P_j)$ in pairs. The cells $\mathsf{E}(P_i, P_j)$ and $\mathsf{E}(P_i, P_k)$ have the same image if and only if $P_i P_j = P_i P_k$. Equivalently, $P_j(|P_i|) = P_k(|P_i|)$, so P_j and P_k are on the same side of the support of P_i.

5.3 The Fundamental Group

In this section our aim is to obtain a presentation for the fundamental group of the complement. It follows from a theorem of Zariski [106], [91] that we may assume that $\mathcal{A}$ is an arrangement of lines in $\mathbb{C}^2$. Zariski and van Kampen [123] developed a general method for computing the fundamental group of the complement of any curve in the complex projective plane. Our curve is a union of lines. This enables us to use the general method to develop a simple algorithm.

In the special case when $\mathcal{A}$ is a complexified real arrangement, a presentation of $\pi_1(M)$ was obtained by Randell [189] and Salvetti [203, 204]. Their main result asserts that a presentation may be obtained from the underlying real arrangement of $\mathcal{A}$, which we refer to as the **real graph** $\Gamma(\mathcal{A})$. In his Ph.D. thesis, W. Arvola [13] gave a presentation for the fundamental group of the complement of an arbitrary arrangement. He constructed a generalization of the real graph called the **admissible** graph, and he proved that a presentation for $\pi_1(M)$ may be obtained from an admissible graph. This section is a modified version of his paper [14]. See the papers of L. Paris [182, 183] for recent work on covering spaces of the complement. It is convenient to continue calling the complex lines $H \in \mathcal{A}$ "hyperplanes" in order to distinguish them from various other "lines" in our constructions. Write $M = M(\mathcal{A})$.

Choose coordinates z_1, z_2 for $\mathbb{C}^2$ and coordinates x_1, y_1, x_2, y_2 for $\mathbb{R}^4$. Identify $\mathbb{C}^2$ with $\mathbb{R}^4$ by

$$z_1 = x_1 + iy_1, \qquad z_2 = x_2 + iy_2.$$

Identify $\mathbb{R}^2$ as the span of the coordinates x_1, x_2 in $\mathbb{R}^4$ and let $\mathbb{R}^3$ be the span of x_1, x_2, y_2. Let ϕ^2, ϕ^3, and ϕ denote the natural projections where $\phi^2 = \phi\phi^3$.

$$\begin{array}{lll} \phi^2 : \mathbb{R}^4 \to \mathbb{R}^2 & \phi^2(x_1, y_1, x_2, y_2) &= (x_1, x_2), \\ \phi^3 : \mathbb{R}^4 \to \mathbb{R}^3 & \phi^3(x_1, y_1, x_2, y_2) &= (x_1, x_2, y_2), \\ \phi : \mathbb{R}^3 \to \mathbb{R}^2 & \phi(x_1, x_2, y_2) &= (x_1, x_2). \end{array}$$

Definition 5.55 *A* **multiple point** *is the nonempty intersection of two or more distinct hyperplanes of $\mathcal{A}$. Let P denote the set of multiple points.*

Definition 5.56 *If u and v are words in a free group, we set $u^v = v^{-1}uv$. If a group Π is given by a set of generators G and a set of relators R, then Π has presentation $\Pi = \langle G \mid R \rangle$. If $w_1, \ldots, w_k$ are words in a free group, we set*

$$[w_1, \ldots, w_k] = \{w_1 \cdots w_k = w_{\sigma(1)} \cdots w_{\sigma(k)} \mid \sigma \in C\},$$

where C is the set of cyclic permutations of the tuple $(1, \ldots, k)$. Thus the symbol on the left represents a **set** *of relators.*

With this notation we can describe Randell's result [189].

Theorem 5.57 *Let $\mathcal{A}$ be a complexified real 2-arrangement. Let M be the complement of $\mathcal{A}$. Let $\Gamma(\mathcal{A})$ be the real graph of $\mathcal{A}$. Let $G = \{g_H \mid H \in \mathcal{A}\}$. Then*

$$\pi_1(M) = \langle G \mid \bigcup_{p \in P} R_p \rangle.$$

Here R_p is a set of relators in Definition 5.56 on suitable conjugates of the generators associated to those hyperplanes which pass through p in the order indicated by $\Gamma(\mathcal{A})$.

Admissible Graphs

Let $\mathcal{A}$ be an arbitrary complex 2–arrangement.

Lemma 5.58 *After a suitable linear transformation, we may assume that $\mathcal{A}$ satisfies the conditions:*
(1) There is no hyperplane in $\mathcal{A}$ of the form $\ker(z_1 - c)$ for any constant $c \in \mathbb{C}$.
(2) If $p, p' \in P$ are distinct multiple points, then $x_1(p) \neq x_1(p')$.

Thus we may introduce a linear order in P by

$$p < p' \iff x_1(p) < x_1(p'), \qquad p, p' \in P.$$

The main tool in the proof of Theorem 5.57 is to sweep $\mathbb{R}^4$ with a family of real 2–planes parametrized by $t \in \mathbb{R}$:

$$K_t = \{q \in \mathbb{C}^2 \mid z_1(q) = t\}.$$

The fact that $\mathcal{A}$ is a complexified real arrangement in Theorem 5.57 implies that the planes K_t pass through all the multiple points. This turns out to be crucial in the argument. For an arbitrary complex 2–arrangement we may not assume that all the multiple points have real first coordinates. Thus the planes K_t may miss some multiple points. If we consider the same family parametrized by $t \in \mathbb{C}$, then we lose the advantages of a 1–dimensional parameter space. The right generalization is to let the parameter space be the graph of a piecewise linear (PL) map $f : \mathbb{R} \to \mathbb{R}$ in the complex line parametrized by z_1. We assign to f the 1–parameter family of 2–planes defined for $t \in \mathbb{R}$ by

$$\begin{aligned} K_t(f) &= \{q \in \mathbb{C}^2 \mid z_1(q) = t + if(t)\} \\ &= \{q \in \mathbb{R}^4 \mid x_1(q) = t, \ y_1(q) = f(t)\}. \end{aligned}$$

We call f the **graphing map**. Choosing $f = 0$, we get $K_t = K_t(0)$. Let $H \in \mathcal{A}$ be a hyperplane. By Lemma 5.58.1, the set $H \cap K_t(f)$ is a single point in $\mathbb{R}^4$ whose coordinates are continuous in the parameter t. Recall that $N = \bigcup_{H \in \mathcal{A}} H$ and $|\mathcal{A}| = n$. Consider the set $N \cap K_t(f)$ for a fixed value of t. This set consists of n distinct points unless there is a multiple point $p \in P$ with $p \in K_t(f)$. This occurs when $t = x_1(p)$ and $f(t) = y_1(p)$. In this case the set consists of $n - v(p) + 1$ points, where $v(p)$ is the number of hyperplanes through p. It follows from Lemma 5.58.2 that for a fixed value of t, at most one multiple point may lie in $N \cap K_t(f)$.

Definition 5.59 *Let $\Gamma_f^4 = \bigcup_{t \in \mathbb{R}} N \cap K_t(f)$ and define its projected images*

$$\Gamma_f^3 = \phi^3(\Gamma_f^4), \qquad \Gamma_f^2 = \phi^2(\Gamma_f^4).$$

Lemma 5.60 *For $H \in \mathcal{A}$, the set $H \cap \Gamma_f^4$ is PL homeomorphic to $\mathbb{R}$.*

Proof. Suppose $H = \ker(a_1 z_1 + a_2 z_2 + c)$ where $a_1, a_2, c \in \mathbb{C}$. Write $a_k = \alpha_k + i\beta_k$ for $k = 1, 2$ and $c = \gamma + i\delta$. Separate real and imaginary parts and substitute $x_1 = t$, $y_1 = f(t)$. By Lemma 5.58.1, we can solve for x_2, y_2. □

As in the case of complexified real arrangements, it is necessary for the family of planes $K_t(f)$ to contain all multiple points.

Definition 5.61 *The graphing map f is* **proper** *if every multiple point is in Γ_f^4. Thus*
$$p \in P \implies y_1(p) = f(x_1(p)).$$

We shall assume that all graphing maps are proper.

Suppose that $\mathcal{A}$ is a complexified real arrangement. Then we may choose the graphing map $f = 0$. Note that f is proper. In this case $(\mathbb{R}^2, \Gamma_f^2)$ is precisely the real graph of $\mathcal{A}$ and $(\mathbb{R}^3, \Gamma_f^3)$ is the natural embedding of the real graph into $\mathbb{R}^3$.

Since f is proper, there is a natural graph structure on Γ_f^4 with vertex set $\mathsf{V}^4 = P$. The edges E^4 are of two types: bounded and unbounded. Each bounded edge joins two distinct vertices with a piecewise linear path. The unbounded edges are paths starting at one vertex and running off to infinity. Addition of a vertex at infinity would make this a graph in the sense of Definition 2.70. Since the roles of the bounded and unbounded edges are different, it is more convenient to think of Γ_f^4 this way. Two distinct edges are either disjoint or meet in one vertex. It follows from Lemma 5.60 that the real line $H \cap \Gamma_f^4$ is a union of edges. Conversely, each edge is contained in a unique $H \in \mathcal{A}$. We call the pair $(\mathbb{R}^4, \Gamma_f^4)$ with this graph structure the **4–graph** of $\mathcal{A}$ relative to f.

Since $y_1 = f(x_1)$ on Γ_f^4, the projection ϕ^3 induces an isomorphism $\phi^3 : \Gamma_f^4 \simeq \Gamma_f^3$. Let $\mathsf{V}^3 = \phi^3(\mathsf{V}^4)$ denote the vertices of Γ_f^3 and let $\mathsf{E}^3 = \phi^3(\mathsf{E}^4)$ denote the edges of Γ_f^3. Thus the pair $\Gamma_f^3 \subseteq \mathbb{R}^3$ is also an embedded combinatorial graph. We call the pair $(\mathbb{R}^3, \Gamma_f^3)$ with this graph structure the **3–graph** of $\mathcal{A}$ relative to f.

It follows from Lemma 5.58.2 that the restriction of the projection $\phi : \Gamma_f^3 \to \Gamma_f^2$ to V^3 is a monomorphism. Call $\mathsf{P} = \phi(\mathsf{V}^3)$ the set of **actual** vertices of Γ_f^2. Note that $\phi^2 : P \to \mathsf{P}$ is a bijection. The projection ϕ is not necessarily a monomorphism on the interiors of the edges of Γ_f^3. If $E \in \mathsf{E}^3$, let int$E$ denote its interior. In order to impose a graph structure on Γ_f^2 best suited for the presentation of $\pi_1(M)$, we want ϕ and f to be well behaved.

Lemma 5.62 *By a suitable choice of coordinates and graphing map, we may assume that ϕ and f satisfy the following conditions:*

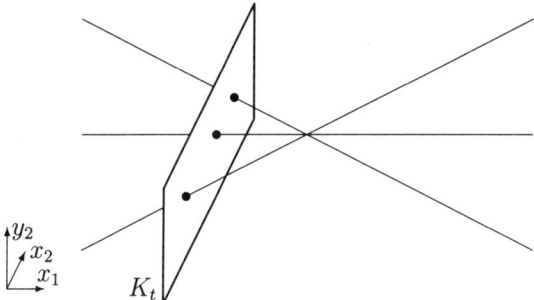

Fig. 5.9. A 3-graph

(1) $|\phi^{-1}(\mathsf{m})| \leq 2$ for all $\mathsf{m} \in \Gamma_f^2$.
(2) If $\mathsf{p} \in \mathsf{P}$, then $|\phi^{-1}(\mathsf{p})| = 1$.
(3) If $E, E' \in \mathsf{E}^3$ and $\mathsf{q} \in \phi(\mathrm{int}\, E) \cap \phi(\mathrm{int}\, E')$, then $\phi(E)$ and $\phi(E')$ meet transversely at q.
(4) The set of **virtual** vertices Q of Γ_f^2 defined by

$$\mathsf{Q} = \{\mathsf{q} \in \Gamma_f^2 \mid |\phi^{-1}(\mathsf{q})| = 2\}$$

is finite.

Proof. The first three assertions follow by transversality. It is clear that there are only a finite number of virtual vertices in the image of the bounded edges. Recall that the multiple points are linearly ordered. If we choose the graphing map to be constant before the smallest multiple point and after the largest multiple point, then only a finite number of additional virtual vertices arise. □

Definition 5.63 *Call Γ_f^2 **regular** provided ϕ, f satisfy the conditions of Lemma 5.62.*

If Γ_f^2 is regular, then we may give $\Gamma_f^2 \subseteq \mathbb{R}^2$ the structure of a planar graph as follows. The vertex set is $\mathsf{V} = \mathsf{P} \cup \mathsf{Q}$. The set of edges E consists of the images of those edges of E^3 where ϕ is a monomorphism, together with the images of the edges where ϕ is not a monomorphism, subdivided by the virtual vertices. Call $(\mathbb{R}^2, \Gamma_f^2)$ with this graph structure the **regular 2-graph** of $\mathcal{A}$ relative to f. Note that if Γ_f^2 has virtual vertices, then ϕ does not respect the graph structures. It could be made into a map of graphs by refining the structure of Γ_f^3, but we will not need this.

Definition 5.64 *For each $H \in \mathcal{A}$, let $\mathsf{H} = \phi^2(H \cap \Gamma_f^4)$ be the **trace** of H in Γ_f^2. Let $\mathsf{A} = \{\mathsf{H} \mid H \in \mathcal{A}\}$ be the set of traces.*

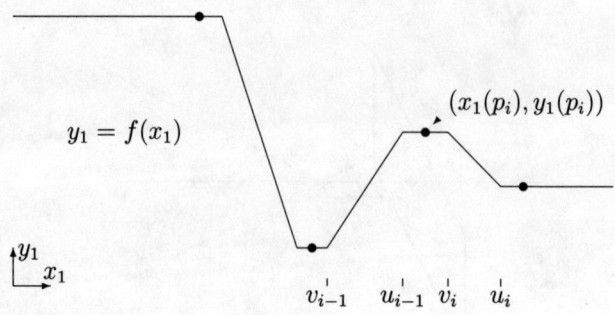

Fig. 5.10. A flat graphing map

It follows from Lemma 5.60 that each trace H is a PL line. Two traces intersect only in a vertex of Γ_f^2. The map $\tau : \mathcal{A} \to \mathsf{A}$ given by $\tau(H) = \phi^2(H \cap \Gamma_f^4)$ is a bijection and $\Gamma_f^2 = \cup_{\mathsf{H} \in \mathcal{A}} \mathsf{H}$.

Corollary 5.65 *If Γ_f^2 is regular, then:*

(1) The actual vertex $\mathsf{p} \in \mathsf{P}$ *is contained in the traces* $\mathsf{H}_1, \ldots, \mathsf{H}_k$ *if and only if the unique multiple point* $p \in P$ *with* $\mathsf{p} = \phi^2(p)$ *is contained in the hyperplanes* $H_1, \ldots, H_k$ *where* $\mathsf{H}_i = \tau(H_i)$.

(2) If $\mathsf{q} \in \mathsf{Q}$ *is a virtual vertex, then it is contained in exactly two traces* H, H' *and these traces meet transversely at* q.

In order to make the construction explicit, we choose a special collection of graphing maps. Since the multiple points play a principal role, we want f to be constant in a neighborhood of each point $x_1(p)$ for $p \in P$. Suppose $|P| = r$. Choose real numbers u_i, v_i for $1 \leq i \leq r-1$ so that

$$x_1(p_i) < v_i < u_i < x_1(p_{i+1}) \quad 1 \leq i \leq r-1.$$

Define a **flat** graphing map $f = f_{\{u_i, v_i\}}$ as follows.

$$\begin{aligned}
-\infty < t \leq v_1 &\Rightarrow f(t) = y_1(p_1), \\
u_{i-1} \leq t \leq v_i &\Rightarrow f(t) = y_1(p_i), \quad 2 \leq i \leq r-1, \\
u_{r-1} \leq t < \infty &\Rightarrow f(t) = y_1(p_r).
\end{aligned}$$

We interpolate linearly on the complementary intervals.

The slope of H changes when the slope of the graph of f changes. In a flat graphing map this occurs at $t = u_i$ and $t = v_i$. Call the corresponding points on the trace H **nodes**. Let N denote the set of nodes in Γ_f^2. The next modification is designed to adjust the nodes and vertices of a regular 2–graph.

Lemma 5.66 *By a suitable choice of coordinates and real numbers u_i, v_i for the flat graphing map f, we may assume that Γ_f^2 is a regular 2-graph whose vertex set V and set of nodes N satisfy the following conditions:*

(1) If $\mathsf{v}, \mathsf{v}' \in \mathsf{V}$ *are distinct vertices, then $x_1(\mathsf{v}) \neq x_1(\mathsf{v}')$.*

(2) *If* $v \in V$ *and* $n \in N$, *then* $x_1(v) \neq x_1(n)$.

Proof. We have already assumed that the actual vertices have distinct x_1 coordinates. Since the set of virtual vertices is also finite, transversality applies. Condition (2) is satisfied by suitable choice of u_i, v_i. □

By Lemma 5.66, we may introduce a linear order in V as follows:
$$v < v' \iff x_1(v) < x_1(v'), \qquad v, v' \in V.$$

Our next aim is to describe the behavior of the edges near a vertex. Let $v \in V$ be a vertex. Since the vertices are linearly ordered, we may choose real numbers $t_1, t_2 \in \mathbb{R}$, $t_1 < x_1(v) < t_2$, so v is the only vertex in the strip $S = \{m \in \mathbb{R}^2 \mid t_1 \leq x_1(m) \leq t_2\}$. Let $C_i = \{m \in \mathbb{R}^2 \mid x_1(m) = t_i\}$ be the boundaries of S. By choice, neither boundary contains a vertex of Γ_f^2. Thus these lines intersect each of the n traces of A exactly once in the interior of an edge. For $i = 1, 2$, let $E(i) = (e_1(i), \ldots, e_n(i))$ be the ordering of those edges which meet C_i, ordered by increasing x_2-coordinates of the intersection points $\{C_i \cap H \mid H \in A\}$. If $e_k(1) \in E(1)$ does not contain v, then $e_k(1)$ intersects both C_1 and C_2, so $e_k(1) \in E(2)$. If $e_k(1)$ contains v, then there is a unique edge $e'_k \in E(2)$ which also contains v and **lies on the same trace** as $e_k(1)$.

Proposition 5.67 *Let j be the first index for which $e_j(1)$ contains v and let k be the last such index. Then*
$$E(2) = (e_1(1), \ldots, e_{j-1}(1), e'_k, e'_{k-1}, \ldots, e'_{j+1}, e'_j, e_{k+1}(1), \ldots, e_n(1)).$$

Proof. The only vertex between C_1 and C_2 is v. Since the graphing map is flat, there is a neighborhood of v in S where the traces which contain v form a linear pencil. Thus $\{e_i(1) \mid j \leq i \leq k\}$ are precisely those edges of E(1) which contain v and the subscripts of the corresponding edges $\{e'_i \mid j \leq i \leq k\}$ which meet C_2 and lie on the same trace occur in reverse order: $e_j(2) = e'_k$, $e_{j+1}(2) = e'_{k-1}$, ..., $e_k(2) = e'_j$. The edges $e_i(1)$ for $i < j$ and $i > k$ do not contain v. Thus they also occur in E(2) and retain the position they had in E(1): $e_i(2) = e_i(1)$ for $i < j$ and $i > k$. In particular, if v is a virtual vertex, then $k = j + 1$, since only two traces contain v. □

A regular 2–graph contains almost all the data necessary to obtain a presentation for $\pi_1(M)$. The only place where information is lost in the projection $\phi : \Gamma_f^3 \to \Gamma_f^2$ occurs at the virtual vertices. Here the images of disjoint edges intersect and we cannot reconstruct their relative positions in the 3–graph from the 2–graph. Suppose $q \in Q$ is a virtual vertex in an admissible 2–graph. Let $E, E' \in E^3$ such that $q \in \phi(\text{int}\,E) \cap \phi(\text{int}\,E')$. We want to mark q to indicate whether it represents an undercrossing or an overcrossing of E and E' in Γ_f^3. Since $\phi^2 = \phi\phi^3$ and ϕ^3 is an isomorphism, we may determine the relative positions of the corresponding edges in Γ_f^4. The advantage is that these edges lie in

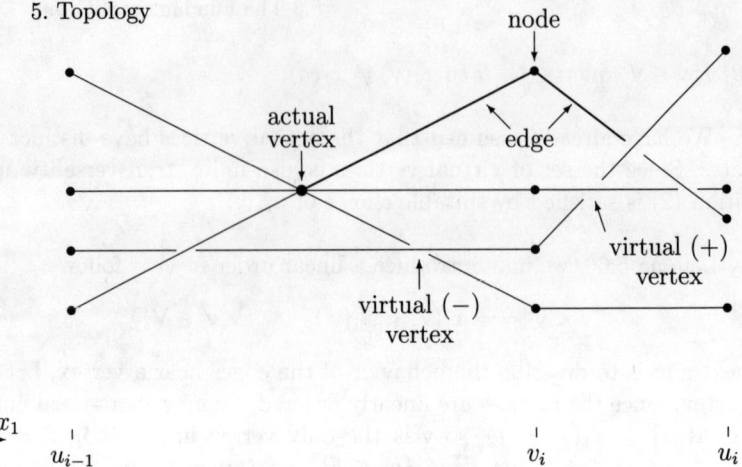

Fig. 5.11. An admissible graph

unique hyperplanes of $\mathcal{A}$. Suppose $\mathsf{q} \in \mathsf{H} \cap \mathsf{H}'$. Let $H = \tau^{-1}(\mathsf{H})$ and $H' = \tau^{-1}(\mathsf{H}')$. Then $H, H' \in \mathcal{A}$ are two hyperplanes such that there exist $q, q' \in \Gamma_f^4$, $q \neq q'$, $q \in H$, $q' \in H'$, and $\phi^2(q) = \mathsf{q} = \phi^2(q')$. It follows that $x_k(q) = x_k(\mathsf{q}) = x_k(q')$ for $k = 1, 2$. Since $y_1(q) = f(x_1(\mathsf{q})) = y_1(q')$, the points q, q' differ only in their y_2 coordinates. We adopt the following convention.

Definition 5.68 *Suppose $\mathsf{q} \in \mathsf{Q}$ is a virtual vertex in a regular 2–graph Γ_f^2. Choose a real number $c < x_1(\mathsf{q})$ sufficiently close to $x_1(\mathsf{q})$. Recall that $H \cap K_c(f)$ and $H' \cap K_c(f)$ are points in Γ_f^4. Assume that H and H' are labeled so $x_2(H \cap K_c(f)) < x_2(H' \cap K_c(f))$. Call the virtual vertex q* **positive** *if $y_2(q) < y_2(q')$, and* **negative** *otherwise.*

Call a regular 2–graph Γ_f^2 **admissible** *if its graphing map f is flat, its vertices and nodes satisfy the conditions of Lemma 5.66, and its virtual vertices are marked by $\pm$ according to this convention.* **We shall assume that all 2–graphs are admissible** *and simplify notation by writing $\Gamma = \Gamma_f^2$.*

Note that in case $\mathcal{A}$ is a complexified real arrangement, the choice of $f = 0$ provides a regular 2–graph with no virtual vertices or nodes. Thus $\Gamma(\mathcal{A}) = \Gamma_0^2$ is an admissible graph.

Arvola's Presentation

First we represent the manifold M as a union of submanifolds suitable to apply van Kampen's theorem. This decomposition is done using Γ. Choose a finite set of real numbers $T \subset \mathbb{R}$ such that

(1) $\{u_j\} \cup \{v_j\} \subset T$, and

(2) if $\mathsf{v}, \mathsf{v}' \in \mathsf{V}$ with $x_1(\mathsf{v}) < x_1(\mathsf{v}')$, then there exists $t \in T$ such that $x_1(\mathsf{v}) < t < x_1(\mathsf{v}')$. Suppose $T = \{t_0, \ldots, t_s\}$, where $t_0 < \cdots < t_s$.

5.3 The Fundamental Group

Definition 5.69 Let $L_i = \{m \in M \mid x_1(m) = t_i\}$ and define submanifolds of M by

$$\begin{aligned} M_{-\infty} &= \{m \in M \mid x_1(m) \leq t_0\} \\ M_i &= \{m \in M \mid t_{i-1} \leq x_1(m) \leq t_i\} \quad 1 \leq i \leq s \\ M_\infty &= \{m \in M \mid t_s \leq x_1(m)\}. \end{aligned}$$

Note that $L_i = M_i \cap M_{i+1}$. Since $M_{-\infty}$ and M_∞ contain no multiple points, we have the following.

Proposition 5.70 L_0 is a strong deformation retract of $M_{-\infty}$ and L_s is a strong deformation retract of M_∞. Thus there is a homotopy equivalence

$$M \sim M_1 \bigcup_{L_1} \cdots \bigcup_{L_{s-1}} M_s.$$

We apply van Kampen's theorem as in [189] to obtain this result.

Corollary 5.71 *The fundamental group is a product*

$$\pi_1(M) \simeq \pi_1(M_1) *_{\pi_1(L_1)} \pi_1(M_2) *_{\pi_1(L_2)} \cdots *_{\pi_1(L_{s-1})} \pi_1(M_s).$$

It remains to compute $\pi_1(M_i)$, $\pi_1(L_i)$, and the maps induced by inclusion. Specify a basepoint for all loops as follows. Choose a real number J sufficiently large so that $-J < y_2(m)$ for all points m in the compact set $\{m \in \Gamma^4 \mid t_0 \leq x_1(m) \leq t_s\}$. Now define

$$B = \{m \in M \mid t_0 \leq x_1(m) \leq t_s, \; y_1(m) = f(x_1(m)), \; y_2(m) = -J\}.$$

The set B is a contractible subspace of the complement M and thus any point in B may serve as basepoint. Let $K_i = K_{t_i}(f)$, $U_i = K_i \cap M$, and $B_i = B \cap U_i$. Then U_i is the complement of the n points $\{K_i \cap H \mid H \in \mathcal{A}\}$ and B_i is a line in U_i.

Lemma 5.72 *The set L_i is the complement of n skew lines in $\mathbb{R}^3$. Its subset U_i is the complement of n points in the real plane K_i. There is a strong deformation retraction $L_i \to U_i$. Thus $\pi_1(L_i)$ is a free group on n generators.*

Proof. The space $L'_i = \{m \in \mathbb{R}^4 \mid x_1(m) = t_i\}$ may be identified with $\mathbb{R}^3$. It contains no multiple point. Thus the lines $\{H \cap L'_i \mid H \in \mathcal{A}\}$ are disjoint. There is a homeomorphism of L'_i onto itself which fixes K_i and makes these lines parallel to each other but still transverse to K_i. Projection onto K_i along the direction of the lines, when restricted to L_i, provides the strong deformation retraction. □

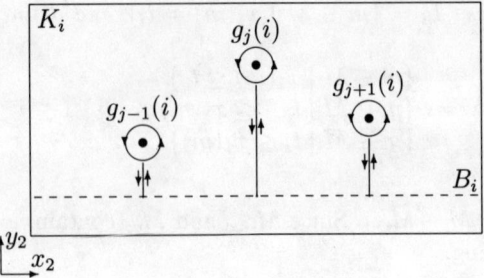

Fig. 5.12. Generators

Choose generators $g_1(i), \ldots, g_n(i)$ for $\pi_1(U_i)$ as indicated in Figure 5.12. Order these generators by increasing x_2-coordinates of the points they loop in K_i. Let $G(i) = (g_1(i), \ldots, g_n(i))$ denote the ordered n-tuple. Identify these loops with their images in $\pi_1(L_i)$ under the homotopy equivalence $U_i \hookrightarrow L_i$. As $K_t(f)$ moves from t_0 to t_s, these loops move around in a continuous fashion. The admissible graph Γ is constructed so knowledge of their order $G(i)$ at the finite number of values $t = t_i$ is sufficient to compute $\pi_1(M)$. In order to find an explicit algorithm, we must determine $G(i)$ from $G(i-1)$ and we must compute $\pi_1(M_i)$ from Γ. Extend the notation of Proposition 5.67 to T. Observe that $\phi^2(K_i) = \mathsf{C}_i$. There is a bijection between the ordered sets $E(i)$ and $G(i)$. The generator $g_k(i)$ loops the point $H \cap K_i$ if and only if $\phi^2(H \cap K_i) \subset \mathsf{e}_k(i)$. Equivalently, the hyperplane $H \in \mathcal{A}$, which is linked by the loop $g_k(i)$, has trace $\mathsf{H} = \tau(H)$, which contains the edge $\mathsf{e}_k(i)$. The key to the algorithm is this correspondence between loops and edges, familiar from knot theory, together with Proposition 5.67. Let $\mathsf{S}_i = \phi^2(M_i)$ be the strip between C_{i-1} and C_i in $\mathbb{R}^2$. By our choice of the set T, this strip may contain at most one vertex of Γ. There are four possibilities: S_i may contain no vertex, a positive virtual vertex, a negative virtual vertex, or an actual vertex. If S_i contains no vertex, then an argument similar to Lemma 5.72 shows that $M_i = L_{i-1} \times [0,1]$ and $G(i) = G(i-1)$. Identify the loops $G(i-1)$ with their images in $\pi_1(M_i)$ under the inclusion $L_{i-1} \hookrightarrow M_i$. When S_i contains the vertex v, we say that $G(i-1)$ is **adapted** to v and write $G(i-1) = (g_1(\mathsf{v}), \ldots, g_n(\mathsf{v}))$. First we need a general result about moving a loop past another loop in the plane.

Lemma 5.73 *Consider the three loops in $\mathbb{R}^2 - \{2 \text{ points}\}$, which are depicted in Figure 5.13. Among these we have the two relations $\gamma = \alpha^\beta$ and $\alpha = \gamma^{\beta^{-1}}$.*

Proof. Recall that the fundamental group consists of based homotopy classes of loops and that the composition of loops is from left to right. Draw the loop $\beta^{-1}\alpha\beta$ and observe that it may be deformed to γ in the complement of the points. □

5.3 The Fundamental Group   187

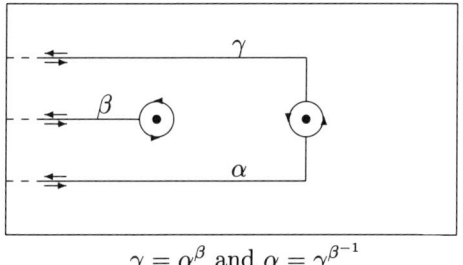

$\gamma = \alpha^\beta$ and $\alpha = \gamma^{\beta^{-1}}$

Fig. 5.13. Loop passing

Lemma 5.74 *Suppose S_i contains the virtual vertex $q \in Q$. Then $\pi_1(M_i)$ is a free group generated by $G(i-1) = (g_1(q), \ldots, g_n(q))$. If q is a positive virtual vertex contained in $e_j(i-1)$ and $e_{j+1}(i-1)$, then we get*

$$G(i) = (g_1(q), \ldots, g_{j-1}(q), g'_{j+1}, g'_j, g_{j+2}(q), \ldots, g_n(q)),$$

where

$$\begin{aligned} g'_j &= g_j(q), \\ g'_{j+1} &= g_{j+1}(q)^{g_j(q)}. \end{aligned}$$

If q is a negative virtual vertex contained in $e_j(i-1)$ and $e_{j+1}(i-1)$, then we get

$$\begin{aligned} g'_j &= g_j(q)^{g_{j+1}(q)^{-1}}, \\ g'_{j+1} &= g_{j+1}(q). \end{aligned}$$

Proof. Let $\mathbb{R}^3_f = \{m \in \mathbb{R}^4 \mid y_1(m) = f(x_1(m))\}$ and define $M' = M_i \cap \mathbb{R}^3_f$. Then M' is a strong deformation retract of M_i. The subspace M' is the complement of a braid which has a single crossing between the strands numbered $j, j+1$. If q is a positive vertex, then as x_1 goes from t_{i-1} to t_i, the loop labeled $g_j(i-1)$ passes under the loop labeled $g_{j+1}(i-1)$ in the x_2, y_2 plane. If q is a negative vertex, then the loop labeled $g_j(i-1)$ passes over the loop labeled $g_{j+1}(i-1)$ in the x_2, y_2 plane. The relationship between $G(i-1)$ and $G(i)$ is a consequence of Lemma 5.73. □

Lemma 5.75 *Suppose S_i contains the actual vertex $p \in P$. Let j be the first index for which $e_j(i-1)$ contains p and let k be the last such index. Let $G(i-1) = (g_1(p), \ldots, g_n(p))$. Then*

$$G(i) = (g_1(p), \ldots, g_{j-1}(p), g'_k, g'_{k-1}, \ldots, g'_{j+1}, g'_j, g_{k+1}(p), \ldots, g_n(p)),$$

where

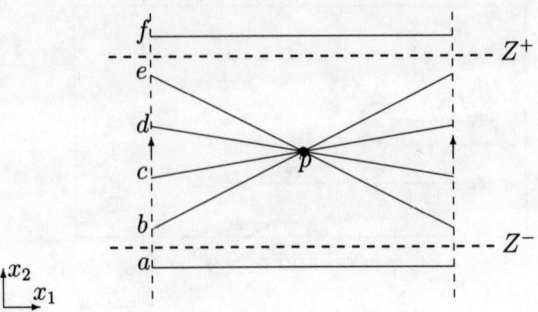

Fig. 5.14. A pencil of lines

$$\begin{aligned}
g'_j &= g_j(\mathsf{p}), \\
g'_{j+1} &= g_{j+1}(\mathsf{p})^{g_j(\mathsf{p})}, \\
g'_{j+2} &= g_{j+2}(\mathsf{p})^{g_{j+1}(\mathsf{p})g_j(\mathsf{p})}, \\
&\vdots \\
g'_k &= g_k(\mathsf{p})^{g_{k-1}(\mathsf{p})\cdots g_j(\mathsf{p})}.
\end{aligned}$$

Let $R_\mathsf{p} = [g_k(\mathsf{p}), \ldots, g_j(\mathsf{p})]$ be the set of relations of Definition 5.56. Observe that the generators adapted to p occur in the relation symbol R_p in the order **reverse** to their order in the generator symbol $G(i-1)$. Then

$$\pi_1(M_i) = \langle g_1(\mathsf{p}), \ldots, g_n(\mathsf{p}) \mid R_\mathsf{p} \rangle.$$

Proof. After a suitable isotopy of the strip, we may assume we have the corresponding strip of a real arrangement. For simplicity of notation, we analyze the situation illustrated in Figure 5.14, where the traces are $\mathsf{H}_a, \ldots, \mathsf{H}_f$ and the generators adapted to p are $G(i-1) = (a, b, c, d, e, f)$. It follows from Proposition 5.67 that the general case differs from it only in the number of traces. Write $G(i) = (w_1, \ldots, w_6)$ and $M = M_i$. We want to show first that

$$w_1 = a, \quad w_2 = e^{dcb}, \quad w_3 = d^{cb}, \quad w_4 = c^b, \quad w_5 = b, \quad w_6 = f.$$

Recall that $\mathbb{R}^3$ is spanned by x_1, x_2, y_2. It follows that $M \cap \mathbb{R}^3$ is the complement of the lines in Figure 5.14. It is evident from this that the loops a and f slide to the right without hindrance. Thus $w_1 = a$ and $w_6 = f$.

Choose real numbers s^- and s^+ so the lines $Z^- = \ker(x_2 - s^-)$ and $Z^+ = \ker(x_2 - s^+)$, together with C_{i-1} and C_i, form a box around the vertex p separating the pencil of traces $\{\mathsf{H}_b, \mathsf{H}_c, \mathsf{H}_d, \mathsf{H}_e\}$ from the traces H_a and H_f as indicated by the dotted lines in Figure 5.14. The subspaces

$$\{m \in M \mid x_2(m) = s^-\}, \qquad \{m \in M \mid x_2(m) = s^+\}$$

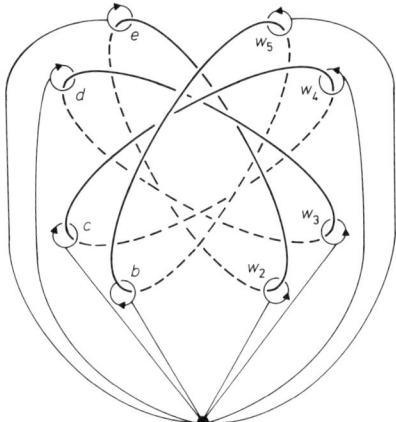

x_1 The y_1-axis points up out of the plane of the figure.

Fig. 5.15. Linking

are contractible. Thus the three sets of loops $\{a\}$, $\{b,c,d,e\}$, and $\{f\}$ are independent in $\pi_1(M)$. Let $p \in P$ be the unique multiple point with $\phi^2(p) = \mathsf{p}$. Set $M_p = \{m \in M \mid s^- \leq x_2(m) \leq s^+\}$. Let S_λ be the 3–sphere centered at p of sufficiently small radius λ so that S_λ is contained in the closure of M_p. Assume without loss of generality that p is the origin and $\lambda = 1$. Consider the map $\psi : \mathbb{R}^4 - \{0\} \to S_\lambda$ given by $\psi(m) = m/|m|$. It induces a strong deformation retraction of M_p onto $M_p \cap S_\lambda$. For $\sigma = b,c,d,e$, let H_σ denote the hyperplane linked by the loop σ, and set $\bar{H}_\sigma = H_\sigma \cap S_\lambda$. Then the $\bar{H}_\sigma$ are mutually disjoint embedded circles in S_λ, and thus $M_p \cap S_\lambda$ is the link complement

$$S_\lambda - (\bar{H}_b \cup \bar{H}_c \cup \bar{H}_d \cup \bar{H}_e).$$

Consider the stereographic projection $S_\lambda - (0,0,0,1) \to \mathbb{R}^3$ given by

$$(x_1, y_1, x_2, y_2) \mapsto \left(\frac{x_1}{1-y_2}, \frac{y_1}{1-y_2}, \frac{x_2}{1-y_2}\right).$$

The point $(0,0,0,1) \in S_\lambda$ may be deleted without changing the fundamental group, and thus we may consider the situation in $\mathbb{R}^3$. This is illustrated in Figure 5.15. The solid half-circles pass above the plane of the figure. The dashed half-circles pass below the plane of the figure.

In order to determine the w_i, we move the loops b,c,d,e along the solid (upper) half-circles. The relations arise by identifying the loops obtained by passing along the dashed (lower) half-circles with the w_i. If we move b along its upper half-circle, it is evident that $w_5 = b$. If we move c along its upper half-circle and compare it to w_4, we are obstructed by $\bar{H}_b$. Thus by Lemma 5.73 we have $w_4 = c^b$. If we move d along its upper half-circle and compare it to w_3, we are obstructed by first $\bar{H}_c$, then $\bar{H}_b$. Thus by Lemma 5.73, we have $w_3 = d^{cb}$. Similarly, $w_2 = e^{dcb}$. This illustrates the general calculation of $G(i)$. Next we want to show that

$$\pi_1(M) = \langle a,b,c,d,e,f \mid [e,d,c,b]\rangle.$$

Move b,c,d,e along their lower half-circles and compare the result with the w_i we found. Again apply Lemma 5.73 as needed to get:

lower		upper
e	$= w_2 =$	e^{dcb}
$d^{e^{-1}}$	$= w_3 =$	d^{cb}
$c^{d^{-1}e^{-1}}$	$= w_4 =$	c^{b}
$b^{c^{-1}d^{-1}e^{-1}}$	$= w_5 =$	$b.$

Move the conjugation to the right side of each equation above. This set of relators is equivalent to the set $[e,d,c,b]$ as required. This illustrates the general calculation of R_{p}. □

We have collected the necessary information to make Corollary 5.71 explicit and give Arvola's presentation.

Theorem 5.76 *Let $\mathcal{A}$ be an arrangement of complex hyperplanes in $\mathbb{C}^2$. Let M be the complement of $\mathcal{A}$. Let Γ be an admissible 2-graph for $\mathcal{A}$. Let $G = \{g_H \mid H \in \mathcal{A}\}$. Then*

$$\pi_1(M) = \langle G \mid \bigcup_{\mathsf{p}\in \mathsf{P}} R_{\mathsf{p}}\rangle,$$

where R_{p} is the set of relators in Lemma 5.75.

Proof. We sweep Γ from t_0 to t_s. There is a bijection between G and the elements of $G(0)$. By successive application of Lemmas 5.74 and 5.75 we can express every $G(i)$ for $1 \leq i \leq s$ as ordered n-tuples of words in G. This shows that G generates $\pi_1(M)$. We apply Lemmas 5.74 and 5.75 again to see that relations arise only at actual vertices. Finally, at every actual vertex $\mathsf{p} \in \mathsf{P}$ we have generators adapted to p, so we can use Lemma 5.75 to calculate the relators. □

Remark. It follows from the algorithm that only virtual vertices which are between actual vertices influence $\pi_1(M)$.

5.4 The Cohomology of $M(\mathcal{A})$

In this section we determine the cohomology groups $H^k(M;\mathbb{Z})$ and the ring structure of $H^*(M;\mathbb{Z})$. These results prove Arnold's conjectures. Brieskorn [41] showed that the groups $H^k(M;\mathbb{Z})$ are free and he obtained the direct sum decomposition of Lemma 5.91 using techniques from algebraic geometry. He also proved that the $\mathbb{Z}$-algebra generated by the forms $(1/2\pi i)\omega_H$ and 1 is isomorphic to the singular cohomology $H^*(M;\mathbb{Z})$. In [171] Brieskorn's Lemma 5.91 was used to show that the Poincaré polynomial of $M(\mathcal{A})$ equals the Poincaré

5.4 The Cohomology of $M(\mathcal{A})$

polynomial of $\mathcal{A}$. The Poincaré polynomial was then used to prove the isomorphism $R(\mathcal{A}) \simeq A(\mathcal{A})$, providing a description of $H^*(M)$ in terms of generators and relators. In [168] we used Brieskorn's Lemma 5.91 to prove that for an inductive triple $(\mathcal{A}, \mathcal{A}', \mathcal{A}'')$, the long exact sequence in cohomology of the pair $(M(\mathcal{A}'), M(\mathcal{A}))$ splits into short exact sequences

$$0 \to H^{k+1}(M(\mathcal{A}')) \to H^{k+1}(M(\mathcal{A})) \to H^k(M(\mathcal{A}'')) \to 0,$$

and that this splitting is equivalent to Brieskorn's Lemma. We remarked there that a direct proof of this splitting would provide a conceptually simpler argument for Brieskorn's Lemma, avoiding references to algebraic geometry. Such an argument appeared in a recent paper by Jozsa and Rice [122]. They use an exact sequence in de Rham cohomology. We give another elementary proof here, based on our work in the algebra $R(\mathcal{A})$. Björner and Ziegler [34] have found an argument which avoids use of the Thom isomorphism. In this section all homology and cohomology groups have integer coefficients.

The Thom Isomorphism

We begin with a topological interpretation of restriction and deletion.

Definition 5.77 *Let $(\mathcal{A}, \mathcal{A}', \mathcal{A}'')$ be a triple of arrangements with distinguished hyperplane $H_0 \in \mathcal{A}$. Let $M = M(\mathcal{A})$, $M' = M(\mathcal{A}')$, $M'' = M(\mathcal{A}'')$.*

Since M and M' are open subsets of V, they are complex manifolds of complex dimension ℓ.

Lemma 5.78 *(1) $M = M' \setminus M''$,*
(2) $M'' = M' \cap H_0$,
(3) M'' is a submanifold of M' of complex codimension 1.

Proof. Assertions (1) and (2) are clear. For (3) note that H_0 has codimension 1 in V. The conclusion follows, since M'' is open in H_0 and M' is open in V. □

Lemma 5.79 *The submanifold $M'' \subset M'$ has a tubular neighborhood $E \subseteq M'$ which has the structure of a trivial $\mathbb{C}$-bundle over M''.*

Proof. View M' as an open smooth manifold of complex dimension ℓ and M'' a smooth submanifold of complex codimension 1. The existence of the tubular neighborhood is a general fact; see [112, p.110]. The bundle is trivial because it is the restriction of a bundle over a contractible space. □

Call the complex line bundle $\xi = (E, M'', p)$ and view E as a subset of M' with inclusion map $q : E \to M'$. Let E_0 be the complement of the zero section in E. We may identify the zero section with M'' and $E_0 = E \setminus M''$. By Lemma 5.78.1, $M = M' \setminus M''$, so we have

$$E \cap M = E \cap M' \setminus E \cap M'' = E \setminus M'' = E_0.$$

Lemma 5.80 *Let $\xi = (E, M'', p)$ be a tubular neighborhood of M'' in M'. Then there are isomorphisms for $k \geq 1$*

$$\tau : H^{k+1}(M', M) \to H^{k-1}(M'').$$

Proof. Since $E_0 = E \cap M$, the embedding $q : E \to M'$ is an inclusion of pairs $q : (E, E_0) \to (M', M)$. This inclusion is excision of the closed subset $M' \setminus E \subseteq M$ and therefore q^* is an isomorphism. Let $z : M'' \to E$ be the zero section. Since $(E, E_0) \simeq M'' \times (\mathbb{C}, \mathbb{C}^*)$, we have isomorphisms $H^{k-1}(M'') \xleftarrow{z^*} H^{k-1}(E) \xrightarrow{r} H^{k+1}(E, E_0)$. Here r is the Thom isomorphism for the trivial bundle ξ. Then $\tau = z^* r^{-1} q^*$. □

Corollary 5.81 *For $k \geq 0$ there is a cohomology long exact sequence*

$$\cdots \to H^k(M') \xrightarrow{i^*} H^k(M) \xrightarrow{\phi} H^{k-1}(M'') \xrightarrow{\psi} H^{k+1}(M') \to \cdots$$

where $\phi = \tau \delta$ and $\psi = j^* \tau^{-1}$.

Proof. Consider the long exact sequence of the pair (M', M) in cohomology:

$$\cdots \to H^k(M') \xrightarrow{i^*} H^k(M) \xrightarrow{\delta} H^{k+1}(M', M) \xrightarrow{j^*} H^{k+1}(M') \to \cdots$$

We may use the isomorphism of Lemma 5.80 to replace $H^{k+1}(M', M)$ by $H^{k-1}(M'')$. This provides the required long exact sequence. □

Definition 5.82 *Let $H = \ker \alpha_H$ and let $M_H = V \setminus H$. The map $\alpha_H : V \to \mathbb{C}$ restricts to $\alpha_H : M_H \to \mathbb{C}^*$. Choose the canonical generator of $H^1(\mathbb{C}^*)$ as $(1/2\pi i)(dz/z)$. Define a rational 1-form*

$$\eta_H = \frac{1}{2\pi i} \frac{d\alpha_H}{\alpha_H}$$

on V. Let $\langle \eta_H \rangle$ be the cohomology class of η_H in $H^1(M_H)$. Then

$$\langle \eta_H \rangle = \alpha_H^* \left(\frac{1}{2\pi i} \frac{dz}{z} \right) \in H^1(M_H).$$

Denote the cohomology class of η_H in $H^1(M)$ by $[\eta_H]$. Let $i_H : M \to M_H$ be the inclusion map. Then $[\eta_H] = i_H^* \langle \eta_H \rangle$.

Lemma 5.83 *The natural orientation of the $\mathbb{C}$-bundle $\xi = (E, M'', p)$ has Thom class $u \in H^2(E, E_0)$ given by $u = q^* \delta[\eta_{H_0}]$.*

Proof. Write $M_0 = M_{H_0}$, $\alpha_0 = \alpha_{H_0}$, $i_0 = i_{H_0}$, and $\eta_0 = \eta_{H_0}$. In the cohomology exact sequence of the pair (V, M_0) we have $H^1(V) = H^2(V) = 0$, so

5.4 The Cohomology of $M(\mathcal{A})$

$\delta : H^1(M_0) \to H^2(V, M_0)$ is an isomorphism. Since $\langle \eta_0 \rangle$ generates $H^1(M_0)$, $\delta \langle \eta_0 \rangle$ generates $H^2(V, M_0)$. Let $x \in M''$ be any point and let $F = p^{-1}(x)$. Let $F_0 = F \cap E_0$. Let $k : (F, F_0) \to (V, M_0)$ be the inclusion of the fiber. Since $\alpha_0 : (F, F_0) \to (\mathbb{C}, \mathbb{C}^*)$ is a homotopy equivalence of pairs, k^* is an isomorphism. If ξ is given the natural orientation, then k^* sends a positive generator to a positive generator. We have the following inclusion of pairs with $k = jqi$:

$$(F, F_0) \xrightarrow{i} (E, E_0) \xrightarrow{q} (M', M) \xrightarrow{j} (V, M_0).$$

In cohomology we get the commutative diagram:

$$\begin{array}{ccccccc}
H^2(V, M_0) & \xrightarrow{j^*} & H^2(M', M) & \xrightarrow{q^*} & H^2(E, E_0) & \xrightarrow{i^*} & H^2(F, F_0) \\
\uparrow \delta & & \uparrow \delta & & & & \\
H^1(M_0) & \xrightarrow{i_0^*} & H^1(M) & & & &
\end{array}$$

Since $i^*q^*j^* = k^*$ is an isomorphism and $\delta \langle \eta_0 \rangle$ generates $H^2(V, M_0)$, it follows that $i^*q^*j^*\delta \langle \eta_0 \rangle = i^*q^*\delta[\eta_0]$ generates $H^2(F, F_0)$. Thus the restriction of $q^*\delta[\eta_0]$ is a generator of $H^2(F, F_0)$ for every fiber. This characterizes the Thom class u of the bundle ξ. $\square$

Let $H \in \mathcal{A}'$. Then η_H gives a cohomology class $[\eta_H]'$ in $H^1(M')$. Naturality implies that for $i : M \to M'$ we have $i^*[\eta_H]' = [\eta_H]$. Let $\alpha_{H_0 \cap H}$ be the restriction of $\alpha_H : V \to \mathbb{C}^*$ to H_0: $\alpha_{H_0 \cap H} = \alpha_H|_{H_0}$. Let $[\eta_{H_0 \cap H}] \in H^1(M'')$ be the cohomology class of the rational 1-form

$$\eta_{H_0 \cap H} = \frac{1}{2\pi i} \frac{d\alpha_{H_0 \cap H}}{\alpha_{H_0 \cap H}}.$$

Lemma 5.84 *Let $H \in \mathcal{A}'$. Then $z^*q^*[\eta_H]' = [\eta_{H_0 \cap H}]$. Here $q : E \to M'$ is the tubular neighborhood and $z : M'' \to E$ is its zero section.*

Proof. Note that $qz : M'' \to M'$ is the inclusion map. Thus the pull-back of η_H by qz is equal to $\eta_{H_0 \cap H}$. $\square$

Lemma 5.85 *Let $H \in \mathcal{A}'$. Then for any $[a] \in H^k(M)$ we have*

$$\delta([a] \cup [\eta_H]) = \delta[a] \cup [\eta_H]'.$$

Proof. Stability [67, p.220] of the diagram

$$\begin{array}{ccccc}
H^k(M) \otimes H^1(M') & \xrightarrow{id \otimes i^*} & H^k(M) \otimes H^1(M) & \xrightarrow{\cup} & H^{k+1}(M) \\
\downarrow \delta \otimes id & & & & \downarrow \delta \\
H^{k+1}(M', M) \otimes H^1(M') & \to & \xrightarrow{\cup} & \to & H^{k+2}(M', M)
\end{array}$$

gives $\delta([a] \cup i^*[\eta_H]') = \delta[a] \cup [\eta_H]'$ and we noted that $i^*[\eta_H]' = [\eta_H]$ by naturality. $\square$

Our next aim is to determine the structure of $H^*(M)$. Recall the modules $R_k(\mathcal{A})$ from Definition 3.116, the residue map from Lemma 3.125, and the exact sequence in Theorem 3.127. Let $\mathcal{K} = \mathbb{Z}$.

Lemma 5.86 *There is a commutative diagram of exact sequences whose vertical maps $\eta : R_k(\mathcal{A}) \to H^k(M)$ are given by $\eta(\omega_H) = [\eta_H]$:*

$$\begin{array}{ccccccccc} 0 & \to & R_{k+1}(\mathcal{A}') & \xrightarrow{i} & R_{k+1}(\mathcal{A}) & \xrightarrow{j} & R_k(\mathcal{A}'') & \to & 0 \\ & & \downarrow \eta' & & \downarrow \eta & & \downarrow \eta'' & & \\ \cdots & \to & H^{k+1}(M') & \xrightarrow{i^*} & H^{k+1}(M) & \xrightarrow{\phi} & H^k(M'') & \to & \cdots \end{array}$$

Proof. The commutativity is clear in the left square. For commutativity in the right square, let $\gamma = \omega_{H_0} \omega_{H_1} \ldots \omega_{H_k}$. We want $\phi \eta(\gamma) = \eta'' j(\gamma)$. Recall that $\phi = z^* r^{-1} q^* \delta$. Thus we have

$$\begin{aligned} \phi \eta(\gamma) &= \phi([\eta_{H_0}] \cup [\eta_{H_1} \ldots \eta_{H_k}]) \\ &= z^* r^{-1} q^* (\delta[\eta_0] \cup [\eta_{H_1} \ldots \eta_{H_k}]') \\ &= z^* r^{-1} (q^* \delta[\eta_0] \cup q^* [\eta_{H_1} \ldots \eta_{H_k}]') \\ &= z^* r^{-1} (u \cup q^* [\eta_{H_1} \ldots \eta_{H_k}]') \\ &= z^* q^* [\eta_{H_1} \ldots \eta_{H_k}]' \\ &= [\eta_{H_0 \cap H_1} \ldots \eta_{H_0 \cap H_k}]. \end{aligned}$$

Here we used naturality and the fact that cupping with u is r. □

Theorem 5.87 *Let $\mathcal{A}$ be a nonempty complex arrangement.*
(1) The map $\eta : R_k(\mathcal{A}) \to H^k(M)$ is an isomorphism for $k \geq 0$.
(2) $H^k(M)$ are free abelian groups.
(3) For $k \geq 0$ there exist split short exact sequences

$$0 \to H^{k+1}(M') \xrightarrow{i^*} H^{k+1}(M) \xrightarrow{\phi} H^k(M'') \to 0.$$

Proof. We argue (1) by a double induction on ℓ and $n = |\mathcal{A}|$. The assertion holds for all 1–arrangements and for all ℓ–arrangements with $|\mathcal{A}| = 1$. The induction step assumes both assertions for all m–arrangements with $m < \ell$ and for all ℓ–arrangements with $|\mathcal{A}| < n$. Thus η' and η'' are isomorphisms. It follows from Lemma 5.86 that η is an isomorphism. This completes the induction step. Corollary 3.128 implies (2). Lemma 5.86 and (1) provide (3). □

Corollary 5.88 *The integral cohomology ring $H^*(M)$ is generated by 1 and the classes $[\eta_H]$ for $H \in \mathcal{A}$.*

Theorem 5.89 *The surjective map $\omega_H \to [(1/2\pi i)\omega_H]$ induces an isomorphism of graded algebras $R(\mathcal{A}) \simeq H^*(M(\mathcal{A}))$.*

This result shows there are no relations in cohomology other than those imposed by the algebraic relations. We showed in Theorem 3.126 that there is an isomorphism of algebras $A(\mathcal{A}) \simeq R(\mathcal{A})$ which sends a_H to ω_H. We may apply this result when $\mathcal{K} = \mathbb{Z}$ to obtain a structure theorem for $H^*(M;\mathbb{Z})$ in terms of generators and the relation ideal.

Theorem 5.90 *Let $\mathcal{A}$ be a complex arrangement and let $A = A(\mathcal{A})$. The map $a_H \to [(1/2\pi i)\omega_H]$ induces an isomorphism $A \to H^*(M)$ of graded $\mathbb{Z}$-algebras.*

This isomorphism is the common thread between the algebra factorization $A(\mathbf{c}\mathcal{A}) \simeq (\mathcal{K} + \mathcal{K}a_0) \otimes A(\mathcal{A})$ of Theorem 3.78 and the topological factorization $M(\mathbf{c}\mathcal{A}) \approx M(\mathcal{A}) \times \mathbb{C}^*$ of Proposition 5.1.

Brieskorn's Lemma

The next result is due to Brieskorn [41]. His proof involves some Lefschetz-type results from algebraic geometry. Alternate arguments have been given by Falk [74], Goresky and MacPherson [91], and Jozsa and Rice [122].

Lemma 5.91 (Brieskorn) *Let $\mathcal{A}$ be a nonempty complex arrangement. For $X \in L(\mathcal{A})$, let $M_X = M(\mathcal{A}_X)$. For $k \geq 0$ there are isomorphisms*

$$\theta_k : \bigoplus_{X \in L_k} H^k(M_X) \to H^k(M)$$

induced by the inclusion maps $i_X : M \to M_X$.

Proof. By Theorem 5.90, there exists an isomorphism $A(\mathcal{A}) \to H^*(M(\mathcal{A}))$ of graded $\mathbb{Z}$-algebras. By Corollary 3.27, we have natural isomorphisms for $k \geq 0$

$$\bigoplus_{X \in L_k} A_k(\mathcal{A}_X) \to A_k(\mathcal{A}).$$

Therefore we have natural isomorphisms for $k \geq 0$

$$\bigoplus_{X \in L_k} H^k(M(\mathcal{A}_X)) \to H^k(M(\mathcal{A}))$$

as required. □

Definition 5.92 *Let $b_p(M) = \mathrm{rank} H^p(M)$ be the Betti numbers of M. The Poincaré polynomial of the complement is*

$$\mathrm{Poin}(M(\mathcal{A}), t) = \sum_{p \geq 0} b_p(M) t^p.$$

Theorem 5.93 *Let $\mathcal{A}$ be a complex arrangement. Then*

$$\mathrm{Poin}(M(\mathcal{A}),t) = \pi(\mathcal{A},t).$$

Proof. If $\mathcal{A}$ is empty, then $M(\mathcal{A}) = V$ and $\mathrm{Poin}(M(\mathcal{A}),t) = 1$. It follows from Theorem 5.87 that $\mathrm{Poin}(M(\mathcal{A}),t)$ satisfies the same recursion under deletion and restriction as $\pi(\mathcal{A},t)$. Thus they are equal. □

Proposition 5.94 *Let $\mathcal{A}$ be a real ℓ-arrangement defined by $Q(\mathcal{A})$. The complement $M(\mathcal{A})$ is an algebraic variety.*

Proof. Consider $\mathbb{R}^{\ell+1}$ with coordinates $x_0, x_1, \ldots, x_\ell$. Then

$$M(\mathcal{A}) \approx \{x \in \mathbb{R}^{\ell+1} \mid x_0 Q(\mathcal{A}) = 1\},$$

so the complement is homeomorphic to a hypersurface. □

Corollary 5.95 (M–property) *Let $(\mathcal{A}_\mathbb{R}, V_\mathbb{R})$ be a real arrangement and let $(\mathcal{A}_\mathbb{C}, V_\mathbb{C})$ be its complexification. Let $M_\mathbb{R} = M(\mathcal{A}_\mathbb{R})$ and $M_\mathbb{C} = M(\mathcal{A}_\mathbb{C})$ be the real and complex complements. Let $b_i(M_\mathbb{R})$ and $b_i(M_\mathbb{C})$ be their respective Betti numbers with coefficients in $\mathbb{Z}/2$. Then $M_\mathbb{R}$ has the M–property*

$$\sum_{i \geq 0} b_i(M_\mathbb{R}) = \sum_{i \geq 0} b_i(M_\mathbb{C}).$$

Proof. Since $M_\mathbb{R}$ is a disjoint union of chambers and each chamber is contractible, $b_i(M_\mathbb{R}) = 0$ for $i > 0$ and $b_0(M_\mathbb{R})$ is the number of chambers. By Theorem 2.68, we have $b_0(M_\mathbb{R}) = \pi(\mathcal{A}_\mathbb{R}, 1)$. By Theorem 5.93, we have

$$\sum_{i \geq 0} b_i(M_\mathbb{C}) = \mathrm{Poin}(M(\mathcal{A}_\mathbb{C}), 1) = \pi(\mathcal{A}_\mathbb{C}, 1).$$

The result follows from the fact that $L(\mathcal{A}_\mathbb{R}) = L(\mathcal{A}_\mathbb{C})$. □

5.5 The Fibration Theorem

Recall strictly linearly fibered arrangements from Definition 5.10 and fiber type arrangements from Definition 5.11. In this section we prove that $\mathcal{A}$ is fiber type if and only if $\mathcal{A}$ is supersolvable. This provides a topological interpretation for the notion of a supersolvable arrangement. Since supersolvable arrangements are central, we assume in this section that $\mathcal{A}$ is a central arrangement. The presentation follows [234].

Let $(\mathcal{A}, V)$ be an ℓ-arrangement and let $Y \subseteq V$ be a subspace which is **not necessarily** in $L = L(\mathcal{A})$. Throughout this section we will use spaces, maps, and properties whose dependence on Y will be suppressed.

Definition 5.96 *Let $p = p_Y : V \to V/Y$ be the natural projection. Extend the definition of $\mathcal{A}_Y$ to subspaces which are not in L by $\mathcal{A}_Y = \{H \in \mathcal{A} \mid Y \subseteq H\}$.*

Then $\mathcal{A}/Y = \{p(H) \mid H \in \mathcal{A}_Y\}$ is an arrangement in V/Y. Let $\bar{Y} = \cap_{H \in \mathcal{A}_Y} H$ be the smallest element of L which contains Y.

Note that $\mathcal{A}_Y = \mathcal{A}_{\bar{Y}}$, so $L(\mathcal{A}_Y) = L(\mathcal{A}_{\bar{Y}})$. There is a natural bijection $\mathcal{A}_Y \to \mathcal{A}/Y$ by $H \to H/Y$, which induces an isomorphism $L(\mathcal{A}_Y) \simeq L(\mathcal{A}/Y)$. The projection p_Y induces a smooth surjection $\pi = \pi_Y : M(\mathcal{A}) \to M(\mathcal{A}/Y)$. We want to determine when π is a bundle map. If $\mathcal{A}$ is not essential, then we may choose $Y = T(\mathcal{A})$ to get a bundle $\pi_{T(\mathcal{A})} : M(\mathcal{A}) \to M(\mathcal{A}/T(\mathcal{A}))$ with fiber $T(\mathcal{A})$. Thus we may assume that $\mathcal{A}$ is an essential central ℓ-arrangement.

Horizontal Subspaces

Definition 5.97 *Call $X \in L$ **horizontal** if $p(X) = V/Y$. Let $\mathrm{Hor} = \mathrm{Hor}_Y$ be the set of horizontal elements of L. Define the **bad set** B by*

$$B = B_Y = \bigcup_{X \in L \setminus \mathrm{Hor}} p(X) \cap M(\mathcal{A}/Y).$$

Example 5.98 Consider $\mathcal{A}$ defined by $Q = xyz(x+y-z)$. Let $H_1 = \ker(x)$, $H_2 = \ker(y)$, $H_3 = \ker(z)$, and $H_4 = \ker(x+y-z)$. Let $Y = H_1 \cap H_2$, so Y is the z-axis. Here $\mathcal{A}_Y = \{H_1, H_2\}$, $\mathrm{Hor} = \{V, H_3, H_4\}$, and $B = p(H_3 \cap H_4)$. We may identify V/Y with the x,y-plane. Then $\mathcal{A}/Y$ consists of the x-axis and the y-axis. The bad set B is the line $x+y = 0$ minus the origin. Note that $\pi^{-1}(u)$ is a twice punctured line for $u \in M(\mathcal{A}/Y) \setminus B$, and a once punctured line for $u \in B$.

Definition 5.99 *Fix $v \in V$. Define $L_v \subseteq L$ by*

$$L_v = \{X \in L \mid (v+Y) \cap X \neq \emptyset\}.$$

Define $A_v \subseteq v + Y$ by

$$A_v = \{(v+Y) \cap X \mid X \in L_v\}.$$

Let $\psi = \psi_v : L_v \to A_v$ be the natural surjection defined for $X \in L_v$ by $\psi(X) = (v+Y) \cap X$.

In Example 5.98 let $v = (1, -1, 1)$. Then $L_v = \{V, H_3, H_4, H_3 \cap H_4\}$. Here $\psi_v(V) = \{(1, -1, c) \mid c \in \mathbb{C}\}$ and $\psi_v(H_3) = \psi_v(H_4) = \psi_v(H_3 \cap H_4) = (1, -1, 0)$.

Definition 5.100 *Define $C = C_Y \subseteq L$ by*

$$C_Y = \{X \in L \mid X \wedge \bar{Y} = V\}.$$

Lemma 5.101 *For all $v \in M(\mathcal{A})$ we have $\mathrm{Hor} \subseteq L_v \subseteq C_Y$.*

Proof. If $X \in$ Hor, then $p(X) = V/Y$, and hence $X + Y = V$. Thus $(v+Y) \cap X \neq \emptyset$ and $X \in L_v$. If $X \in L_v$, then
$$v \in X + Y \subseteq X + \bar{Y} \subseteq X \wedge \bar{Y}.$$
Since $X \wedge \bar{Y} \in L$ and the only element of L which contains a point of $M(\mathcal{A})$ is V, we have $X \wedge \bar{Y} = V$. Therefore $X \in C_Y$. □

In Example 5.98 we have $C_Y = \{V, H_3, H_4, H_3 \cap H_4\} = L_v$. Note that the inclusion Hor $\subset L_v$ is proper.

Lemma 5.102 $C_Y = \{X \in L \mid p(X) \cap M(\mathcal{A}/Y) \neq \emptyset\}$.

Proof. We have
$$\begin{aligned} p(X) \cap M(\mathcal{A}/Y) = \emptyset &\iff p(X) \subseteq p(H) \quad \text{for some } H \in \mathcal{A}_Y \\ &\iff X + Y \subseteq H \quad \text{for some } H \in \mathcal{A}_Y \\ &\iff X + \bar{Y} \subseteq H \quad \text{for some } H \in \mathcal{A}_Y \\ &\iff X \wedge \bar{Y} \neq V. \end{aligned}$$
This completes the proof. □

Proposition 5.103 *The following four conditions on Y are equivalent.*
(1) $L_v =$ Hor for all $v \in M(\mathcal{A})$.
(2) L_v is independent of $v \in M(\mathcal{A})$.
(3) The bad set is empty, $B = \emptyset$.
(4) Hor $= C_Y$.

Proof. (1) $\Rightarrow$ (2) is clear. We argue (2) $\Rightarrow$ (3) by contradiction. If $B \neq \emptyset$, then there exists $v \in M(\mathcal{A})$ such that $\pi(v) \in p(X)$ for some nonhorizontal $X \in L$. Since X is not horizontal, there exists $w \in M(\mathcal{A})$ such that $\pi(w) \notin p(X)$. Then $X \in L_v$, but $X \notin L_w$. This contradicts (2). For (3) $\Rightarrow$ (4), note that $B = \emptyset$ implies
$$\{X \in L \mid p(X) \cap M(\mathcal{A}/Y) \neq \emptyset\} \subseteq \text{Hor}.$$
The reverse inclusion follows from Lemmas 5.101 and 5.102. Finally, (4) $\Rightarrow$ (1) follows from Lemma 5.101. □

Good Subspaces

Definition 5.104 *Call Y a **good** subspace of V if it satisfies the conditions of Proposition 5.103.*

Proposition 5.105 *Let $v \in M(\mathcal{A})$. If $\pi(v) \in M(\mathcal{A}/Y) \setminus B$, then*
(1) Hor $= L_v$, and
(2) the map $\psi_v : L_v \to \mathcal{A}_v$ is bijective.

Proof. (1) It follows from Lemma 5.101 that Hor $\subseteq L_v$. If $X \in L_v$, then $v \in X + Y$ and $\pi(v) \in p(X)$. It follows from $\pi(v) \notin B$ that $X \in$ Hor. Thus $L_v \subseteq$ Hor.

(2) The map is surjective by Definition 5.99. Let $X_1, X_2 \in L_v$ with $\psi(X_1) = \psi(X_2)$. Then

$$(v + Y) \cap X_1 = (v + Y) \cap X_2 = (v + Y) \cap (X_1 \cap X_2) \neq \emptyset.$$

Thus $X_1 \cap X_2 \in L_v =$ Hor and $Y + (X_1 \cap X_2) = V$. We also have

$$\dim(Y \cap X_1) = \dim(Y \cap X_2) = \dim(Y \cap X_1 \cap X_2).$$

Hence

$$\begin{aligned}
\dim(X_1 \cap X_2) &= \dim Y + \dim(Y \cap X_1 \cap X_2) - \dim(Y + (X_1 \cap X_2)) \\
&= \dim Y + \dim(Y \cap X_1) - \dim V \\
&= \dim Y + \dim(Y \cap X_1) - \dim(Y + X_1) \\
&= \dim X_1.
\end{aligned}$$

This implies that $X_1 = X_1 \cap X_2$. A similar argument shows that $X_2 = X_1 \cap X_2$ and hence $X_1 = X_2$. Thus ψ is injective. □

Corollary 5.106 *If Y is good, then $\psi_v : L_v \to A_v$ is a bijection for all $v \in M(\mathcal{A})$.*

Proposition 5.107 *Suppose $Y \in L$. Then Y is good if and only if Y is a modular element.*

Proof. Since $Y \in L$, we have $Y = \bar{Y}$. Suppose $X \in C_Y$. If Y is good, then by Lemma 5.101 we have $X \in$ Hor and $V = X + Y$. Thus $X \wedge Y = V = X + Y$. It follows from Lemma 2.24 that (X, Y) is a modular pair. Since this holds for all $X \in C_Y$, it follows from Lemma 2.30 that Y is a modular element.

If Y is modular, then $X + Y = X \wedge Y$ for all $X \in L$. In particular, Hor $= C_Y$ and Y is good by Lemma 5.101. □

Good Lines

In the rest of this section assume that $\dim Y = 1$.

Lemma 5.108 *For all $v \in M(\mathcal{A})$*

$$\text{Hor} = \{V\} \cup \{\mathcal{A} \setminus \mathcal{A}_Y\} = \{X \in L_v \mid r(X) \leq 1\}.$$

Proof. We have

$$X \in L_v \text{ and } r(X) \leq 1 \iff v \in X+Y \text{ and } \dim X \geq \ell-1$$
$$\iff X = V \text{ or } (Y \not\subseteq X \text{ and } \dim X = \ell-1)$$
$$\iff X = V \text{ or } X \in \mathcal{A} \setminus \mathcal{A}_Y$$
$$\iff X + Y = V$$
$$\iff X \in \text{Hor}.$$

This completes the argument. □

Lemma 5.109 *For all $v \in M(\mathcal{A})$ we have $\psi(\text{Hor}) = A_v$.*

Proof. Suppose $X \in L_v$. Then $\psi(X) = (v+Y) \cap X \in A_v$. There exists $H \in \mathcal{A}$ such that $X \subseteq H$. If $H \in \mathcal{A}_Y$, then $v \in X+Y \subseteq H$. This is a contradiction because $v \in M(\mathcal{A})$. Thus $H \in \mathcal{A} \setminus \mathcal{A}_Y$. It follows from Lemma 5.108 that $H \in \text{Hor}$ and $\psi(H) \supseteq \psi(X) \neq \emptyset$. Since $\dim Y = 1$ and $Y \not\subseteq H$, we see that $\psi(H)$ is a point. It follows that $\psi(H) = \psi(X)$. □

Proposition 5.110 *Let $C_k = \mathbb{C} \setminus \{0, 1, \ldots, (k-1)\}$ be the complex line with k integer points removed. It is known that C_k is homeomorphic to the complement of any other k points in $\mathbb{C}$, thus we may abuse notation and use C_k to denote any one of these spaces. It is also known that C_k is homeomorphic to C_m if and only if $k = m$.*

Theorem 5.111 *Let $(\mathcal{A}, V)$ be an essential central ℓ-arrangement and let Y be a subspace of V with $\dim Y = 1$. The following four conditions are equivalent:*
  (1) The map $\pi : M(\mathcal{A}) \to M(\mathcal{A}/Y)$ is a fiber bundle projection.
  (2) Each fiber of π is homeomorphic to C_k where $k = |\mathcal{A} \setminus \mathcal{A}_Y|$.
  (3) Y is good.
  (4) $\bar{Y}$ is a modular element with $r(\bar{Y}) = \ell - 1$. In particular, $Y = \bar{Y} \in L$.

Proof. (1) ⇔ (2) Since $\dim Y = 1$, if $X \in L_v \setminus \{V\}$, then $(v+Y) \cap X$ is a point. Thus the fiber $F_v = \pi^{-1}\pi(v)$ is C_m where $m = |A_v| - 1$. It follows from Proposition 5.105 and Lemma 5.108 that if $v \in M(\mathcal{A})$ and $\pi(V) \in M(\mathcal{A}/Y) \setminus B$, then

(∗) $$|A_v| = |L_v| = |\text{Hor}| = |\mathcal{A} \setminus \mathcal{A}_Y| + 1.$$

Since B is a proper Zariski closed subset of $M(\mathcal{A}/Y)$, the generic fiber is C_k where $k = |\mathcal{A} \setminus \mathcal{A}_Y|$. Proposition 5.110 completes the argument.

  (3) ⇒ (2) This follows from (∗) and Corollary 5.106.
  (2) ⇒ (3) By (2) we have $|A_v| = |\text{Hor}|$ for all $v \in M(\mathcal{A})$. It follows from Lemma 5.109 that the restriction map $\psi|_{\text{Hor}} : \text{Hor} \to A_v$ is bijective. Suppose $X \in L_v$. If $r(X) = 1$, then Lemma 5.108 implies that $X \in \text{Hor}$. If $r(X) > 1$, then there exist distinct hyperplanes $H_1, H_2 \in \mathcal{A}$ containing X. First note that $H_i \in \mathcal{A} \setminus \mathcal{A}_Y$ for $i = 1, 2$ by the argument used in the proof of Lemma 5.109 and thus $H_i \in \text{Hor}$. But then we have

$$(v+Y) \cap H_1 = (v+Y) \cap X = (v+Y) \cap H_2,$$

contradicting the bijectivity of $\psi|_{\text{Hor}}$. This shows that $r(X) = 1$ and $L_v = \text{Hor}$. It follows from Proposition 5.103 that Y is good.

(3) $\Rightarrow$ (4) Suppose $X \in C_Y$. Then Proposition 5.103 implies that $X \in \text{Hor}$ and hence $X \wedge \bar{Y} = V = X + Y \subseteq X + \bar{Y}$. Thus $X \wedge \bar{Y} = X + \bar{Y}$ and $(X, \bar{Y})$ is a modular pair. It follows from Lemma 2.30 that $\bar{Y}$ is a modular element. Now suppose that $X \in L$ is a complement of $\bar{Y}$. Then $V = X \wedge \bar{Y} = X + Y = X + \bar{Y}$ and $\{0\} = X \vee \bar{Y} = X \cap \bar{Y} \supseteq X \cap Y$. It follows that $\dim X + \dim \bar{Y} = \ell = \dim X + \dim Y$ and $\dim \bar{Y} = 1$. This shows that $Y = \bar{Y} \in L$.

(4) $\Rightarrow$ (3) Suppose $Y = \bar{Y}$ is a modular element with $r(Y) = \ell - 1$. Then $r(X \wedge Y) + r(X \vee Y) = r(X) + r(Y)$. If $X \in C_Y$, then either $X = V$ and $X \in \text{Hor}$, or $X \vee Y = \{0\}$ and hence $r(X) = 1$. It follows that $Y \not\subseteq X$ and hence $\dim(X + Y) > \dim X = \ell - 1$. Therefore $X + Y = V$ and $X \in \text{Hor}$. It follows from Proposition 5.103 that Y is good. □

Corollary 5.112 *Let $\mathcal{A}$ be an essential central ℓ-arrangement. Then $\mathcal{A}$ is strictly linearly fibered if and only if there is a modular element $Y \in L(\mathcal{A})$ with $r(Y) = \ell - 1$.*

Theorem 5.113 (Fibration) *Let $\mathcal{A}$ be an essential central arrangement. Then $\mathcal{A}$ is fiber type if and only if $\mathcal{A}$ is supersolvable.*

Proof. Suppose $\mathcal{A}$ is fiber type. Then there is a tower of fibrations

$$M(\mathcal{A}) = M_\ell \xrightarrow{\pi_{\ell-1}} M_{\ell-2} \xrightarrow{\pi_{\ell-1}} \cdots \to M_2 \xrightarrow{\pi_1} M_1 = \mathbb{C}^*$$

such that the fiber of π_k is C_{m_k}. This tower is the restriction of a tower of vector space projections

$$V = V_\ell \xrightarrow{p_{\ell-1}} V_{\ell-2} \xrightarrow{p_{\ell-1}} \cdots \to V_2 \xrightarrow{p_1} V_1 = \mathbb{C}.$$

For $1 \leq k \leq \ell - 1$, let $Y_k = \ker(p_k)$ and let $X_k = \ker(p_{\ell-1} p_{\ell-2} \ldots p_k)$. Let $X_0 = V$ and $X_\ell = \{0\}$. Then we have a chain in $L(\mathcal{A})$

(1) $\qquad V = X_0 < X_1 < \cdots < X_{\ell-1} < X_\ell = \{0\}.$

For $1 \leq k \leq \ell - 1$, we have $V_k = V_{k+1}/Y_k = V/X_k$. Define k-arrangements $(\mathcal{A}_k, V_k)$ inductively by $\mathcal{A}_\ell = \mathcal{A}$ and $\mathcal{A}_k = \mathcal{A}_{k+1}/Y_k$ for $1 \leq k \leq \ell - 1$. Then $\mathcal{A}_k$ is an essential k-arrangement and $L(\mathcal{A}_k) \simeq L(\mathcal{A}_{k+1})_{Y_k} \simeq L(\mathcal{A}_{X_k})$. It follows from Corollary 5.112 that Y_k is modular in $L(\mathcal{A}_{k+1})$. Thus X_k is modular in $L(\mathcal{A}_{k+1})$. It follows from Lemma 2.31 that $\mathcal{A}$ is supersolvable.

If $\mathcal{A}$ is supersolvable, then we have a maximal chain (1) of modular elements. Choose coordinates so $X_k = \ker(p_{\ell-1} p_{\ell-2} \ldots p_k)$ for $1 \leq k \leq \ell - 1$. Repeated application of Corollary 5.112 shows that each restriction map π_k is a strictly linear fibration. Thus $\mathcal{A}$ is fiber type. □

This result provides a topological interpretation of the algebra factorization of a supersolvable arrangement $\mathcal{A}$ of rank r and maximal chain of modular elements $V = Y_0 < Y_1 < \ldots < Y_r = T$ proved in Theorem 3.81:

$$(\mathcal{K} + B_1) \otimes \ldots \otimes (\mathcal{K} + B_r) \simeq A(\mathcal{A}).$$

Here $\mathcal{B}_i = \mathcal{A}_{Y_i} \setminus \mathcal{A}_{Y_{i-1}}$ and $B_i = \sum_{H \in \mathcal{B}_i} \mathcal{K} a_H$. It follows from Theorem 3.81 that $\mathcal{A}$ is fiber type. Let $\mathcal{K} = \mathbb{Z}$ and let $b_i = |\mathcal{B}_i|$. The fiber F_k is the complex line with b_k points removed. It follows from Theorem 5.13 that the cohomology of the complement is the tensor product of the $H^*(F_k)$. The latter is isomorphic to $\mathbb{Z} + B_k$. The naturality of the construction is a consequence of Theorem 5.90.

Proposition 5.114 *Let $\mathcal{A}$ be a real central 3-arrangement. Its complexification $\mathcal{A}_\mathbb{C}$ is fiber type if either*

(1) there is a direction so all multiple points in the affine part of $\mathbf{d}\mathcal{A}$ are contained in lines of $\mathbf{d}\mathcal{A}$ parallel to this direction, or

(2) there is a multiple point P of $\mathbf{d}\mathcal{A}$ so the pencil of lines of $\mathbf{d}\mathcal{A}$ which contains P contains all the remaining multiple points of $\mathbf{d}\mathcal{A}$, including those on the line at infinity.

These conditions are equivalent and correspond to different deconings of $\mathcal{A}$. Condition (1) is convenient to see that the complexification of $\mathcal{A}$ defined by

$$Q(\mathcal{A}) = xyz(x - z)(x + z)(y - z)(y + z)$$

is fiber type. Figure 2.8 shows $\mathbf{d}\mathcal{A}$. Both vertical and horizontal directions satisfy (1). In fact, $M(\mathcal{A}_\mathbb{C}) = C_3 \times C_3 \times C_1$. The B_3-arrangement provides a more interesting example. Figure 1.3 shows its decone. Both vertical and horizontal directions satisfy (1). The fiber is C_5 but the total space is not a product.

Condition (2) is convenient to see that the arrangement $\mathcal{B}$ defined in Example 4.59 is supersolvable. Figure 4.1 shows $\mathbf{d}\mathcal{B}$ containing two dotted lines. Let P be their intersection. These dotted lines were added precisely to satisfy condition (2). Thus $\mathcal{B}_\mathbb{C}$ is fiber type. It follows from Theorem 5.114 that $\mathcal{B}$ is supersolvable.

5.6 Related Research

Minimal Models

We start with work of Kohno [127, 129] and Falk [75], who use the theory of minimal models to establish a connection between $\pi_1(M(\mathcal{A}))$ and $H^*(M(\mathcal{A}))$. First we define the terminology and describe some general results of Sullivan [225] and Morgan [160]. Griffiths and Morgan [97] have given a detailed exposition of this work.

All vector spaces and algebras are over the rational numbers $\mathbb{Q}$. A differential graded algebra (DG algebra) A is a graded vector space $A = \oplus_{i \geq 0} A^i$ with a

degree 1 coboundary operator $d: A^i \to A^{i+1}$ and a product $\wedge : A^i \otimes A^j \to A^{i+j}$. These satisfy:
(1) $d^2 = 0$,
(2) $d(x \wedge y) = dx \wedge y + (-1)^i x \wedge dy$ for $x \in A^i$,
(3) $x \wedge y = (-1)^{ij} y \wedge x$ for $x \in A^i$ and $y \in A^j$,
(4) $\wedge$ makes A an associative algebra with unit $1 \in A^0$.

If $A^0 = \mathbb{Q}$, then A is called connected. The augmentation ideal of a connected algebra is $\mathcal{I}(A) = \oplus_{i>0} A^i$. The quotient, $I(A) = \mathcal{I}(A)/\mathcal{I}(A) \wedge \mathcal{I}(A)$, is the set of **indecomposable elements**.

Let V be a finite dimensional vector space. Let $\Lambda_r(V)$ be the graded–commutative algebra over V where $\deg v = r$ for $v \in V$. Thus $\Lambda_r(V)$ is a polynomial algebra if r is even and an exterior algebra if r is odd. Note that in this case $I(A) = V$.

Let B be a DG algebra. Let A be a DG subalgebra of B. The inclusion $A \subseteq B$ is called a **Hirsch extension** of degree r if for some V there is an isomorphism of graded–commutative algebras $B \simeq A \otimes \Lambda_r(V)$ and the differential d of B satisfies $dV \subseteq A^{r+1}$.

Let (M, d) be a DG algebra. It is called **minimal** if
(1) $M^0 = \mathbb{Q}$,
(2) there is an increasing filtration of DG subalgebras

$$\mathbb{Q} = M_0 \subseteq M_1 \subseteq M_2 \subseteq \cdots$$

such that $M = \cup M_i$ and each inclusion $M_i \subseteq M_{i+1}$ is a Hirsch extension,
(3) d is decomposable: $dM \subseteq \mathcal{I}(M) \wedge \mathcal{I}(M)$.

Let A be a DG algebra. An i–**minimal** model for A is a map $\rho: M \to A$ of DG algebras such that
(1) M is minimal,
(2) M is generated by elements of degree $\leq i$,
(3) $\rho^* : H^p(M) \to H^p(A)$ is an isomorphism for $p \leq i$ and injective for $p = i + 1$.

In case $i = \infty$, $\rho : M \to A$ is called a **minimal model** for A. It follows from the work of Sullivan [225] and Morgan [160] that if A is a connected DG algebra, then an i–minimal model for A exists for each i, and it is unique up to isomorphism.

Suppose X is a connected polyhedron. Sullivan defined the DG algebra $A = A(X)$ of $\mathbb{Q}$–polynomial forms on X. The 1–minimal model $\mathcal{M} \to A$ is an increasing union of Hirsch extensions of degree 1:

$$\mathbb{Q} = \mathcal{M}_0 \subseteq \mathcal{M}_1 \subseteq \mathcal{M}_2 \subseteq \ldots$$

$$\mathcal{M} = \cup_{n \geq 0} \mathcal{M}_n.$$

Let V_n be the degree 1 part of $\mathcal{M}_n$. The differential in $\mathcal{M}_n$ is determined by its restriction to V_n. Since $\mathcal{M}$ is minimal, we have

$$d|_{V_n} : V_n \to V_n \wedge V_n.$$

The Lie algebra $\mathcal{L}_n$ dual to $\mathcal{M}_n$ has underlying vector space V_n^*. The bracket is dual to $d|_{V_n}$
$$[\,,\,] : V_n^* \wedge V_n^* \to V_n^*.$$
The inclusions $\mathcal{M}_n \subseteq \mathcal{M}_{n+1}$ give rise to maps $\mathcal{L}_n \leftarrow \mathcal{L}_{n+1}$. Induction shows that the $\mathcal{L}_n$ are nilpotent. This constructs from the filtration of the 1–minimal model $\mathcal{M}$ a tower of nilpotent Lie algebras:
$$0 \leftarrow \mathcal{L}_1 \leftarrow \mathcal{L}_2 \leftarrow \cdots$$

Let G be a finitely presented group. The **lower central series** G_n of G is defined by setting $G_0 = G$ and $G_n = [G_{n-1}, G]$ for $n \geq 1$. Here $[G_k, G]$ denotes the subgroup generated by elements of the form $xyx^{-1}y^{-1}$ with $x \in G_k$, $y \in G$. The quotients G_{n-1}/G_n are finitely generated abelian groups. Let $\phi_n(G) = \operatorname{rank}(G_{n-1}/G_n)$. The quotients G/G_n are nilpotent groups. By a construction of Malcev, see [97, pp.142–145], it is possible to "tensor" these nilpotent groups by $\mathbb{Q}$ and use the central extensions
$$0 \to G_{n-1}/G_n \to G/G_n \to G/G_{n-1} \to 1$$
to define a Lie algebra structure on $(G/G_n) \otimes \mathbb{Q}$, called $\mathcal{L}_n(G)$. If $G = \pi_1(X)$, this leads to Sullivan's theorem; see [225], [160, 5.11], [97, p.145].

Theorem 5.115 *Let X be a connected polyhedron, let $\rho : \mathcal{M} \to A(X)$ be a 1-minimal model, and let*
$$0 \leftarrow \mathcal{L}_1 \leftarrow \mathcal{L}_2 \leftarrow \cdots$$
be the tower of dual nilpotent Lie algebras. Let $G = \pi_1(X)$. Then $\mathcal{L}_n \simeq \mathcal{L}_n(G)$ for $n \geq 0$.

Since $\mathcal{M}_n$ is a degree 1 Hirsch extension of $\mathcal{M}_{n-1}$, we have $\mathcal{M}_n \simeq \mathcal{M}_{n-1} \otimes \Lambda_1(W_n)$ for some vector space W_n. The following is a direct consequence of Theorem 5.115.

Corollary 5.116 *Let X be a connected polyhedron with finitely generated rational cohomology. Let $\rho : \mathcal{M} \to A(X)$ be a 1-minimal model and let $G = \pi_1(X)$. Then $\phi_n(G) = \dim W_n$.*

Now suppose $\mathcal{A}$ is an arrangement and $M = M(\mathcal{A})$. Since M is a formal space in the sense of Sullivan [225, p.315], we may replace the algebra $A(M)$ of $\mathbb{Q}$–polynomial forms on M with the algebra $A = A(\mathcal{A})$ defined in Section 3.2. We may view A as a DG algebra with zero differential, so $H^*(A) = A$. Let $\rho : \mathcal{M} \to A$ be a 1-minimal model.

Definition 5.117 *Call $\mathcal{A}$ a **rational $K(\pi,1)$-arrangement** if the map $\rho^* : H^*(\mathcal{M}) \to A$ is an isomorphism.*

Falk [75] and Kohno [133] used different methods to prove the following:

Theorem 5.118 *Let $\mathcal{A}$ be a rational $K(\pi,1)$-arrangement and write $\phi_n = \phi_n(\pi_1(M))$. Then*
$$\prod_{n\geq 1}(1-t^n)^{\phi_n} = \text{Poin}(M,-t).$$

This formula is called the LCS (lower central series) formula. It connects the ranks of the successive quotients in the lower central series of the fundamental group of M with the Poincaré polynomial of M. It is natural to ask for the largest class of arrangements for which the LCS formula is valid. Falk and Randell [80] showed that the LCS formula holds for fiber type arrangements. It is also known to hold for certain reflection arrangements and to be false for other reflection arrangements. See also [117, 75].

Discriminantal Arrangements

Next we describe work of Manin and Schechtman [149], who constructed a family of arrangements which generalize the braid arrangements. Let $W = \mathbb{K}^k$ and let $\mathcal{A}^0 = \{H_1^0, \ldots, H_n^0\}$ with $n > k$ be a general position arrangement in W. Let $U(n,k)$ be the set of k-arrangements $\mathcal{A} = \{H_1, \ldots, H_n\}$ in W which satisfy

(1) H_i is parallel to H_i^0 for $1 \leq i \leq n$,
(2) $\mathcal{A}$ is a general position arrangement.

Manin and Schechtman showed that $U(n,k)$ is itself the complement of an arrangement. Let $V = \mathbb{K}^n$ be the space of parallel translations of the hyperplanes of $\mathcal{A}^0$. Denote by $C(n,a)$ the set of subsets of $\{1,\ldots,n\}$ of cardinality a. For $K \in C(n,a)$ let D_K be the set of parallel translations $(H_1,\ldots,H_n) \in V$ such that $\cap_{i \in K} H_i \neq \emptyset$. If $|K| \leq k$, then $D_K = V$, and if $|K| \geq k+1$, then $\text{codim} D_K = |K| - k$. In particular, for $J \in C(n, k+1)$ the set D_J is a hyperplane in V and these hyperplanes are pairwise distinct. Let $\mathcal{B}(n,k) = \{D_J \mid J \in C(n,k+1)\}$. The arrangement $\mathcal{B}(n,1)$ is the braid arrangement. Manin and Schechtman called $\mathcal{B}(n,k)$ a **discriminantal arrangement**. In our terminology, they proved that
$$U(n,k) = M(\mathcal{B}(n,k)).$$

Although $\mathcal{B}(n,k)$ also depends on $\mathcal{A}^0$, its combinatorial properties do not. Let $L(n,k) = L(\mathcal{B}(n,k))$ and let $\ell = n - k$. It is easy to see that $L(n,k)$ has a unique maximal element $D_{(1,\ldots,n)}$ of codimension ℓ and hence $r(B(k+\ell,k)) = \ell$. In fact we may identify $D_{(1,\ldots,n)}$ with W. Assume that $H_1^0,\ldots,H_k^0$ contain the origin of W. For every $w \in W$ there is a parallel translation such that every hyperplane contains the endpoint of w. Thus for $\ell = 1$ and $\ell = 2$, the lattices $L(k+1,k)$ and $L(k+2,k)$ are easy to describe.

Manin and Schechtman gave the following description of $L(k+3,k)$. For $J = (1,\ldots,k+3) - (i,j)$, write $D_J = (i,j)$. For $K = (1,\ldots,k+3) - (i)$, write

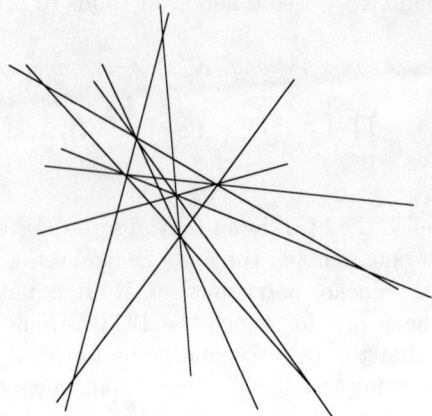

Fig. 5.16. The arrangement $C^*(5)$

$D_K = (i)$. If i, j, l, m are distinct indices, write $(ij, lm) = (i, j) \cap (l, m)$. The hyperplanes of $\mathcal{B}(k+3, k)$ are the $(k+2)(k+3)/2$ sets (i, j). The elements of $L(k+3, k)$ of rank 2 are the $(k+3)$ sets (i) and the $k(k+1)(k+2)(k+3)/8$ sets (ij, lm). Each (i) is contained in $(k+2)$ hyperplanes and each (ij, lm) is contained in two hyperplanes. Let T denote the unique maximal element of rank three. This gives:

$$\begin{aligned}
\mu(V) &= 1, \\
\mu((i,j)) &= -1, \\
\mu((i)) &= k+1, \\
\mu((ij,lm)) &= 1, \\
\mu(T) &= -(1/8)k(k+3)(k^2+3k+6) - 1.
\end{aligned}$$

Proposition 5.119 *The Poincaré polynomial of* $\mathcal{B}(k+3, k)$ *is*

$$\pi(\mathcal{B}(k+3,k), t) = 1 + (1/2)(k+2)(k+3)t + (1/8)(k+1)(k+3)(k^2+2k+8)t^2$$
$$+ ((1/8)k(k+3)(k^2+3k+6) + 1)t^3.$$

The first nontrivial cases are the arrangements $\mathcal{B}(k+3, k)$ for $k \geq 2$. It is immediate from the description above that they may be visualized as follows. Let $C^*(n)$ be an affine real 2-arrangement consisting of n points in the plane labeled (i) for $1 \leq i \leq n$ with the following properties:

(1) no three points are collinear,

(2) if $(i), (j), (l), (m)$ are distinct points, then the line through $(i), (j)$ is not parallel to the line through $(l), (m)$.

Let $C(n)$ be the central 3-arrangement obtained by embedding $C^*(n)$ as the affine set $z = 1$ and coning over the origin. (This is not the cone over $C^*(n)$ in the sense of Definition 1.15. The defining polynomial of $\mathcal{C}(n)$ is the homogenization of the defining polynomial of $\mathcal{C}^*(n)$. The difference is that here we do not add the hyperplane $\ker(x_0)$.)

It follows from the description that

$$\mathcal{B}(k+3,k) \simeq \mathcal{C}(k+3) \times \Phi_k.$$

The arrangement $\mathcal{C}^*(5)$ is illustrated in Figure 5.16. It is not known whether the arrangements $\mathcal{B}(n,k)$ are $K(\pi,1)$.

Proposition 5.120 *The discriminantal arrangements $\mathcal{B}(k+3,k)$ are not free for $k \geq 2$.*

Proof. For $k=1$ we have a braid arrangement which is free. Write $\mathcal{B}_k = \mathcal{B}(k+3,k)$. It follows from the Factorization Theorem 4.137 that it suffices to show that if there exist integers a,b such that

$$\pi(\mathcal{B}_k, t) = (1+t)(1+at)(1+bt),$$

then $k=1$. If we factor the Poincaré polynomial given in Proposition 5.119, we get $\pi(\mathcal{B}_k) = (1+t)p_k(t)$ where

$$p_k(t) = 1 + (1/2)(k^2 + 5k + 4)t + (1/8)(k^4 + 6k^3 + 15k^2 + 18k + 8)t^2.$$

The discriminant of $p_k(t)$ is

$$D(k) = -(1/4)k(k+1)(k^2 + k - 4).$$

Thus $D(1) = 1$ and $D(k) < 0$ for $k \geq 2$. □

Alexander Duality

The algebra $A(\mathcal{A})$ describes the cohomology of $M(\mathcal{A})$. Its classes are "torical" in the sense that they are products of 1–dimensional classes. The elements of the algebra $B(\mathcal{A})$ are "spherical" in the sense of Theorem 4.116. It is natural to ask whether these classes also have a topological interpretation.

Consider the unit sphere $S^{2\ell-1} \subset V$ and let $\hat{M} = S^{2\ell-1} \cap M$, $\hat{N} = S^{2\ell-1} \cap N$. Clearly, $\hat{M}$ is a strong deformation retract of M. Alexander duality [215, p.296] in the compact $(2\ell - 1)$-manifold $S^{2\ell-1}$ for the compact polyhedron $\hat{N}$ gives:

$$H_{q+1}(S^{2\ell-1}, \hat{M}) \simeq H^{2\ell-q-2}(\hat{N}).$$

Thus for $1 \leq q \leq 2\ell - 3$ we have

$$H_q(\hat{M}) \simeq H^{2\ell-q-2}(\hat{N}).$$

This gives rise to a linking pairing in $S^{2\ell-1}$, which Falk [78] interpreted as a geometric linking. He constructed an embedding of the complex $\mathsf{F}(\mathcal{A})$ in $\hat{N}$

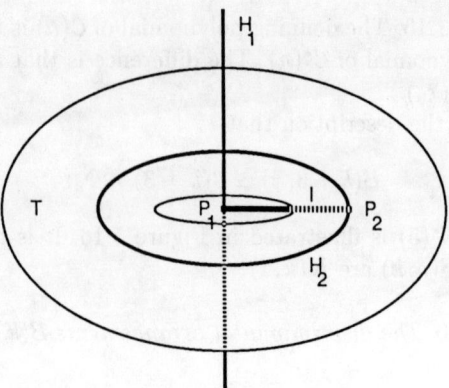

Fig. 5.17. Falk's Linking

which induces isomorphism in homology and in homotopy through dimension $(\ell - 2)$. It follows from Theorems 4.111 and 4.116 that this is a geometric representation of $\mathcal{H}_\ell(\mathcal{A}) = B_\ell(\mathcal{A})$. He also constructed embeddings of ℓ-tori in $\hat{M}$ which represent the ℓ-dimensional homology of $\hat{M}$. Then he showed that these classes link appropriately.

Theorem 5.121 *Suppose $S = (H_1, \ldots H_\ell)$ is independent, so the cycle*

$$z_S = \sum_{\pi \in Sym(\ell)} (-1)^{\ell-1}(\mathrm{sign}\pi)(H_{\pi 1}, H_{\pi 1} \cap H_{\pi 2}, \ldots, H_{\pi 1} \cap \cdots \cap H_{\pi(\ell-1)})$$

is a generator of $\tilde{H}_{\ell-2}(\hat{N})$. Let $\tau_S \in H_\ell(\hat{M})$ be the algebraic dual of $\omega_S \in H^\ell(\hat{M})$. The linking number of z_S and τ_S is ± 1.

The case $\ell = 2$ is illustrated in Figure 5.17. Let $H_i = \ker(x_i)$ for $i = 1, 2$. Stereographic projection from $(0, 1)$ sends S^3 onto R^3. The image of $\hat{N}$ consists of the vertical axis and the unit circle in the horizontal plane. The embedded torus

$$T = \{(x_1, x_2) \in S^3 \mid |x_1| = 1, |x_2| = 1\}$$

is a generator of $H_2(\hat{M}) \simeq \mathbb{Z}$. The Folkman complex consists of one point for each hyperplane, $\mathsf{F}(\mathcal{A}) = \{P_1, P_2\}$. Embed $\mathsf{F}(\mathcal{A}) \to \hat{N}$ by letting $P_i \in H_i \cap S^3$. The group $\tilde{H}_0(\hat{N}) \simeq \mathbb{Z}$ is generated by the cycle $z = (P_2 - P_1)$. The fact that T and z have linking number ± 1 follows because we can choose an embedded 1-chain I with $\partial I = z$ such that the geometric intersection $I \cap T$ is a single point.

Alexander duality in the open 2ℓ-manifold $V = \mathbb{C}^\ell = \mathbb{R}^{2\ell}$ for the closed polyhedron N and its complement M gives for $1 \leq q \leq 2\ell - 1$

$$H_q(M) \simeq H_c^{2\ell-q-1}(N)$$

where H_c denotes cohomology with compact supports. It would be interesting to give a geometric interpretation of Alexander duality for all q. Björner and Ziegler [34] give a combinatorial interpretation of Alexander duality.

The Milnor Fiber of a Generic Arrangement

Let $\mathcal{A}$ be a central complex arrangement defined by $Q = Q(\mathcal{A})$. Recall the Milnor fibration $Q : M \to \mathbb{C}^*$ and the Milnor fiber $F = Q^{-1}(1)$ from Proposition 5.2. It admits a free action by the cyclic group $G(n)$ of order $n = |\mathcal{A}|$. Let $\zeta = e^{2\pi i/n}$ be a generator of $G(n)$. The map $F \to F$ induced by multiplication by ζ is called the **monodromy** of the Milnor fiber. It follows that $B = M/\mathbb{C}^* = F/G(n)$. If we use cohomology with complex coefficients, we get $[H^k(F)]^{G(n)} = H^k(B)$. This describes the 1–eigenspace of the monodromy. The eigenspaces of the other n-th roots of unity are harder to detect in general, but we get lower bounds
$$b_k(F) \geq b_k(B).$$

Recall general position arrangements from Definition 5.19, generic arrangements from Definition 5.22, and Hattori's Theorem 5.21. We compute the cohomology groups and the monodromy of the Milnor fiber of a generic arrangement following [170].

Lemma 5.122 *Let $\mathcal{B}$ be a general position $(\ell-1)$–arrangement with $|\mathcal{B}| = n-1$ where $n > \ell \geq 3$, and let $B = M(\mathcal{B})$. Then*
(1) $\pi_1(B)$ is free abelian of rank $(n-1)$,
(2) $\pi_k(B) = 0$ for $2 \leq k \leq \ell - 2$,
(3) for $0 \leq k \leq \ell - 1$
$$b_k(B) = \binom{n-1}{k},$$
(4) the Euler characteristic of B is
$$\chi(B) = (-1)^{\ell-1}\binom{n-2}{\ell-2}.$$

Proof. We think of T^{n-1} as the $(n-1)$–dimensional hypercube with opposite faces identified. Then Hattori's subspace B_0 is obtained from T^{n-1} by removing cells in dimensions $n-1, n-2, \ldots, \ell$ corresponding to the interior of the cube and to pairs of faces of the cube. Thus B_0 has the same $(\ell-1)$–skeleton as T^{n-1}. The boundaries of the removed ℓ–cells give rise to nonvanishing homotopy classes, but they are nullhomologous. This proves parts (1), (2), and (3). Part (4) follows from Lemma 5.123 below. □

Lemma 5.123 *For $m > k$ we have*
$$\binom{m-1}{k} = \binom{m}{k} - \binom{m}{k-1} + \cdots + (-1)^k\binom{m}{0}.$$

Proof. We use induction on k. The formula holds for $k = 1$. If we assume it for $k - 1$, then it follows for k because we have

$$\binom{m}{k} = \binom{m-1}{k} + \binom{m-1}{k-1}.$$

This proves the assertion. □

Theorem 5.124 *Assume $\ell \geq 3$. Let $\mathcal{A}$ be a generic arrangement with total space $M = M(\mathcal{A})$. Let $p_M : M \to B$ be the restriction of the Hopf bundle. Let $Q : M \to \mathbb{C}^*$ be the Milnor fibration and let F be the associated Milnor fiber. Let $\xi = \exp(2\pi i/n)$. Let $h^* : H^*(F) \to H^*(F)$ be the monodromy induced by $h(z_1, \ldots, z_\ell) = (\xi z_1, \ldots, \xi z_\ell)$. Let $u = \binom{n-2}{\ell-2}$ and let $v = \binom{n-2}{\ell-1}$. Then*
(1) $\pi_1(F)$ is a free abelian group of rank $(n-1)$,
(2) $b_k(F) = b_k(B)$ for $0 \leq k \leq \ell - 2$ and hence the monodromy is trivial in this range,
(3) $b_{\ell-1}(F) = u + nv$,
(4) the characteristic polynomial of the monodromy on $H^{\ell-1}(F)$ is

$$\Delta_{\ell-1}(t) = (1-t)^u (1-t^n)^v.$$

Proof. If we think of the universal cover of T^{n-1} as $\mathbb{R}^{n-1}$ subdivided into hypercubes by the integer lattice, then the universal cover of B is a giant "Swiss cheese," since in each hypercube the same cells are removed as those removed to get B_0. Since the restriction of the Hopf map $p_F : F \to B$ is an n-fold covering, F has the homotopy type of the union of n such hypercubes with the appropriate identifications. This proves (1) and (2). Part (3) follows from (2) together with the formula for the Euler characteristic of a covering $\chi(F) = n\chi(B)$ and the calculation of $\chi(B)$ in Lemma 5.122.4.

To prove (4), we use Milnor's work [155, pp.76–77]. The Weil ζ function of the mapping h can be expressed as a product

$$\zeta(t) = \prod_{d|n} (1-t^d)^{-r_d}$$

where the exponents $-r_d$ can be computed from the formula

$$\chi_j = \sum_{d|j} d r_d.$$

Here χ_j is the Lefschetz number of the mapping h^j, the j-fold iterate of h. Milnor showed that χ_j is the Euler characteristic of the fixed point manifold of h^j. Since h^j has no fixed points for $1 \leq j < n$ and $\chi(F) = n\chi(B)$, we conclude that

(i) $$\zeta(t) = (1-t^n)^{-\chi(B)}.$$

The zeta function can be expressed as an alternating product of polynomials

(ii) $$\zeta(t) = \Delta_0(t)^{-1}\Delta_1(t)\Delta_2(t)^{-1}\cdots\Delta_{\ell-1}(t)^{\pm 1}$$

where $\Delta_k(t)$ is the characteristic polynomial of the monodromy on $H^k(F)$. Part (4) now follows from equations (i), (ii), and the fact that $\Delta_k(t) = (1-t)^{b_k(F)}$ for $0 \leq k \leq \ell - 2$, which is a consequence of (2). □

Proposition 5.125 *If $\ell = 2$, then $\pi_1(F)$ is a free group of rank $(n-1)^2$. Conclusions (2)–(4) of Theorem 5.124 remain valid.*

Proof. A central 2-arrangement is always generic. In this case Q has an isolated singularity at the origin. Thus $b_0(F) = 1$ and $b_1(F) = (n-1)^2$. In fact, F has the homotopy type of a wedge of $(n-1)^2$ circles. In this case B is the complex line with $(n-1)$ points removed. Thus $b_0(B) = 1$ and $b_1(B) = n-1$. This agrees with assertions (2) and (3). The characteristic polynomial of the monodromy on $H^1(F)$ may be computed using the divisor formula in [156]:

$$\delta(h) = (nE_n - 1)^2 = n(n-2)E_n + 1 = (n-2)\Lambda_n + 1.$$

Thus $\Delta_1(t) = (1-t)(1-t^n)^{n-2}$, which agrees with (4). □

Remark. It is shown in [11] that the complexification of the D_3 arrangement defined by $Q = (x-y)(x+y)(x-z)(x+z)(y-z)(y+z)$ has $b_1(F) = 7$ while $b_1(B) = 5$. Thus the Milnor fiber is a more subtle invariant of a nongeneric arrangement.

Arrangements of Subspaces

Goresky and MacPherson [91] considered real arrangements of affine subspaces of possibly various dimensions as an example of their general results in stratified Morse theory. We call these **subspace** arrangements. In [91, pp.237-244] they computed the groups $H_*(M(\mathcal{A}); \mathbb{Z})$ using the order complex of Definition 4.93. They partially order $L(\mathcal{A})$ by inclusion. Our statement of their theorem is adjusted to agree with our conventions. Here complex arrangements are also considered as real arrangements.

Definition 5.126 *Let $(\mathcal{A}, V)$ be a real arrangement of subspaces. Let $L = L(\mathcal{A})$ be the set of all intersections partially ordered by reverse inclusion with rank function $r(X) = \text{codim} X$. Given $X, Y \in L$ with $X < Y$, recall the segment $[X, Y] = \{Z \in L \mid X \leq Z < Y\}$ from Definition 2.10. Define the segment*

$$(X, Y) = \{Z \in L \mid X < Z < Y\}.$$

Write $\mathsf{K}_* = \mathsf{K}(*)$ for the corresponding order complexes.

Table 5.1. A subspace arrangement

X	$r(X)$	$\mathsf{K}_{[V,X)}$	$\mathsf{K}_{(V,X)}$	b_0	b_1	b_2
V	0	$\emptyset$	$\emptyset$	1		
A_1	2	D^0	$\emptyset$		1	
A_2	3	D^0	$\emptyset$			1
A_3	3	D^0	$\emptyset$			1
B_1	4	D^1	∂D^1			1
B_2	4	D^1	∂D^1			1
O	5	D^2	C			

Thus if $\mathcal{A}$ is a complex arrangement, then the definition of the poset $L(\mathcal{A})$ is the same as in Definition 2.1, but the rank function in Definition 2.2 is complex codimension, while here it is real codimension.

Theorem 5.127 *The homology of the complement $M = M(\mathcal{A})$ is given by*

$$H_i(M;\mathbb{Z}) = \bigoplus_{X \in L} H^{r(X)-i-1}(\mathsf{K}_{[V,X)}, \mathsf{K}_{(V,X)}; \mathbb{Z})$$

with the convention that $H^{-1}(\emptyset,\emptyset) = \mathbb{Z}$. Thus V contributes a copy of $\mathbb{Z}$ to $H_0(M)$.

Example 5.128 Let $\mathcal{A}$ consist of a 3–plane and two 2–planes in $\mathbb{R}^5$:

$$A_1: \quad x_1 = x_5 = 0$$
$$A_2: \quad x_1 = x_2 = x_3 = 0$$
$$A_3: \quad x_3 = x_4 = x_5 = 0.$$

These subspaces are not in general position. In addition to $V = \mathbb{R}^5$ and the elements of $\mathcal{A}$, the intersection poset contains the subspaces:

$$B_1: \quad x_1 = x_2 = x_3 = x_5 = 0$$
$$B_2: \quad x_1 = x_3 = x_4 = x_5 = 0$$
$$O: \quad x_1 = x_2 = x_3 = x_4 = x_5 = 0.$$

Table 5.1 indicates their contributions. Here D^k is a k-disk with boundary ∂D^k and C is an arc in ∂D^2. It follows that $\mathrm{Poin}(M(\mathcal{A}),t) = 1 + t + 4t^2$.

It is natural to ask which properties of hyperplane arrangements hold for subspace arrangements. The calculation of the fundamental group of the complement of a subspace arrangement has not been carried out. Suppose $A \in \mathcal{A}$ has codimension d. We may form $\mathcal{A}'$ and $\mathcal{A}''$ as before. Let $M' = M(\mathcal{A}')$ and $M'' = M(\mathcal{A}'')$. Then M'' has a tubular neighborhood in M' which is a trivial

d–plane bundle. The Thom isomorphism arguments of Section 5.4 hold and we get the long exact sequence

$$\cdots \to H^k(M') \xrightarrow{i^*} H^k(M) \xrightarrow{\phi} H^{k+1-d}(M'') \xrightarrow{\psi} H^{k+1}(M') \to \cdots$$

We showed in Theorem 5.87 that for arrangements of complex hyperplanes this sequence splits into short exact sequences. A similar splitting for arbitrary subspace arrangements would result in the formula

(1) $$\text{Poin}(M,t) = \text{Poin}(M',t) + t^{d-1}\text{Poin}(M'',t).$$

Example 5.128 illustrates that not all choices of $A \in \mathcal{A}$ result in a similar splitting. The choice of $A_1 \in \mathcal{A}$ as the distinguished element gives $d = 3$, $\text{Poin}(M',t) = 1 + 2t^2 + t^3$, and $\text{Poin}(M'',t) = 1 + 3t$. Thus (1) does not hold in this case. It is interesting to note that (1) holds with either A_2 or A_3 as distinguished element.

6. Reflection Arrangements

In this chapter we study the equivariant theory of arrangements. Let $GL(V)$ denote the general linear group of V. Suppose $G \subseteq GL(V)$ is a finite linear group and $\mathcal{A}$ is an arrangement which is stable under the action of G. *We assume that the order of G is not divisible by the characteristic of the field $\mathbb{K}$.* Section 6.1 contains definitions, examples, and generalities about the action of G on the polynomial algebra S and on the S–modules $\operatorname{Der}_{\mathbb{K}}(S)$ and $\Omega[V]$. We also study the action of G on the algebra $A(\mathcal{A})$ and on the cohomology ring $H^*(M)$ in this section.

There is a collection of linear groups and associated arrangements with particularly nice properties. The groups G are generated by reflections and the arrangements $\mathcal{A}(G)$ are the fixed hyperplanes of the reflections, called reflection arrangements. These are the objects we study in the rest of the chapter. In Section 6.2 we discuss some basic results. Chevalley's theorem 6.19 asserts that finite groups generated by reflections are distinguished by the fact that their algebra of G–invariant polynomials is a polynomial algebra. This result is the key to their nice properties. Irreducible complex reflection groups were classified by Shephard and Todd [210]. It is customary to call a proof "case free" if it does not depend on this classification.

In Section 6.3 we prove that every reflection arrangement $\mathcal{A}(G)$ is free. This was first proved for Coxeter groups independently by V. I. Arnold [8, 10] and by K. Saito [199, 201]. It was shown for complex reflection arrangements in [227] and for general reflection arrangements in [235]. The argument here is case free. It proves existence of a basis for $D(\mathcal{A}(G))$, but it does not construct a basis for $D(\mathcal{A}(G))$. It does not even determine the exponents of $\mathcal{A}(G)$. We show that the exponents of $\mathcal{A}(G)$ have an interpretation in the invariant theory of G, which allows determination of the exponents of $\mathcal{A}(G)$ using character tables for G [172]. The results are listed in Appendix B. At this stage, the situation is similar to Example 4.54. Construction of a basis for $D(\mathcal{A}(G))$ involves additional work [166, 237]. We also present these results in Appendix B.

The group G acts by permutation on the poset $L(\mathcal{A})$. We discuss this action in Section 6.4. A complete set of orbit types and related information is presented for the irreducible groups in Appendix C. We study the structure of $L(\mathcal{A})$ and determine for all $X \in L(\mathcal{A})$ the structure of the restriction $L(\mathcal{A}^X)$. A study of the restrictions $L(\mathcal{A}^X)$ shows that if $\mathcal{A} = \mathcal{A}(G)$ is a reflection arrangement and $X \in \mathcal{A}$ with $p = \dim X$, then there exist integers $\{b_1^X, \ldots, b_p^X\}$ such that

$$\pi(\mathcal{A}^X, t) = (1 + b_1^X) \cdots (1 + b_p^X).$$

This observation is the source of Conjecture 6.90. The values of b_i^X are tabulated in Appendix C. In Section 6.5 we consider restrictions of reflection arrangements. If Conjecture 6.90 is true, then these restrictions are free. The only known general result, proved in [180], asserts that if $\mathcal{A}$ is a Coxeter arrangement and $H \in \mathcal{A}$, then $\mathcal{A}^H$ is free. If $\exp \mathcal{A} = \{m_1, \ldots, m_{\ell-1}, m_\ell\}$ where $m_1 \leq \ldots \leq m_\ell$, then $\exp \mathcal{A}^H = \{m_1, \ldots, m_{\ell-1}\}$. We prove this in Section 6.5. Unfortunately, this does not settle Conjecture 6.90 for Coxeter arrangements $\mathcal{A}$ because $\mathcal{A}^H$ is not in general a Coxeter arrangement. In Section 6.6 we consider the orbit space V/G. It follows from Chevalley's theorem 6.19 that V/G has the structure of affine space. The image N/G of the variety of $\mathcal{A}$ is the discriminant locus. We study the stratification of V/G induced by the stratification of V in Definition 5.16. In Section 6.6 we also define a subclass of complex reflection groups called Shephard groups. We follow [177, 178] and associate to each Shephard group G an irreducible Coxeter group W. We show that, with suitable choices, the discriminant loci of G and W are equal. It follows from Deligne's theorem 5.15 that the complement of the discriminant of W is a $K(\pi, 1)$ space. This identification proves that the complement of the discriminant of G is a $K(\pi, 1)$ space. It follows that $\mathcal{A}(G)$ is a $K(\pi, 1)$-arrangement for all Shephard groups.

6.1 Equivariant Theory

The Action of G

Definition 6.1 *Let $G \subset GL(V)$ be any finite group. Let $\langle x, v \rangle = x(v)$ denote the usual pairing $V^* \times V \to \mathbb{K}$. The **contragradient action** of G on V^* is defined by $(gx)(v) = x(g^{-1}v)$. Then we have*

$$\langle gx, gv \rangle = (gx)(gv) = x(v) = \langle x, v \rangle.$$

If $\mathbb{K} = \mathbb{R}$, then we may choose a G–invariant positive definite quadratic form in V. Thus the elements of G are orthogonal transformations in V. If $\mathbb{K} = \mathbb{C}$, then we may choose a G–invariant positive definite Hermitian form in V. Thus the elements of G are unitary transformations in V. We may also choose a G–invariant positive definite orthogonal or Hermitian form in the dual space V^*.

Definition 6.2 *If $g \in G$, let $\mathrm{Fix}(g) = \{v \in V \mid gv = v\}$ be the fixed subspace of g. If $v \in V$, let $G_v = \{g \in G \mid gv = v\}$ be the **fixer** of v. If U is a subspace of V, let $G_U = \cap_{v \in U} G_v$. Note that $G_U = \{g \in G \mid U \subseteq \mathrm{Fix}(g)\}$.*

Recall the S-module of derivations $\mathrm{Der}_S = \mathrm{Der}_\mathbb{K}(S)$ from Definition 1.18 and the S-module of differential 1-forms $\Omega_S = \Omega^1[V]$ from Section 4.6. If

$v \in V$, let $D_v \in \mathrm{Der}_S$ be the derivation defined by $D_v(x) = \langle x, v \rangle$ for $x \in V^*$. The spaces S, Der_S, and Ω_S have G–module structures.

Definition 6.3 *Let $g \in G$, $v \in V$, $a \in S$, $\theta \in \mathrm{Der}_S$, and $\omega \in \Omega_S$. Define the G–module structure*
(1) in S by $(ga)(v) = a(g^{-1}v)$,
(2) in Der_S by $(g\theta)(a) = g(\theta(g^{-1}a))$,
(3) in Ω_S by $(g\omega)(\theta) = g(\omega(g^{-1}\theta))$.

Proposition 6.4 *The following transformation formulas hold:*
(1) $g(D_v a) = D_{gv}(ga)$,
(2) $d(ga) = g(da)$,
(3) $g(D_v) = D_{gv}$,
(4) $g(a\theta) = (ga)(g\theta)$,
(5) $g(a\omega) = (ga)(g\omega)$,
(6) $\langle g\omega, g\theta \rangle = g\langle \omega, \theta \rangle$.

Proof. (1) Since D_v is a derivation, we may take $a = x$ to get
$$g(D_v x) = g(\langle x, v \rangle) = \langle x, v \rangle = \langle gx, gv \rangle = D_{gv}(gx).$$

(2) Recall from Lemma 4.73 that $df(\theta) = \langle \theta, df \rangle = \theta(f)$. We use Definition 6.3 twice.
$$(gda)(D_v) = g[da(g^{-1}D_v)] = g[(g^{-1}D_v)(a)] =$$
$$g[g^{-1}(D_v(ga))] = D_v(ga) = d(ga)(D_v).$$

For (3) we use Definition 6.3.2 and (1):
$$(gD_v)(a) = g[D_v(g^{-1}a)] = D_{gv}(a).$$

In (4), let $s \in S$. Then
$$[g(a\theta)](s) = g[(a\theta)(g^{-1}s)] = g[a\theta(g^{-1}s)] = (ga)g[\theta(g^{-1}s)] = (ga)(g\theta)(s).$$

The argument for (5) is similar. Assertion (6) follows from Definition 6.3.3. □

Definition 6.5 *Let $R = S^G$ be the algebra of G–**invariant polynomials**. Let Der_S^G be the R–module of G–**invariant derivations** and let Ω_S^G be the R–module of G–**invariant differential forms**.*

Recall the grading of S from Section 4.1, the two gradings of Der_S from Definitions 4.2 and 4.3, and the two gradings of Ω_S from Definitions 4.62 and 4.63. These induce a grading in R, and two gradings in Der_S^G and in Ω_S^G, one by polynomial degree and one by total degree.

Matrices

Let $M_\ell(S)$ denote the set of $\ell \times \ell$ matrices with polynomial entries. We give $M_\ell(S)$ the structure of a G-module as follows:

$$(g\mathsf{P})_{i,j} = g(\mathsf{P})_{i,j} \quad \mathsf{P} \in M_\ell(S).$$

One example is the coefficient matrix $\mathsf{M}(\theta_1, \ldots, \theta_\ell)$ of $\theta_1, \ldots, \theta_\ell \in \mathrm{Der}_S$ from Definition 4.11. Two more examples are important here.

Definition 6.6 *Given $f_1, \ldots, f_\ell \in S$, we define their **Jacobian matrix** $\mathsf{J}(f_1, \ldots, f_\ell)$ by*

$$\mathsf{J}(f_1, \ldots, f_\ell)_{i,j} = D_i f_j.$$

Definition 6.7 *If $f \in S$, define $\mathrm{Hess}(f) : \mathrm{Der}_S \to \Omega_S$ by*

$$\mathrm{Hess}(f)\theta = \sum_{i=1}^{\ell} \theta(D_i f) dx_i \quad \theta \in \mathrm{Der}_S.$$

Let $\mathsf{H}(f)$ denote the matrix of $\mathrm{Hess}(f)$ with respect to the pair of bases $\{D_i\}$ and $\{dx_i\}$. It is the usual Hessian matrix of second partial derivatives of f.

Lemma 6.8 *If $g \in G$, let $[g]$ denote the matrix for g in the basis $\{e_i\}$ of V so $[g]_{i,j} = \langle x_j, g e_i \rangle$. Let $[g^*] \in GL(V^*)$ be the matrix of the contragredient action of G on V^*. Let $[\ ]^T$ denote the transpose.*
(1) $[g^*] = [(g)^{-1}]^T$.
(2) If $\theta_1, \ldots, \theta_\ell \in \mathrm{Der}_S^G$ and $\mathsf{M} = \mathsf{M}(\theta_1, \ldots, \theta_\ell)$, then $g\mathsf{M} = [g^{-1}]\mathsf{M}$.
(3) If $f_1, \ldots, f_\ell \in R$ and $\mathsf{J} = \mathsf{J}(f_1, \ldots, f_\ell)$, then $g\mathsf{J} = [g]^T \mathsf{J}$.
(4) If $f \in R$, then $g\mathsf{H}(f) = [g]^T \mathsf{H}(f)[g]$.

Proof. We show (1).

$$[g^*]_{ij} = \langle gx_i, e_j \rangle = \langle x_i, g^{-1} e_j \rangle = [g^{-1}]_{ji}.$$

The other assertions follow from Proposition 6.4. $\square$

Lemma 6.9 *If $f \in R$, then $\mathrm{Hess}(f) : \mathrm{Der}_S \to \Omega_S$ is a G-module homomorphism and thus induces a map $\mathrm{Hess}(f) : \mathrm{Der}_S^G \to \Omega_S^G$.*

Proof. Let $g \in G$ and let $f \in S$ be any polynomial. Write $h = \mathrm{Hess}(f)$. Then

$$hD_v = \sum D_v(D_i f) dx_i = \sum D_i(D_v f) dx_i = d(D_v f).$$

Using this and Proposition 6.4, we get

$$\begin{aligned} g(hD_v) &= d(g(D_v f)) = d[D_{gv}(gf)], \\ h(gD_v) &= h(D_{gv}) = d(D_{gv} f). \end{aligned}$$

Thus if $f \in R$, then $g(hD_v) = h(gD_v)$. Let $a \in S$. Since h is an S–module map, we have

$$\begin{aligned} g(h(aD_v)) &= g(a(hD_v)) = (ga)(g[hD_v]) = (ga)(h[gD_v]) \\ &= h[(ga)(gD_v)] = h(g(aD_v)). \end{aligned}$$

The result follows because $\text{Der}_S = \sum SD_i$. □

Corollary 6.10 *If $f \in R$ is a homogeneous polynomial of degree p, then* $\text{Hess}(f) : \text{Der}_S^G \to \Omega_S^G$ *is a graded R–module map. If Der_S^G and Ω_S^G are graded by polynomial degree, then $\text{Hess}(f)$ has degree $p - 2$. If the modules are graded by total degree, then the map has degree p.*

Character Formulas

Next suppose $\mathcal{A}$ is an arrangement in V which is stable under the action of G. Then G operates on $\mathcal{A}$ by permutation and the induced action on $L(\mathcal{A})$ is order preserving. This induces a G–action on the algebras $A(\mathcal{A})$, $B(\mathcal{A})$, $R(\mathcal{A})$, and on the Folkman complex $\mathsf{F}(\mathcal{A})$. We compute the character of the representation of G on $A(\mathcal{A})$ following [171]. The proofs of Theorem 3.110 and Theorem 4.116 yield the following results.

Proposition 6.11 *The map $\theta : A(\mathcal{A}) \to B(\mathcal{A})$ is a G–isomorphism of graded $\mathcal{K}$–algebras.*

Proposition 6.12 *Let $\ell \geq 2$. Then the map $\tau : B_T \to \tilde{H}_{\ell-2}(\mathsf{F};\mathcal{K})$ is a G–isomorphism.*

Let $\mathbb{K} = \mathcal{K} = \mathbb{C}$. Let $R(G)$ be the representation ring of G. If M is a G–module, we let $[M]$ denote its image in $R(G)$. If H is a subgroup of G and N is an H–module, we let Ind_H^G denote the induced G–module and write $\text{Ind}_H^G[N] = [\text{Ind}_H^G N]$. For simplicity of notation, we write $[M] = [G/H]$ if $M = \mathbb{C}[G/H]$. For $\ell \geq 2$, let $\mathsf{F}_0(\mathcal{A})$ be the augmented Folkman complex obtained from $\mathsf{F}(\mathcal{A})$ by adjoining a simplex of dimension -1 on which G acts trivially. Then the reduced homology $\tilde{H}(\mathsf{F})$ of F is the homology $H(\mathsf{F}_0)$ of F_0. If $X \in L$, let $S_X = \{g \in G \mid gX = X\}$ be the **stabilizer** of X. Note that in general this is a larger group than the fixer G_X of Definition 6.2. If σ is a simplex of F_0, let $d(\sigma)$ denote its dimension, let $S_\sigma = \{g \in G \mid g\sigma = \sigma\}$, and let $S_{X,\sigma} = S_X \cap S_\sigma$.

Theorem 6.13 *Let $\mathcal{A}$ be an arrangement which is stable under the action of the finite linear group G. Then*

$$\begin{aligned} [A_0] &= [\mathbb{C}], \\ [A_1] &= \sum_{X \in L_1} |G : S_X|^{-1} [G/S_X], \end{aligned}$$

and for $2 \leq p \leq \ell$

$$[A_p] = (-1)^p \sum_{X \in L_p} \sum_{\sigma \in \mathsf{F}_0(\mathcal{A}_X)} (-1)^{d(\sigma)} |G : S_{X,\sigma}|^{-1} [G/S_{X,\sigma}].$$

Proof. The assertion is clear for A_0 and A_1, since $A_1 = E_1$ is the G-module defined by permuting the hyperplanes. This proves the theorem for ℓ-arrangements with $\ell \leq 1$. We may assume that $2 \leq p \leq \ell$. The Hopf trace formula says

$$\sum_{q=0}^{\ell-2} (-1)^q [H_q(\mathsf{F})] = \sum_{q=0}^{\ell-2} (-1)^q [C_q(\mathsf{F})]$$

where $C_q(\mathsf{F})$ is the group of q-chains of F. From Theorem 4.106 and Proposition 6.12, we get

$$\sum_{q=0}^{\ell-2} (-1)^q [H_q(\mathsf{F})] = [\mathbb{C}] + (-1)^\ell [B_T].$$

Since the action of G is order preserving in $L(\mathcal{A})$, G preserves the orientation of each simplex $\sigma \in \mathsf{F}$. Thus the modules $\mathbb{C}[G\sigma]$ and $\mathbb{C}[G/S_\sigma]$ are isomorphic. This gives

$$\sum_{q=0}^{\ell-2} (-1)^q [C_q(\mathsf{F})] = \sum_{\sigma \in \mathsf{F}} (-1)^{d(\sigma)} |G : S_\sigma|^{-1} [G/S_\sigma].$$

We apply the Hopf trace formula to get

$$[\mathbb{C}] + (-1)^\ell [B_T] = \sum_{\sigma \in \mathsf{F}} (-1)^{d(\sigma)} |G : S_\sigma|^{-1} [G/S_\sigma].$$

Replace F by F_0. This absorbs the term $[\mathbb{C}]$ on the right side, leaving

$$(1) \qquad (-1)^\ell [B_T] = \sum_{\sigma \in \mathsf{F}_0} (-1)^{d(\sigma)} |G : S_\sigma|^{-1} [G/S_\sigma].$$

If $X \in L$ and $r(X) \geq 2$, then we may apply this formula to $\mathcal{A}_X$ and the group S_X to conclude that

$$(-1)^{r(X)} [B_X(\mathcal{A}_X)] = \sum_{\sigma \in \mathsf{F}_0(\mathcal{A}_X)} (-1)^{d(\sigma)} |S_X : S_{X,\sigma}|^{-1} [S_X/S_{X,\sigma}]$$

in the representation ring $R(S_X)$. Let $\mathcal{O}$ be a G-orbit on L_p. Let $B_\mathcal{O} = \oplus_{X \in \mathcal{O}} B_X$. Thus $B_p = \oplus B_\mathcal{O}$, sum over all orbits. Since $gB_X(\mathcal{A}) = B_{gX}(\mathcal{A})$, we have $B_\mathcal{O} \simeq \mathrm{Ind}_{S_X}^G B_X(\mathcal{A})$ for any fixed $X \in \mathcal{O}$. Since $B_X(\mathcal{A}) \simeq B_X(\mathcal{A}_X)$ as S_X-modules, we have $[B_\mathcal{O}] = \mathrm{Ind}_{S_X}^G [B_X(\mathcal{A}_X)]$. It follows from the last formula and transitivity of induction that

$$[B_p] = (-1)^p \sum_{X \in L_p} \sum_{\sigma \in \mathsf{F}_0(\mathcal{A}_X)} (-1)^{d(\sigma)} |G : S_{X,\sigma}|^{-1} [G/S_{X,\sigma}].$$

The assertion of the theorem follows from Proposition 6.11. □

By choosing representatives for the orbits, we may write $[A_p]$ as a $\mathbb{Z}$-linear combination of certain $[G/S_{X,\sigma}]$ and thereby put the formulas of Theorem 6.13 in a form more suitable for calculation. Let T_p be a set of representatives for the G-orbits on L_p. For each $X \in T_p$ let U_X be the set of representatives for G-orbits on the set of simplexes of $\mathsf{F}_0(\mathcal{A}_X)$. Then $[A_1] = \sum_{X \in T_1}[G/S_X]$ and for $2 \leq p \leq \ell$

(2) $$[A_p] = (-1)^p \sum_{X \in T_p} \sum_{\sigma \in U_X} (-1)^{d(\sigma)}[G/S_{X,\sigma}].$$

Theorem 6.14 *Let G be a finite subgroup of $GL(V)$ and let $\mathcal{A}$ be an arrangement in V which is stable under G. Let $L = L(\mathcal{A})$. If $g \in G$, let $L^g = \{X \in L \mid gX = X\}$ and let μ_g be the Möbius function of the poset L^g. Then*

$$\sum_{p=0}^{\ell} \mathrm{tr}(g|A_p)t^p = \sum_{X \in L^g} \mu_g(X)(-t)^{r(X)}.$$

Proof. The statement is clear for $p = 0, 1$. We assume that $p \geq 2$ and compute $\mathrm{tr}(g|B_p)$. Since $gB_X(\mathcal{A}) = B_{gX}(\mathcal{A})$, the only terms which contribute to the trace have $gX \subseteq X$. Let $L_p^g = L^g \cap L_p$. Then

$$\mathrm{tr}(g|B_p) = \sum_{X \in L_p^g} \mathrm{tr}(g|B_X(\mathcal{A})) = \sum_{X \in L_p^g} \mathrm{tr}(g|B_X(\mathcal{A}_X)).$$

Choose $X \in L_p^g$. Let $\mathcal{O}_1, \ldots, \mathcal{O}_s$ be the orbits of S_X on F_0 and choose $\sigma_j \in \mathcal{O}_j$. Let $M_j = \mathbb{C}[\mathcal{O}_j] \simeq \mathbb{C}[S_X/S_{X,\sigma_j}]$. Now (1) gives

$$(-1)^p \mathrm{tr}(g|B_X(\mathcal{A}_X)) = \sum_{j=1}^{s}(-1)^{d(\sigma_j)}\mathrm{tr}(g|M_j)$$
$$= \sum_{j=1}^{s}(-1)^{d(\sigma_j)}|\mathcal{O}_j^g|$$
$$= -1 + e(\mathsf{F}(\mathcal{A}_X)^g)$$

where e denotes Euler characteristic. If L is any finite poset with minimal element V and maximal element T, then Rota [196, Cor. 2] has shown that $e(\mathsf{F}) = 1 + \mu(T)$. Since $\mathsf{F}(\mathcal{A}_X)^g = \mathsf{F}(L_X^g)$, this gives $(-1)^p \mathrm{tr}(g|B_X(\mathcal{A}_X)) = \mu_g(X)$. □

Remark. Note that L^g is a lattice which contains V and T. Joins of elements of L^g are the same as in L, but L^g need not satisfy the chain condition and the elements of L^g need not be joins of atoms of L^g.

Example 6.15 Let $\mathcal{A}$ be the braid arrangement on 4 strands and let $G = Sym(4)$ be the symmetric group on 4 letters. It follows from Proposition 2.9 that we may identify $L(\mathcal{A})$ with the partition lattice of the set $\{1, 2, 3, 4\}$, shown in Figure 6.1. Choose representatives $g = (1), (12), (123), (12)(34), (1234)$ for the conjugacy classes of G. We give the Hasse diagrams of L^g for $g \neq (1)$ in Figure 6.2.

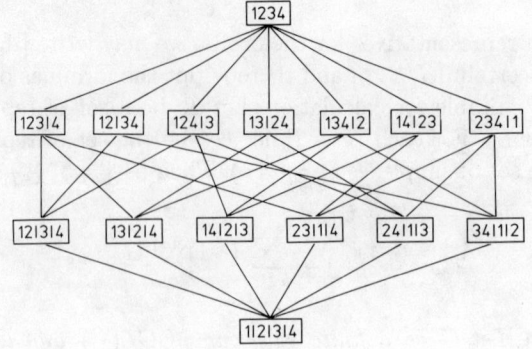

Fig. 6.1. The lattice $L = L^{(1)}$

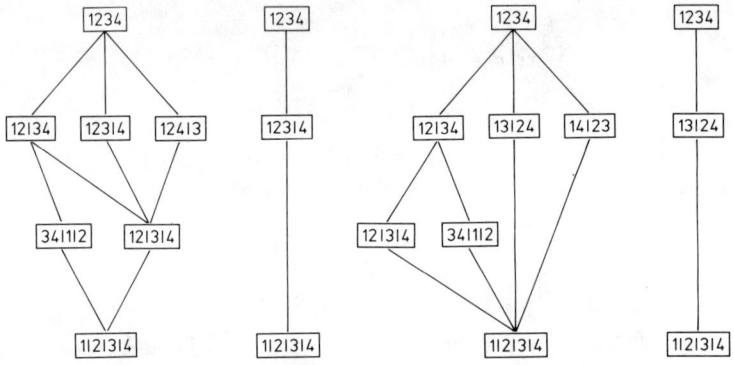

Fig. 6.2. The lattices $L^{(12)}$, $L^{(123)}$, $L^{(12)(34)}$, $L^{(1234)}$

Topological Interpretation

Since the isomorphism in Theorem 5.90 is G equivariant, it follows that Theorem 6.14 has a topological interpretation.

Corollary 6.16 *Let G be a finite subgroup of $GL(V)$ and let $\mathcal{A}$ be an arrangement in V which is stable under G. Let $M = M(\mathcal{A})$ and let* $\mathrm{Poin}(M,g,t) = \sum_{p \geq 0} \mathrm{tr}(g|H^p(M,\mathbb{C}))t^p$. *Then*

$$\mathrm{Poin}(M,g,t) = \sum_{X \in L^g} \mu_g(X)(-t)^{r(X)}.$$

When we compute the values of $\mu_g(X)$ and use Corollary 6.16 to compute $\mathrm{Poin}(M,g,t)$, we find the following remarkable factorizations. The first line of

Table 6.1. Poincaré polynomials

g	$\text{Poin}(M,g,t)$
(1)	$(1+t)(1+2t)(1+3t)$
(12)	$(1+t)(1+t)$
(123)	$(1-t)(1+t)$
(12)(34)	$(1-t)(1+t)(1+2t)$
(1234)	$(1-t)(1+t)$

Table 6.1 is a special case of Arnold's theorem [6] quoted in the introduction; see Proposition 2.54.

Corollary 6.17 *Let M/G be the orbit space of the action of G on M. Let $b_i = b_i(M/G)$ be its Betti numbers. We have $b_0 = 1$, $b_1 = |T_1|$, and for $2 \leq p \leq \ell$*

$$b_p = (-1)^p \sum_{X \in T_p} \sum_{\sigma \in U_X} (-1)^{d(\sigma)},$$

$$\text{Poin}(M/G, t) = \frac{1}{|G|} \sum_{g \in G} \sum_{X \in L^g} \mu_g(X)(-t)^{r(X)}.$$

Proof. Since each induced module $\mathbb{C}[G/S_{X,\sigma}]$ contains the trivial module with multiplicity 1, the Betti numbers follow from (2). The Poincaré polynomial formula is a consequence of Corollary 6.16. □

6.2 Reflection Arrangements

Basic Properties

Definition 6.18 *An element $s \in GL(V)$ is a **reflection** if it has finite order and its fixed point set is a hyperplane H_s. We call H_s the **reflecting hyperplane** of s. A finite subgroup $G \subset GL(V)$ is called a **reflection group** if it is generated by reflections. When $\mathbb{K} = \mathbb{R}$, the group G is called a **Coxeter group**.*

If $G \subset GL(V)$ is a reflection group, then Chevalley's theorem [49] describes $R = S^G$.

Theorem 6.19 *Let $G \subset GL(V)$ be a finite reflection group.*
 (1) There exist homogeneous polynomials $f_1, \ldots, f_\ell \in R$ such that $R = \mathbb{K}[f_1, \ldots, f_\ell]$.

(2) There exists a finite-dimensional G-stable graded vector subspace U of S such that $S \simeq U \otimes_{\mathbb{K}} R$.
(3) The graded vector space U affords the regular representation of G.

Definition 6.20 The invariant algebra R is both a subalgebra of S and a polynomial algebra without reference to S. A polynomial in the invariants may be viewed both ways. In order to eliminate this ambiguity, we introduce indeterminates $T_1, \ldots, T_\ell$ and sometimes view $R = \mathbb{K}[T_1, \ldots, T_\ell]$. The embedding of R into S is obtained by the substitution $T_i = f_i$ for $1 \leq i \leq \ell$.

Definition 6.21 The reflection group G is **reducible** if $V = V_1 \oplus V_2$ where V_i is stable under G. The restriction G_i of G to V_i is a reflection group in V_i. The group is called **irreducible** if it is not reducible.

Definition 6.22 A set $\mathcal{F} = \{f_1, \ldots, f_\ell\}$ which satisfies Theorem 6.19 is called a set of **basic invariants** for G. The polynomials f_i are not unique, but their degrees $d_i = \deg f_i$ are determined uniquely by G, and are called the basic degrees. The integers $m_i = d_i - 1$ are the **exponents** of G. It is customary to label them in increasing order:

$$m_1 \leq \ldots \leq m_\ell.$$

The following result is called the duality of the exponents of irreducible Coxeter groups [38, p.118].

Theorem 6.23 If G is an irreducible Coxeter group, then $m_i + m_{\ell-i+1}$ is independent of i.

Definition 6.24 Let $G \subset GL(V)$ be a finite reflection group. The set $\mathcal{A} = \mathcal{A}(G)$ of reflecting hyperplanes of G is called the **reflection arrangement** of G. In particular, when the group G is a Coxeter group, the reflection arrangement $\mathcal{A}(G)$ is called a **Coxeter arrangement**.

Steinberg [223, Theorem 1.5] proved the following result.

Theorem 6.25 Let $G \subset GL(V)$ be a finite complex reflection group and let T be a subspace of V. Then G_T is a reflection group generated by reflections in the hyperplanes of $\mathcal{A}(G)$ which contain T.

The following result is essentially due to Springer [216].

Theorem 6.26 Let G be an irreducible Coxeter group and let $\mathcal{A}(G)$ be its Coxeter arrangement. Then the common zeros of the first $\ell - 1$ basic invariants is the union of 1-dimensional subspaces not contained in $\bigcup_{H \in \mathcal{A}(G)} H$.

Theorem 6.27 *Let $G \subset GL(V)$ be a finite complex reflection group. Write $\mathcal{A} = \mathcal{A}(G)$ and $L = L(\mathcal{A})$.*
 (1) If $g \in G$, then $\mathrm{Fix}(g) \in L$.
 (2) If $X \in L$, then there exists $g \in G$ with $\mathrm{Fix}(g) = X$.

Proof. Let $g \in G$ and let $U = \mathrm{Fix}(g)$. To prove that $U \in L$, we argue by induction on $|G|$. The assertion is clear if G is cyclic. Now assume that the statement holds for all reflection groups of order $< |G|$. It follows from Theorem 6.25 that G_U is a reflection group. If $G_U \neq G$, then we are done by induction, since $g \in G_U$ and $|G_U| < |G|$. If $G_U = G$, then $U \subseteq \cap_{H \in \mathcal{A}} H = T$, so $U = T \in L$. This proves (1).

Suppose $X \in L$ and $r(X) = p$. There exist p linearly independent hyperplanes in $\mathcal{A}$ so $X = H_1 \cap \cdots \cap H_p$. The groups G_{H_i} are cyclic. For $1 \leq i \leq p$, let s_i be a generator for G_{H_i}. We prove that $X = \mathrm{Fix}(s_1 \cdots s_p)$. The inclusion $X \subseteq \mathrm{Fix}(s_1 \cdots s_p)$ is clear. Let $(\,,\,)$ be a G–invariant positive definite Hermitian form in V. Let ϵ_i be the unique eigenvalue of s_i different from 1. Let $w_i \in V$ be a nonzero vector perpendicular to H_i. Coxeter [56, 8.73] gives the following formula for a reflection

$$(1) \qquad s_i v = v + (\epsilon_i - 1) \frac{(v, w_i)}{(w_i, w_i)} w_i.$$

It follows that

$$(2) \qquad s_1 \cdots s_p v \equiv v \quad \mod \mathbb{C} w_1 + \cdots + \mathbb{C} w_p \quad v \in V.$$

We prove $X \supseteq \mathrm{Fix}(s_1 \cdots s_p)$ by induction on p. This is clear for $p = 1$. Suppose $v \in \mathrm{Fix}(s_1 \cdots s_p)$. By (2) we may write $s_2 \cdots s_p v = v + v'$ where $v' \in \mathbb{C} w_2 + \cdots + \mathbb{C} w_p$. On the other hand, we use (1) to get

$$s_2 \cdots s_p v = s_1^{-1} v = v + (\epsilon_1^{-1} - 1) \frac{(v, w_1)}{(w_1, w_1)} w_1.$$

Since $r(X) = p$, the w_i are linearly independent and hence $w_1 \notin \mathbb{C} w_2 + \cdots + \mathbb{C} w_p$. It follows that $(v, w_1) = 0$ and $v \in H_1$. Furthermore, $s_1 v = v$, so $s_2 \cdots s_p v = v$, and thus $v \in H_2 \cap \cdots \cap H_p$ by induction. This shows that $X \supseteq \mathrm{Fix}(s_1 \cdots s_p)$ and proves (2). □

Corollary 6.28 *Let $G \subset GL(V)$ be a complex reflection group. Write $\mathcal{A} = \mathcal{A}(G)$ and $L = L(\mathcal{A})$. If $X \in L$, then*
 (1) $G_X = \{g \in G \mid \mathrm{Fix}(g) \leq X\}$, and
 (2) $\mathcal{A}_X = \mathcal{A}(G_X)$.

Examples

We introduced the B_3–arrangement in Example 1.7. The corresponding reflection group is called the Coxeter group of type B_3. It is generated by reflections about the symmetry planes of the cube. The braid arrangement of Example 1.9 is also a reflection arrangement corresponding to the symmetric group. It is generated by the transpositions (i,j) with fixed set $H_{i,j} = \ker(x_i - x_j)$. These are examples of Coxeter groups. Shephard and Todd [210] classified finite irreducible complex reflection groups. Every real reflection group may be viewed as a complex reflection group. We give three examples of complex reflection groups which are not complexified real reflection groups.

Example 6.29 Let $r \geq 2$ be an integer and let $C(r)$ be the cyclic group of order r generated by $\theta = \exp(2\pi i/r)$. The **full monomial group** $G(r,1,\ell)$ is the wreath product of $C(r)$ and $Sym(\ell)$, consisting of all $\ell \times \ell$ monomial matrices with entries in $C(r)$. Its reflection arrangement is defined by

$$Q = x_1 \ldots x_\ell \prod_{1 \leq i < j \leq \ell} (x_i^r - x_j^r).$$

Its lattice is the Dowling lattice $Q_\ell(\mathbb{Z}_r)$; see [68]. See Section 6.4 for a complete description of these arrangements. In particular, $G(r,1,2)$ is generated by the matrices

$$s_1 = \begin{bmatrix} \theta & 0 \\ 0 & 1 \end{bmatrix}, \quad s_2 = \begin{bmatrix} 0 & 1 \\ 1 & 0 \end{bmatrix}.$$

Example 6.30 Every nonsingular cubic in $\mathbb{C}P^2$ is projectively equivalent to a nonsingular cubic defined by $f(x,y,z) = x^3 + y^3 + z^3 - 3axyz$ where $a^3 \neq 1$ and $a \neq \infty$. The inflection points of a cubic are the solutions of $f = 0 = H(f)$ where $H(f)$ is the Hessian determinant of second partials. Direct calculation shows that for this family of curves, the nine inflection points are independent of a. Set $\omega = e^{2\pi i/3}$. The projective coordinates of the nine inflection points are

$$\begin{array}{lll}
(0,1,-1) & (0,1,-\omega) & (0,1,-\omega^2) \\
(1,0,-1) & (1,0,-\omega) & (1,0,-\omega^2) \\
(1,-1,0) & (1,-\omega,0) & (1,-\omega^2,0)
\end{array}$$

These nine points lie on 12 projective lines, which are the four degenerate cubics corresponding to the parameter values $a = \infty$ and $a^3 = 1$:

(1) $\quad x = 0, \ y = 0, \ z = 0, \ x + \omega^i y + \omega^j z = 0,$

where $i,j = 0,1,2$. These 12 projective lines meet in 12 additional points. This configuration of 12 lines and 21 points is called the **Hessian configuration**. Each of the first nine points is contained in four lines, so we refer to them as quadruple points. Each of the second 12 points is contained in two lines, so we refer to them as double points. Each line contains three quadruple points and two double points. Figure 6.3 illustrates the configuration.

The Hessian configuration has a distinguished history. For a complete account see Brieskorn and Knörrer's book [42, pp.289–305]. It appeared recently in the work of Hirzebruch [113]. Its group of symmetries was determined by

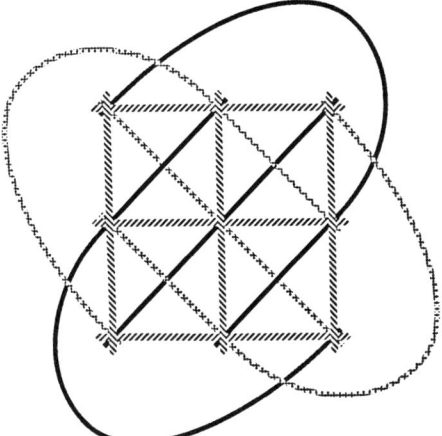

Fig. 6.3. The Hessian configuration

Jordan in 1878 as a subgroup of $PGL(3,\mathbb{C})$ of order 216. It is generated by projective transformations of order 3 which leave one of the 12 projective lines pointwise fixed. In addition, the group contains nine projective transformations of order 2 which fix projective lines. This group gives rise to two complex reflection groups. One, say G, has order 648. It is generated by reflections of order 3. Its reflection arrangement is defined by the 12 lines of (1)

$$Q(G) = xyz \prod_{i,j=0,1,2} (x + \omega^i y + \omega^j z).$$

See Examples 6.92 and 6.127 for more on this arrangement. The other, say G', has order 1296 and contains additional reflections of order 2. Its reflection arrangement is defined by

$$Q(G') = (x^3 - y^3)(x^3 - z^3)(y^3 - z^3)Q(G).$$

Example 6.31 Klein's simple group of order 168 is a subgroup of $PGL(3,\mathbb{C})$. It gives rise to a complex reflection group $G \subset GL(3,\mathbb{C})$ of order 336. The corresponding reflection arrangement has 21 hyperplanes. For the many important properties of this group and arrangement see [125] and [21, pp.88–96]. Here we note that all reflections of G are of order 2, but the group is not the complexification of a real group. See Examples 6.69 and 6.118 for more on this arrangement.

Example 6.32 If $\mathbb{K} = \mathbb{F}_q$ and q is odd, then the orthogonal groups $O^\pm(\ell,q)$ are generated by reflections. Their reflection arrangement $\mathcal{A}$ consists of all hyperplanes through the origin. This arrangement was defined in Example 1.10. We determined its structure in Example 2.12 and computed its Poincaré polynomial in Proposition 2.53. We showed in Example 4.24 that $\mathcal{A}$ is free with

$$\exp \mathcal{A} = \{1, q, \ldots, q^{\ell-1}\}.$$

Relative Invariants

Consider a reflecting hyperplane $H \in \mathcal{A}(G)$. Recall from Definition 6.2 that the set of all elements of G which fix H pointwise is denoted G_H. Since G_H is a subgroup of G, the characteristic of the field $\mathbb{K}$ does not divide $|G_H|$. By Maschke's theorem, there exists a 1–dimensional G_H-stable subspace L_H with $V = H \oplus L_H$. Since the representation $G_H \to GL(L_H) \simeq \mathbb{K} \setminus \{0\}$ is faithful, G_H is a cyclic group. Let $e_H = |G_H|$ and denote by s_H a fixed generator of G_H. Then $\det[s_H]$ is a primitive e_H-th root of unity. Let $\chi : G \to \mathbb{K} \setminus \{0\}$ be a fixed linear character. For each $H \in \mathcal{A}(G)$ define an integer a_H by the condition that a_H is the least nonnegative integer satisfying $\chi(s_H) = (\det[s_H^*])^{a_H}$. Then a_H depends only on G_H and not on the generator s_H. Let $H = \ker(\alpha_H)$. Define $d_\chi \in S$ by

$$d_\chi = \prod_{H \in \mathcal{A}(G)} \alpha_H^{a_H},$$

and let

$$S_\chi^G = \{f \in S \mid gf = \chi(g)f \text{ for all } g \in G\}.$$

Elements of S_χ^G are called **relative invariants**. Our next aim is to show that $S_\chi^G = S^G d_\chi$. The following lemma is due to R. Stanley [220, 2.2].

Lemma 6.33 *If $f \in S_\chi^G$, then f is divisible by d_χ.*

Proof. Choose $H \in \mathcal{A}$. Let s_H be the fixed generator of G_H as above. Put $\rho = \det[s_H]$. Let e_1 be a basis for the 1–dimensional space L_H. Then $s_H e_1 = \rho e_1$. Let $e_2, \ldots, e_\ell$ be a basis for H. Let $x_1, \ldots, x_\ell$ be the dual basis. Then $s_H x_1 = \rho^{-1} x_1$, and for $2 \leq i \leq \ell$ we have $s_H x_i = x_i$ by Proposition 6.8. Thus

$$\begin{aligned} f(\rho^{-1} x_1, x_2, \ldots, x_\ell) &= f(s_H x_1, x_2, \ldots, x_\ell) \\ &= \chi(s_H) f(x_1, \ldots, x_\ell) \\ &= \rho^{-a_H} f(x_1, \ldots, x_\ell), \end{aligned}$$

which shows that f must be divisible by $x_1^{a_H}$ and thus by $\alpha_H^{a_H}$. Since this is true for all $H \in \mathcal{A}$, we have the desired result. □

Lemma 6.34 *If $g(H) = K$ for $H, K \in \mathcal{A}(G)$ and $g \in G$, then $a_H = a_K$.*

Proof. Since $g^{-1} G_K g = G_H$, we have $|G_K| = |G_H| = o(s_H)$. This shows that $g s_H g^{-1}$ is a generator for G_K. We can choose $s_K = g s_H g^{-1}$. Then $\det[s_H^*] = \det[s_K^*]$ and $\chi(s_H) = \chi(s_K)$. Thus $a_H = a_K$. □

Lemma 6.35 *Let $\mathcal{A}(G) = \{H_0, H_1, \ldots, H_n\}$ and let $H = H_0$. Write $\alpha_i = \alpha_{H_i}$ and $a_i = a_{H_i}$. Then $\prod_{i>0} \alpha_i^{a_i}$ is a G_H-invariant polynomial.*

Proof. First note that we can replace α_i, if necessary, by a constant multiple. By considering the G_H–orbit decomposition of $\{H_1, \ldots, H_n\}$, we may assume without loss of generality that $\{H_1, \ldots, H_n\}$ itself is a G_H–orbit. Then the G_H–orbit of α_1 is a set of linear forms defining $H_1, \ldots, H_n$. Therefore we may assume that $\{\alpha_1, \ldots, \alpha_n\}$ is a G_H–orbit. So the product $\prod_{i>0} \alpha_i$ is G_H–invariant. Since $a_i = a_j$ for $1 \leq i < j \leq n$ by Lemma 6.34, this completes the proof. □

Lemma 6.36 $d_\chi \in S_\chi^G$.

Proof. It suffices to show that
$$s_H(d_\chi) = \det(s_H^*)^{a_H} d_\chi$$
for each $H \in \mathcal{A}(G)$. Fix $H \in \mathcal{A}(G)$. Write $a = a_H, s = s_H$, and $\alpha = \alpha_H$. Since $s(\alpha) = \det(s^*)\alpha$, we have
$$s(\alpha^a) = \det(s^*)^a \alpha^a.$$
The result follows from Lemma 6.35. □

The next result follows from Lemmas 6.33 and 6.36.

Theorem 6.37 $S_\chi^G = S^G d_\chi$.

This theorem was shown by R. Stanley [220, 3.1] when $\mathbb{K} = \mathbb{C}$. The following three examples of d_χ are important.

Example 6.38 When χ is the trivial character, we have $d_\chi = 1$. In this case $S_\chi^G = S^G$ is the invariant subring of S under G.

Example 6.39 When $\chi = \det$, we have $a_H = e_H - 1$. In this case we have
$$d_\chi = \prod_{H \in \mathcal{A}(G)} \alpha_H^{e_H - 1}.$$
Let m denote the number of reflections in G. Then $\deg d_{\det} = m$.

Example 6.40 When $\chi = \det^{-1}$, we have $a_H = 1$. In this case we have
$$d_\chi = \prod_{H \in \mathcal{A}(G)} \alpha_H = Q(\mathcal{A}(G)).$$
Recall that $|\mathcal{A}| = n$ is the number of reflecting hyperplanes in $\mathcal{A}(G)$. Then $\deg d_{\det^{-1}} = n$.

Jacobian and Discriminant

Let $\mathcal{F} = \{f_1, \ldots, f_\ell\}$ be a set of basic invariants for G. Let $J(\mathcal{F})$ be the determinant of the Jacobian matrix $\mathsf{J}(f_1, \ldots, f_\ell)$ in Definition 6.6:

$$J(\mathcal{F}) = \det \mathsf{J}(f_1,\ldots,f_\ell) = \det[D_i f_j].$$

The polynomial $J(\mathcal{F})$ is called the **Jacobian** of the reflection group G. Steinberg [223], [38, Ch.5, Sect.5.5, Prop.6] proved that $J(\mathcal{F}) \doteq \prod_{H \in \mathcal{A}(G)} \alpha_H^{e_H - 1}$. Thus $J(\mathcal{F})$ is independent of the choice of basic invariants $\mathcal{F}$. In order to prove this result we need a preliminary lemma, which was proved by K. Saito [198, 3.4] and by G. Scheja and U. Storch [208, 1.2]. We refer to Appendix A for the definition and basic properties of a regular sequence.

Lemma 6.41 *Suppose $h_1,\ldots,h_\ell \in S = \mathbb{K}[x_1,\ldots,x_\ell]$ form a homogeneous regular sequence. Assume that $e_i = \deg h_i$ is invertible in $\mathbb{K}$. Then the determinant $\det[\partial h_i/\partial x_j]_{1 \leq i,j \leq \ell}$ does not belong to the ideal of S generated by $h_1,\ldots,h_\ell$.*

Proof. We argue by induction on ℓ. The assertion is clear for $\ell = 1$. Assume $\ell > 1$. Since the height of the ideal generated by $h_2,\ldots,h_\ell$ is equal to $\ell - 1$ by Theorem A.33 and S is Cohen-Macaulay by Theorem A.31, each associated prime of the ideal is also of height $\ell - 1$ by Theorem A.32. Thus the ideal $(x_1,\ldots,x_\ell)S$, whose height is ℓ, is not contained in any associated prime $\wp$ of $(h_2,\ldots,h_\ell)S$. By Proposition A.28, $(h_2,\ldots,h_\ell)S : x_1 S = (h_2,\ldots,h_\ell)S$ and the sequence $h_2,\ldots,h_\ell,x_1$ is a regular sequence. By Euler's formula,

$$e_j h_j = \sum_{i=1}^{\ell} x_i (\partial h_i/\partial x_j) \quad (j = 1,\ldots,\ell).$$

Let $\Delta = \det[\partial h_i/\partial x_j]_{1 \leq i,j \leq \ell}$. It follows from Cramer's rule that

$$x_1 \Delta = \det \begin{bmatrix} e_1 h_1 & \partial h_1/\partial x_2 & \cdots & \partial h_1/\partial x_\ell \\ \cdot & \cdot & \cdots & \cdot \\ \cdot & \cdot & \cdots & \cdot \\ e_\ell h_\ell & \partial h_\ell/\partial x_2 & \cdots & \partial h_\ell/\partial x_\ell \end{bmatrix}.$$

Let $\Delta_2 = \det[\partial h_i/\partial x_j]_{2 \leq i,j \leq \ell}$. In the rest of the proof, the congruence $\equiv$ always means modulo the ideal $(h_2,\ldots,h_\ell)S$. Thus $x_1 \Delta \equiv e_1 h_1 \Delta_2$. If $\Delta \in (h_1,\ldots,h_\ell)S$, then $\Delta \equiv g_1 h_1$ for some $g_1 \in S$. Thus

$$x_1 g_1 h_1 \equiv e_1 h_1 \Delta_2.$$

Since $h_1,\ldots,h_\ell$ form a regular sequence, we have

$$x_1 g_1 \equiv e_1 \Delta_2.$$

Thus $\Delta_2 \in (x_1, h_2,\ldots,h_\ell)S$. For $p \in S$, let $\bar{p}$ denote $p(0,x_2,\ldots,x_\ell) \in \bar{S} = \mathbb{K}[x_2,\ldots,x_\ell]$. Then $\bar{h}_2,\ldots,\bar{h}_\ell$ form a regular sequence in $\bar{S}$ because $x_1,h_2,\ldots,h_\ell$ form a regular sequence in S. Note that $\overline{(\partial p/\partial x_i)} = \partial \bar{p}/\partial x_i$ for $i > 1$. Thus

$$\det[\partial \bar{h}_i/\partial x_j]_{2 \leq i,j \leq \ell} = \bar{\Delta}_2 \in (\bar{h}_2,\ldots,\bar{h}_\ell)\bar{S}.$$

This contradicts the induction assumption because $\bar{h}_2,\ldots,\bar{h}_\ell$ form a regular sequence in $\mathbb{K}[x_2,\ldots,x_\ell]$. □

6.2 Reflection Arrangements

Theorem 6.42 *Let G be a reflection group and let $\mathcal{F}$ be a set of basic invariants for G. Then*
$$J(\mathcal{F}) \doteq \prod_{H \in \mathcal{A}(G)} \alpha_H^{e_H - 1}$$
and $m = m_1 + \cdots + m_\ell$.

Proof. Consider the linear character $\chi(g) = \det[g]$ in Example 6.39. Write $J = J(\mathcal{F})$. Then
$$d_\chi = \prod_{H \in \mathcal{A}(G)} \alpha_H^{e_H - 1}.$$
We have $g(J) = (\det[g])J = \chi(g)$ by Proposition 6.8.3. Thus $J \in S_\chi^G$. By Theorem 6.37, we have $J \in d_\chi S^G$. Note that J is homogeneous in S and that $S^G = \mathbb{K}[f_1, \ldots, f_\ell]$. If J/d_χ has positive degree, then J/d_χ belongs to the ideal $(f_1, \ldots, f_\ell)S$. This contradicts Lemma 6.41. Thus $J \doteq d_\chi$. By comparing degrees, we find the number of reflections equals the sum of the exponents. □

Corollary 6.43 *Let $J = \prod_{H \in \mathcal{A}(G)} \alpha_H^{e_H - 1}$. If $a \in S$ and $ga = \det(g)a$ for all $g \in G$, then $a \in RJ$.*

Definition 6.44 *Define the **discriminant** $\delta(x_1, \ldots, x_\ell)$ of the group G by*
$$\delta(x_1, \ldots, x_\ell) = JQ \doteq \prod_{H \in \mathcal{A}(G)} \alpha_H^{e_H}.$$
It follows from Examples 6.39 and 6.40 *that $\delta = d_{\det} d_{\det^{-1}} \in R$ is a homogeneous invariant of degree $m+n$. Thus δ is also a polynomial in the basic invariants. We define a polynomial $\Delta(T_1, \ldots, T_\ell; \mathcal{F})$, also called the **discriminant**, in the indeterminates $T_1, \ldots, T_\ell$ and depending on $\mathcal{F}$ by $\Delta(f_1, \ldots, f_\ell; \mathcal{F}) = \delta(x_1, \ldots, x_\ell)$.*

Classification

Shephard and Todd [210] classified finite unitary reflection groups. Since a finite group of unitary transformations is either irreducible or completely reducible [47, p.263], it suffices to determine the irreducible groups. Every real reflection group (Coxeter group) is a complex reflection group. An irreducible complex reflection group has a real form if and only if it has a quadratic invariant, $d_1 = 2$. The classification of finite Coxeter groups is well known [38, p.193]. We refer to these groups by their classification symbols.

The classification of finite unitary reflection groups contains three infinite families and 34 exceptional groups. The infinite families are:

(1) The symmetric groups. These are real groups of type A_ℓ.

(2) The family denoted $G(r, p, \ell)$. These are subgroups of the full monomial group $G(r, 1, \ell)$ of Example 6.29. The Coxeter groups of type B_ℓ are the groups $G(2, 1, \ell)$. The Coxeter groups of type D_ℓ are the groups $G(2, 2, \ell)$. The Coxeter groups of type $I_2(n)$ are the groups $G(n, n, 2)$.

(3) The cyclic groups $C(r)$.

It is customary to refer to the exceptional groups by their number in [210, Table VII] as G_4–G_{37}. For example, the two groups in Example 6.30 are G_{25} and G_{26}. The group in Example 6.31 is G_{24}. The icosahedral groups H_3, H_4 are G_{23}, G_{30}. The Coxeter group of type F_4 is G_{28}. The Coxeter groups of type E_6, E_7, E_8 are G_{35}, G_{36}, G_{37}.

Basic invariants for unitary reflection groups are found in [46, 51, 125, 150, 151, 251].

6.3 Free Arrangements

Invariant Theory

Lemma 6.45 *Let G be a finite reflection group. Suppose M is a G–module of dimension r. The R–module $(S \otimes M)^G$ is free of rank r.*

Proof. Recall from Theorem 6.19 that U affords the regular representation. Let $U_p = U \cap S_p$. If M is an irreducible representation, then it occurs with multiplicity $\dim M = r$ in U. Note the isomorphism $(U_p \otimes M)^G \simeq \operatorname{Hom}_G(M^*, U_p)$. We may choose a homogeneous $\mathbb{K}$–basis $u_1, \ldots, u_r$ for $(U \otimes M)^G$. Since G–modules are semisimple, such choice may be made for any G–module M. Thus

$$(S \otimes M)^G \simeq (R \otimes U \otimes M)^G \simeq R \otimes (U \otimes M)^G \simeq Ru_1 \oplus \cdots \oplus Ru_r$$

as required. $\square$

Let ΛM denote the exterior algebra over M. Next we will study the G–invariant part of $S \otimes \Lambda M$. Let $q_1, \ldots, q_r$ be a basis for M. Let $[g_M]$ be the matrix for the action of $g \in G$ on M in this basis. Define a linear character $\chi : G \to \mathbb{K} \setminus \{0\}$ by $\chi(g) = \det[g_M]^{-1}$. Recall the definition of relative invariant d_χ from Section 6.2.

Lemma 6.46 $(S \otimes \Lambda^r M)^G = d_\chi R \otimes (\Lambda^r M)$.

Proof. Note that $\Lambda^r M$ is a 1–dimensional space on which $g \in G$ acts as multiplication by $\det[g_M]$. The result is a direct consequence of Theorem 6.37, since $S_\chi^G = d_\chi R$. $\square$

Proposition 6.47 *Let $q_1, \ldots, q_r$ be a basis for M. If there exist elements $u_1, \ldots, u_r \in (S \otimes M)^G$ such that*

$$u_1 \wedge \cdots \wedge u_r \doteq d_\chi \otimes (q_1 \wedge \cdots \wedge q_r),$$

then $(S \otimes \Lambda M)^G$ is an exterior algebra over R:

$$(S \otimes \Lambda M)^G = \bigoplus_{p=0}^{\ell} \bigoplus_{i_1 < \cdots < i_p} R u_{i_1} \wedge \cdots \wedge u_{i_p}.$$

Proof. Given a multiindex $K = (k_1, \ldots, k_s)$, define

$$u_K = u_{k_1} \wedge \cdots \wedge u_{k_s} \in S \otimes \Lambda^s M,$$
$$q_K = q_{k_1} \wedge \cdots \wedge q_{k_s} \in \Lambda^s M.$$

Let $I = (i_1, \ldots, i_p)$ be an ordered multiindex of length p, $1 \leq i_1 < \cdots < i_p \leq r$. Let $I^c = (j_1, \ldots, j_{r-p})$ be the ordered complement of I, $1 \leq j_1 < \cdots < j_{r-p} \leq r$. Thus

$$\{i_1, \ldots, i_p\} \cup \{j_1, \ldots, j_{r-p}\} = \{1, \ldots, r\}.$$

Let $\sigma(I)$ be the sign of the permutation $(i_1, \ldots, i_p, j_1, \ldots, j_{r-p})$. Let $\mathcal{I}$ be the set of ordered multiindices of length p.

Let $u \in (S \otimes \Lambda^p M)^G$ and let $I \in \mathcal{I}$. Then

$$u \wedge u_{I^c} \in (S \otimes \Lambda^r M)^G = d_\chi R \otimes (q_1, \ldots, q_r)$$

by Lemma 6.46. Define $f_I \in R$ by

$$u \wedge u_{I^c} = d_\chi f_I \otimes (q_1, \ldots, q_r).$$

Let $\eta = u - \sum_{I \in \mathcal{I}} \sigma(I) f_I u_I$. Then $\eta \wedge u_{I^c} = 0$ for all $I \in \mathcal{I}$. Since $u_1 \wedge \cdots \wedge u_r \doteq d_\chi \otimes (q_1 \wedge \cdots \wedge q_r)$, Cramer's formula implies that $d_\chi q_i \in Su_1 + \cdots + Su_r$. Thus for any $K = (k_1, \ldots, k_{r-p})$, $d_\chi^{r-p} q_K$ is a linear combination of $\{u_{I^c} \mid I \in \mathcal{I}\}$ over S. This implies that $\eta \wedge q_K = 0$. Therefore $q = 0$ and

$$u = \sum_{I \in \mathcal{I}} \sigma(I) f_I u_I.$$

This shows that $\{u_{i_1} \wedge \ldots \wedge u_{i_p} \mid 1 \leq i_1 < \cdots < i_p \leq r\}$ spans $(S \otimes \Lambda^p M)^G$. To show that this set is R–independent, assume $\sum_{I \in \mathcal{I}} f_I u_I = 0$. By taking exterior product with u_{I^c}, we get $f_I = 0$. □

Lemma 6.48 *The R–modules Der_S^G and Ω_S^G are free of rank ℓ.*

Proof. Proposition 6.4 shows that the map $D_v \mapsto v$ induces a G–equivariant isomorphism $\mathrm{Der}_S \simeq S \otimes V$ and the map $dx \mapsto x$ induces a G–equivariant isomorphism $\Omega_S \simeq S \otimes V^*$. These maps induce R–module isomorphisms $\mathrm{Der}_S^G \simeq (S \otimes V)^G$ and $\Omega_S^G \simeq (S \otimes V^*)^G$. The conclusion follows from Lemma 6.45 setting $M = V$ and $M = V^*$. □

Consider a set $\mathcal{F} = \{f_1, \ldots, f_\ell\}$ of basic invariants for G. Then $df_1, \ldots, df_\ell$ belong to $\Omega_S^G \simeq (S \otimes V^*)^G$. We have

$$df_1 \wedge \cdots \wedge df_\ell = J(\mathcal{F})(dx_1 \wedge \cdots \wedge dx_\ell) = \prod_{H \in \mathcal{A}(G)} \alpha_H^{e_H - 1}(dx_1 \wedge \cdots \wedge dx_\ell)$$

by Theorem 6.42. Since $\det[g_{V^*}]^{-1} = \det[g]$, we can apply Proposition 6.47 replacing M by V^* to obtain a theorem of Solomon [211].

Theorem 6.49 *If $\mathcal{F} = \{f_1, \ldots, f_\ell\}$ is a set of basic invariants for G, then*

$$(S \otimes \Lambda V^*)^G \simeq \left(\bigoplus_{p=0}^{\ell} \Omega_S^p\right)^G = \bigoplus_{p=0}^{\ell} \bigoplus_{i_1 < \cdots < i_p} R(df_{i_1} \wedge \cdots \wedge df_{i_p}).$$

In particular, $df_1, \ldots, df_\ell$ is a basis for the R-module Ω_S^G. Thus the exponents $m_1, \ldots, m_\ell$ are the degrees where G-modules isomorphic to V occur in U.

Definition 6.50 *A homogeneous basis $\Theta = \{\theta_1, \ldots, \theta_\ell\}$ of Der_S^G is called a set of **basic derivations** for G. Let $\{n_1, \ldots, n_\ell\}$ be the polynomial degrees of a homogeneous basis of Der_S^G over R. These integers are called the **coexponents** of G. It follows that the coexponents are the degrees where G-modules isomorphic to V^* occur in U. It is customary to label the coexponents in increasing order*

$$n_1 \leq \ldots \leq n_\ell.$$

The fact that $n_1 = 1$ is seen in this interpretation as $V^* = S_1 \subseteq U$. If G is a Coxeter group, then V and V^* are isomorphic G-modules and $n_i = m_i$ for $1 \leq i \leq \ell$.

The Hessian

We showed in Corollary 6.10 that $f \in R$ induces an R-module map $\mathrm{Hess}(f) : \mathrm{Der}_S^G \to \Omega_S^G$. It follows from Lemma 6.48 that we may choose bases and study the matrix of the map. If we choose the basis given in Theorem 6.49 for Ω_S^G, then the resulting matrix equation connects the Jacobian matrix with the coefficient matrix of the basic derivations and the Hessian matrix. If $a_1, \ldots, a_\ell$ are the columns of a matrix A, we write $\mathsf{A} = [a_1 | \cdots | a_\ell]$.

Lemma 6.51 *Let $\mathsf{A} = [a_1 | \cdots | a_\ell]$ and $\mathsf{B} = [b_1 | \cdots | b_\ell]$ be $\ell \times \ell$ matrices with coefficients in $\mathbb{K}$. If B is invertible, then*

$$(\mathsf{B}^{-1}\mathsf{A})_{i,j} = (\det \mathsf{B})^{-1} \det[b_1 | \cdots | b_{i-1} | a_j | b_{i+1} | \cdots | b_\ell].$$

Proof. Let $c_{i,j} = (\mathsf{B}^{-1}\mathsf{A})_{i,j}$ and let $c_j = [c_{1,j}, \ldots, c_{\ell,j}]^T$. Then $[c_1 | \cdots | c_\ell] = \mathsf{B}^{-1}\mathsf{A} = [\mathsf{B}^{-1}a_1 | \cdots | \mathsf{B}^{-1}a_\ell]$. Thus $\mathsf{B}c_j = a_j$. Fix j and view this as a system of linear equations with given $a_{i,j}$ and unknown $c_{i,j}$ for $1 \leq i \leq \ell$. The assertion of the lemma is Cramer's rule. □

Lemma 6.52 *Let G be a unitary reflection group. Let $\mathcal{F}$ be a set of basic invariants for G and let $\mathsf{J} = \mathsf{J}(\mathcal{F})$ be the Jacobian matrix. Suppose $\mathsf{A} \in M_\ell(S)$ and $g\mathsf{A} = [g]^T \mathsf{A}$ for all $g \in G$. Then there exists a matrix $\mathsf{C} \in M_\ell(R)$ such that $\mathsf{A} = \mathsf{J}\mathsf{C}$.*

Proof. Write $A = [a_1|\cdots|a_\ell]$ and $J = [b_1|\cdots|b_\ell]$. Fix $g \in G$ and define ga_i by $gA = [ga_1|\cdots|ga_\ell]$. Write $P = [g]^T$. Since $gA = PA$ by assumption, and $gJ = PJ$ by Lemma 6.8.3, we have $ga_i = Pa_i$ and $gb_i = Pb_i$. For each $1 \leq i, j \leq \ell$ define $T(i,j) \in M_\ell(S)$ by

$$T(i,j) = [b_1|\cdots|b_{i-1}|a_j|b_{i+1}|\cdots|b_\ell].$$

Then $gT(i,j) = PT(i,j)$. Thus

$$g(\det T(i,j)) = (\det P)(\det T(i,j)) = (\det g)(\det T(i,j)).$$

It follows from Corollary 6.43 that $\det T(i,j) \in RJ$. Theorem 6.42 implies that $\det J \doteq J \neq 0$, so J is an invertible matrix over the quotient field $F(S)$. Let $C = J^{-1}A$. It follows from Lemma 6.51 that $C_{i,j} \doteq J^{-1} \det T(i,j)$. Since $\det T(i,j) \in RJ$, we conclude that $C_{i,j} \in R$. □

Theorem 6.53 *Let G be a unitary reflection group. Let $\mathcal{F}$ be a set of basic invariants and let Θ be a set of basic derivations for G. If $f \in R$, then there exists $C \in M_\ell(R)$ such that*

$$H(f)M(\Theta) = J(\mathcal{F})C.$$

If f is homogeneous, then the nonzero entries of C are homogeneous.

Proof. Write $J = J(\mathcal{F})$, $M = M(\Theta)$, and $H = H(f)$. It follows from Lemma 6.8 that $g(HM) = (gH)(gM) = [g]^T HM$. We may apply Lemma 6.52 to find a matrix $C \in M_\ell(R)$ which satisfies $HM = JC$. The nonzero entries of a given column of J are homogeneous polynomials of the same degree. The same holds for M. If f is homogeneous, then all nonzero entries of H have the same degree. Thus the nonzero entries of C are homogeneous invariants. □

The polynomials which may be chosen as basic invariants for the irreducible unitary reflection groups were constructed by invariant theorists in the last century. The concept of a basic derivation is new. It is reasonable to try to use knowledge of basic invariants to construct basic derivations. Theorem 6.53 will be used in Appendix B to define standard basic derivations associated to a set of basic invariants.

$D_R(\delta)$ Is Free

Lemma 6.54 *Let $A = \mathbb{K}[z_1,\ldots,z_\ell]$ be a graded polynomial algebra. Assume that $h_1,\ldots,h_\ell$ are algebraically independent homogeneous polynomials. Let $B = \mathbb{K}[h_1,\ldots,h_\ell] \subseteq A$. Denote the quotient fields of A and B by $F(A)$ and $F(B)$. Suppose that $F(A)/F(B)$ is a separable extension. Let $J_{A/B}$ be the Jacobian $\det[\partial h_j/\partial z_i]_{1\leq i,j\leq \ell}$. Given $\theta \in \mathrm{Der}_B$, there exists a unique derivation $\theta_A : A \to F(A)$ which extends θ such that $J_{A/B}\theta_A \in \mathrm{Der}_A$.*

6. Reflection Arrangements

Proof. Since $F(A)/F(B)$ is a finite separable extension, θ is uniquely extendable to a derivation $\theta_{F(A)} : F(A) \to F(A)$. Let θ_A be the restriction of $\theta_{F(A)}$ to A. It follows from $\theta(h_j) = \sum_{i=1}^{\ell}(\partial h_j/\partial z_i)\theta_A(z_i)$ by Cramer's formula that $J_{A/B}\theta_A(z_j) \in A$. □

The following special case is of particular interest. Let $\mathcal{F} = \{f_1, \ldots, f_\ell\}$ be a set of basic invariants for G. Define derivations $D_{f_i} = \partial/\partial f_i \in \mathrm{Der}_R$ characterized by $D_{f_i}(f_j) = \delta_{i,j}$. It follows from Lemma 6.54 that D_{f_i} is extendable to a derivation $S \to F(S)$. We agree to call this extension again D_{f_i}. If $\theta \in \mathrm{Der}_S^G$, then $\theta R \subseteq R$. Let $\bar{\theta}$ be the restriction of θ to R. Then

$$\bar{\theta} = \sum_{i=1}^{\ell} \theta(f_i) D_{f_i}. \tag{1}$$

Let Der_R^0 be the R–module of derivations of R extendable to a derivation $S \to S$:

$$\mathrm{Der}_R^0 = \{\theta \in \mathrm{Der}_R \mid \theta_S(S) \subseteq S\}.$$

Lemma 6.55 *The map $\theta \mapsto \theta_S$ induces an R–module isomorphism*

$$\mathrm{Der}_R^0 \to \mathrm{Der}_S^G.$$

Proof. The correspondence is clearly injective and R–linear. It remains to show surjectivity. If $\eta \in \mathrm{Der}_S^G$ and $f \in R = S^G$, then $\eta(f) \in R$. Thus $\eta(R) \subseteq R$ and $\bar{\eta} \in \mathrm{Der}_R^0$. It follows that the extension of $\bar{\eta}$ to S is η. □

Recall from Definition 4.4 the S–module $D(f)$ for any $f \in S$:

$$D_S(f) = D(f) = \{\theta \in \mathrm{Der}_S \mid \theta(f) \in fS\}.$$

Proposition 6.56 *Let χ be a linear character of G. Then $\mathrm{Der}_S^G \subseteq D(d_\chi)$.*

Proof. Given $\theta \in \mathrm{Der}_S^G$ and $g \in G$, we have

$$g(\theta(d_\chi)) = (g\theta)(gd_\chi) = \theta(\chi(g)d_\chi) = \chi(g)\theta(d_\chi).$$

Thus $\theta(d_\chi) \in S_\chi^G = Rd_\chi$ by Theorem 6.37. □

Theorem 6.57 *Let $D_R(\delta) = \{\theta \in \mathrm{Der}_R \mid \theta(\delta) \in \delta R\}$ where δ is the discriminant. Then $\mathrm{Der}_R^0 = D_R(\delta)$.*

Proof. If $\theta \in \mathrm{Der}_R^0$, then $\theta_S \in \mathrm{Der}_S^G$ by Lemma 6.55. Let $\chi(g) = \det[g^*]$ for $g \in G$. Then $d_\chi = J$ by Example 6.39. Let $\chi'(g) = \det[g]$ for $g \in G$. Then $d_{\chi'} = Q$ by Example 6.40. It follows from Proposition 6.56 that $\theta_S \in D_S(J) \cap D_S(Q)$. Recall that $\delta = JQ$. Then $\theta(\delta) = \theta_S(QJ) = Q\theta_S(J) + J\theta_S(Q) \in \delta R$. This shows $\mathrm{Der}_R^0 \subseteq D_R(\delta)$.

To show $\operatorname{Der}_R^0 \supseteq D_R(\delta)$, let $H \in \mathcal{A}(G)$ and recall from Definition 6.2 that G_H is the pointwise fixer of H. Let $R_1 = S^{G_H}$. It follows from Theorem 6.25 that G_H is a finite reflection group. It follows from Chevalley's theorem that R_1 is a graded polynomial $\mathbb{K}$–algebra with $R \subseteq R_1 \subseteq S$. We may assume that $\alpha_H = x_1$ and that $x_2, \ldots, x_\ell \in R_1$. Write $e = e_H = |G_H|$. Then the discriminant of the reflection group G_H is given by $\delta_1 = x_1^e \in R_1$. Let $\delta_2 = \delta/\delta_1 \in R_1$. Choose an arbitrary $\theta \in D_R(\delta)$. It is proved in [38, Ch.5, Sect.5.2] that $F(S)/F(R)$ is a Galois extension with Galois group G. Thus $F(R_1)/F(R)$ is a finite separable extension. By Lemma 6.54, θ is extendable to $\theta_{R_1} : R_1 \to F(R_1)$ in such a way that $\psi = J_{R/R_1}\theta_{R_1} \in \operatorname{Der}_{R_1}$. Then we have

$$\delta_1 \psi(\delta_2) + \delta_2 \psi(\delta_1) = \psi(\delta) = \delta R_1 = \delta_1 \delta_2 R_1.$$

Since δ_1 and δ_2 are coprime, we have $\psi(\delta_1) \in \delta_1 R_1$. By Lemma 6.54 again, we extend ψ to $\psi_S : S \to F(S)$. Then

$$e x_1^{e-1} \psi_S(x_1) = \psi(\delta_1) \in \delta_1 R_1 = x_1^e R_1.$$

Since e is invertible in the field $\mathbb{K}$, we have $\psi_S(x_1) \in S$. Since $\psi_S(x_i) = \psi(x_i) \in R_1$ $(i > 1)$, we have $J_{R_1/R}\theta_S = \psi_S \in \operatorname{Der}_S$. This is the case for all $H \in \mathcal{A}(G)$. Since the polynomials $J_{R_1/R} = J/\alpha_H^{e-1}$ for all $H \in \mathcal{A}(G)$ have no common factor, we obtain $\theta_S \in \operatorname{Der}_S$. □

Corollary 6.58 *The map $\beta : \operatorname{Der}_S^G \to D_R(\delta)$ defined by $\beta(\theta) = \bar{\theta}$ is an R–module isomorphism. Thus $D_R(\delta)$ is a free R–module of rank ℓ. If $\theta_1, \ldots, \theta_\ell$ is a basis for Der_S^G, then $\bar{\theta}_1, \ldots, \bar{\theta}_\ell$ is a basis for $D_R(\delta)$.*

$D(\mathcal{A})$ Is Free

Theorem 6.59 *If $\mathcal{A}$ is a reflection arrangement, then*

$$D(\mathcal{A}) \simeq S \otimes_R \operatorname{Der}_S^G.$$

Proof. Let $Q = Q(\mathcal{A})$ and let $\theta \in \operatorname{Der}_S^G$. We have $d_\chi = Q$ for the character $\chi(g) = \det[g]$. It follows from Proposition 6.56 that $\operatorname{Der}_S^G \subseteq D(\mathcal{A})$. Recall the vector subspace $U \subset S$ provided by Theorem 6.19. Thus

$$S \otimes_R \operatorname{Der}_S^G \simeq U \otimes_{\mathbb{K}} R \otimes_R \operatorname{Der}_S^G \simeq U \otimes_{\mathbb{K}} \operatorname{Der}_S^G$$

is naturally embedded in $D(\mathcal{A})$. The image of the embedding is $S(\operatorname{Der}_S^G) = U(\operatorname{Der}_S^G)$.

To show the reverse inclusion, let $\phi \in D(\mathcal{A})$. Choose a $\mathbb{K}$-basis $u_1, \ldots, u_n$ for U. Let $f_1, \ldots, f_\ell$ be a set of basic invariants of G. Then there exist $r_{ij} \in R$ such that for $1 \leq i \leq \ell$

$$\phi(f_i) = \sum_{j=1}^n u_j r_{i,j}.$$

Define $\phi_j \in \mathrm{Der}_R$ by $\phi_j(f_i) = r_{i,j}$. Then $\sum_j u_j \phi_j$ is a derivation in R. We may extend it to a derivation $S \to F(S)$. When we apply it to the algebraically independent polynomials $f_1, \ldots, f_\ell$, we note that its values are the same as ϕ applied to these polynomials. Thus $\phi = \sum_j u_j \phi_j$. Recall from Proposition 4.8 that $\phi(\alpha_H) \in \alpha_H S$. Thus

$$\phi(\delta) = \phi(\prod_{H \in \mathcal{A}} \alpha_H^{e_H}) = \sum_{H \in \mathcal{A}} (e_H \delta / \alpha_H) \phi(\alpha_H) \in \delta S.$$

There exist $h_j \in R$ ($1 \leq j \leq n$) such that $\phi(\delta) = \sum_j u_j h_j \delta$. We have

$$\sum_j u_j(h_j \delta) = \phi(\delta) = \sum_j u_j \phi_j(\delta).$$

Since the expression is unique, $h_j \delta = \phi_j(\delta)$, and therefore $\phi_j \in D_R(\delta) \simeq \mathrm{Der}_S^G$ for each j. Thus $\phi = \sum_j u_j \phi_j \in U(\mathrm{Der}_S^G) = S(\mathrm{Der}_S^G)$. □

Theorems 6.48 and 6.59 give the following result.

Theorem 6.60 *If G is a finite reflection group, then its reflection arrangement $\mathcal{A} = \mathcal{A}(G)$ is free with $\exp \mathcal{A} = \{n_1, \ldots, n_\ell\}$. If $\Theta = \{\theta_1, \ldots, \theta_\ell\}$ is a set of basic derivations for G, and hence an R–module basis for Der_S^G, then Θ is also an S–module basis for $D(\mathcal{A})$.*

Saito's criterion 4.19 has the following consequence.

Corollary 6.61 *Let G be a finite reflection group with reflection arrangement $\mathcal{A} = \mathcal{A}(G)$. Let Q be a defining polynomial for $\mathcal{A}(G)$. If $\Theta = \{\theta_1, \ldots, \theta_\ell\}$ is a set of basic derivations for G, then*

$$\det \mathsf{M}(\Theta) = \det[\theta_i(x_j)]_{1 \leq i,j \leq \ell} \doteq Q.$$

Corollary 6.62 *Let G be a finite reflection group with coexponents $n_1, \ldots, n_\ell$ and let $\mathcal{A} = \mathcal{A}(G)$ be its reflection arrangement. Then*

$$\pi(\mathcal{A}, t) = (1 + n_1 t) \cdots (1 + n_\ell t).$$

Corollary 6.63 *The number of reflecting hyperplanes of $\mathcal{A}(G)$ is the sum of the coexponents of G*

$$n = n_1 + \cdots + n_\ell.$$

The Discriminant Matrix

The discriminant is a polynomial which received considerable attention around the turn of the century. It gained renewed importance in the theory of singularities in work of Arnold [7], Brieskorn [40], and others.

The ring of invariants $R = \mathbb{K}[f_1, \ldots, f_\ell]$ has a natural grading $R = \oplus R_p$ inherited from S, in which $\deg f_i = d_i$ and $R_p = R \cap S_p$. With this grading, the discriminant $\delta \in R$ is homogeneous of degree $m + n$. If $\theta(R_p) \subseteq R_{p+r}$, we call $\theta \in \mathrm{Der}_R(d_1, \ldots, d_\ell)$–homogeneous of degree r. Thus D_{f_i} is $(d_1, \ldots, d_\ell)$–homogeneous of degree $-d_i$. Recall the grading of Der_S by total degree in Definition 4.3. If $\theta \in \mathrm{Der}_S^G$ has total degree r, then (1) shows that $\bar\theta$ is $(d_1, \ldots, d_\ell)$–homogeneous of degree r.

If we view R as a polynomial algebra $R = \mathbb{C}[T_1, \ldots, T_\ell]$ without reference to S, then it has no distinguished grading. We show next that the requirement that the discriminant is homogeneous distinguishes the natural grading. If $a = (a_1, \ldots, a_\ell)$ is any ℓ-tuple of positive integers, we may grade R by letting $\deg T_i = a_i$. Let $R = \oplus R_p^a$. If $f \in R_p^a$, we say that f is $(a_1, \ldots, a_\ell)$–homogeneous of degree p. If $f \in R_p^a$, then the Euler formula says

$$\sum_{i=1}^{\ell} a_i T_i D_{T_i}(f) = pf.$$

The corresponding grading of Der_R is defined similarly.

Proposition 6.64 *Let G be a finite irreducible unitary reflection group. Let $a = (a_1, \ldots, a_\ell)$ be an ℓ-tuple of positive integers. Suppose Δ is $(a_1, \ldots, a_\ell)$–homogeneous of degree p. Then there exists k such that $p = k(m+n)$ and $a_i = k d_i$ for $1 \leq i \leq \ell$.*

Proof. Let $\theta_1, \ldots, \theta_\ell \in \mathrm{Der}_S^G$ be a set of basic derivations. Their total degrees are $n_1 - 1, \ldots, n_\ell - 1$. It follows from Corollary 6.58 that $\bar\theta_1, \ldots, \bar\theta_\ell$ is a basis for $D_R(\Delta)$. Thus $\bar\theta_i$ is $(d_1, \ldots, d_\ell)$–homogeneous of degree $n_i - 1$. Since G is irreducible, it follows from Corollary 4.30 that $1 = n_1 < n_2$. Thus

$$\bar\theta_1 = d_1 T_1 D_{T_1} + \cdots + d_\ell T_\ell D_{T_\ell}$$

is, up to constant, the unique $(d_1, \ldots, d_\ell)$–homogeneous element of degree 0 in $D_R(\Delta)$. Define

$$\eta = a_1 T_1 D_{T_1} + \cdots + a_\ell T_\ell D_{T_\ell}.$$

Since Δ is $(a_1, \ldots, a_\ell)$–homogeneous of degree p, it follows from the Euler formula that $\eta(\Delta) = p\Delta$, so $\eta \in D_R(\Delta)$. Clearly, η is $(d_1, \ldots, d_\ell)$–homogeneous of degree 0. Thus there exists $k \neq 0$ such that $\eta = k\bar\theta_1$. □

Definition 6.65 *A polynomial $P \in \mathbb{C}[T_1, \ldots, T_\ell]$ is called **weighted homogeneous** with weights $w_1, \ldots, w_\ell$ if it is a sum of monomials $T_1^{a_1} \cdots T_\ell^{a_\ell}$ such that $\sum a_i/w_i = 1$.*

Corollary 6.66 *Let G be an irreducible unitary reflection group. Its discriminant $\Delta(T_1, \ldots, T_\ell; \mathcal{F})$ is weighted homogeneous with weights $(m+n)/d_1, \ldots, (m+n)/d_\ell$. These weights are unique and independent of $\mathcal{F}$.*

This assertion need not hold for reducible groups. For example, if G is of type $A_1 \times A_1$ acting naturally in $\mathbb{C}^2$, then $Q = x_1 x_2$. If we choose $\mathcal{F} = \{x_1^2, x_2^2\}$, then $\Delta(T_1, T_2; \mathcal{F}) = T_1 T_2$, which is (a_1, a_2)–homogeneous of degree $a_1 + a_2$ for any positive integers a_1, a_2. The weights are $(a_1 + a_2)/a_1, (a_1 + a_2)/a_2$.

Let $\mathcal{F} = \{f_1, \ldots, f_\ell\}$ be a set of basic invariants and let $\Theta = \{\theta_1, \ldots, \theta_\ell\}$ be a set of basic derivations for G. Let $\mathsf{J}(\mathcal{F}) = \mathsf{J}(f_1, \ldots, f_\ell)$ be the Jacobian matrix of Definition 6.6 and let $\mathsf{M}(\Theta) = \mathsf{M}(\theta_1, \ldots, \theta_\ell)$ be the coefficient matrix of Definition 4.11. Since

(2) $$(\mathsf{J}^T \mathsf{M})_{i,j} = \sum (\theta_j x_k)(D_k f_i) = \theta_j f_i,$$

it follows that $\mathsf{J}^T \mathsf{M} \in M_\ell(R)$. Let $T_1, \ldots, T_\ell$ be indeterminates. Since the basic invariants are algebraically independent, there exist unique polynomials $\psi_{i,j} \in \mathbb{C}[T_1, \ldots, T_\ell]$ such that $(\mathsf{J}^T \mathsf{M})_{i,j} = \psi_{i,j}(f_1, \ldots, f_\ell)$.

Definition 6.67 *Let $\mathcal{F} = \{f_1, \ldots, f_\ell\}$ be a set of basic invariants and let $\Theta = \{\theta_1, \ldots, \theta_\ell\}$ be a set of basic derivations for G. Define the* **discriminant matrix** $\mathsf{M}_\Delta(T_1, \ldots, T_\ell; \mathcal{F}, \Theta)$ *by* $\mathsf{M}_\Delta(T_1, \ldots, T_\ell; \mathcal{F}, \Theta)_{i,j} = \psi_{i,j}(T_1, \ldots, T_\ell)$. *Thus*

$$\mathsf{M}_\Delta(f_1, \ldots, f_\ell; \mathcal{F}, \Theta) = \mathsf{J}(\mathcal{F})^T \mathsf{M}(\Theta).$$

Recall that the columns of the matrix $\mathsf{M}(\Theta)$ are the coefficients of $\theta_1, \ldots, \theta_\ell$. In view of formulas (1) and (2), the columns of the matrix $\mathsf{M}_\Delta(f_1, \ldots, f_\ell; \mathcal{F}, \Theta)$ are the coefficients of $\bar{\theta}_1, \ldots, \bar{\theta}_\ell$ when written as R–linear combinations of $D_{f_1}, \ldots, D_{f_\ell}$.

Proposition 6.68

$$\det \mathsf{M}_\Delta(T_1, \ldots, T_\ell; \mathcal{F}, \Theta) \doteq \Delta(T_1, \ldots, T_\ell; \mathcal{F}).$$

Thus the discriminant matrix depends on $\mathcal{F}$ and Θ, but its determinant depends only on $\mathcal{F}$.

Proof. We have

$$\begin{aligned}
\det \mathsf{M}_\Delta(f_1, \ldots, f_\ell; \mathcal{F}, \Theta) &= \det(\mathsf{J}(\mathcal{F})^T \mathsf{M}(\Theta)) \\
&= (\det \mathsf{J}(\mathcal{F}))(\det \mathsf{M}(\Theta)) \\
&\doteq JQ = \Delta(f_1, \ldots, f_\ell; \mathcal{F})
\end{aligned}$$

as required. $\square$

Example 6.69 Recall the definition of G_{24} from Example 6.31. We compute its discriminant matrix. Basic invariants are derived from Klein's famous quartic [125], whose automorphism group is the simple group of order 168. The notation is simplified by the following convention: in formulas where the subscripts i,

$i+1$, $i+2$ appear, they represent the integers 1,2,3 in cyclic permutation, and summation is over 1,2,3.

$$f_1 = z_1^3 z_2 + z_2^3 z_3 + z_3^3 z_1 = \sum z_i^3 z_{i+1},$$
$$f_2 = 5 z_1^2 z_2^2 z_3^2 - \sum z_i^5 z_{i+2},$$
$$f_3 = \sum [z_i^{14} - 34 z_i^{11} z_{i+1}^2 z_{i+2} + 18 z_i^7 z_{i+1}^7 - 250 z_i^4 z_{i+1}^9 z_{i+2}$$
$$+ 375 z_i^8 z_{i+i}^4 z_{i+2}^2 - 126 z_i^5 z_{i+1}^6 z_{i+2}^3].$$

We provide a method for computing basic derivations for all unitary reflection groups in Appendix B. Applied in this case, our method yields

$$\theta_1 = \sum z_i D_i,$$
$$\theta_2 = (1/2) \sum [z_{i+1}^9 - z_i^7 z_{i+1}^2 + 25 z_i^4 z_{i+1}^4 z_{i+2} - 19 z_i z_{i+1}^6 z_{i+2}^2$$
$$- 38 z_i^5 z_{i+1} z_{i+2}^3 + 25 z_i^2 z_{i+1}^3 z_{i+2}^4 + 2 z_i^3 z_{i+2}^6 + 9 z_{i+1}^2 z_{i+2}^7] D_i,$$
$$\theta_3 = \sum [212 z_i^5 z_{i+1}^6 - 96 z_i^9 z_{i+1} z_{i+2} - 404 z_i^2 z_{i+1}^8 z_{i+2} + 1308 z_i^6 z_{i+1}^3 z_{i+2}^2$$
$$- 848 z_i^3 z_{i+1}^5 z_{i+2}^3 - 250 z_i^7 z_{i+2}^4 + 154 z_i^7 z_{i+1}^4 z_{i+2}^4 + 230 z_i^4 z_{i+1}^2 z_{i+2}^5$$
$$- 866 z_i z_{i+1}^4 z_{i+2}^6 + 58 z_i^2 z_{i+1} z_{i+2}^8 + 14 z_{i+2}^{14}] D_i.$$

The rest of the calculation consists of expressing each entry of $J^T M$ as a polynomial in the f_i. We get $\mathsf{M}_\Delta(T_1, T_2, T_3; \mathcal{F}, \Theta) =$

$$\begin{bmatrix} 4T_1 & 6T_2^2 & 14T_3 - 288 T_1^2 T_2 \\ 6T_2 & (-1/2) T_3 & 1024 T_1 T_2^2 - 224 T_1^4 \\ 14T_3 & \begin{matrix} 1024 T_1 T_2^3 - 24 T_1^2 T_3 \\ -224 T_1^4 T_2 \end{matrix} & \begin{matrix} 2296 T_1 T_2 T_3 - 2240 T_1^3 T_2^2 \\ +1792 T_1^6 - 4704 T_2^4 \end{matrix} \end{bmatrix}.$$

See Example 6.118 for more on this discriminant.

A Character Formula

Suppose $\mathbb{K} = \mathbb{C}$. Let $e_1, \ldots, e_\ell$ be a basis for V.

Proposition 6.70

$$(S \otimes \Lambda V)^G = \oplus_{p=0}^\ell \bigoplus_{i_1 < \cdots < i_p} R u_{i_1} \wedge \cdots \wedge u_{i_p}.$$

Proof. It follows from Lemma 6.48 that

$$(S \otimes V)^G = R u_1 \oplus \ldots \oplus R u_\ell$$

where $u_1, \ldots, u_\ell$ is a homogeneous basis for $(U \otimes V)^G$. The u_i are the images of a set of basic derivations under the map $D_v \mapsto v$. Let Θ be a basis for Der_S^G.

Since $u_1 \wedge \cdots \wedge u_\ell \doteq \mathsf{M}(\Theta)(e_1 \wedge \cdots \wedge e_\ell) \doteq Q(e_1 \wedge \cdots \wedge e_\ell)$, the assertion follows from Proposition 6.47. □

For $g \in G$, let $k(g) = \dim \text{Fix}(g)$. Shephard and Todd [210] found a formula which involves the exponents of G:

$$\sum_{g \in G} t^{k(g)} = \prod_{i=1}^{\ell}(t + m_i).$$

Setting $t = 1$ in this formula shows that $|G| = d_1 \cdots d_\ell$. Proposition 6.70 may be used to obtain a related character formula; see [172, 3.10].

Theorem 6.71 *Let G be a unitary reflection group with coexponents $n_1, \ldots, n_\ell$. Then*

$$\sum_{g \in G}(-1)^{\ell-k(g)} \det(g) t^{k(g)} = \prod_{i=1}^{\ell}(t + n_i).$$

Proof. Recall from Lemma 6.8 that if $[g]$ denotes the matrix for g in its action on V, then its action on V^* has matrix $[g^{-1}]^T$. Let $\lambda_1(g), \ldots, \lambda_\ell(g)$ be the eigenvalues of $[g]$. Let x and y be indeterminates. Since $\deg u_i = n_i$, Proposition 6.70 gives the Poincaré series

$$(3) \qquad \sum_{n \geq 0} \sum_{p \geq 0} \dim(S_n \otimes \Lambda^p V)^G x^n y^p = \prod_{i=1}^{\ell} \frac{1 + x^{n_i} y}{1 - x^{d_i}}.$$

The eigenvalues of $g|(\Lambda^p V)$ are $\lambda_{i_1}(g) \cdots \lambda_{i_p}(g)$ where $i_1 < \cdots < i_p$. Thus

$$\sum_{p \geq 0} \text{tr}(g|\Lambda^p V) y^p = \det(1 + gy)$$

where det denotes the determinant of the endomorphism $1 + gy$ of $\mathbb{C}[y] \otimes V$. Let $E_{n,p} = S_n \otimes \Lambda^p V$. Then

$$\sum_{n,p \geq 0} \text{tr}(g|E_{n,p}) x^n y^p = \left(\sum_{n \geq 0} \text{tr}(g|S_n) x^n\right)\left(\sum_{p \geq 0} \text{tr}(g|\Lambda^p V) y^p\right)$$

$$= \frac{\det(1 + gy)}{\det(1 - g^{-1}x)}.$$

Recall that if N is any finite dimensional G–module, then the dimension of the space of G–invariants is given by $|G| \dim(N^G) = \sum_{g \in G} \text{tr}(g|N)$. We substitute (3) in the last equation to get

$$(4) \qquad \sum_{g \in G} \frac{\det(1 + gy)}{\det(1 - g^{-1}x)} = |G| \prod_{i=1}^{\ell} \frac{1 + x^{n_i} y}{1 - x^{d_i}}.$$

Let L and R denote the left and right sides of (4). We compare their values at the pole $x = 1$. Fix $g \in G$ and number the eigenvalues so $\lambda_i(g^{-1}) = \lambda_i(g)^{-1}$. Then

$$(5) \qquad \frac{\det(1+gy)}{\det(1-g^{-1}x)} = \prod_{i=1}^{\ell} \frac{1+\lambda_i(g)y}{1-\lambda_i(g)^{-1}x}.$$

Recall the substitution $y = t(1-x) - 1$ used in the proof of Proposition 4.131. The same substitution gives

$$(6) \qquad \frac{1+\lambda_i(g)y}{1-\lambda_i(g)^{-1}x} = \frac{1-\lambda_i(g)}{1-\lambda_i(g)^{-1}x} + \frac{t(1-x)\lambda_i(g)}{1-\lambda_i(g)^{-1}x}.$$

If $\lambda_i(g) = 1$, then (6) contributes t to the product (5) at $x = 1$. If $\lambda_i(g) \neq 1$, then (6) contributes $(1-\lambda_i(g))/(1-\lambda_i(g)^{-1}) = -\lambda_i(g)$ to the product (5) at $x = 1$. This gives $L|_{x=1} = \sum_{g \in G}(-1)^{\ell-k(g)}\det(g)t^{k(g)}$.

Make the same substitution in R. Then $1 + x^{n_i}y = 1 - x^{n_i} + t(1-x)x^{n_i}$. This shows that both the numerator and the denominator of R have zeros of order ℓ at $x = 1$. We know from above that $|G| = d_1 \cdots d_\ell$. Thus we get $R|_{x=1} = \prod(t + n_i)$. □

6.4 The Structure of $L(\mathcal{A})$

The group G acts on the lattice $L(\mathcal{A})$. The orbits of this action were computed for irreducible finite Coxeter groups in [175] and for the remaining unitary reflection groups in [174]. We present the results in this section. It follows from Corollary 6.28 that for $X \in L$ the subarrangement $\mathcal{A}_X$ is the arrangement of the reflection group G_X. Thus $\mathcal{A}_X$ may be considered known. On the other hand, the restriction $\mathcal{A}^X$ is not necessarily a reflection arrangement. We also study these restrictions in this section. Let $\rho : V^* \to X^*$ be the restriction map. The dependence on X will be clear from the context. If $\mathcal{A}$ is the set of kernels of linear forms α, then $\mathcal{A}^X$ is the set of kernels of the nonzero linear forms $\rho(\alpha)$. Write $y_i = \rho(x_i)$ for all i.

If G is reducible and $V = V_1 \oplus V_2$ where V_i is stable under G, then the restriction G_i of G to V_i is a unitary reflection group in V_i and defines an arrangement $\mathcal{A}_i$ in V_i. Then $\mathcal{A} = \mathcal{A}_1 \times \mathcal{A}_2$ and all relevant information about $\mathcal{A}$ may be obtained from the corresponding information about the $\mathcal{A}_i$. Thus it suffices to consider the irreducible groups and their arrangements.

The Symmetric Group

The braid arrangement of Example 1.9 is the product of the irreducible reflection representation of the Coxeter group of type $A_{\ell-1}$ with an empty 1–arrangement. Recall that for $1 \leq i < j \leq \ell$, the hyperplanes are $H_{i,j} = \ker(x_i - x_j)$ and

$$Q(\mathcal{A}) = \prod_{1 \leq i < j \leq \ell}(x_i - x_j).$$

We showed in Proposition 2.9 that $L(\mathcal{A})$ is isomorphic to the partition lattice. Recall our notation. Let $I = \{1, \ldots, \ell\}$. Let $\mathcal{P}(\ell)$ be the set of partitions of

I. An element of $\mathcal{P}(\ell)$ is a collection $\Lambda = \{\Lambda_1, \ldots, \Lambda_r\}$ of nonempty pairwise disjoint subsets of I, called the blocks of Λ, whose union is I. Let $\lambda_q = |\Lambda_q|$ and agree to number so $\lambda_1 \geq \ldots \geq \lambda_p > 0$. Let $\|\Lambda\| = (\lambda_1, \ldots, \lambda_p)$ be the corresponding partition of ℓ. For simplicity we write $\lambda = \|\Lambda\|$. Let $b_i > 0$ be the number of blocks of Λ with cardinality i, and let $b_\lambda = b_1! b_2! \ldots (1!)^{b_1} (2!)^{b_2} \ldots$. The number of partitions of I of type $(b_1, \ldots, b_\ell)$ is $\ell!/b_\lambda$. The sum $\sum \ell!/b_\lambda$ over all partitions of ℓ into p parts is the Stirling number $S(\ell, p)$; see [1, p.70]. Let $P(\ell)$ denote the number of partitions of ℓ.

Proposition 6.72 *Let $\mathcal{A}$ be the braid arrangement and let $L = L(\mathcal{A})$ be the partition lattice. Two subspaces $X = \Lambda$ and $X' = \Lambda'$ lie in the same G orbit if and only if $\|\Lambda\| = \|\Lambda'\|$. The cardinality of the orbit of $X = \Lambda$ with $\lambda = \|\Lambda\|$ is*

$$|\mathcal{O}_X| = \frac{\ell!}{b_\lambda}.$$

The number of p-dimensional subspaces in L is $S(\ell, p)$. The number of G orbits on L is $P(\ell)$.

Proposition 6.73 *Let $\mathcal{A}$ be the braid arrangement and let $L = L(\mathcal{A})$ be the partition lattice. If $X \in L$ and $\dim X = p$, then $\mathcal{A}^X$ is isomorphic to the braid arrangement on p strands. Thus $\mathcal{A}^X$ is free with $\exp \mathcal{A}^X = \{0, 1, \ldots, p-1\}$. In particular,*

$$\pi(\mathcal{A}^X, t) = (1+t)(1+2t) \cdots (1+(p-1)t).$$

Proof. Arguing by induction, it suffices to show that $\mathcal{A}^H$ is isomorphic to the braid arrangement on $\ell - 1$ strands for all $H \in \mathcal{A}$. Since G acts transitively on $\mathcal{A}$, we may assume that $H = H_{\ell-1,\ell}$. Then $\rho(x_i - x_j) = y_i - y_j$ for $1 \leq i < j \leq \ell - 1$, $\rho(x_i - x_\ell) = y_i - y_{\ell-1}$ for $1 \leq i < \ell - 1$, and $\rho(x_{\ell-1} - x_\ell) = 0$. □

The Full Monomial Group

In describing the monomial groups and their arrangements, we use the notation for the symmetric group Sym_ℓ and the partition lattice above. Let $r \geq 2$ be an integer, let $\theta = \exp(2\pi i/r)$, and let $C(r)$ be the cyclic group generated by θ. Let $\sigma \in Sym_\ell$ and let $\epsilon : I \to C(r)$ be any function. The full monomial group $G \subset GL(V)$ of Example 6.29 is the group of all transformations

$$g(\sigma, \epsilon) e_i = \epsilon(i) e_{\sigma(i)}.$$

Its reflecting hyperplanes are $H_i = \ker(x_i)$ where $1 \leq i \leq \ell$ and $H_{i,j}(\zeta) = \ker(x_i - \zeta x_j)$ where $1 \leq i \neq j \leq \ell$ and $\zeta \in C(r)$. Let $\mathcal{A}_\ell(r)$ be the arrangement consisting of these reflecting hyperplanes. We fix r and write $\mathcal{A}_\ell = \mathcal{A}_\ell(r)$.

Let Ω be a nonempty subset of I and let $\nu : \Omega \to C(r)$. Define a subspace $Y(\Omega, \nu)$ by

$$Y(\Omega, \nu) = \bigcap_{i,j \in \Omega} H_{i,j}(\nu(i)\nu(j)^{-1}).$$

We agree that $H_{i,i}(1) = V$. Since $H_{i,j}(\nu(i)\nu(j)^{-1}) \cap H_{j,k}(\nu(j)\nu(k)^{-1}) \subseteq H_{i,k}(\nu(i)\nu(k)^{-1})$, we see that if $\Omega = \{1, \ldots, m\}$, then $Y(\Omega, \nu)$ is the space defined by

$$\nu(1)^{-1}x_1 = \cdots = \nu(m)^{-1}x_m.$$

If $W \subseteq V$ is a subspace of V, let W^0 be its annihilator in V^*. Since $Y(\Omega, \nu)^0 = \sum_{i,j \in \Omega} \mathbb{C}(\nu(j)x_i - \nu(i)x_j)$, we have $\mathrm{codim} Y(\Omega, \nu) = |\Omega| - 1$. Note that $Y(\Omega, \nu) = Y(\Omega, \zeta\nu)$ for any $\zeta \in C(r)$. Let $J \subseteq I$ be any subset and let $\nu : J \to C(r)$. Let $\Lambda = \{\Lambda_1, \ldots, \Lambda_p\}$ be a partition of J and let $\nu_q = \nu|\Lambda_q$ be the restriction of ν to Λ_q for $1 \leq q \leq p$. Define

$$Y(\Lambda, \nu) = \bigcap_{q=1}^{p} Y(\Lambda_q, \nu_q).$$

Since the Λ_q are disjoint, we have $Y(\Lambda, \nu)^0 = \oplus_{q=1}^{p} Y(\Lambda_q, \nu_q)^0$, so $\mathrm{codim} Y(\Lambda, \nu) = |J| - p$. If $\Delta \subseteq I$, define

$$Z(\Delta) = \bigcap_{i \in \Delta} H_i.$$

If Δ is empty, we agree that $Z(\Delta) = V$. Since $Z(\Delta)^0 = \sum_{i \in \Delta} \mathbb{C}x_i$, we have $\mathrm{codim} Z(\Delta) = |\Delta|$.

Let $\mathcal{T}(I)$ be the set of all triples (Δ, Λ, ν) where $\Delta \subseteq I$, Λ is a partition of $J = I \setminus \Delta$ and $\nu : J \to C(r)$. If $(\Delta, \Lambda, \nu) \in \mathcal{T}$, define a subspace $X(\Delta, \Lambda, \nu) = Y(\Lambda, \nu) \cap Z(\Delta)$. Since $Y(\Lambda, \nu)$ and $Z(\Delta)$ are defined as intersections of elements of $\mathcal{A}_\ell$, we have $X(\Delta, \Lambda, \nu) \in \mathcal{A}_\ell$. Moreover, if Λ has p blocks, then $\mathrm{codim} X(\Delta, \Lambda, \nu) = \mathrm{codim} Y(\Lambda, \nu) + \mathrm{codim} Z(\Delta) = \ell - p$.

If $J \subseteq I$, $\psi : J \to C(r)$, and $\sigma \in \mathrm{Sym}_\ell$, define $\sigma\psi : \sigma J \to C(r)$ by $(\sigma\psi)(i) = \psi(\sigma^{-1}i)$. Then $\sigma(\psi\psi') = \sigma\psi \cdot \sigma\psi'$ for all ψ, ψ'. The action of G on the space $X(\Delta, \Lambda, \nu)$ is given by

(1) $$g(\sigma, \epsilon) X(\Delta, \Lambda, \nu) = X(\sigma\Delta, \sigma\Lambda, \sigma(\epsilon\nu))$$

where $\sigma\Lambda = \{\sigma\Lambda_1, \ldots, \sigma\Lambda_p\}$. In order to check (1), it suffices to show that $g(\sigma, \epsilon) H_{i,j}(\zeta) = H_{\sigma i, \sigma j}(\zeta \epsilon(i) \epsilon(j)^{-1})$.

Proposition 6.74 *The map $\pi : \mathcal{T}(I) \to L(\mathcal{A}_\ell)$ defined by $\pi(\Delta, \Lambda, \nu) = X(\Delta, \Lambda, \nu)$ is surjective. We have $\pi(\Delta, \Lambda, \nu) = \pi(\Delta', \Lambda', \nu')$ if and only if $\Delta = \Delta'$, $\Lambda = \Lambda'$, and for each block Λ_q of Λ there exists $\zeta_q \in C(r)$ such that $\nu'_q = \zeta_q \nu_q$. Thus if Λ has p blocks, then $|\pi^{-1} X(\Delta, \Lambda, \nu)| = r^p$.*

Proof. Choose $X \in L(\mathcal{A}_\ell)$. We construct a triple (Δ, Λ, ν) such that $\pi(\Delta, \Lambda, \nu) = X$. Let $\Delta = \{i \in I \mid X \subseteq H_i\}$. Define an equivalence relation on $J = I \setminus \Delta$ as follows. Agree that $i \sim i$ and for $i \neq j$ that $i \sim j$ if there exists $\zeta \in C(r)$ such that $X \subseteq H_{i,j}(\zeta)$. This is an equivalence relation because $H_{i,j}(\zeta) = H_{j,i}(\zeta^{-1})$ and $H_{i,j}(\zeta) \cap H_{j,k}(\zeta') \subseteq H_{i,k}(\zeta\zeta')$ for $i \neq k$. If $i = k$, then $x_i(1 - \zeta\zeta')$ vanishes on X.

Since $i \notin \Delta$, this implies that $\zeta' = \zeta^{-1}$, and thus $H_{i,j}(\zeta) \cap H_{j,i}(\zeta') \subseteq H_{i,i}(1) = V$. This also shows that if $i, j \in J$, $i \neq j$, and $i \sim j$, then $X \subseteq H_{i,j}(\zeta)$ for a unique $\zeta \in C(r)$ which we denote by $\eta(i,j)$.

Let $\Lambda = \{\Lambda_1, \ldots, \Lambda_p\}$ be the partition of J defined by $\sim$. Let η_q be the restriction of η to Λ_q. Define $\eta_q(i,i) = 1$. Since $\eta_q(i,j)\eta_q(j,k)\eta_q(k,i) = 1$, there exists $\nu_q : \Lambda_q \to C(r)$ such that $\eta_q(i,j) = \nu_q(i)\nu_q(j)^{-1}$. Define $\nu : J \to C(r)$ by $\nu|\Lambda_q = \nu_q$. By construction, X and $X(\Delta, \Lambda, \nu)$ are contained in the same hyperplanes of $\mathcal{A}_\ell$. Since they both lie in $L(\mathcal{A}_\ell)$, they are equal. This proves that π is surjective. The function ν_q is determined uniquely up to a replacement by $\zeta_q \nu_q$ for some $\zeta_q \in C(r)$. The r^p triples which map to X are obtained this way. □

Proposition 6.75 *Two subspaces $X(\Delta, \Lambda, \nu)$ and $X(\Delta', \Lambda', \nu')$ lie in the same G orbit if and only if $|\Delta| = |\Delta'|$ and $\|\Lambda\| = \|\Lambda'\|$. The cardinality of the orbit of $X = X(\Delta, \Lambda, \nu)$ is*

$$|\mathcal{O}_X| = r^{j-p} \binom{\ell}{j} \frac{j!}{b_\lambda}$$

where $j = |I \setminus \Delta|$ and $\lambda = \|\Lambda\|$.

Proof. Formula (1) shows that these conditions are necessary. Conversely, if $|\Delta| = |\Delta'|$ and $\|\Lambda\| = \|\Lambda'\|$, then there exists $\sigma \in Sym_\ell$ with $\sigma\Delta = \Delta'$ and $\sigma\Lambda = \Lambda'$. Let $J = I \setminus \Delta$ and let $J' = I \setminus \Delta'$. Then $J' = \sigma J$. Choose $\delta : J \to C(r)$ so $\sigma(\delta\nu) = \nu'$ and let $\epsilon : I \to C(r)$ be any extension of δ. Then (1) shows that $g(\sigma, \epsilon) X(\Delta, \Lambda, \nu) = X(\Delta', \Lambda', \nu')$. It follows from Proposition 6.74 that $|\mathcal{O}_X|$ is r^{-p} times the number of triples (Δ, Λ, ν) with $|\Delta| = |\Delta'|$ and $\|\Lambda\| = \|\Lambda'\|$. There are $\binom{\ell}{j}$ choices for Δ'. Once Δ' is chosen, there are r^j choices for ν'. The number of partitions Λ' with $\|\Lambda\| = \|\Lambda'\|$ is $j!/b_\lambda$ by Proposition 6.72. □

Corollary 6.76 *The number of p-dimensional subspaces of $L(\mathcal{A}_\ell)$ is*

$$\sum_{j=0}^{\ell} r^{j-p} \binom{\ell}{j} S(j,p).$$

The number of G orbits on $L(\mathcal{A}_\ell)$ is $P(0) + \cdots + P(\ell)$.

Proposition 6.77 *The arrangement $\mathcal{A}_\ell$ is free with*

$$\exp \mathcal{A}_\ell = \{1, r+1, 2r+1, \ldots, (\ell-1)r+1\}.$$

If $X \in L(\mathcal{A}_\ell)$ has dimension p, then $\mathcal{A}_\ell^X$ is isomorphic to $\mathcal{A}_p$. Thus $\mathcal{A}^X$ is free with $\exp \mathcal{A}^X = \{1, r+1, \ldots, (p-1)r+1\}$. In particular,

$$\pi(\mathcal{A}^X, t) = (1+t)(1+(r+1)t) \cdots (1 + [(p-1)r+1]t).$$

Proof. It follows from Theorem 6.60 that $\mathcal{A}_\ell$ is free. We have

$$Q(\mathcal{A}) = x_1 \cdots x_\ell \prod_{1 \leq i < j \leq \ell} (x_i^r - x_j^r).$$

For $1 \leq i \leq \ell$ we may choose the derivations

$$\theta_i = \sum_{j=1}^{\ell} x_j^{(i-1)r+1} D_j.$$

Direct calculation shows that $\theta_i \in D(\mathcal{A})$. It follows from Saito's criterion 4.19 that θ_i form a basis for $D(\mathcal{A})$. For the restriction it suffices to show that if $H \in \mathcal{A}_\ell$, then $\mathcal{A}_\ell^H$ is isomorphic to $\mathcal{A}_{\ell-1}$. Since G has two orbits on $\mathcal{A}_\ell$, we may assume that $H = H_\ell$ or $H = H_{\ell-1,\ell}(1)$. If $H = H_\ell$, then the set of restrictions to H of the forms which define $\mathcal{A}_\ell$ consists of the forms $0, y_i$, and $y_i - \zeta y_j$, so the assertion is clear. If $H = H_{\ell-1,\ell}(1)$, then $\rho(x_{\ell-1} - \zeta x_\ell) = (1 - \zeta)y_{\ell-1}$ also occurs as a restriction. It is superfluous, however, since it is 0 or it defines the same hyperplane as $y_{\ell-1}$. □

The Monomial Group $G(r, r, \ell)$

The full monomial group has certain irreducible subgroups generated by reflections. Let p be a divisor of r. Let $G(r, p, \ell)$ be the subgroup of the full monomial group consisting of all $g(\sigma, \epsilon)$ where $\prod \epsilon(i)$ is a power of θ^p. These groups are generated by reflections and are irreducible since $r \geq 2$. If $p < r$, then $G(r, p, \ell)$ contains the reflections $e_i \mapsto \theta^p e_i$ and $e_i \mapsto e_j$ for $i \neq j$. The corresponding arrangement is the same as the arrangement of the full monomial group $G = G(r, 1, \ell)$, which we have described already. However, if $p = r$, then the arrangement $\mathcal{A}_\ell^0(r)$ corresponding to the subgroup $G^0 = G(r, r, \ell)$ consists of the hyperplanes $H_{i,j}(\zeta)$ where $1 \leq i \neq j \leq \ell$ and $\zeta \in C(r)$. For $r = 2$, the full monomial group $G(2, 1, \ell)$ is the Weyl group of type B_ℓ and $G(2, 2, \ell)$ is the Weyl group of type D_ℓ.

We fix r and write $\mathcal{A}_\ell^0 = \mathcal{A}_\ell^0(r)$. Since $\mathcal{A}_\ell^0 \subseteq \mathcal{A}_\ell$, $L(\mathcal{A}_\ell^0) \subseteq L(\mathcal{A}_\ell)$, so every $X \in L(\mathcal{A}_\ell^0)$ may be written as $X = X(\Delta, \Lambda, \nu) = Y(\Lambda, \nu) \cap Z(\Delta)$ for some subset Δ of I, partition Λ of $J = I \setminus \Delta$, and function $\nu : J \to C(r)$. Let $\mathcal{S}(I)$ be the subset of $\mathcal{T}(I)$ consisting of all triples (Δ, Λ, ν) with $|\Delta| \neq 1$.

Proposition 6.78 *The map* $\pi : \mathcal{S}(I) \to L(\mathcal{A}_\ell^0)$ *defined by* $\pi(\Delta, \Lambda, \nu) = X(\Delta, \Lambda, \nu)$ *is surjective. We have* $\pi(\Delta, \Lambda, \nu) = \pi(\Delta', \Lambda', \nu')$ *if and only if* $\Delta = \Delta'$, $\Lambda = \Lambda'$, *and for each block* Λ_q *of* Λ *there exists* $\zeta_q \in C(r)$ *such that* $\nu'_q = \zeta_q \nu_q$. *Thus if* Λ *has* p *blocks, then* $|\pi^{-1} X(\Delta, \Lambda, \nu)| = r^p$.

Proof. Clearly, $Y(\Lambda, \nu) \in L(\mathcal{A}_\ell^0)$ for all Λ, ν. If $i \neq j$ and $\zeta \neq \zeta'$ are in $C(r)$, then $H_i \cap H_j = H_{i,j}(\zeta) \cap H_{i,j}(\zeta') \in L(\mathcal{A}_\ell^0)$. Thus $Z(\Delta) \in L(\mathcal{A}_\ell^0)$ if $|\Delta| \geq 2$. This proves that $\pi \mathcal{S}(I) \subseteq L(\mathcal{A}_\ell^0)$. To prove that π is surjective, choose $X \in L(\mathcal{A}_\ell^0)$. Recall that the subset Δ associated with X is $\Delta = \{i \in I \mid X \subseteq H_i\}$. To prove that $X \in \pi \mathcal{S}(I)$ we must show that $|\Delta| \neq 1$. Suppose $|\Delta| \geq 1$. Then $X \subseteq H_i$ for some i. Let S be the set of all linear forms $x_k - \zeta x_j$ such that $X \subseteq H_{k,j}(\zeta)$. Since

$H_i \not\subseteq L(\mathcal{A}_\ell^0)$, we must have $x_i \in \sum_S \mathbb{C}(x_k - \zeta x_j)$. Since $H_{k,j}(\zeta) = H_{j,k}(\zeta^{-1})$, we may assume that some k is i. Then both x_i and $x_i - \zeta x_j$ vanish on X for some ζ, j so $X \subseteq H_j$. Thus $|\Delta| \geq 2$. This proves that π is surjective. The rest of the argument follows as in Proposition 6.74. □

The next result is a corrected version of [174, 2.9].

Proposition 6.79 *If $X(\Delta, \Lambda, \nu)$ and $X(\Delta', \Lambda', \nu')$ lie in the same G^0-orbit, then $|\Delta| = |\Delta'|$ and $\|\Lambda\| = \|\Lambda'\|$. Conversely, suppose $|\Delta| = |\Delta'|$ and $\|\Lambda\| = \|\Lambda'\|$.*
(1) If $|\Delta| \geq 2$, then $X(\Delta, \Lambda, \nu)$ and $X(\Delta', \Lambda', \nu')$ lie in the same G^0-orbit. The cardinality of this orbit is as in Proposition 6.75.
(2) If $|\Delta| = 0$, then the G-orbit determined by $\|\Lambda\| = (\lambda_1, \ldots, \lambda_p)$ is the union of d orbits of G^0 where d is the greatest common divisor $d = (r, \lambda_1, \ldots, \lambda_p)$. The cardinality of this orbit is

$$|\mathcal{O}_X| = \frac{r^{\ell-p}\ell!}{d\, b_\lambda}.$$

Proof. It follows from (1) that the conditions $|\Delta| = |\Delta'|$ and $\|\Lambda\| = \|\Lambda'\|$ are necessary. For sufficiency we argue as in Proposition 6.75 up to the point where we choose $\epsilon : I \to C(r)$ to extend the map $\delta : J \to C(r)$. The full monomial group G contains all $g(\sigma, \nu)$, so any extension of δ will do. In the group G^0 there is the restriction $\epsilon(1) \cdots \epsilon(\ell) = 1$. If $|\Delta| \geq 2$, then there is sufficient freedom to choose a suitable ϵ. If $|\Delta| = 0$, then $J = I$ and there is a unique map $\delta : I \to C(r)$ such that $\delta\nu = \sigma^{-1}\nu'$. If $\delta(1) \cdots \delta(\ell) = 1$, then $g(\sigma, \delta) \in G^0$ and thus (1) shows that $Y(\Lambda, \nu)$ and $Y(\Lambda', \nu')$ are in the same G^0-orbit.

The remaining possibility is that we may change the map ν which represents the space $Y(\Lambda, \nu)$. Recall from Proposition 6.74 that this may be done in r^p ways. If $\nu_q = \nu|\Lambda_q$, then the function ν_q is determined uniquely up to a replacement by $\zeta_q \nu_q$ for some $\zeta_q \in C(r)$. Define $\tau : I \to C(r)$ by $\tau(i) = \zeta_q$ if $i \in \Lambda_q$. We may replace ν by $\tau\nu$. It follows that $Y(\Lambda, \nu)$ and $Y(\Lambda', \nu')$ are in the same G^0-orbit if and only if there exists τ such that $\epsilon = \delta\tau \in G^0$. Let $\zeta_q = \theta^{a_q}$. The integers a_q are the free parameters in the choice of τ. We may choose a_q so

$$\tau(1) \cdots \tau(\ell) = \theta^{a_1\lambda_1 + \cdots + a_p\lambda_p}$$

is any power of θ^d. Since it is always divisible by θ^d, we conclude that $\epsilon \in G^0$ if and only if $\delta(1) \cdots \delta(\ell)$ is a power of θ^d. In particular, if we fix $\|\Lambda\|$, then the single G-orbit represented by $Y(\Lambda, \nu)$ is the union of d orbits of G^0 corresponding to

$$\nu(1) \cdots \nu(\ell) = 1, \theta, \theta^2, \ldots, \theta^{d-1}$$

as required. □

Example 6.80 Let $r = \ell = 4$ and consider $\|\Lambda\| = (2, 2)$. The single G-orbit represented by $\{x_1 = x_2, x_3 = x_4\}$ is the union of two G^0-orbits represented

Table 6.2. Restrictions

k	H		Type $\mathcal{A}^H$
0	Arbitrary		$\mathcal{A}_{\ell-1}^1(r)$
$1,\ldots,\ell-1$	$H_{i,j}(\zeta)$	$1 \leq i < j \leq k < \ell$	$\mathcal{A}_{\ell-1}^{k-1}(r)$
$1,\ldots,\ell-1$	$H_{i,j}(\zeta)$	$1 \leq i \leq k < j < \ell$	$\mathcal{A}_{\ell-1}^{k}(r)$
$1,\ldots,\ell-1$	$H_{i,j}(\zeta)$	$1 < k < i < j \leq \ell$	$\mathcal{A}_{\ell-1}^{k+1}(r)$
$1,\ldots,\ell-1$	H_i	$1 \leq i \leq \ell$	$\mathcal{A}_{\ell-1}^{\ell-1}(r)$
ℓ	Arbitrary		$\mathcal{A}_{\ell-1}^{\ell-1}(r)$

by $\{x_1 = x_2, x_3 = x_4\}$ and $\{x_1 = ix_2, x_3 = x_4\}$. Note that $\{x_1 = x_2, x_3 = x_4\}$ and $\{x_1 = -x_2, x_3 = x_4\}$ are in the same G^0-orbit. Here $\delta(1)\cdots\delta(4) = -1$ and we may choose τ to be one of several maps, for example $\tau(1) = i$, $\tau(2) = i$, $\tau(3) = 1$, $\tau(4) = 1$. In this case $\epsilon = \delta\tau$ is the diagonal matrix $(i, -i, 1, 1) \in G^0$.

Corollary 6.81 *The number of p-dimensional subspaces in $L(\mathcal{A}_\ell^0)$ is*

$$\sum_{\substack{j=p \\ j \neq \ell-1}} r^{j-p} \binom{\ell}{j} S(j,p).$$

The number of G^0-orbits on $L(\mathcal{A}_\ell^0)$ is $P(0) + P(1) + \cdots + P(\ell-2) + \sum_{d|r} dP_d(\ell)$ where $P_d(\ell)$ is the number of partitions of ℓ such that the greatest common divisor of the parts is d.

Next we describe the restrictions $\mathcal{A}^X$. To do this, we define arrangements $\mathcal{A}_\ell^k = \mathcal{A}_\ell^k(r)$, which interpolate between $\mathcal{A}_\ell^0$ and $\mathcal{A}_\ell^\ell$. For each $\ell \geq 2$ and $0 \leq k \leq \ell$, let $\mathcal{A}_\ell^k$ be the arrangement consisting of all hyperplanes $H_{i,j}(\zeta)$ where $1 \leq i \neq j \leq \ell$ and $\zeta \in C(r)$ together with the hyperplanes H_i for $1 \leq i \leq k$. The set of arrangements $\mathcal{A}_\ell^k$ is closed under restriction to a hyperplane. Arguing by induction, we show that if $\mathcal{A} = \mathcal{A}_\ell^0$, then $\mathcal{A}^X = \mathcal{A}_p^n$ where $p = \dim X$ and $0 \leq n \leq p$. We determine n in terms of X in Proposition 6.84.

Proposition 6.82 *Let $\mathcal{A} = \mathcal{A}_\ell^k(r)$ and let $H \in \mathcal{A}$. The type of $\mathcal{A}^H$ is given in Table 6.2.*

Proof. If $k = 0$, then $\mathcal{A} = \mathcal{A}_\ell^0$. Since the group $G^0 = G(r, r, \ell)$ acts transitively on $\mathcal{A}$, we may assume that $H = H_{\ell-1,\ell}(1)$. The set of restrictions to H of the linear forms defining $\mathcal{A}$ consists of the forms $y_i - \zeta y_j$ for $1 \leq i < j \leq \ell - 1$ together with $(1 - \zeta)y_{\ell-1}$. Thus $\mathcal{A}^H \simeq \mathcal{A}_{\ell-1}^1$. If $k = \ell$, then it follows from the argument in Proposition 6.77 that $\mathcal{A}^H \simeq \mathcal{A}_{\ell-1}^{\ell-1}$.

Suppose $1 \leq k \leq \ell - 1$. All four cases are argued in similar fashion. We do the case where $H = H_{i,j}(\zeta)$ and $1 < k < i < j \leq \ell$. We may assume that

$H = H_{\ell-1,\ell}(1)$. Then $\rho(x_i) = y_i$ for $1 \leq i \leq k$; $\rho(x_i - \zeta x_j) = y_i - \zeta y_j$ for $1 \leq i < j \leq \ell - 1$; $\rho(x_i - \zeta x_\ell) = y_i - \zeta y_{\ell-1}$ for $1 \leq i < \ell - 1$; $\rho(x_{\ell-1} - x_\ell) = 0$; and $\rho(x_{\ell-1} - \zeta x_\ell) = (1-\zeta)y_{\ell-1}$ for $\zeta \neq 1$. Thus $\mathcal{A}^H \simeq \mathcal{A}_{\ell-1}^{k+1}$. □

Among the arrangements $\mathcal{A}_\ell^k(r)$, only $\mathcal{A}_\ell^0(r)$ and $\mathcal{A}_\ell^\ell(r)$ are reflection arrangements. Proposition 6.82 provides examples where the restriction of a reflection arrangement is not a reflection arrangement.

Example 6.83 The arrangement of the Coxeter group $D_4 = G(2,2,4)$ is $\mathcal{A}_4^0(2)$. Its restriction to every hyperplane is $\mathcal{A}_3^1(2)$, sometimes denoted D_3^1. This restriction is not a reflection arrangement:

$$Q(D_3^1) = x(x-y)(x+y)(x-z)(x+z)(y-z)(y+z).$$

Proposition 6.84 Let $\mathcal{A} = \mathcal{A}_\ell^0$. Suppose $X \in L(\mathcal{A})$ and $X = (\Delta, \Lambda, \nu)$. If $|\Delta| \geq 2$, then $\mathcal{A}^X \simeq \mathcal{A}_p^p = \mathcal{A}_p$ where $p = \dim X$. If $|\Delta| = 0$ and Λ has n blocks Λ_i with $|\Lambda_i| > 1$, then $\mathcal{A}^X \simeq \mathcal{A}_p^n$.

Proof. Suppose Λ has p blocks. We choose notation so $i \in \Lambda_i$ for $1 \leq i \leq p$. Then $y_1, \ldots, y_p$ is a basis for X^* and $\dim X = p$. Recall that $\mathcal{A}^X$ consists of the nonzero linear forms $y_i - \zeta y_j$. It follows that if $1 \leq i \leq \ell$, then either $y_i = 0$ or $y_i \in C(r)y_1 \cup \ldots \cup C(r)y_p$. Thus $\mathcal{A}^X$ consists of the kernels of the forms $y_i - \zeta y_j$ with $1 \leq i \neq j \leq p$ and $\zeta \in C(r)$, and perhaps some forms $c_i y_i$ where $c_i \in \mathbb{C}$ is not 0.

Suppose $|\Delta| \geq 2$. Choose $m \in \Delta$. Then $\rho(x_i - x_m) = y_i$ for $1 \leq i \leq p$, so all $y_1, \ldots, y_p$ occur among the forms defining $\mathcal{A}^X$ and hence $\mathcal{A}^X \simeq \mathcal{A}_p$.

If $|\Delta| = 0$, label the blocks Λ_i so $|\Lambda_i| > 1$ for $1 \leq i \leq n$ and $|\Lambda_i| = 1$ for $n+1 \leq i \leq p$. Suppose $1 \leq i \leq n$. Since $i \in \Lambda_i$, we may choose $j \in \Lambda_i$ with $i < j$. Then $X \subseteq H_{i,j}(\zeta)$ for some $\zeta \in C(r)$. We remarked in the proof of Proposition 6.74 that this ζ is unique. Then $\rho(x_i - \zeta x_j) = 0$ and $\rho(x_i - \zeta' x_j) = (1 - \zeta^{-1}\zeta')y_i \neq 0$ for $\zeta' \neq \zeta$. Thus $c_1 y_1, \ldots, c_n y_n$ occur among the forms defining $\mathcal{A}^X$, where $c_i \in \mathbb{C}$ are not zero. Since $y_{n+1}, \ldots, y_\ell$ do not occur, we have $\mathcal{A}^X \simeq \mathcal{A}_p^n$. □

Proposition 6.85 The arrangement $\mathcal{A}_\ell^k(r)$ is free with

$$\exp \mathcal{A}_\ell^k(r) = \{1, r+1, \ldots, (\ell-2)r+1, (\ell-1)r - \ell + k + 1\}.$$

In particular,

$$\pi(\mathcal{A}_\ell^k(r), t) = (1 + [(\ell-1)r - \ell + k + 1]t) \prod_{i=0}^{\ell-2}(1 + [ir+1]t).$$

Proof. Recall that

$$Q(\mathcal{A}_\ell^k) = x_1 \cdots x_k \prod_{1 \leq i < j \leq \ell} (x_i^r - x_j^r).$$

Consider the derivations

$$\theta_i = \sum_{j=1}^{\ell} x_j^{(i-1)r+1} D_j \quad 1 \leq i \leq \ell-1,$$

$$\theta_\ell = x_1 \cdots x_k (x_1 \cdots x_\ell)^{r-1} \sum_{i=1}^{\ell} x_i^{-(r-1)} D_i.$$

Direct calculation shows that $\theta_i \in D(\mathcal{A}_\ell^k)$. It follows from Saito's criterion 4.19 that θ_i form a basis for $D(\mathcal{A}_\ell^k)$. □

Corollary 6.86 *The arrangement $\mathcal{A}_\ell^0$ has exponents*

$$\exp \mathcal{A}_\ell = \{1, r+1, 2r+1, \ldots, (\ell-2)r+1, (\ell-1)(r-1)\}.$$

If $X \in \mathcal{A}^0$, then $\mathcal{A}^X$ is free. Its exponents are determined by Propositions 6.84 and 6.85.

The Exceptional Groups

For simplicity, assume all arrangements are essential. Thus $r(\mathcal{A}^X) = \dim X$. We compute $\pi(\mathcal{A}^X, t)$ recursively starting with $X = 0$, where $\pi(\mathcal{A}^X, t) = 1$. The recursion uses the following result.

Lemma 6.87 *Let $\mathcal{A}$ be any central arrangement. Then*

$$\sum_{Y \in L^X} (-t)^{r(Y)-r(X)} \pi(\mathcal{A}^Y, t) = 1.$$

Proof. It suffices to prove the assertion for $X = V$.

$$\sum_{Y \in L} (-t)^{r(Y)} \pi(\mathcal{A}^Y, t) = \sum_{Y \in L} (-t)^{r(Y)} \sum_{Y \leq Z} \mu(Y, Z)(-t)^{r(Z)-r(Y)}$$

$$= \sum_{Z \in L} (-t)^{r(Z)} \Big(\sum_{V \leq Y \leq Z} \mu(Y, Z) \Big)$$

$$= 1.$$

The second equality is an interchange of summation. The third follows from Lemma 2.38. □

We rewrite the sum in Lemma 6.87 as a sum over G-orbits. If $X \in L$, let $\mathcal{O}_X$ be the G-orbit of X. If $X, Y \in L$, let $u(X, Y)$ be the number of $Z \in L$ such that $Z \in \mathcal{O}_Y$ and $Z \geq X$. This number depends only on $\mathcal{O}_X$ and $\mathcal{O}_Y$. The polynomial $\pi(\mathcal{A}^Y, t)$ depends only on $\mathcal{O}_Y$. The sum on the left side of Lemma 6.87 depends only on $\mathcal{O}_X$. Let $\Xi = \{X_1, \ldots, X_r\}$ be a set of representatives for the G-orbits on L numbered so $\dim X_i \geq \dim X_j$ if $i \leq j$. Write $\mathcal{O}_j = \mathcal{O}_{X_j}$ and $r(j) = r(X_j)$. Let $u_{i,j} = u(X_i, X_j)$ and let $\pi_i(t) = \pi(\mathcal{A}^{X_i}, t)$. Then $U = U(G) = (u_{i,j})$ is an upper triangular $r \times r$ matrix with $u_{i,i} = 1$. Lemma 6.87 gives

$$(2) \qquad \sum_{j \geq i} u_{i,j}(-t)^{r(j)-r(i)} \pi_j(t) = 1.$$

Thus to compute the Poincaré polynomials $\pi(\mathcal{A}^X, t)$, we must compute the matrices U. The first step is to find the G-orbits. If G_0 is a subgroup of G, let $N(G_0)$ be its normalizer in G. Recall the definition of the fixer G_X and the stabilizer S_X from Section 6.1.

Lemma 6.88 *Let G be a unitary reflection group, let $\mathcal{A} = \mathcal{A}(G)$, and let $L = L(\mathcal{A})$. If $X \in L$, then $\mathrm{Fix}(G_X) = X$. Two elements $X, Y \in L$ lie in the same G-orbit if and only if G_X and G_Y are conjugate in G. Moreover, $S_X = N(G_X)$ and $|\mathcal{O}_X| = |G : N(G_X)|$.*

Proof. The first assertion follows from Corollary 6.28. If $Y = gX$ for some $g \in G$, then $G_Y = gG_Xg^{-1}$. Conversely, if $G_Y = gG_Xg^{-1}$, then $Y = \mathrm{Fix}(G_Y) = \mathrm{Fix}(gG_Xg^{-1}) = \mathrm{Fix}(G_{gX}) = gX$. Finally, take $Y = X$. □

The explicit calculation of the orbits and of the matrices U is fairly complicated. We refer to [174, 175] for details. The results are tabulated in Appendix C. We show the use of (2) for the group G_{25} defined in Example 6.30. The group G_{25} has five orbits on $L(\mathcal{A})$:

$\mathcal{O}_1 = V$ has fixer the identity group, called A_0,

$\mathcal{O}_2$ consists of the 12 hyperplanes, with fixer $C(3)$,

$\mathcal{O}_3$ consists of the 12 lines which are in only two planes, with fixer $C(3) \times C(3)$,

$\mathcal{O}_4$ consists of the nine lines which are in four planes, with fixer the group called G_4 in the classification,

$\mathcal{O}_5$ is the origin, with fixer G_{25}.

Table 6.3 of $U = U(G_{25})$ summarizes what we know about the lattice. The orbits are labeled by their fixers. This table also appears in Appendix C.

Now use formula (2) inductively. We get $\pi_5(t) = 1$ and $\pi_4(t) = \pi_3(t) = 1+t$. The second line of the table gives

$$\pi_2(t) - 2t\pi_3(t) - 3t\pi_4(t) + t^2\pi_5(t) = 1,$$
$$\pi_2(t) = 1 + 5t + 4t^2 = (1+t)(1+4t).$$

The first line of the table gives

$$\pi_1(t) - 12t(1+t)(1+4t) + 12t^2(1+t) + 9t^2(1+t) - t^3 = 1,$$
$$\pi_1(t) = 1 + 12t + 39t^2 + 28t^3 = (1+t)(1+4t)(1+7t).$$

Table 6.3. Orbits in G_{25}

	A_0	$C(3)$	$C(3)^2$	G_4	G_{25}
A_0	1	12	12	9	1
$C(3)$		1	2	3	1
$C(3)^2$			1	0	1
G_4				1	1
G_{25}					1

The factorization of $\pi_1(t)$ is no surprise, since we know that every reflection arrangement is free, but this factorization reveals that the coexponents of G are $\{1, 4, 7\}$. We had no reason to anticipate the factorization of $\pi_2(t)$.

When these calculations are done for all restrictions of all irreducible complex reflection groups we get the following.

Proposition 6.89 *Let $G \subset GL(V)$ be a finite reflection group with reflection arrangement $\mathcal{A} = \mathcal{A}(G)$. For each $X \in L(\mathcal{A})$ with $\dim(X) = p$, there exist integers $b_1^X, \ldots, b_p^X$ such that*

$$\pi(\mathcal{A}^X, t) = \prod_{i=1}^{p}(1 + b_i^X t).$$

The values of b_i^X are tabulated in Appendix C for all exceptional groups and all orbit types. N. Spaltenstein [214] observed that the b_i^X occur in the representation theory of G in case G is a Weyl group of exceptional type. He conjectured that the same holds for all Weyl groups. This was proved by Lehrer and Shoji [143]. Propositions 6.73, 6.77, 6.86, and 6.89 lead to the following.

Conjecture 6.90 *Every reflection arrangement is hereditarily free.*

K. Brandt has worked out many examples and they indicate even more may be true. Say that $\mathcal{A}$ is inductively k–free if $\mathcal{A}^X$ is inductively free for all $X \in L(\mathcal{A})$ with $r(X) \leq k$. Similarly, $\mathcal{A}$ is hereditarily inductively free if it is inductively k–free for all k.

Conjecture 6.91 *Every reflection arrangement is hereditarily inductively free.*

Example 6.92 Table 6.4 shows that the arrangement of the group G_{25} defined in Example 6.30 is inductively free. Experimentation with this example will reveal how sensitive it is to the order of the hyperplanes.

Table 6.4. Induction table for G_{25}

exp $\mathcal{A}'$	α_H	exp $\mathcal{A}''$
0,0,0	$x+y+z$	0,0
0,0,1	$x+y+\omega z$	0,1
0,1,1	$x+y+\omega^2 z$	0,1
0,1,2	z	0,1
0,1,3	x	1,3
1,1,3	$x+\omega y+z$	1,3
1,2,3	$x+\omega y+\omega z$	1,3
1,3,3	$x+\omega y+\omega^2 z$	1,3
1,3,4	y	1,4
1,4,4	$x+\omega^2 y+z$	1,4
1,4,5	$x+\omega^2 y+\omega z$	1,4
1,4,6	$x+\omega^2 y+\omega^2 z$	1,4
1,4,7		

6.5 Restrictions

The Cardinality of $\mathcal{A}^H$

Let V be an ℓ–dimensional complex vector space. Let $G \subset GL(V)$ be an irreducible finite unitary reflection group with reflection arrangement $\mathcal{A} = \mathcal{A}(G)$. If $H, K \in \mathcal{A}$ are in the same G–orbit, then $\mathcal{A}^H$ and $\mathcal{A}^K$ are isomorphic. However, if the hyperplanes are in different G–orbits, then there is no reason to expect any connection between $\mathcal{A}^H$ and $\mathcal{A}^K$. We show next that the cardinality of $\mathcal{A}^H$ is independent of the choice of H. More precisely, let $n = |\mathcal{A}|$ and let $t = 2n/\ell$. Then for all $H \in \mathcal{A}$

$$|\mathcal{A}^H| = n + 1 - t.$$

Thus for an irreducible unitary reflection group G, the number $2|\mathcal{A}|/\ell$ is an integer.

Proposition 6.93 *Choose a G–invariant Hermitian form $(\ ,\ )$ on V. Let $K \in \mathcal{A}$. Denote the orthogonal complement of K by $K^\perp$. Then $K^\perp$ has dimension one. Choose $w_K \in K^\perp$ so $|w_K|^2 = (w_K, w_K) = 1$. For any $v \in V$ we have*

$$2 \sum_{K \in \mathcal{A}} (v, w_K) w_K = tv.$$

Proof. Define a map $f : V \to V$ by

$$f(v) = \sum_{K \in \mathcal{A}} (v, w_K) w_K.$$

Then f is $\mathbb{C}$–linear. Note that the definition does not depend on the choice of w_K ($K \in \mathcal{A}$) as long as $(w_K, w_K) = 1$. Let $g \in G$ and $v \in V$. We have

$$f(g(v)) = \sum_{K \in \mathcal{A}} (g(v), w_K) w_K = \sum_{K \in \mathcal{A}} (v, g^{-1}(w_K)) w_K$$
$$= g[\sum_{K \in \mathcal{A}} (v, g^{-1}(w_K)) g^{-1}(w_K)] = g(f(v)).$$

By Schur's lemma, there exists a constant C such that $f(v) = Cv$ for all $v \in V$. Let $e_1, \ldots, e_\ell$ be an orthonormal basis for V. To determine C, we compute

$$C\ell = \sum_{i=1}^{\ell} (f(e_i), e_i) = \sum_{i=1}^{\ell} \sum_{K \in \mathcal{A}} |(e_i, w_K)|^2 = \sum_{K \in \mathcal{A}} |w_K|^2 = n.$$

Thus $C = n/\ell = t/2$. □

Proposition 6.94 *Let $H \in \mathcal{A}$. Then*

$$t = 2 \sum_{K \in \mathcal{A}} |(w_H, w_K)|^2.$$

Proof. By Proposition 6.93, we have

$$t|v|^2 = (tv, v) = 2 \sum_{K \in \mathcal{A}} |(v, w_K)|^2.$$

Set $v = w_H$. □

Fix $H \in \mathcal{A}$ and recall that

$$\mathcal{A}^H = \{H \cap K \mid K \in \mathcal{A} \setminus \{H\}\}.$$

Then $\mathcal{A}^H$ is a finite collection of $(\ell-2)$–dimensional vector subspaces of H. Let $X \in \mathcal{A}^H$. Denote by $X^\perp$ the orthogonal complement of X in V. Then $X^\perp$ has dimension 2. Define

$$\mathcal{A}_X = \{K \in \mathcal{A} \mid X \subset K\}, \quad \mathcal{A}_X^\perp = \{X^\perp \cap K \mid K \in \mathcal{A}_X\}.$$

Then $\mathcal{A}_X^\perp$ is a finite collection of 1–dimensional vector subspaces of $X^\perp$. Recall from Theorem 6.25 that G_X is a unitary reflection group. Since $X^\perp$ is G_X–stable, there is a natural injection $G_X \to GL(X^\perp)$. Denote the image of this injection by $G_X^\perp$. Then $G_X^\perp$ is a unitary reflection group, and the arrangement defined by $G_X^\perp$ is $\mathcal{A}_X^\perp$ by Corollary 6.28.

Lemma 6.95 *Let $H \in \mathcal{A}$ and let $X \in \mathcal{A}^H$. Then*

$$|\mathcal{A}_X| = 2 \sum_{K \in \mathcal{A}_X} |(w_H, w_K)|^2.$$

Proof. Note that $w_K \in X^\perp$ for $K \in \mathcal{A}_X$.

Case 1. If $G_X^\perp$ is irreducible, we can apply Proposition 6.94 to get the desired result.

Case 2. If $G_X^\perp$ is reducible, then
$$|\mathcal{A}_X| = |\mathcal{A}_X^\perp| = 2.$$
Let $\mathcal{A}_X = \{H, K\}$. Then $(w_H, w_K) = 0$. So, in this case,
$$|\mathcal{A}_X| = 2 = 2 \sum_{K \in \mathcal{A}_X} |(w_H, w_K)|^2$$
as required. □

Lemma 6.96 *Let $\mathcal{A}' = \mathcal{A} \setminus \{H\}$ and $\mathcal{A}'_X = \mathcal{A}_X \setminus \{H\}$ for $X \in \mathcal{A}^H$. Then we have a disjoint union*
$$\mathcal{A}' = \bigcup_{X \in \mathcal{A}^H} \mathcal{A}'_X.$$

Theorem 6.97 *Let G be an irreducible unitary reflection group and let $\mathcal{A} = \mathcal{A}(G)$. Let $n = |\mathcal{A}|$ and let $t = 2n/\ell$. Then for all $H \in \mathcal{A}$*
$$|\mathcal{A}^H| = n + 1 - t.$$

Proof. We apply Proposition 6.94, Lemma 6.96, and Lemma 6.95:
$$\begin{aligned}
t &= 2 \sum_{K \in \mathcal{A}} |(w_H, w_K)|^2 = 2 + 2 \sum_{K \in \mathcal{A}'} |(w_H, w_K)|^2 \\
&= 2 + 2 \sum_{X \in \mathcal{A}^H} \sum_{K \in \mathcal{A}'_X} |(w_H, w_K)|^2 \\
&= 2 + \sum_{X \in \mathcal{A}^H} (2 \sum_{K \in \mathcal{A}_X} |(w_H, w_K)|^2 - 2) \\
&= 2 + \sum_{X \in \mathcal{A}^H} (|\mathcal{A}_X| - 2) = 2 + \sum_{X \in \mathcal{A}^H} (|\mathcal{A}_X| - 1) - |\mathcal{A}^H| \\
&= 2 + \sum_{X \in \mathcal{A}^H} |\mathcal{A}'_X| - |\mathcal{A}^H| = 2 + |\mathcal{A}'| - |\mathcal{A}^H| = 1 + n - |\mathcal{A}^H|.
\end{aligned}$$
This completes the proof. □

Corollary 6.98 *If G is an irreducible unitary reflection group, then $2|\mathcal{A}|/\ell$ is an integer.*

$\mathcal{A}^H$ Is Free In Coxeter Arrangements

Let V be an ℓ-dimensional real vector space. Let $G \subset GL(V)$ be an irreducible finite real reflection group. Then $\mathcal{A} = \mathcal{A}(G)$ is a Coxeter arrangement. Recall from Definition 6.50 that $m_i = n_i$. Thus $\mathcal{A}$ is free with $\exp(\mathcal{A}) =$

$\{m_1, \ldots, m_{\ell-1}, m_\ell\}$. Since $\mathcal{A}$ is irreducible, $1 = n_1 = m_1 < m_2$. It follows from Theorem 6.23 that $m_{\ell-1} < m_\ell$. Recall from Definition 6.22 that $m_i = d_i - 1$, where d_i $(i = 1, \ldots, \ell)$ are the basic degrees.

Definition 6.99 *The integer $h = h(G) = m_\ell + 1 = d_\ell$ is the **Coxeter number** of G. It is the unique maximal basic degree.*

The duality of Theorem 6.23 takes the form

$$m_i + m_{\ell+1-i} = h \quad (i = 1, \ldots, \ell).$$

We have

$$t = 2|\mathcal{A}|/\ell = (2 \sum_{i=1}^{\ell} m_i)/\ell = h\ell/\ell = h.$$

Thus the integer $t = t(G)$, which was defined for any finite irreducible unitary reflection group, is equal to the Coxeter number h when G is real. In this case Theorem 6.97 can be written as follows.

Theorem 6.100 *Let G be a finite irreducible real reflection group. Let h be the Coxeter number of G and let $\mathcal{A} = \mathcal{A}(G)$. Then for all $H \in \mathcal{A}$*

$$|\mathcal{A}^H| = |\mathcal{A}| + 1 - h.$$

Next we follow [180] to show that $\mathcal{A}^H$ is free. Unfortunately, induction may not be used in combination with this result because in general $\mathcal{A}^H$ is not the arrangement of any Coxeter group; see Example 6.83.

Lemma 6.101 *Let $\Delta = \Delta(T_1, \ldots, T_\ell; \mathcal{F})$ be the discriminant of G with respect to the basic invariants $\mathcal{F}$. Then*

$$\Delta \doteq T_\ell^\ell + A_1 T_\ell^{\ell-1} + \cdots + A_\ell$$

where $A_i \in \mathbb{R}[T_1, \ldots, T_{\ell-1}]$.

Proof. Recall that Δ is weighted homogeneous with weights $(m+n)/d_1, \ldots, (m+n)/d_\ell$ by Corollary 6.66. Since G is a Coxeter group, we have $m = n$ and $d_\ell = 2n/\ell$. So $(m+n)/d_\ell = \ell$. Recall from Theorem 6.23 that the maximal degree d_ℓ is unique, $d_{\ell-1} < d_\ell$. We have to show that $\Delta \notin (T_1, \ldots, T_{\ell-1})$. Choose $0 \neq v \in V$ such that $f_1(v) = \cdots = f_{\ell-1}(v) = 0$. It follows from Theorem 6.26 that v is not on any hyperplane in $\mathcal{A}$. Thus $\delta(v) \neq 0$. This shows $\delta \notin (f_1, \ldots, f_{\ell-1})R$, and thus $\Delta \notin (T_1, \ldots, T_{\ell-1})$. □

Fix $H_0 \in \mathcal{A} = \mathcal{A}(G)$. Choose an orthonormal basis $x_1, \ldots, x_\ell$ for V^* such that $\ker(x_\ell) = H_0$. Fix a set $\mathcal{F} = \{f_1, \ldots, f_\ell\}$ of basic invariants with

$$2 = d_1 = \deg f_1 \leq \cdots \leq h = d_\ell = \deg f_\ell.$$

In Appendix B we associate to these basic invariants the set of basic derivations

$$\{\theta_{f_1},\ldots,\theta_{f_\ell}\}.$$

For simplicity, we sometimes write $\theta_i = \theta_{f_i}$. Then $\theta_i(x_j) = \partial f_i/\partial x_j$. Define $\bar{S} = S/x_\ell S$. We can naturally identify $\bar{S}$ with the symmetric algebra of the dual space H_0^* of H_0. Identify $\bar{S}$ with the polynomial ring $\mathbb{R}[\bar{x}_1,\ldots,\bar{x}_{\ell-1}]$. Let $\pi : S \to \bar{S}$ be the natural projection. We denote by $\bar{q}$ the residue class of $q \in S$ in $\bar{S}$. Then $\bar{q}(\bar{x}_1,\ldots,\bar{x}_{\ell-1}) = q(\bar{x}_1,\ldots,\bar{x}_{\ell-1},0)$.

Lemma 6.102 *The polynomials $\bar{f}_1,\ldots,\bar{f}_{\ell-1}$ in $\mathbb{R}[\bar{x}_1,\ldots,\bar{x}_{\ell-1}]$ are algebraically independent.*

Proof. Since $\ker(x_\ell) \in \mathcal{A}$, we have $\bar{\delta} = 0$. Thus Lemma 6.101 implies that $\bar{f}_\ell$ is algebraic over the field $F = \mathbb{R}(\bar{f}_1,\ldots,\bar{f}_{\ell-1})$. Thus the field $F(\bar{f}_\ell)$ is algebraic over F. Let $E = \mathbb{R}(\bar{x}_1,\ldots,\bar{x}_\ell) = \mathbb{R}(\bar{x}_1,\ldots,\bar{x}_{\ell-1})$. Then E is algebraic over $F(\bar{f}_\ell)$ because $\mathbb{R}(x_1,\ldots,x_\ell)$ is algebraic over $\mathbb{R}(f_1,\ldots,f_\ell)$ with Galois group G. Therefore E/F is an algebraic extension. Since $\bar{x}_1,\ldots,\bar{x}_{\ell-1}$ are algebraically independent, it follows that $\bar{f}_1,\ldots,\bar{f}_{\ell-1}$ are algebraically independent. □

Since $\theta_i \in D(\mathcal{A})$, we have $\theta_i(x_\ell) \in x_\ell S$ $(1 \leq i \leq \ell)$. Therefore the derivations $\theta_1,\ldots,\theta_\ell$ on S induce the derivations $\bar{\theta}_1,\ldots,\bar{\theta}_\ell$ on $\bar{S}$:

$$\begin{array}{ccc} S & \xrightarrow{\theta_i} & S \\ \pi \downarrow & & \downarrow \pi \\ \bar{S} & \xrightarrow{\bar{\theta}_i} & \bar{S}. \end{array}$$

Theorem 6.103 *Let G be a finite irreducible Coxeter group. Let $\mathcal{A}'' = \mathcal{A}^{H_0}$ be the restriction of the Coxeter arrangement $\mathcal{A}$ to the hyperplane H_0. Let $\mathcal{F} = \{f_1,\ldots,f_\ell\}$ be a set of basic invariants with*

$$\deg f_1 \leq \cdots \leq \deg f_\ell$$

and let $\theta_1,\ldots,\theta_\ell$ be the associated derivations which form a basis for $D(\mathcal{A})$. Then the induced derivations $\bar{\theta}_1,\ldots,\bar{\theta}_{\ell-1}$ on $\bar{S}$ form a basis for $D(\mathcal{A}'')$.

Proof. We will show first that the derivations $\bar{\theta}_1,\ldots,\bar{\theta}_{\ell-1}$ are linearly independent over $\bar{S}$. It suffices to show that

$$\det[\bar{\theta}_i(\bar{x}_j)]_{1 \leq i,j \leq \ell-1} \neq 0.$$

Since

$$\bar{\theta}_i(\bar{x}_j) = \overline{\theta_i(x_j)} = \overline{\partial f_i/\partial x_j} = \partial \bar{f}_i/\partial \bar{x}_j$$

for $1 \leq i,j \leq \ell-1$, it is enough to show that the Jacobian of $\bar{f}_1,\ldots,\bar{f}_{\ell-1}$ does not vanish. This is equivalent to the algebraic independence of $\bar{f}_1,\ldots,\bar{f}_{\ell-1}$ by a well-known theorem in field theory [138, Ch.I, 11.4]. The algebraic independence was shown in Lemma 6.102.

Next we have

$$\deg \bar{\theta}_1 + \cdots + \deg \bar{\theta}_{\ell-1} = \deg \theta_1 + \cdots + \deg \theta_{\ell-1}$$
$$= (\sum_{i=1}^{\ell} m_i) - m_\ell = |\mathcal{A}| - (h-1) = |\mathcal{A}''|$$

by Theorem 6.100. Theorem 4.23 completes the proof. □

Theorem 6.104 *Let G be a finite irreducible Coxeter group with exponents $m_1 \leq \ldots \leq m_\ell$. Let $\mathcal{A} = \mathcal{A}(G)$ be its reflection arrangement. For $H \in \mathcal{A}$, the restriction $\mathcal{A}^H$ is free with*

$$\exp \mathcal{A}^H = \{m_1, \ldots, m_{\ell-1}\}.$$

6.6 Topology

In this section we study the topological properties of the discriminant locus and its complement. The discriminant locus is the image of all orbits of dimension $\leq \ell - 1$. It is the zero set of the discriminant. The orbit stratification of the discriminant is determined by finding its subvarieties which are the images of the lower dimensional orbits. We show that the discriminant matrix encodes this information in the determinants of its minors.

It is conjectured that M and M/G are $K(\pi, 1)$ spaces for all complex reflection groups. We observed in Proposition 5.6 that every 2–arrangement is $K(\pi, 1)$. Nakamura [162] showed that if G is an imprimitive complex reflection group, then its arrangement is $K(\pi, 1)$. This holds in particular for the groups defined in Example 6.29. The fundamental groups $\pi_1(M/G)$ are not known for all G. They were computed for $\ell = 2$ by Bannai [17]. We define a subclass of complex reflection groups called Shephard groups. These groups are closely related to Coxeter groups. In particular, the fact that $M(G)$ is a $K(\pi, 1)$ space if G is a Coxeter group enables us to show that $M(G)$ is a $K(\pi, 1)$ space if G is a Shephard group. Only the six groups numbered 24,27,29,31,33,34 are not known to have $K(\pi, 1)$ complements.

Stratification of the Discriminant

The construction given in Example 5.8 for the braid space and the symmetric group was generalized by Brieskorn [41] to Coxeter groups. If G is a complex reflection group, the construction is similar. We follow [167].

Definition 6.105 *Let $G \subset GL(V)$ be a complex reflection group and let $\mathcal{A} = \mathcal{A}(G)$. Let $\gamma : V \to V/G$ be the orbit map. Given a set of basic invariants $\mathcal{F} = \{f_1, \ldots, f_\ell\}$, define $\tau : V/G \to \tilde{V} = \mathbb{C}^\ell$ by*

$$\tau(Gv) = (f_1(v), \ldots, f_\ell(v)).$$

Proposition 6.106
*(1) The map $\gamma : V \to V/G$ is a branched covering. The branch locus is the variety $N = \cup_{H \in \mathcal{A}} H$. The image N/G is the **discriminant locus**. The restriction $\gamma : M \to M/G$ is a $|G|$-fold covering.*

(2) The map τ is a bijection. Let $\pi = \tau\gamma : V \to \tilde{V}$ and let $T_1, \ldots, T_\ell$ be the coordinate functions in $\tilde{V}$. Let $\tilde{N} = \pi(N)$. Then the discriminant locus is the zero set of the discriminant:

$$\tilde{N} = \{(z_1, \ldots, z_\ell) \in \mathbb{C}^\ell \mid \Delta(z_1, \ldots, z_\ell; \mathcal{F}) = 0\}.$$

Proof. (1) If $v \in N$, then there exists $H \in \mathcal{A}$ such that $v \in H$. Then $G_v \supseteq G_H \neq \{1\}$. If $v \in M$, then consider the line T spanned by v. Since the action is linear, $G_v = G_T$. It follows from Theorem 6.25 that $G_T = \{1\}$.

(2) Chevalley's theorem 6.19 implies that τ is a bijection. By definition, δ is the invariant of minimal degree which is divisible by Q. This implies the assertion about the discriminant locus. □

The discriminant locus is the image under π of all the orbits of dimension $\leq \ell - 1$. The orbit stratification of the discriminant is determined by finding the subvarieties of $\tilde{N}$ which are the images under π of the lower dimensional orbits. For $0 \leq p \leq \ell$ let

$$N^p = \cup\{X \in L \mid \dim X = p\}$$

and let

$$\tilde{N}^p = \pi(N^p).$$

In general, $\tilde{N}^p$ is a reducible variety whose components are the images of the p-dimensional orbits of G on L. The discriminant provides a defining polynomial for $\tilde{N}$. Our aim is to find defining polynomials for the varieties $\tilde{N}^p$ for all p. Recall the open submanifold $M^X = M(\mathcal{A}^X)$ of X from Definition 5.16 and the stratification of V

$$V = \bigcup_{X \in L} M^X.$$

Lemma 6.107 *Suppose $v \in M^X$. Let $\tilde{M}^X = \pi(M^X)$. The restriction $\pi : M^X \to \tilde{M}^X$ is a smooth covering. Thus for all $v \in M^X$, the differential $d\pi_{|v} : TM_v^X \to T\tilde{M}_{\pi v}^X$ is an isomorphism.*

Proof. We claim that $G_v = G_X$ for all $v \in M^X$. Clearly, $G_X \subset G_v$. For the converse, recall that we proved in Theorem 6.27 that $\text{Fix}(g) \in L$ for all $g \in G$. If $v \in M^X$ and $g \in G_v$, this implies that $\text{Fix}(g)$ contains the smallest element of L which contains v. Thus $X \subset \text{Fix}(g)$, so $g \in G_X$ and $G_v \subset G_X$. We showed in

6.6 Topology   261

Lemma 6.88 that the stabilizer of X is the normalizer of G_X in G. Thus $\pi|M^X$ is the orbit map of the free action of the group $N(G_X)/G_X$. The orbit map of a free action is a covering. Since it is smooth, it is a local diffeomorphism. □

Recall the evaluation map from Proposition 5.17. Given $v \in V$ and $\theta \in \mathrm{Der}_S^G$, write $\theta = \sum u_i D_i$ and let

$$\rho_v(\theta) = \sum u_i(v) D_i.$$

Then $\rho_v(\theta)$ is a vector in $TV_v = \sum \mathbb{C} D_i$. Thus we get a map $\rho_v : \mathrm{Der}_S^G \to TV_v$. Let $(\mathrm{Der}_S^G)_v = \mathrm{im}\rho_v$. There is a similar map in the orbit space. Given $w \in \tilde{V}$ and $\eta \in D_R(\delta)$, write $\eta = \sum h_i D_{f_i}$ and let

$$\rho_w(\eta) = \sum h_i(w) D_{f_i}.$$

Then $\rho_w(\eta)$ is a vector in $T\tilde{V}_w = \sum \mathbb{C} D_{f_i}$. Thus we get a map $\rho_w : D_R(\delta) \to T\tilde{V}_w$. Let $(D_R(\delta))_w = \mathrm{im}\rho_w$.

Lemma 6.108 *There is a commutative diagram of surjective maps*

$$\begin{array}{ccc} \mathrm{Der}_S^G & \xrightarrow{\beta} & D_R(\delta) \\ \rho_v \downarrow & & \downarrow \rho_{\pi v} \\ (\mathrm{Der}_S^G)_v & \xrightarrow{d\pi|_v} & (D_R(\delta))_{\pi v} \end{array}$$

Proof. Recall that $\pi(v) = (f_1(v), \ldots, f_\ell(v))$. Relative to the bases $\{D_i\}$, $\{D_{f_j}\}$, the matrix of $d\pi|_v$ is $\{\partial f_j / \partial x_i\}_v$. Let $\theta \in \mathrm{Der}_S^G$ and write $\theta = \sum u_i D_i$. Then $d\pi|_v \rho_v(\theta) = \sum\sum u_i(v)(\partial f_j / \partial x_i)(v) D_{f_j}$, $\rho_{\pi v} \beta(\theta) = \sum\sum (u_i(\partial f_j/\partial x_i))(\pi v) D_{f_j}$. Thus the diagram commutes. The maps ρ_v and $\rho_{\pi v}$ are surjective by definition and β is an isomorphism by Corollary 6.58. Thus $d\pi|_v$ is surjective. □

Lemma 6.109 *For all $v \in V$, the map $d\pi|_v : (\mathrm{Der}_S^G)_v \to (D_R(\delta))_{\pi v}$ is an isomorphism of vector spaces.*

Proof. Suppose $v \in M^X$. Clearly, $(\mathrm{Der}_S^G)_v \subseteq D(\mathcal{A})_v$. By Proposition 5.17, $D(\mathcal{A})_v = TM_v^X$. Thus $\rho_v \theta \in TM_v^X$. The map is injective by Lemma 6.107 and surjective by Lemma 6.108. □

Corollary 6.110 *If $v \in M^X$, then $(D_R(\delta))_{\pi v} \subseteq T\tilde{M}_{\pi v}^X$. Thus we have the commutative diagram*

$$\begin{array}{ccc} (\mathrm{Der}_S^G)_v & \xrightarrow{d\pi|_v} & (D_R(\delta))_{\pi v} \\ \nu \downarrow & & \downarrow \tilde{\nu} \\ TM_v^X & \xrightarrow{d\pi|_v} & T\tilde{M}_{\pi v}^X \end{array}$$

where the horizontal maps are isomorphisms and the vertical maps ν, $\tilde{\nu}$ are inclusions.

Let $\mathcal{F}$ be a set of basic invariants for G and let Θ be a set of basic derivations. In order to find the orbit stratification, we consider next the matrices $\mathsf{J}(\mathcal{F})$ and $\mathsf{M}(\Theta)$. If P is a matrix of polynomials and $v \in V$, let $\mathsf{P}(v)$ be the scalar matrix obtained by evaluating P at v. Steinberg [223] proved the next result.

Theorem 6.111 *For all $\mathcal{F}$ and for all $v \in V$, the maximum number of linearly independent reflecting hyperplanes which contain v equals the nullity of $\mathsf{J}(\mathcal{F})(v)$.*

In order to prove the analog for $\mathsf{M}(\Theta)$ we need the following lemma.

Lemma 6.112 *Fix a p-tuple $(i_1, \ldots, i_p)$, let $W = \mathbb{C}x_{i_1} \oplus \ldots \oplus \mathbb{C}x_{i_p}$, and let $\mathcal{A}' = \{H \in \mathcal{A} \mid \alpha_H \in W\}$. Let $M(i_1, \ldots, i_p; j_1, \ldots, j_p)$ denote the determinant of the $p \times p$ minor of $\mathsf{M}(\Theta)$ consisting of rows $i_1, \ldots, i_p$ and columns $j_1, \ldots, j_p$. Let $H \in \mathcal{A}$. The linear form α_H divides $M(i_1, \ldots, i_p; j_1, \ldots, j_p)$ for all $(j_1, \ldots, j_p)$ if and only if $H \in \mathcal{A}'$.*

Proof. We may assume that $(i_1, \ldots, i_p) = (1, \ldots, p)$ and write $M(j_1, \ldots, j_p) = M(1, \ldots, p; j_1, \ldots, j_p)$. We show that if $H \in \mathcal{A}'$, then α_H divides $M(j_1, \ldots, j_p)$. Let G' be the reflection group generated by all reflections in G which fix hyperplanes in $\mathcal{A}'$. Every G-invariant derivation is G'-invariant. Since G' acts trivially in W, the derivations $D_{p+1}, \ldots, D_\ell$ are also G'-invariant. For any $\theta_{j_1}, \ldots, \theta_{j_p} \in \mathrm{Der}_S^G$, it follows from Proposition 4.12 that

$$\det \mathsf{M}(\theta_{j_1}, \ldots, \theta_{j_p}, D_{p+1}, \ldots, D_\ell) = M(j_1, \ldots, j_p)$$

is divisible by $\prod_{H \in \mathcal{A}'} \alpha_H$. For the converse, assume that $H \notin \mathcal{A}'$, but α_H divides each $M(j_1, \ldots, j_p)$. Choose coordinates so that $\alpha_H = x_{p+1}$. Apply the first part of the argument to the 1-dimensional subspace $\mathbb{C}x_{p+1}$ to conclude that every entry in row $p+1$ is divisible by x_{p+1}. By assumption, x_{p+1} also divides every $M(j_1, \ldots, j_p)$. Thus the determinant of every $(p+1) \times (p+1)$ minor which involves the first $p+1$ rows is divisible by x_{p+1}^2. Expanding $\det \mathsf{M}(\theta_1, \ldots, \theta_\ell)$ by these minors shows that $\det \mathsf{M}(\theta_1, \ldots, \theta_\ell)$ is divisible by x_{p+1}^2. This contradicts the fact that $Q \doteq \det \mathsf{M}(\Theta)$ is square free. $\square$

Theorem 6.113 *For all Θ and for all $v \in V$, the maximum number of linearly independent reflecting hyperplanes which contain v equals the nullity of $\mathsf{M}(\Theta)(v)$.*

Proof. Let p be the number of linearly independent reflecting hyperplanes which contain v and let q be the nullity of $\mathsf{M}(\Theta)(v)$. Choose coordinates so the kernels of $x_1, \ldots, x_p$ are linearly independent reflecting hyperplanes containing v. By Lemma 6.112, each entry of row i is divisible by x_i for $1 \leq i \leq p$. Thus $p \leq q$. For the converse, expand $Q \doteq \det \mathsf{M}(\Theta)$ by the $p \times p$ minors of the first p rows. Write

(1) $$Q \doteq \sum \pm M(j_1, \ldots, j_p) N(j_{p+1}, \ldots, j_\ell)$$

where $N(j_{p+1},\ldots,j_\ell)$ is the determinant of the minor with rows $p+1,\ldots,\ell$ and columns $j_{p+1},\ldots,j_\ell$. The sum is over all permutations $j_1,\ldots,j_\ell$ with $j_1 < \ldots < j_p, j_{p+1} < \ldots < j_\ell$. As in Lemma 6.112, let $\mathcal{A}' = \{H \in \mathcal{A} \mid \alpha_H \in \mathbb{C}x_1 \oplus \ldots \oplus \mathbb{C}x_p\}$. Let $Q' = \prod_{H \in \mathcal{A}'} \alpha_H$ and write $Q = Q'Q''$. Note that $Q''(v) \neq 0$. By Lemma 6.112, every $M(j_1,\ldots,j_p)$ is divisible by Q'. Write $M(j_1,\ldots,j_p) = R(j_1,\ldots,j_p)Q'$. Divide both sides of (1) by Q'. We get

$$Q'' \doteq \sum \pm R(j_1,\ldots,j_p)N(j_{p+1},\ldots,j_\ell).$$

Since the left side is not 0 at v, there is at least one summand on the right side which is not 0 at v. In particular, some $N(j_{p+1},\ldots,j_\ell)$ is not 0 at v. Thus rank$\mathsf{M}(\Theta)(v) \geq \ell - p$ and hence $q \leq p$. □

Corollary 6.114 *Let $v \in M^X$. Then the maps $\nu : (\mathrm{Der}_S^G)_v \to TM_v^X$ and $\tilde{\nu} : (D_R(\delta))_{\pi v} \to T\tilde{M}_{\pi v}^X$ are isomorphisms.*

Proof. Since ν and $\tilde{\nu}$ are inclusions, it suffices to show that $\dim(\mathrm{Der}_S^G)_v = \dim TM_v^X$. Suppose $\dim(\mathrm{Der}_S^G)_v = p$. Let $\Theta = \{\theta_1,\ldots,\theta_\ell\}$ be a set of basic derivations. Since $\theta_j = \sum(\theta_j x_i)D_i = \sum \mathsf{M}(\Theta)_{i,j}D_i$, we have $p = \dim(\mathrm{Der}_S^G)_v = \mathrm{rank}\mathsf{M}(\Theta)(v)$. Thus the nullity of $\mathsf{M}(\Theta)(v)$ is $\ell - p$. By Theorem 6.113, the maximum number of linearly independent hyperplanes containing v is also $\ell - p$. Thus $\ell - p = \mathrm{codim} M^X$ and hence $p = \dim M^X = \dim TM_v^X$, showing that ν is an isomorphism. It follows from Corollary 6.110 that $\tilde{\nu}$ is also an isomorphism. □

Theorem 6.115 *For all $\mathcal{F}$, Θ, and $v \in V$, the maximum number of linearly independent reflecting hyperplanes which contain v is equal to the nullity of $\mathsf{M}_\Delta(f_1,\ldots,f_\ell;\mathcal{F},\Theta)(\pi v)$.*

Proof. Suppose $v \in M^X$ and $\dim M^X = \dim \tilde{M}^X = p$. Then the maximum number of linearly independent reflecting hyperplanes containing v is $\ell - p$. Since $\tilde{\nu}$ is an isomorphism, it follows from Corollary 6.110 that $\dim(D_R(\delta))_{\pi v} = p$. Given $\mathcal{F} = \{f_1,\ldots,f_\ell\}$ and $\Theta = \{\theta_1,\ldots,\theta_\ell\}$, it follows from Theorem 6.58 that for $1 \leq i \leq \ell$

$$\bar{\theta}_j = \sum \theta_j(f_i)D_{f_i} = \sum \mathsf{M}_\Delta(f_1,\ldots,f_\ell;\mathcal{F},\Theta)_{i,j}D_{f_i}$$

is a basis for $D_R(\delta)$. Therefore

$$p = \dim(D_R(\delta))_{\pi v} = \mathrm{rank}\mathsf{M}_\Delta(f_1,\ldots,f_\ell;\mathcal{F},\Theta)(\pi v)$$

as required. □

There are convenient reformulations of these results. For $0 \leq p \leq \ell - 1$, let $\mathcal{I}_p(\Theta)$ be the ideal in $\mathbb{C}[x_1,\ldots,x_\ell]$ generated by the determinants of the $(p+1) \times (p+1)$ minors of $\mathsf{M}(\Theta)$. Let $V(\mathcal{I})$ denote the variety of the ideal $\mathcal{I}$. Then Theorem 6.113 gives the next result.

Theorem 6.116 *For* $0 \leq p \leq \ell - 1$ *and for all* $\Theta = \{\theta_1, \ldots, \theta_\ell\}$, *we have*
$$N^p = V(\mathcal{I}_p(\Theta)).$$

For $0 \leq p \leq \ell - 1$, let $\mathcal{J}_p(\mathcal{F}, \Theta)$ be the ideal in $\mathbb{C}[T_1, \ldots, T_\ell]$ generated by the determinants of the $(p+1) \times (p+1)$ minors of $\mathsf{M}_\Delta(T_1, \ldots, T_\ell; \mathcal{F}, \Theta)$. The following reformulation of Theorem 6.115 shows that the discriminant matrix completely determines the orbit stratification of the discriminant.

Theorem 6.117 *For* $0 \leq p \leq \ell - 1$ *and for all* $\mathcal{F} = \{f_1, \ldots, f_\ell\}$ *and* $\Theta = \{\theta_1, \ldots, \theta_\ell\}$, *we have*
$$\tilde{N}^p = V(\mathcal{J}_p(\mathcal{F}, \Theta)).$$

Example 6.118 We computed the discriminant matrix $\mathsf{M}_\Delta(T_1, T_2, T_3; \mathcal{F}, \Theta)$ of G_{24} in Example 6.69. We determine its stratification here. The only point in $\tilde{V}$ where every entry of M_Δ vanishes is the origin, so $\tilde{N}^0 = \{0\}$. For $\tilde{N}^1$ we compute the determinants of the 2×2 minors of M_Δ. We know from Appendix C that G has two orbits of lines on L. Thus $\tilde{N}^1$ has two irreducible components. Explicit calculation yields the following common solutions of the determinants of the 2×2 minors of M_Δ:

$$T_1 = -u^2, \quad T_2 = 2u^3, \quad T_3 = 144u^7,$$
$$T_1 = 3u^2, \quad T_2 = 2u^3, \quad T_3 = -48u^7.$$

For $\tilde{N}^2$ we compute the discriminant $\Delta \doteq \det \mathsf{M}_\Delta$:

$$\Delta = -2048\, T_1^9 T_2 + 22016\, T_1^6 T_2^3 - 60032\, T_1^3 T_2^5 + 1728\, T_2^7 + 256\, T_1^7 T_3$$
$$+ 1088\, T_1^4 T_2^2 T_3 + 1008\, T_1 T_2^4 T_3 - 88\, T_1^2 T_2 T_3^2 + T_3^3.$$

We know from Appendix C that G is transitive on $\mathcal{A}$. Thus the discriminant is irreducible. Setting $T_1 = 0$, we see that

$$\Delta(T_1, T_2, T_3; \mathcal{F}) \equiv T_3^3 + 1728\, T_2^7 \mod (T_1).$$

This discriminant was well known to Klein as the discriminant of his simple group. Using our notation, he showed [125, p.449] that

$$(-f_2)^7 \equiv \left(\frac{f_3}{12}\right)^3 - 27\left(\frac{J}{216}\right)^2 \mod (f_1).$$

In this group, $J^2 \doteq \delta$ because all reflections are of order 2. Thus these congruences amount to the same assertion.

Naruki [163] computed the fundamental group $\pi_1(\tilde{M})$ of the complement of this discriminant. This is the smallest example where it is not known whether M and $\tilde{M}$ are $K(\pi, 1)$ spaces.

Shephard Groups

Regular complex polytopes were introduced by Shephard [209].

Definition 6.119 *A* **Shephard group** *is the symmetry group of a regular complex polytope.*

Shephard groups are irreducible complex reflection groups. There are irreducible complex reflection groups which are not Shephard groups. Coxeter groups which are also Shephard groups are the symmetry groups of (real) regular polytopes. We indicate in Appendix B which irreducible complex reflection groups are Shephard groups.

Recall the Hessian map from Definition 6.7. We showed in Lemma 6.9 that $f \in R$ induces a map $\mathrm{Hess}(f) : \mathrm{Der}_S^G \to \Omega_S^G$. If $\det \mathsf{H}(f_1) \neq 0$, then the map is a monomorphism.

Lemma 6.120 *Suppose G is irreducible and $f \in R$. If $\mathrm{Hess}(f) : \mathrm{Der}_S^G \to \Omega_S^G$ is surjective, then f is an invariant of minimal positive degree.*

Proof. Use total degrees in Der_S^G and in Ω_S^G. Suppose $f \in R_p$ and suppose that $R_q \neq 0$ for some $q > 0$. It follows from Theorem 6.49 that $(\Omega_S^G)_q \supseteq dR_q \neq 0$. Since $\mathrm{Hess}(f)$ is a surjective map of degree p, we have $(\mathrm{Der}_S^G)_{q-p} \neq 0$. Since G is an irreducible group, invariant derivations have minimal total degree $n_1 - 1 = 0$. Thus $q - p \geq 0$. □

The following characterization of Shephard groups was obtained in [177] using case by case arguments.

Theorem 6.121 *Let $G \subset GL(V)$ be a finite irreducible unitary reflection group with m reflections and n reflecting hyperplanes. Let f_1 be an invariant of minimal positive degree. The following statements are equivalent:*

(1) G is a Coxeter group or Shephard group,
(2) $\mathrm{Hess}(f_1) : \mathrm{Der}_S^G \to \Omega_S^G$ is an isomorphism,
(3) $m_i - n_i = d_1 - 2$ for $1 \leq i \leq \ell$,
(4) $m - n = \ell(d_1 - 2)$,
(5) $\det \mathsf{H}(f_1) \doteq \prod_{H \in \mathcal{A}} \alpha_H^{e_H - 2}$.

Definition 6.122 *Let G be an irreducible Coxeter group or Shephard group. Let $\mathcal{F} = \{f_1, \ldots, f_\ell\}$ be a set of basic invariants. Define the set of* **associated basic derivations** $\Theta = \{\theta_1, \ldots, \theta_\ell\}$ *by* $\mathrm{Hess}(f_1)\theta_i = m_1 df_i$. *Thus* $\mathsf{M}(\Theta)$ *is defined by the matrix equation*

$$\mathsf{H}(f_1)\mathsf{M}(\Theta) = m_1 \mathsf{J}(\mathcal{F}).$$

The coefficient m_1 normalizes the equation, so $\theta_1 = \theta_E$ is the Euler derivation. This equation is used to construct basic derivations in Appendix B.

Definition 6.123 *Let G be an irreducible Coxeter group or Shephard group. Let $\mathcal{F} = \{f_1, \ldots, f_\ell\}$ be a set of basic invariants. Let $\Theta = \{\theta_1, \ldots, \theta_\ell\}$ be the associated basic derivations. Define*

$$\mathsf{M}_\Delta(T_1, \ldots, T_\ell; \mathcal{F}) = \mathsf{M}_\Delta(T_1, \ldots, T_\ell; \mathcal{F}, \Theta).$$

If associated basic derivations are used, then the discriminant matrix of an irreducible Coxeter or Shephard group is symmetric because $\mathsf{M}_\Delta(f_1, \ldots, f_\ell; \mathcal{F}) \doteq \mathsf{M}(\Theta)^T \mathsf{H}(f_1) \mathsf{M}(\Theta)$.

Coxeter [56] showed that every Shephard group G admits a presentation with generating reflections $s_1, \ldots, s_\ell$ and relations

(2) $$s_i^{p_i} = 1 \qquad 1 \leq i \leq \ell,$$

(3) $$s_i s_j s_i \ldots = s_j s_i s_j \ldots \qquad i \neq j,$$

where $p_i \geq 2$ are integers and there are a certain number $q_{i,j} = q_{j,i} \geq 2$ terms on both sides of equation (3). Coxeter associated the symbol $p_1[q_1]p_2 \cdots p_{\ell-1}[q_{\ell-1}]p_\ell$ to the group G. It follows from the classification of Shephard groups that the symbol is uniquely determined by G up to reversal. Thus we may make the following definition.

Definition 6.124 *If $G \subset GL(V)$ is the Shephard group $p_1[q_1]p_2 \cdots p_{\ell-1}[q_{\ell-1}]p_\ell$, let $W \subset GL(V)$ be the Coxeter group $2[q_1]2 \cdots 2[q_{\ell-1}]2$. We call W the **Coxeter group associated** to G. The group W is uniquely determined by G up to conjugacy in $GL(V)$. If G is a Coxeter group, then $W = G$.*

The following is a complete list of Shephard groups G and the associated Coxeter groups W, where $G \neq W$. In the infinite families, $G(r, 1, \ell)$ has symbol $r[4]2[3] \cdots [3]2$. Its basic degrees are $r, 2r, \ldots, \ell r$. The associated Coxeter group has type B_ℓ with symbol $2[4]2[3] \cdots [3]2$. Its basic degrees are $2, 4, \ldots, 2\ell$. The group $C(r)$ has symbol r. Its basic degree is r. The associated Coxeter group has symbol 2. Its basic degree is 2. The exceptional groups are collected in Table 6.5. The first column indicates the classification number.

Both G and W are isomorphic to quotients of the same Artin group defined by the relators in (3). Since W is also a reflection group, it has basic invariants, basic derivations, discriminant, etc. When both groups are in consideration, we use notation like $M(G), M(W), d_i^G, d_i^W$, etc. to indicate the dependence on G, W. Given a Shephard group G and the associated Coxeter group W, inspection of the values of d_i^G and d_i^W listed above and in Table 6.5 reveals the existence of a constant κ such that

$$\frac{d_1^G}{d_1^W} = \cdots = \frac{d_\ell^G}{d_\ell^W} = \kappa.$$

6.6 Topology 267

Table 6.5. The associated pairs (G, W)

ST	G	d_i^G	W	d_i^W
4	3[3]3	4,6	2[3]2	2,3
8	4[3]4	8,12	2[3]2	2,3
16	5[3]5	20,30	2[3]2	2,3
5	3[4]3	6,12	2[4]2	2,4
10	4[4]3	12,24	2[4]2	2,4
18	5[4]3	30,60	2[4]2	2,4
20	3[5]3	12,30	2[5]2	2,5
6	3[6]2	4,12	2[6]2	2,6
9	4[6]2	8,24	2[6]2	2,6
17	5[6]2	20,60	2[6]2	2,6
14	3[8]2	6,24	2[8]2	2,8
21	3[10]2	12,60	2[10]2	2,10
25	3[3]3[3]3	6,9,12	2[3]2[3]2	2,3,4
26	2[4]3[3]3	6,12,18	2[4]2[3]2	2,4,6
32	3[3]3[3]3[3]3	12,18,24,30	2[3]2[3]2[3]2	2,3,4,5

It was shown in [178] that the invariant theories of G and W have the following connection.

Theorem 6.125 *Let G be a Shephard group and let W be the associated Coxeter group. There exist basic invariants $\mathcal{F}^G = \{f_1^G, \ldots, f_\ell^G\}$ and $\mathcal{F}^W = \{f_1^W, \ldots, f_\ell^W\}$ such that*

$$\mathsf{M}_\Delta^G(T_1, \ldots, T_\ell; \mathcal{F}^G) = \mathsf{M}_\Delta^W(\kappa T_1, \ldots, \kappa T_\ell; \mathcal{F}^W).$$

The $K(\pi, 1)$ Problem

Theorem 6.125 and the fact that Coxeter arrangements are $K(\pi, 1)$ may be used to show that the arrangements of Shephard groups are $K(\pi, 1)$.

Corollary 6.126 *Let G be a Shephard group and let W be the associated Coxeter group. There exist basic sets $\mathcal{F}^G = \{f_1^G, \ldots, f_\ell^G\}$ and $\mathcal{F}^W = \{f_1^W, \ldots, f_\ell^W\}$ such that their discriminant polynomials are equal:*

$$\Delta^G(T_1, \ldots, T_\ell; \mathcal{F}^G) \doteq \Delta^W(T_1, \ldots, T_\ell; \mathcal{F}^W).$$

Proof. Choose $\bar{\mathcal{F}}^G = \{f_1^G, \ldots, f_\ell^G\}$ and $\bar{\mathcal{F}}^W = \{f_1^W, \ldots, f_\ell^W\}$ which satisfy Theorem 6.125. Thus

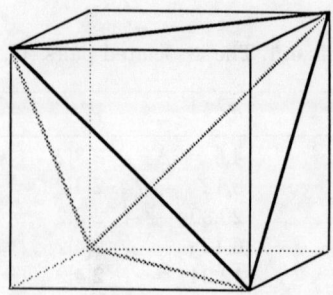

Fig. 6.4. A tetrahedron in the cube

$$\mathsf{M}^G_\Delta(T_1/d_1^G, \ldots, T_\ell/d_\ell^G; \bar{\mathcal{F}}^G) = \mathsf{M}^W_\Delta(T_1/d_1^W, \ldots, T_\ell/d_\ell^W; \bar{\mathcal{F}}^W).$$

Let $\mathcal{F}^G = \{d_1^G f_1^G, \ldots, d_\ell^G f_\ell^G\}$ and $\mathcal{F}^W = \{d_1^W f_1^W, \ldots, d_\ell^W f_\ell^W\}$. The discriminant matrices with respect to $\mathcal{F}^G$ and $\mathcal{F}^W$ change considerably, but their determinants satisfy the assertion. $\square$

Example 6.127 The group G_{25} introduced in Example 6.30 is a Shephard group. The associated Coxeter group W is of type $A_3 = D_3$. We illustrate Corollary 6.126 with the pair (G_{25}, D_3). Maschke [150, p.326] constructed certain homogeneous polynomials C_6, C_9, C_{12}, C_{12} of degrees 6,9,12 and 12, where $C_{12} = Q(G)$ defines $\mathcal{A}(G)$; see Example 6.30. Shephard and Todd [210, p.286] remarked that we may choose $\mathcal{F}^G = \{C_6, C_9, C_{12}\}$ as basic invariants for G. It follows from Maschke's work [150, p.326] that

$$\mathcal{C}_{12}^3 \doteq (432 C_9^2 - C_6^3 + 3 C_6 C_{12})^2 - 4 C_{12}^3.$$

Since every reflection in G has order 3, $\prod \alpha_H^{e_H} = \mathcal{C}_{12}^3$. Thus we have

(4) $$\Delta^G(T_1, T_2, T_3; \mathcal{F}^G) = (432 T_2^2 - T_1^3 + 3 T_1 T_3)^2 - 4 T_3^3.$$

Next we compute the discriminant for $W = D_3$. Since D_3 is a subgroup of B_3 of index 2, we may think of W as the group of symmetries of one of the two tetrahedra inscribed in a cube; see Figure 6.4. Thus we may choose a basis x, y, z for V^* such that

$$\prod_{H \in \mathcal{A}(W)} \alpha_H = (x^2 - y^2)(x^2 - z^2)(y^2 - z^2).$$

In this coordinate system $p_1 = x^2 + y^2 + z^2$, $p_2 = xyz$, and $p_3 = x^2y^2 + x^2z^2 + y^2z^2$ is a set of basic invariants for W. Consider a cubic polynomial with roots x^2, y^2, z^2. The formula for the discriminant of this cubic, expressed in terms of the elementary symmetric functions of the roots, gives the identity

$$\prod_{H \in \mathcal{A}(W)} \alpha_H^2 = (2p_1^3 - 9p_1 p_3 + 27 p_2^2)^2 - 4(p_1^2 - 3p_3)^3.$$

Let $f_1 = p_1$, $f_2 = p_2/4$, and $f_3 = p_1^2 - 3p_3$. Then $\mathcal{F}^W = \{f_1, f_2, f_3\}$ is also a set of basic invariants for W and we have

(5) $\qquad \Delta^W(T_1, T_2, T_3; \mathcal{F}^W) = (432T_2^2 - T_1^3 + 3T_1T_3)^2 - 4T_3^3.$

Comparing (4) and (5) illustrates Corollary 6.126 for the pair (G_{25}, D_3).

Theorem 6.128 *If $G \subset GL(V)$ is a Shephard group, then $M(G)$ is a $K(\pi, 1)$ space.*

Proof. Let W be the Coxeter group associated to G. It follows from Theorem 5.15 of Deligne that $M(W)$ is a $K(\pi, 1)$ space. Thus $\tilde{M}(W)$ is a $K(\pi, 1)$ space. By Corollary 6.126, $\tilde{M}(W) = \tilde{M}(G)$ and therefore $\tilde{M}(G)$ and $M(G)$ are also $K(\pi, 1)$ spaces. □

A. Some Commutative Algebra

In this appendix we collect some definitions and facts in commutative algebra. These facts concern free modules, Krull dimension of rings and dimension of modules, graded rings and modules, associated primes, and regular sequences.

A.1 Free Modules

Let R be a commutative Noetherian ring with unit and let M, N be R–modules. We assume that **all our modules are finitely generated**. The following lemma is one of the most fundamental results in linear algebra when R is a field.

Lemma A.1 *Let A be an $m \times n$ matrix with entries in R. Suppose that all n–minors are zero (or $m < n$). Then there exists a nonzero vector $\mathbf{a}$ such that $\mathsf{A}\mathbf{a} = \mathbf{0}$.*

Proof. This is obvious when $\mathsf{A} = 0$. Let $r > 0$ be the number such that there is a nonzero r–minor of A and that all the $(r+1)$–minors are 0. (In linear algebra r is the rank of A.) We can assume without loss of generality that $n = r+1$ and that the principal r–minor of A is not 0. Let $\tilde{\mathsf{A}}$ be a square matrix consisting of the first $r+1$ rows of A. The cofactor expansion formula gives

$$\mathsf{A}\,\mathrm{ad}(\tilde{\mathsf{A}}) = 0$$

where ad stands for the adjoint matrix. Since the $(r+1, r+1)$–entry of $\mathrm{ad}(\tilde{\mathsf{A}})$ is not 0, we may take the $(r+1)$–column of $\mathrm{ad}(\tilde{\mathsf{A}})$ as $\mathbf{a}$. □

Definition A.2 *An R–module M is **free** if M has a linearly independent set of generators over R. We call such a set a **basis**.*

Proposition A.3 *Let M be a free R–module with basis $\{f_1, \ldots, f_\ell\}$. Let $G = \{m_1, \ldots, m_k\}$ be a finite subset of M.*
  (1) If G is linearly independent, then $k \leq \ell$.
  (2) If G generates M, then $k \geq \ell$. Moreover, if equality holds, then G is a basis for M.

Proof. There exists an $\ell \times k$ matrix A such that
$$[m_1, \ldots, m_k] = [f_1, \ldots, f_\ell]\, \mathsf{A}.$$

(1) If $k > \ell$, then by Lemma A.1 there exists a nonzero vector $\mathbf{a}$ with $\mathsf{A}\mathbf{a} = \mathbf{0}$. Thus $[m_1, \ldots, m_k]\,\mathbf{a} = 0$, which implies that G is linearly dependent.

(2) Since G generates M, there exists a $k \times \ell$ matrix B such that
$$[f_1, \ldots, f_\ell] = [m_1, \ldots, m_k]\, \mathsf{B}.$$
If $k < \ell$, then there exists a nonzero vector $\mathbf{b}$ with $\mathsf{B}\mathbf{b} = \mathbf{0}$ by Lemma A.1. Thus $[f_1, \ldots, f_\ell]\,\mathbf{b} = 0$. This contradicts the fact that $f_1, \ldots, f_\ell$ are linearly independent. Finally, assume that $k = \ell$. Then we get $\mathsf{AB} = \mathsf{I}_\ell$. Thus B is an invertible matrix and G is a basis. □

Definition A.4 *Suppose M is a free R-module. It follows from Proposition A.3 that the number of elements in a basis $\{m_1, \ldots, m_\ell\}$ for M over R is independent of the choice of basis. We call this number the* **rank** *of M and write* $\mathrm{rank}_R M = \ell$.

A.2 Krull Dimension

If $m \in M$, let
$$\mathrm{Ann}\, m = \mathrm{Ann}_R\, m = \{r \in R \mid rm = 0\}$$
be the **annihilator** of m. Let
$$\mathrm{Ann}\, M = \mathrm{Ann}_R M = \{r \in R \mid rM = 0\}$$
be the **annihilator** of M. Let $\mathrm{Spec}\, R$ be the set of all prime ideals of R. If I is an ideal of R, let
$$V(I) = \{\wp \in \mathrm{Spec}\, R \mid \wp \supseteq I\}.$$
If $\wp \in \mathrm{Spec}\, R$, let $R_\wp$ be the **localization** of R at $\wp$ and let $M_\wp$ be the **localization** of M at $\wp$ [152, 1.G]. Note that $M_\wp$ is an $R_\wp$-module.

Lemma A.5 *Let $\wp \in \mathrm{Spec}\, R$. If $m \in M$, then $m/1 \neq 0$ in $M_\wp$ if and only if $\wp \supseteq \mathrm{Ann}\, m$.*

Proof. We have $m/1 = 0$ if and only if there exists $s \in R \setminus \wp$ with $sm = 0$. □

Proposition A.6 *Let $\wp \in \mathrm{Spec}\, R$. Then $M_\wp \neq 0$ if and only if $\wp \supseteq \mathrm{Ann}\, M$.*

Proof. Since M is finitely generated, we may write $M = Rm_1 + \ldots + Rm_k$. Lemma A.5 gives
$$\begin{aligned} M_\wp \neq 0 &\iff \text{there exists } i \text{ such that } m_i/1 \neq 0 \\ &\iff \text{there exists } i \text{ such that } \mathrm{Ann}\, m_i \subseteq \wp \\ &\iff \mathrm{Ann}\, M = \bigcap_{i=1}^k \mathrm{Ann}\, m_i \subseteq \wp. \end{aligned}$$
This completes the proof. □

Definition A.7 *The* **support** *of M is defined by*

$$\operatorname{Supp} M = \operatorname{Supp}_R M = \{\wp \in \operatorname{Spec} R \mid M_\wp \neq 0\}.$$

The next result follows from Proposition A.6 and Definition A.7.

Proposition A.8 $\operatorname{Supp} M = V(\operatorname{Ann} M)$.

See [152, 3.D] for the next result.

Theorem A.9 *If $\wp \in \operatorname{Spec} R$ and $M' \to M \to M''$ is an exact sequence of R-modules, then $M'_\wp \to M_\wp \to M''_\wp$ is an exact sequence of $R_\wp$-modules.*

Definition A.10 *If $\wp \in \operatorname{Spec} R$, the* **height** *of $\wp$, $\operatorname{ht} \wp$, is the largest integer n for which there exists a chain $\wp = \wp_0 \supset \wp_1 \supset \ldots \supset \wp_n$ with $\wp_i \in \operatorname{Spec} R$ for $0 \leq i \leq n$.*

Definition A.11 *The* **Krull dimension** *of R is defined by*

$$\operatorname{Krull\ dim} R = \sup\{\operatorname{ht} \wp \mid \wp \in \operatorname{Spec} R\}.$$

Thus Krull $\dim R$ is the largest integer n for which there exists a chain $\wp_0 \supset \wp_1 \supset \ldots \supset \wp_n$ with $\wp_i \in \operatorname{Spec} R$ for $0 \leq i \leq n$. For the next result see [152].

Proposition A.12 *If $\varphi : R \to S$ is an epimorphism of rings, then the correspondence $\wp \in \operatorname{Spec} S \mapsto \varphi^{-1}(\wp) \in \operatorname{Spec} R$ gives a bijection from $\operatorname{Spec} S$ to $\operatorname{Spec} R \cap V(\ker(\varphi))$.*

Proposition A.13 *Let I be an ideal of R. Then*

$$\operatorname{Krull\ dim}(R/I) = \sup_{\wp \in V(I)} \operatorname{Krull\ dim}(R/\wp).$$

Proof. This is clear from Definition A.11 and Proposition A.12. □

Definition A.14 *If $M \neq 0$, then the* **dimension** *of M is defined by*

$$\dim M = \dim_R M = \operatorname{Krull\ dim}(R/\operatorname{Ann} M).$$

Proposition A.15 *If $M \neq 0$, then*

$$\dim M = \sup_{\wp \in \operatorname{Supp} M} \operatorname{Krull\ dim}(R/\wp).$$

274   A. Some Commutative Algebra

Proof. This is clear from Definition A.14, Proposition A.13, and Proposition A.8. □

In Propositions A.16–A.18, let $\mathbb{K}$ be a field and let $R = \mathbb{K}[x_1, \ldots, x_\ell]$ where $x_1, \ldots, x_\ell$ are indeterminates. The next results are in [152, 14.A, Cor. and 14.H, Cor.3].

Proposition A.16 *The ideal $(x_1, \ldots, x_k)$ is a prime ideal of height k for $1 \leq k \leq \ell$ and Krull dim $R = \ell$.*

Proposition A.17 *If $\wp \in \text{Spec } R$, then Krull $\dim(R/\wp) = \ell - \text{ht } \wp$.*

Proposition A.18
$$\dim M = \ell - \min_{\wp \in \text{Supp}(M)} \text{ht } \wp.$$

Proof. This follows from Proposition A.15 and Proposition A.17. □

A.3 Graded Modules

In this section we assume that $R = \oplus_{p \in \mathbb{Z}} R_p$ is a **finitely generated $\mathbb{K}$-algebra with $R_p = 0$ for $p < 0$ and $R_0 = \mathbb{K}$. Assume that M is a graded R-module**, $M = \oplus_{p \in \mathbb{Z}} M_p$. Let $R_+ = \oplus_{p > 0} R_p$. Then R_+ is a maximal ideal of R.

Theorem A.19 *Let M be a free graded R-module of rank ℓ. A finite subset of M consisting of homogeneous elements is a basis if and only if it is a minimal set of generators.*

Proof. Suppose that $\{m_1, \ldots, m_k\}$ is a minimal set of generators. Let $\bar{M} = M/R_+M$. Then $\bar{M}$ is naturally a $\mathbb{K} = R_0 \simeq R/R_+$-vector space. Denote the class of x in $\bar{M}$ by $\bar{x}$. By Proposition A.3, it is sufficient to show that $k \leq \ell$. Assume $k > \ell$. Since ℓ elements (the classes of members of a basis) generate $\bar{M}$, $\bar{m}_1, \ldots, \bar{m}_k$ are linearly dependent over $\mathbb{K}$. Thus there exist $c_1, \ldots, c_k \in \mathbb{K}$, not all 0, such that
$$c_1 \bar{m}_1 + \ldots + c_k \bar{m}_k = 0.$$
This implies that
$$c_1 m_1 + \ldots + c_k m_k = R_+ M.$$
There exist $r_1, \ldots r_k \in R_+$ such that
$$c_1 m_1 + \ldots + c_k m_k = r_1 m_1 + \ldots + r_k m_k.$$

Suppose $c_1 \neq 0$ and consider the homogeneous component of degree $\deg m_1$ in the last equation. Since $r_1 \in R_+$, $\deg r_1 m_1 > \deg c_1 m_1$. Thus

$$c_1 m_1 \in Rm_2 + \ldots + Rm_k.$$

This contradicts the minimality of $m_1, \ldots, m_k$. The converse is obvious. □

Theorem A.20 *If a graded R–module M is free of rank ℓ, then it has a homogeneous basis $\{m_1, \ldots, m_\ell\}$.*

Proof. Consider all homogeneous components of members of a basis. They form a set of homogeneous generators. Choose a minimal set of generators among them. This set is a basis by Theorem A.19. □

Recall the Poincaré series of a finitely generated graded module from Definition 4.117.

Theorem A.21 *Let $0 \to M_1 \to M_2 \to M_3 \to 0$ be an exact sequence of finitely generated graded R–modules and S–homomorphisms of degree 0. Then*

$$\mathrm{Poin}(M_1, t) - \mathrm{Poin}(M_2, t) + \mathrm{Poin}(M_3, t) = 0.$$

Theorem A.22 *Suppose $M \neq 0$. Then $\mathrm{Poin}(M, t)$ has a pole at $t = 1$ of order at most $\dim_R M$.*

Proof. It is proved in [165, p.346, Th.19] that $\dim M_n$ is a polynomial in n of degree $\dim M - 1$ for sufficiently large n. The weaker result here follows from the fact that every polynomial of degree $d - 1$ in n is a linear combination of $\{n^{[i]}\}$, where

$$n^{[i]} = (n+1)(n+2)\ldots(n+i) \quad 0 \leq i \leq d-1,$$

and

$$\frac{1}{(1-t)^i} = (-1)^{i-1} \sum_{n \geq 0} n^{[i-1]} t^n \quad 0 \leq i \leq d-1.$$

This completes the argument. □

Example A.23

(1) Regard the polynomial ring $S = \mathbb{K}[x_1, \ldots, x_\ell]$ as a graded $\mathbb{K}$–algebra by defining $\deg x_i = 1$ for $1 \leq i \leq \ell$. Then

$$\mathrm{Poin}(S, t) = (1 - t)^{-\ell}.$$

(2) Let d be an integer. The graded R–module $R(-d)$ is defined by shifting the grading of R by d so that $R(-d)_m = R_{m-d}$. Then

$$\mathrm{Poin}(R(-d), t) = t^d \mathrm{Poin}(R, t).$$

(3) Let d_i be integers for $1 \leq i \leq \ell$. If a graded R–module M is isomorphic to $\bigoplus_{i=1}^\ell R(-d_i)$, then by Theorem A.21

$$\text{Poin}(M, t) = (t^{d_1} + \ldots + t^{d_\ell})\text{Poin}(R, t).$$

(4) If the set $m_1, \ldots, m_\ell$ is a basis for the free graded R–module M with $m_i \in M_{d_i}$, then $M \cong \bigoplus_{i=1}^{\ell} R(-d_i)$ and thus by (3)

$$\text{Poin}(M, t) = (t^{d_1} + \cdots + t^{d_\ell})\text{Poin}(R, t).$$

Proposition A.24 *Assume the graded R–module M is free with a homogeneous basis $\{m_1, \ldots, m_\ell\}$. Then the degrees $\{\deg m_1, \ldots, \deg m_\ell\}$ (with multiplicity but neglecting the order) depend only on M.*

Proof. This follows from Example A.23.4. □

Proposition A.25 *The following three conditions are equivalent:*
(1) M is finite dimensional over $\mathbb{K}$,
(2) $R_+ \subseteq \sqrt{\text{Ann}\, M}$,
(3) there exists a positive integer μ such that $(R_+)^\mu M = 0$.

Proof. Since R is Noetherian, (2) and (3) are equivalent.

(1) $\Rightarrow$ (2) There exists a positive integer μ such that $M_n = 0$ for $n \geq \mu$. Thus we have $(R_+)^\mu M \subseteq \bigoplus_{n \geq \mu} M_n = 0$.

(3) $\Rightarrow$ (1) Let $x_1, \cdots, x_m$ be a finite set of generators for M over R. Let $r_1, \ldots, r_n$ be a basis for $\bigoplus_{p=0}^{\mu} R_p$ over $\mathbb{K}$. Then $\{r_i x_j \mid 1 \leq i \leq m, 1 \leq j \leq n\}$ generates M over $\mathbb{K}$. □

A.4 Associated Primes and Regular Sequences

Let R be a Noetherian ring with 1 and let M be a finite R–module.

Definition A.26 *A prime ideal of R of type $\text{ann}(x) = \{r \in R \mid rx = 0\}$ ($0 \neq x \in M$) is called an **associated prime** of M. Given an ideal I of R, we say that $\wp \in \text{Spec}\, R$ is an **associated prime** of I if $\wp$ is an associated prime of the module R/I.*

Lemma A.27 *The maximal elements of the set $\{\text{ann}(x) \mid 0 \neq x \in M\}$ are associated primes of M.*

Proof. Let $\text{ann}(y)$ be a maximal element. Suppose that $\text{ann}(y)$ is not prime. Choose $a \notin \text{ann}(y)$ and $b \notin \text{ann}(y)$ with $ab \in \text{ann}(y)$. Then

$$a \notin \text{ann}(y) \subseteq \text{ann}(by) \ni a.$$

This contradicts the maximality of $\text{ann}(y)$. □

Let I and J be two ideals of R. The ideal $I : J$ is defined by

A.4 Associated Primes and Regular Sequences   277

$$I : J = \{x \in R \mid xJ \subseteq I\}.$$

Proposition A.28 *Let I and J be two ideals of R. Then $I : J = I$ if and only if $J \not\subseteq \wp$ for each associated prime $\wp$ of I.*

Proof. We apply Lemma A.27 to get

$$\begin{aligned} I : J \neq I &\iff xJ \subseteq I \text{ for some } x \notin I \\ &\iff J \subseteq \operatorname{ann}(x+I) \text{ for some } x+I \in R/I, x+I \neq I \\ &\iff J \subseteq \wp \text{ for some associated prime } \wp \text{ of } I \end{aligned}$$

as required.   □

Definition A.29 *A sequence $a_1, \ldots, a_n$ of elements of a ring R is called an **M-regular sequence** if for every $x \in M$*

$$a_i x \in a_1 M + \cdots + a_{i-1} M \implies x \in a_1 M + \cdots + a_{i-1} M$$

*for $1 \leq i \leq n$. In the special case when $M = R$, we simply call it a regular sequence. In other words, the sequence $a_1, \ldots, a_n$ of elements of a ring R is called a **regular sequence** if*

$$(a_1, \ldots, a_{i-1}) : (a_i) = (a_1, \ldots, a_{i-1})$$

for $1 \leq i \leq n$.

Definition A.30 *A local ring R is called **Cohen-Macaulay** if there exists a regular sequence in the maximal ideal whose length is equal to the Krull dimension. A ring R is said to be **Cohen-Macaulay** if the locaization at every maximal ideal is Cohen-Macaulay.*

The following three results [153, Thms. 17.7, 17.6, 17.4] are fundamental concerning Cohen-Macaulay rings.

Theorem A.31 *A polynomial ring over a field is Cohen-Macaulay.*

Theorem A.32 *Let R be a Cohen-Macaulay ring. Suppose an ideal $(r_1, \ldots, r_k)R$ is of height k. Then the height of each associated prime of $(r_1, \ldots, r_k)R$ is equal to k.*

Theorem A.33 *Let R be a Cohen-Macaulay ring. Suppose that $r_1, \ldots, r_k$ are in a maximal ideal. Then the following three conditions are equivalent:*
(1) $r_1, \ldots, r_k$ is a regular sequence,
(2) $\operatorname{ht}(r_1, \ldots, r_k)R = k$,
(3) $\operatorname{ht}(r_1, \ldots, r_i)R = i$ for $1 \leq i \leq k$.

B. Basic Derivations

In this appendix we construct basic derivations for all unitary reflection groups G. We follow our work in [166, 237]. If G is reducible, then $V = V_1 \oplus V_2$, where V_1, V_2 are G–modules and the restrictions G_1, G_2 of G to V_1, V_2 are either the identity or unitary reflection groups. Let $\mathcal{A}_1 = (\mathcal{A}(G_1), V_1)$, $\mathcal{A}_2 = (\mathcal{A}(G_2), V_2)$. Then $\mathcal{A}(G) = \mathcal{A}_1 \times \mathcal{A}_2$. It follows from Theorem 4.28 and Theorem 6.59 that it suffices to find bases for $\mathrm{Der}_S^{G_1}$ and $\mathrm{Der}_S^{G_2}$. Thus we may assume that G acts irreducibly and use the classification; see Section 6.2. Various methods are used in the constructions.

B.1 The Infinite Families

(1) We showed in Example 4.22 that the derivations defined for $1 \leq k \leq \ell$ by

$$\theta_k = \sum_{j=1}^{\ell} x_j^{k-1} D_j$$

form a basis for the braid arrangement. Direct calculation shows that $\theta_k \in \mathrm{Der}_S^G$. In order to get a set of basic derivations for the irreducible reflection representation of the Coxeter group of type $A_{\ell-1}$ we set $\sum x_j = 0$.

(2) The monomial groups $G(r, p, \ell)$ are divided into two families with different reflection arrangements; see Section 6.4.

(2′) For the monomial groups $G(r, p, \ell)$ with $p < r$ we have

$$Q(\mathcal{A}) = x_1 \cdots x_\ell \prod_{1 \leq i < j \leq \ell} (x_i^r - x_j^r).$$

For $1 \leq i \leq \ell$ we may choose the derivations

(∗) $$\theta_i = \sum_{j=1}^{\ell} x_j^{(i-1)r+1} D_j.$$

Direct calculation shows that $\theta_i \in \mathrm{Der}_S^G$. It follows from Saito's criterion 4.19 that the θ_i form a basis for Der_S^G.

(2″) For the monomial groups $G(r, r, \ell)$ we have

$$Q(\mathcal{A}) = \prod_{1 \leq i < j \leq \ell} (x_i^r - x_j^r).$$

We may choose θ_i as in $(*)$ for $1 \leq i \leq \ell - 1$ together with

$$\theta_\ell = \sum_{j=1}^{\ell} (x_1 \cdots x_{j-1} x_{j+1} \cdots x_\ell)^{r-1} D_j.$$

Direct calculation shows that $\theta_i \in \text{Der}_S^G$. It follows from Saito's criterion 4.19 that the θ_i form a basis for Der_S^G.

(3) Here the basis is the Euler derivation in one variable.

B.2 Exceptional Groups of Rank 2

Here we may choose

$$\begin{aligned} \theta_1 &= x_1 D_1 + x_2 D_2, \\ \theta_2 &= (-D_2 Q) D_1 + (D_1 Q) D_2. \end{aligned}$$

Direct calculation shows that $\theta_1, \theta_2 \in \text{Der}_S^G$ and that $\det \mathsf{M}(\theta_1, \theta_2) = (\deg Q) Q$. It follows from Saito's criterion 4.19 that θ_1, θ_2 form a basis for Der_S^G.

Recall that we showed in Example 4.20 that every 2–arrangement is free by constructing derivations $\theta_E, \theta \in D(\mathcal{A})$ such that $\det \mathsf{M}(\theta_E, \theta) = Q$. Note that $\theta_1 = \theta_E$, but $\theta_2 \neq \theta$. Neither could be replaced by the other. In general, $\theta \notin \text{Der}_S^G$, so we could not replace θ_2 by θ here. The argument in Example 4.20 is claimed for all $\mathbb{K}$. If the characteristic of $\mathbb{K}$ divides $\deg Q$, then θ_1 and θ_2 are linearly dependent. Hence we could not replace θ by θ_2 in Example 4.20.

B.3 Exceptional Groups of Rank ≥ 3

The groups numbered 23–37 remain. The essential tool in the construction of basic derivations for these groups is the matrix equation of Theorem 6.53:

(1) $$\mathsf{H}(f) \mathsf{M}(\Theta) = m_1 \mathsf{J}(\mathcal{F}) \mathsf{C}.$$

Here $\mathsf{H}(f)$ is the Hessian matrix of the G–invariant polynomial $f \in R$, $\Theta = \{\theta_1, \ldots, \theta_\ell\}$ is a set of basic derivations, $\mathsf{M}(\Theta)$ is its coefficient matrix, $\mathcal{F} = \{f_1, \ldots, f_\ell\}$ is a set of basic invariants, $\mathsf{J}(\mathcal{F})$ is the Jacobian, and C is a nonsingular $\ell \times \ell$ matrix with entries in R. If f is homogeneous, then the nonzero entries of C are homogeneous. The coefficient m_1 normalizes the equation, so $\theta_1 = \theta_E$ is the Euler derivation.

Basic invariants for unitary reflection groups are found in [46, 51, 125, 150, 151, 251]. Choose f with $\det \mathsf{H}(f) \neq 0$. We may view (1) as an equation which contains two known matrices $\mathsf{H}(f)$, $\mathsf{J}(\mathcal{F})$ and two unknown matrices $\mathsf{M}(\Theta)$, C.

B.3 Exceptional Groups of Rank ≥ 3   281

It follows from (1) that either of the unknown matrices determines the other. The amount of work is reduced by considering C first, because it is possible to choose a suitable C with very few nonzero entries. We treat the groups in sets with increasing difficulty of finding a suitable matrix C. In each case $\theta_1 = \theta_E$ is the Euler derivation and we omit it from the list of basic derivations.

Coxeter Groups

The remaining Coxeter groups are numbered 23, 28, 30, 35, 36, 37. Each group has an invariant real quadratic form $f = f_1$ with $\det \mathsf{H}(f) \neq 0$. Thus $\mathsf{H}(f)$ is an invertible scalar matrix. If we choose an orthonormal basis in V, then $\mathsf{H}(f) = 2\,\mathsf{I}$. It follows from Theorem 6.122 that we may choose $\mathsf{C} = \mathsf{I}$. Thus (1) gives in this special case that the coefficient matrix is a constant multiple of the Jacobian:
$$\mathsf{M}(\Theta) = (1/2)\mathsf{J}(\mathcal{F}).$$

We make this explicit. Let $x_1, \ldots, x_\ell$ be an orthonormal basis for V^*. Let $\mathcal{F} = \{f_1, \ldots, f_\ell\}$ be basic invariants. For $1 \leq i \leq \ell$ define

$$\theta_{f_i} = \frac{1}{2}\sum_{j=1}^{\ell}\left(\frac{\partial f_i}{\partial x_j}\right)D_j.$$

Then $\Theta = \{\theta_{f_1}, \ldots, \theta_{f_\ell}\}$ is the set of basic derivations associated to $\mathcal{F}$.

The notion of **flat** basic invariants was introduced in [202]. Recall from Definition 6.99 that $d_{\ell-1} < d_\ell$. Thus D_{T_ℓ} is a well defined derivation in $R = \mathbb{R}[T_1, \ldots, T_\ell]$. In our terminology, $\mathcal{F}$ is flat if $D_{T_\ell}(\mathsf{M}_\Delta(T_1, \ldots, T_\ell; \mathcal{F})) \in M_\ell(\mathbb{R})$. Saito [200] proved the existence of flat basic invariants for all Coxeter groups.

Shephard Groups

The remaining Shephard groups are numbered 25, 26, 32. If $f = f_1$ is an invariant of minimal positive degree, then $\det \mathsf{H}(f) \neq 0$. It follows from Theorem 6.122 that we may choose $\mathsf{C} = \mathsf{I}$. Thus

(2) $$\mathsf{M}(\Theta) = m_1[\mathsf{H}(f)]^{-1}\mathsf{J}(\mathcal{F}).$$

Even with the aid of fast computers, there is some difficulty in calculating the right side of (2). This calculation involves division by the Hessian determinant which, in G_{32}, is a homogeneous polynomial of degree 40 in four variables. The notion of **flat** basic invariants was extended to Shephard groups in [178]. We have $d_{\ell-1} < d_\ell$. Thus D_{T_ℓ} is a well defined derivation in $R = \mathbb{C}[T_1, \ldots, T_\ell]$ and $\mathcal{F}$ is flat if $D_{T_\ell}(\mathsf{M}_\Delta(T_1, \ldots, T_\ell; \mathcal{F})) \in M_\ell(\mathbb{C})$. We choose flat basic invariants and reproduce the associated basic derivations and the resulting discriminant matrix from [178].

In order to be able to use Maschke's polynomials C_6, C_9, C_{12}, C_{18} defined in [151, p.326], we agree to let the basis of V^* be z_1, z_2, z_3 for $\ell = 3$ and $z_0, z_1, z_2,$

z_3 for $\ell = 4$. This allows us to use Maschke's convention that the subscripts i, $i+1$, $i+2$, when they appear, represent the integers 1,2,3 in cyclic permutation. Summation is over 1,2,3 in G_{25}, G_{26} and over 0,1,2,3 in G_{32}.

$$\begin{aligned} C_6 &= z_1^6 + z_2^6 + z_3^6 - 10(z_1^3 z_2^3 + z_2^3 z_3^3 + z_3^3 z_1^3), \\ C_9 &= (z_1^3 - z_2^3)(z_2^3 - z_3^3)(z_3^3 - z_1^3), \\ C_{12} &= (z_1^3 + z_2^3 + z_3^3)[(z_1^3 + z_2^3 + z_3^3)^3 + 216 z_1^3 z_2^3 z_3^3], \\ C_{18} &= (z_1^3 + z_2^3 + z_3^3)^6 - 540 z_1^3 z_2^3 z_3^3 (z_1^3 + z_2^3 + z_3^3)^3 - 5832 z_1^6 z_2^6 z_3^6. \end{aligned}$$

The Group G_{25}

$$\begin{aligned} f_1 &= C_6, \\ f_2 &= 32\sqrt{3} C_9, \\ f_3 &= 5 C_6^2 - 8 C_{12}, \\ \theta_2 &= 8\sqrt{3} \sum [z_{i+2}^3 - z_{i+1}^3] z_i D_i, \\ \theta_3 &= 6 \sum [-z_i^6 + 7(z_{i+1}^6 + z_{i+2}^6) - 14 z_i^3 (z_{i+1}^3 + z_{i+2}^3) + 42 z_{i+1}^3 z_{i+2}^3] z_i D_i. \end{aligned}$$

$\mathsf{M}_\Delta(T_1, T_2, T_3; \mathcal{F}) =$

$$\begin{bmatrix} 6 T_1 & 9 T_2 & 12 T_3 \\ 9 T_2 & 36 T_1^2 + 12 T_3 & 90 T_1 T_2 \\ 12 T_3 & 90 T_1 T_2 & 216 T_1^3 + 54 T_2^2 \end{bmatrix}.$$

The Group G_{26}

$$\begin{aligned} f_1 &= C_6, \\ f_2 &= 12 C_{12} - 3 C_6^2, \\ f_3 &= 96 C_{18} + 18 C_6^3 - 72 C_6 C_{12}, \\ \theta_2 &= 18 \sum [z_i^6 - 3(z_{i+1}^6 + z_{i+2}^6) + 2 z_i^3 (z_{i+1}^3 + z_{i+2}^3) - 26 z_{i+1}^3 z_{i+2}^3] z_i D_i, \\ \theta_3 &= 18 \sum [7 z_i^{12} + 28 z_i^9 (z_{i+1}^3 + z_{i+2}^3) + 162 z_i^6 (z_{i+1}^6 + z_{i+2}^6) \\ &\quad - 1236 z_i^6 z_{i+1}^3 z_{i+2}^3 + 2580 z_i^3 (z_{i+1}^6 z_{i+2}^3 + z_{i+1}^3 z_{i+2}^6) \\ &\quad - 308 z_i^3 (z_{i+1}^9 + z_{i+2}^9) - 17(z_{i+1}^{12} + z_{i+2}^{12}) \\ &\quad + 676 (z_{i+1}^9 z_{i+2}^3 + z_{i+1}^3 z_{i+2}^9) + 90 z_{i+1}^6 z_{i+2}^6] z_i D_i. \end{aligned}$$

$\mathsf{M}_\Delta(T_1, T_2, T_3; \mathcal{F}) =$

$$\begin{bmatrix} 6 T_1 & 12 T_2 & 18 T_3 \\ 12 T_2 & 216 T_1^3 + 108 T_1 T_2 + 18 T_3 & 864 T_1^2 T_2 + 72 T_2^2 \\ 18 T_3 & 864 T_1^2 T_2 + 72 T_2^2 & 7776 T_1^5 + 1080 T_1 T_2^2 \end{bmatrix}.$$

The Group G_{32}

In the description of the basic derivations we need the following polynomials:
$$C_6(0) = C_6, \quad C_9(0) = C_9, \quad C_{12}(0) = C_{12}.$$

For $i = 1, 2, 3$ define:
$$\begin{aligned}
C_6(i) &= z_0^6 + z_{i+1}^6 + z_{i+2}^6 + 10(z_{i+1}^3 z_{i+2}^3 - z_0^3 z_{i+1}^3 + z_0^3 z_{i+2}^3), \\
C_9(i) &= z_{i+1}^3 z_{i+2}^6 + z_{i+1}^6 z_{i+2}^3 + z_0^3 z_{i+1}^6 - z_0^6 z_{i+1}^3 - z_0^3 z_{i+2}^6 - z_0^6 z_{i+2}^3, \\
C_{12}(i) &= z_0^{12} + z_{i+1}^{12} + z_{i+2}^{12} \\
&\quad - 4(z_{i+1}^9 z_{i+2}^3 + z_{i+1}^3 z_{i+2}^9 + z_0^9 z_{i+2}^3 + z_0^3 z_{i+2}^9 - z_0^9 z_{i+1}^3 - z_0^3 z_{i+1}^9) \\
&\quad + 6(z_0^6 z_{i+1}^6 + z_0^6 z_{i+2}^6 + z_{i+1}^6 z_{i+2}^6) \\
&\quad - 228(z_0^6 z_{i+1}^3 z_{i+2}^3 + z_0^3 z_{i+1}^6 z_{i+2}^3 - z_0^3 z_{i+1}^3 z_{i+2}^6).
\end{aligned}$$

We use polynomials in [151, p.337] to define basic invariants:
$$\begin{aligned}
f_1 &= F_{12}, \\
f_2 &= \frac{4}{3} F_{18}, \\
f_3 &= 21 F_{12}^2 - 25 F_{24}, \\
f_4 &= \frac{8}{5}(11 F_{12} F_{18} - 25 F_{30}), \\
\theta_2 &= 6\sum [-3 z_i^6 + 7 C_6(i)] z_i D_i, \\
\theta_3 &= 36 \sum [7 z_i^{12} - 26 z_i^6 C_6(i) + 2080 z_i^3 C_9(i) \\
&\quad + \frac{26}{3} C_6(i)^2 - \frac{65}{3} C_{12}(i)] z_i D_i, \\
\theta_4 &= 216 \sum [-11 z_i^{18} + 57 z_i^{12} C_6(i) - 3040 z_i^9 C_9(i) + 722 z_i^6 C_6(i)^2 \\
&\quad - 1235 z_i^6 C_{12}(i) + 4560 z_i^3 C_6(i) C_9(i) - \frac{38}{3} C_6(i)^3 \\
&\quad + \frac{95}{3} C_6(i) C_{12}(i) + 7600 C_9(i)^2] z_i D_i.
\end{aligned}$$

$\mathsf{M}_\Delta(T_1, T_2, T_3, T_4; \mathcal{F}) =$

$$\begin{bmatrix}
12 T_1 & 18 T_2 & 24 T_3 & 30 T_4 \\
18 T_2 & 24 T_3 + 144 T_1^2 & 30 T_4 + 360 T_1 T_2 & 432 T_1 T_3 + 216 T_2^2 \\
24 T_3 & 30 T_4 + 360 T_1 T_2 & \begin{array}{c} 432 T_1 T_3 + 432 T_2^2 \\ + 1728 T_1^3 \end{array} & 504 T_2 T_3 + 6048 T_1^2 T_2 \\
30 T_4 & 432 T_1 T_3 + 216 T_2^2 & 504 T_2 T_3 + 6048 T_1^2 T_2 & \begin{array}{c} 288 T_3^2 + 6912 T_1 T_2^2 \\ + 20736 T_1^4 \end{array}
\end{bmatrix}.$$

The Last Six Groups

It remains to consider the groups numbered 24, 27, 29, 31, 33, 34. We follow [166, 237]. These groups have the following common properties:

(1) $d_1 < d_2$, so up to constant there is a unique invariant of minimal positive degree,
(2) $\det \mathsf{H}(f_1) \neq 0$ and may be taken as one of the basic invariants,
(3) if we take $f = f_1$, then $\det \mathsf{C} \doteq \det \mathsf{H}(f_1)$.

If we choose basic invariants so $f_k \doteq \det \mathsf{H}(f_1)$, then there is an ordering of the coexponents so there is a **unique** matrix C with the following properties:
(1) C is upper triangular,
(2) its diagonal elements are: $c_{q,q} = f_k$ and $c_{i,i} = 1$ for $i \neq q$,
(3) $c_{i,j} = 0$ for $i \neq q$,
(4) $c_{q,j} = 0$ for $j < q$ and $c_{q,j} \in \mathbb{C}[f_1, \ldots, f_{k-1}, f_{k+1}, \ldots, f_\ell]$ for $j > q$.

Thus basic derivations for G are determined by the following data:
(1) a set of basic invariants which includes $f_k \doteq \det \mathsf{H}(f_1)$,
(2) the elements $c_{q,j}$ for $j > q$.

In explicit calculations with a symbolic manipulation package like MACSYMA or MATHEMATICA, we found it most efficient to compute the matrix $\mathsf{P} = m_1 \mathsf{AJ}(\mathcal{F})\mathsf{C}$ where A is the adjoint of $\mathsf{H}(f_1)$, then divide each term of P by $\det \mathsf{H}(f_1)$. In the description of basic sets for G_{24} and G_{27} it is convenient to have the following definition. Let $u, v \in \mathbb{C}[z_1, z_2, z_3]$. Define

$$\mathrm{bord}(u,v) = \det \begin{bmatrix} \partial^2 u/\partial^2 z_1 & \partial^2 u/\partial z_1 \partial z_2 & \partial^2 u/\partial z_1 \partial z_3 & \partial v/\partial z_1 \\ \partial^2 u/\partial z_1 \partial z_2 & \partial^2 u/\partial^2 z_2 & \partial^2 u/\partial z_2 \partial z_3 & \partial v/\partial z_2 \\ \partial^2 u/\partial z_1 \partial z_3 & \partial^2 u/\partial z_2 \partial z_3 & \partial^2 u/\partial^2 z_3 & \partial v/\partial z_3 \\ \partial v/\partial z_1 & \partial v/\partial z_2 & \partial v/\partial z_3 & 0 \end{bmatrix}.$$

The Group G_{24}

We introduced this group in Example 6.31. Basic invariants are derived from Klein's famous quartic [125], whose automorphism group is the simple group of order 168.

$$\begin{aligned} f_1 &= z_1^3 z_2 + z_2^3 z_3 + z_3^3 z_1, \\ f_2 &= (1/54) \det \mathsf{H}(f_1), \\ f_3 &= (1/9)\mathrm{bord}(f_1, f_2), \\ c_{2,3} &= -48 f_1^2. \end{aligned}$$

These basic invariants and the resulting basic derivations appear in Example 6.69.

The Group G_{27}

Here we use the invariants of Wiman [251].

$$\begin{aligned} f_1 &= 10 z_1^3 z_2^3 + 9 z_3(z_1^5 + z_2^5) - 45 z_1^2 z_2^2 z_3^2 - 135 z_1 z_2 z_3^4 + 27 z_3^6, \\ f_2 &= (1/20250) \det \mathsf{H}(f_1), \\ f_3 &= (1/24300)\mathrm{bord}(f_1, f_2), \\ c_{2,3} &= (13/9) f_1^3. \end{aligned}$$

The Group G_{29}

Here we use the invariants of Maschke [150, pp.504–5].

$$\begin{aligned}
f_1 &= \Phi_1, \\
f_2 &= (-1/20736)\det \mathsf{H}(f_1), \\
f_3 &= F_{12}, \\
f_4 &= F_{20}, \\
c_{2,3} &= (-63/80)f_1, \\
c_{2,4} &= (-57/320)f_1^3 - (3/5)f_3.
\end{aligned}$$

The Group G_{31}

Here we use the invariants of Maschke [150, pp.504–5].

$$\begin{aligned}
f_1 &= F_8, \\
f_2 &= F_{12}, \\
f_3 &= F_{20}, \\
f_4 &= (1/265531392)\det \mathsf{H}(f_1), \\
c_{2,3} &= (-1/5)f_1, \\
c_{2,4} &= (-1/1620)f_2.
\end{aligned}$$

The Group G_{33}

Here we use the invariants of Burkhardt [46, p.208].

$$\begin{aligned}
f_1 &= J_4, \\
f_2 &= J_6, \\
f_3 &= (1/63700992)\det \mathsf{H}(f_1), \\
f_4 &= J_{12}, \\
f_5 &= J_{18}, \\
c_{2,3} &= (-1/576)f_1, \\
c_{2,4} &= (-1/192)f_2, \\
c_{2,5} &= (-1/192)f_4.
\end{aligned}$$

The Group G_{34}

Here we use the invariants of Conway and Sloane [51, p.438].

$$\begin{aligned}
f_1 &= (1/1944)\mu_6, \\
f_2 &= (1/3888)\mu_{12},
\end{aligned}$$

$$f_3 = (1/1944)\mu_{18},$$
$$f_4 = (1/23328000000)\det \mathsf{H}(f_1),$$
$$f_5 = (1/1944)\mu_{30},$$
$$f_6 = (1/1944)\mu_{42},$$
$$c_{2,3} = -\frac{25623}{175}f_1,$$
$$c_{2,4} = \frac{29}{437500}f_1^2 - \frac{2}{984375}f_2,$$
$$c_{2,5} = -\frac{18958101273}{28700}f_1^3 + \frac{4383422469}{315700}f_1 f_2 - \frac{1197907}{6314}f_3,$$
$$\begin{aligned}c_{2,6} = & -\frac{78307327099903642416708516}{12141111960078125}f_1^5 \\ & + \frac{566404141998190388920389}{1184498727812500}f_1^3 f_2 \\ & - \frac{119778687088911246 7288053}{267104463121718750}f_1 f_2^2 \\ & - \frac{4578546012088741561134}{2428222392015625}f_1^2 f_3 \\ & + \frac{17293466952248756 67163}{106841785248687500}f_2 f_3 - \frac{2079266672313}{8239991150}f_5.\end{aligned}$$

B.4 The Coexponents

The values of the coexponents n_i were first computed in [172] before the construction of explicit basic derivations. The calculations were case by case. For convenience, we list the results in Table B.1. The table also includes the exponents m_i because there are interesting relations between the m_i and the n_i. The first column lists the Shephard–Todd classification number. If the group is a Coxeter group, then $m_i = n_i$ and we omit the n_i. In the right column we mark Coxeter groups by C and Shephard groups by S. Coxeter groups and Shephard groups satisfy these equations for $1 \leq i \leq \ell$:

$$m_i - n_i = d_1 - 2, \qquad m_i + n_{\ell-i+1} = d_\ell.$$

Additional groups which satisfy the second equation are called duality groups and marked by D in the tables. The monomial groups $G(r,p,\ell)$ divide into several subclasses with respect to these properties. They are listed in Table B.2.

Table B.1. Exponents and coexponents

Group	Exponents	Coexponents	
1	$1, 2, \ldots, \ell - 1$		C, S
2' $p < r$	$r - 1, 2r - 1, \ldots,$ $(\ell - 1)r - 1, \ell q - 1$	$1, r + 1, \ldots,$ $(\ell - 1)r + 1$	
2'' $p = r$	$r - 1, 2r - 1, \ldots,$ $(\ell - 1)r - 1, \ell - 1$	$1, r + 1, \ldots,$ $(\ell - 2)r + 1, (\ell - 1)(r - 1)$	
3	$r - 1$	1	S
4	3,5	1,3	S
5	5,11	1,7	S
6	3,11	1,9	S
7	11,11	1,13	
8	7,11	1,5	S
9	7,23	1,17	S
10	11,23	1,13	S
11	23,23	1,25	
12	5,7	1,11	
13	7,11	1,17	
14	5,23	1,19	S
15	11,23	1,25	
16	19,29	1,11	S
17	19,59	1,41	S
18	29,59	1,31	S
19	59,59	1,61	
20	11,29	1,19	S
21	11,59	1,49	S
22	11,19	1,29	
23	1,5,9		C, S
24	3,5,13	1,9,11	D
25	5,8,11	1,4,7	S
26	5,11,17	1,7,13	S
27	5,11,29	1,19,25	D
28	1,5,7,11		C, S
29	3,7,11,19	1,9,13,17	D
30	1,11,19,29		C, S
31	7,11,19,23	1,13,17,29	
32	11,17,23,29	1,7,13,19	S
33	3,5,9,11,17	1,7,9,13,15	D
34	5,11,17,23,29,41	1,13,19,25,31,37	D
35	1,4,5,7,8,11		C
36	1,5,7,9,11,13,17		C
37	1,7,11,13,17,19,23,29		C

Table B.2. The monomial groups

	$G(2,1;\ell)$	C,S
2'	$G(r,1,\ell), r \geq 3$	S
	$G(r,p,\ell), 1 < p < r$	
	$G(2,2,\ell), \ell \geq 4$	C
2''	$G(r,r,2), r \geq 5$	C,S
	$G(r,r,\ell), r,\ell \geq 3$	

C. Orbit Types

In this appendix we present the tables of the orbits of G on $L(\mathcal{A})$ and the values of b_i^X. These results are reproduced from [175, 174]. Construction of the matrix $U(G)$ is outlined in Section 6.4. In a complete matrix $U(G)$, the rows index the types T of the orbits. We use the symbol A_0 for the trivial group, the symbols in [38, p.193] for irreducible Coxeter groups, $G(r,p,\ell)$ for the monomial groups, and G_m where m is in [210, Table VII] for the remaining irreducible unitary reflection groups. If two orbits have type T, we label them T', T''. The columns have the same indices in the same order, but these indices are omitted. The columns to the right of the matrix $U(G)$ give the values of b_i^X in Theorem 6.89. The b_i^X are computed recursively using the matrix $U(G)$ and the formula in Lemma 6.87. We list the orbits and their sizes for the exceptional groups of rank 2 in Tables C.1 and C.2. The rows index the groups. The columns index the types T of the orbits. This information is sufficient to construct the matrix $U(G)$ in each case. For example, Table C.2 shows that in G_{15} there are two orbits A_1', A_1'' of type A_1 with cardinalities 12, 6 and one orbit of type $C(3)$ with cardinality 8. The matrix $U(G_{15})$ and the values of b_i^X are given in Table C.3. Groups of rank ≥ 3 comprise the remaining tables.

Table C.1. The rank 2 groups (I)

G	A_1'	A_1''	$C(3)'$	$C(3)''$	$C(4)$	$C(5)$
4			4			
5			4	4		
6	6		4			
7	6		4	4		
8					6	
9	12				6	

Table C.2. The rank 2 groups (II)

G	A_1'	A_1''	$C(3)'$	$C(3)''$	$C(4)$	$C(5)$
10			8		6	
11	12		8		6	
12	12					
13	12	6				
14	12		8			
15	12	6	8			
16						12
17	30					12
18				20		12
19	30			20		12
20				20		
21	30			20		
22	30					

Table C.3. Orbits in G_{15}

A_0	1	12	6	8	1	1	25
A_1'		1	0	0	1	1	
A_1''			1	0	1	1	
$C(3)$				1	1	1	
G_{15}					1		

Table C.4. Orbits in H_3

A_0	1	15	15	10	6	1	1	5	9
A_1		1	2	2	2	1	1	5	
A_1^2			1	0	0	1	1		
A_2				1	0	1	1		
$I_2(5)$					1	1	1		
H_3						1			

Table C.5. Orbits in G_{24}

	1	21	28	21	1	1	9	11
A_0	1	21	28	21	1	1	9	11
A_1		1	4	4	1	1	7	
A_2			1	0	1	1		
B_2				1	1	1		
G_{24}					1			

Table C.6. Orbits in G_{25}

	1	12	12	9	1	1	4	7
A_0	1	12	12	9	1	1	4	7
$C(3)$		1	2	3	1	1	4	
$C(3)^2$			1	0	1	1		
G_4				1	1	1		
G_{25}					1			

Table C.7. Orbits in G_{26}

	1	9	12	36	9	12	1	1	7	13
A_0	1	9	12	36	9	12	1	1	7	13
A_1		1	0	4	0	4	1	1	7	
$C(3)$			1	3	3	2	1	1	7	
$A_1 \times C(3)$				1	0	0	1	1		
G_4					1	0	1	1		
$G(3,1,2)$							1	1		
G_{26}							1			

Table C.8. Orbits in G_{27}

	1	45	60	60	45	36	1	1	19	25
A_0	1	45	60	60	45	36	1	1	19	25
A_1		1	4	4	4	4	1	1	15	
A_2'			1	0	0	0	1	1		
A_2''				1	0	0	1	1		
B_2					1	0	1	1		
$I_2(5)$						1	1	1		
G_{27}							1			

Table C.9. Orbits in F_4

A_0	1	12	12	72	16	16	18	12	12	48	48	1	1	5	7	11
A_1		1	0	6	4	0	3	3	6	4	12	1	1	5	7	
$\tilde{A}_1$			1	6	0	4	3	6	3	12	4	1	1	5	7	
$A_1 \times \tilde{A}_1$				1	0	0	0	1	1	2	2	1	1	5		
A_2					1	0	0	0	3	0	3	1	1	5		
$\tilde{A}_2$						1	0	3	0	3	0	1	1	5		
B_2							1	2	2	0	0	1	1	3		
C_3								1	0	0	0	1	1			
B_3									1	0	0	1	1			
$A_1 \times \tilde{A}_2$										1	0	1	1			
$\tilde{A}_1 \times A_2$											1	1	1			
F_4												1				

Table C.10. Orbits in G_{29}

A_0	1	40	120	160	30	160	80	80	40	20	1	1	9	13	17
A_1		1	6	12	3	16	12	12	9	6	1	1	9	11	
A_1^2			1	0	0	4	2	2	2	0	1	1	9		
A_2				1	0	1	2	2	1	2	1	1	7		
B_2					1	0	0	0	4	2	1	1	5		
$A_1 \times A_2$						1	0	0	0	0	1	1			
A_3'							1	0	0	0	1	1			
A_3''								1	0	0	1	1			
B_3									1	0	1	1			
$G(4,4,3)$										1	1	1			
G_{29}											1				

Table C.11. Orbits in H_4

A_0	1	60	450	200	72	600	360	300	60	1	1	11	19	29
A_1		1	15	10	6	40	36	30	15	1	1	11	19	
A_1^2			1	0	0	4	4	2	2	1	1	11		
A_2				1	0	3	0	6	3	1	1	11		
$I_2(5)$					1	0	5	0	5	1	1	9		
$A_1 \times A_2$						1	0	0	0	1	1			
$A_1 \times I_2(5)$							1	0	0	1	1			
A_3								1	0	1	1			
H_3									1	1	1			
H_4										1				

C. Orbit Types

Table C.12. Orbits in G_{31}

A_0	1	60	360	320	30	960	480	60	1	1	13	17	29
A_1		1	12	16	3	64	48	15	1	1	13	17	
A_1^2			1	0	0	8	4	2	1	1	13		
A_2				1	0	3	6	3	1	1	11		
$G(4,2,2)$					1	0	0	6	1	1	5		
$A_1 \times A_2$						1	0	0	1	1			
A_3							1	0	1	1			
$G(4,2,3)$								1	1	1			
G_{31}									1				

Table C.13. Orbits in G_{32}

A_0	1	40	90	240	360	40	1	1	7	13	19
$C(3)$		1	9	12	45	12	1	1	7	13	
G_4			1	0	4	4	1	1	7		
$C(3)^2$				1	6	2	1	1	7		
$C(3) \times G_4$					1	0	1	1			
G_{25}						1	1	1			
G_{32}							1				

Table C.14. Orbits in G_{33}

A_0	1	45	270	240	270	720	540	40	540	216	45	40	1	1	7	9	13	15
A_1		1	12	16	18	64	72	8	84	48	12	16	1	1	7	9	11
A_1^2			1	0	3	8	6	0	18	12	3	4	1	1	7	9	
A_2				1	0	3	9	2	9	9	3	7	1	1	6	7	
A_1^3					1	0	0	0	6	0	2	0	1	1	7		
$A_1 \times A_2$						1	0	0	3	3	0	1	1	1	6		
A_3							1	0	1	2	1	2	1	1	5		
$G(3,3,3)$								1	0	0	0	4	1	1	3		
$A_1 \times A_3$									1	0	0	0	1	1			
A_4										1	0	0	1	1			
D_4											1	0	1	1			
$G(3,3,4)$												1	1	1			
G_{33}													1				

C. Orbit Types

Table C.15. Orbits in G_{34} (I)

A_0	1	126	2835	1680	11340	30240	11340	560	45360	30240	68040	27216
A_1		1	45	40	270	960	540	40	1800	1440	3780	2160
A_1^2			1	0	12	32	12	0	112	96	216	144
A_2				1	0	18	27	4	27	36	162	162
A_1^3					1	0	0	0	12	0	18	0
$A_1\times A_2$						1	0	0	3	6	9	9
A_3							1	0	0	0	6	12
$G(3,3,3)$								1	0	0	0	0
$A_1^2\times A_2$									1	0	0	0
A_2^2										1	0	0
$A_1\times A_3$											1	0
A_4												1
D_4												
$A_1\times G(3,3,3)$												
$G(3,3,4)$												
$A_2\times A_3$												
$A_1\times A_4$												
A_5'												
A_5''												
D_5												
$A_1\times G(3,3,4)$												
$G(3,3,5)$												
G_{33}												
G_{34}												

Table C.16. Orbits in E_6 (I)

A_0	1	36	270	120	540	720	270	1080	120	540	216	45	360	216
A_1		1	15	10	45	80	45	150	20	105	60	15	70	66
A_1^2			1	0	6	8	3	28	4	18	12	3	20	20
A_2				1	0	6	9	9	2	18	18	6	6	18
A_1^3					1	0	0	6	0	3	0	1	6	6
$A_1\times A_2$						1	0	3	1	3	3	0	4	6
A_3							1	0	0	2	4	2	0	4
$A_1^2\times A_2$								1	0	0	0	0	2	2
A_2^2									1	0	0	0	3	0
$A_1\times A_3$										1	0	0	0	2
A_4											1	0	0	1
D_4												1	0	0
$A_1\times A_2^2$													1	0
$A_1\times A_4$														1
A_5														
D_5														
E_6														

Table C.17. Orbits in G_{34} (II)

2835	5040	1680	45360	27216	9072	9072	3402	5040	672	126	1	1	13	19	25	31	37
270	400	240	3240	2376	1080	1080	540	760	160	45	1	1	13	19	25	27	
18	16	16	336	240	144	144	84	80	32	12	1	1	13	19	23		
27	36	42	135	162	108	108	81	126	40	18	1	1	13	16	19		
3	0	0	36	36	12	12	18	12	0	3	1	1	13	19			
0	2	1	27	18	18	18	9	10	8	3	1	1	13	16			
3	0	4	4	12	12	12	15	12	8	6	1	1	11	13			
0	9	12	0	0	0	0	0	36	12	9	1	1	7	13			
0	0	0	3	6	0	0	3	2	0	0	1	1	13				
0	0	0	6	0	3	3	0	0	2	0	1	1	13				
0	0	0	2	2	2	2	1	2	0	1	1	1	11				
0	0	0	0	1	2	2	2	0	2	1	1	1	9				
1	0	0	0	0	0	0	6	0	0	2	1	1	7				
	1	0	0	0	0	0	0	4	4	0	1	1	7				
		1	0	0	0	0	0	3	2	3	1	1	7				
			1	0	0	0	0	0	0	0	1	1					
				1	0	0	0	0	0	0	1	1					
					1	0	0	0	0	0	1	1					
						1	0	0	0	0	1	1					
							1	0	0	0	1	1					
								1	0	0	1	1					
									1	0	1	1					
										1	1	1					
											1						

Table C.18. Orbits in E_6 (II)

36	27	1	1	4	5	7	8	11
15	15	1	1	4	5	7	8	
6	7	1	1	4	5	7		
6	9	1	1	4	5	5		
1	3	1	1	4	5			
3	3	1	1	4	5			
2	5	1	1	3	4			
0	1	1	1	4				
3	0	1	1	5				
1	1	1	1	3				
1	2	1	1	3				
0	3	1	1	2				
0	0	1	1					
0	0	1	1					
1	0	1	1					
	1	1	1					
		1						

Table C.19. Orbits in E_7 (I)

A_0	1	63	945	336	315	3780	5040	1260	3780	15120	3360	1260	7560	2016	315	5040
A_1		1	30	16	15	180	320	120	240	1200	320	140	840	320	60	480
A_1^2			1	0	1	12	16	4	24	112	32	12	72	32	6	64
A_2				1	0	0	15	15	0	45	20	15	90	60	15	15
$(A_1^3)'$					1	0	0	0	12	0	0	12	0	0	0	16
$(A_1^3)''$						1	0	0	3	12	0	0	6	0	1	12
$A_1 \times A_2$							1	0	0	6	4	1	6	4	0	3
A_3								1	0	0	0	1	6	8	3	0
A_1^4									1	0	0	0	0	0	0	4
$A_1^2 \times A_2$										1	0	0	0	0	0	1
A_2^2											1	0	0	0	0	0
$(A_1 \times A_3)'$												1	0	0	0	0
$(A_1 \times A_3)''$													1	0	0	0
A_4														1	0	0
D_4															1	0
$A_1^3 \times A_2$																1
$A_1 \times A_2^2$																
$A_1^2 \times A_3$																
$A_2 \times A_3$																
$A_1 \times A_4$																
A_5'																
A_5''																
$A_1 \times D_4$																
D_5																
$A_1 \times A_2 \times A_3$																
$A_2 \times A_4$																
$A_1 \times A_5$																
A_6																
$A_1 \times D_5$																
D_6																
E_6																
E_7																

C. Orbit Types   297

Table C.20. Orbits in E_7 (II)

10080	7560	5040	6048	336	1008	945	378	5040	2016	1008	288	378	63	28	1	1	5	7	9	11	13 17
1120	960	720	1056	80	240	195	120	800	416	256	96	126	30	16	1	1	5	7	9	11 13	
160	128	112	160	16	48	30	28	160	96	64	32	36	13	8	1	1	5	7	9	11	
60	90	75	180	20	60	45	45	75	66	60	30	45	15	10	1	1	5	7	8	9	
0	72	0	0	16	0	18	0	48	0	48	0	12	12	0	1	1	5	7	11		
24	18	12	24	0	4	6	6	36	24	12	8	12	4	4	1	1	5	7	9		
16	12	18	24	4	12	3	6	23	20	16	12	9	6	4	1	1	5	7	8		
0	6	4	24	4	12	9	15	4	8	12	8	15	7	6	1	1	5	5	7		
0	6	0	0	0	0	3	0	12	0	4	0	6	3	0	1	1	5	7			
4	2	1	4	0	0	0	1	7	6	4	2	3	1	2	1	1	5	7			
3	0	6	0	1	3	0	0	6	6	3	6	0	3	1	1	1	5	7			
0	6	0	0	4	0	3	0	4	0	12	0	3	6	0	1	1	5	7			
0	1	2	4	0	2	1	1	2	4	2	4	3	2	2	1	1	5	5			
0	0	0	3	1	3	0	3	0	1	3	3	3	3	3	1	1	4	5			
0	0	0	0	0	0	3	6	0	0	0	0	6	3	4	1	1	3	5			
0	0	0	0	0	0	0	0	3	0	0	0	3	0	0	1	1	5				
1	0	0	0	0	0	0	0	2	2	1	0	0	0	1	1	1	5				
	1	0	0	0	0	0	0	2	0	2	0	1	1	0	1	1	5				
		1	0	0	0	0	0	1	2	0	2	0	1	0	1	1	5				
			1	0	0	0	0	0	1	1	1	1	0	1	1	1	4				
				1	0	0	0	0	0	3	0	0	3	0	1	1	5				
					1	0	0	0	0	0	2	0	1	1	1	1	3				
						1	0	0	0	0	0	2	2	0	1	1	3				
							1	0	0	0	0	1	1	2	1	1	3				
								1	0	0	0	0	0	0	1	1					
									1	0	0	0	0	0	1	1					
										1	0	0	0	0	1	1					
											1	0	0	0	1	1					
												1	0	0	1	1					
													1	0	1	1					
														1	1	1					
															1						

Table C.21. Orbits in E_8 (I)

	1	120	3780	1120	37800	40320	7560	113400	302400	67200	151200	24192	3150	604800
A_0	1	120	3780	1120	37800	40320	7560	113400	302400	67200	151200	24192	3150	604800
A_1		1	63	28	945	1344	378	3780	12600	3360	8820	2016	315	30240
A_1^2			1	0	30	32	6	180	560	160	360	96	15	1920
A_2				1	0	36	27	0	270	120	540	216	45	540
A_1^3					1	0	0	12	24	0	12	0	1	160
$A_1 \times A_2$						1	0	0	15	10	15	6	0	45
A_3							1	0	0	0	20	16	5	0
A_4								1	0	0	0	0	0	16
$A_1^2 \times A_2$									1	0	0	0	0	6
A_2^2										1	0	0	0	0
$A_1 \times A_3$											1	0	0	0
A_4												1	0	0
D_4													1	0
$A_1^3 \times A_2$														1
$A_1 \times A_2^2$														
$A_1^2 \times A_3$														
$A_2 \times A_3$														
$A_1 \times A_4$														
A_5														
$A_1 \times D_4$														
D_5														
$A_1^2 \times A_2^2$														
$A_1 \times A_2 \times A_3$														
$A_1^2 \times A_4$														
A_3^2														
$A_2 \times A_4$														
$A_1 \times A_5$														
A_6														
$A_2 \times D_4$														
$A_1 \times D_5$														
D_6														
E_6														
$A_1 \times A_2 \times A_4$														
$A_3 \times A_4$														
$A_1 \times A_6$														
A_7														
$A_2 \times D_5$														
D_7														
$A_1 \times E_6$														
E_7														
E_8														

Table C.22. Orbits in E_8 (II)

403200	453600	302400	241920	40320	37800	7560	604800	604800	362880	151200	241920	120960	34560
23520	30240	22680	22176	5040	4095	1260	40320	50400	36288	15120	26208	16128	6048
1600	1920	1680	1600	480	300	140	3520	4800	3456	1680	2880	1920	960
720	1620	1350	2160	720	540	270	1080	2700	3240	1080	2376	2160	1080
96	144	72	96	16	30	12	384	480	384	144	288	192	96
80	90	135	120	60	15	15	150	345	270	180	300	240	180
0	60	40	160	80	60	50	0	80	240	40	160	240	160
0	12	0	0	0	4	0	48	48	48	12	0	16	0
8	6	3	8	0	0	1	28	42	36	12	36	24	12
6	0	18	0	6	0	0	9	36	0	36	36	18	36
0	6	6	8	4	3	1	0	16	24	12	24	24	24
0	0	0	10	10	0	5	0	0	15	0	10	30	30
0	0	0	0	0	12	12	0	0	0	0	0	0	0
0	0	0	0	0	0	0	6	3	6	0	0	0	0
1	0	0	0	0	0	0	3	6	0	0	6	3	0
	1	0	0	0	0	0	0	4	4	2	0	4	0
		1	0	0	0	0	0	2	0	4	4	0	4
			1	0	0	0	0	0	3	0	3	3	3
				1	0	0	0	0	0	0	0	3	6
					1	0	0	0	0	0	0	0	0
						1	0	0	0	0	0	0	0
							1	0	0	0	0	0	0
								1	0	0	0	0	0
									1	0	0	0	0
										1	0	0	0
											1	0	0
												1	0
													1

Table C.23. Orbits in E_8 (III)

50400	45360	3780	1120	241920	120960	34560	8640	30240	1080	3360	120	1	1	7	11	13	17	19	23	29
6300	7938	945	336	28224	16128	6336	2016	5796	378	1036	63	1	1	7	11	13	17	19	23	
720	1080	195	80	3712	2496	1152	480	1040	126	272	30	1	1	7	11	13	17	19		
765	1620	270	120	2376	1512	1080	432	1107	135	360	36	1	1	7	11	13	14	17		
88	156	30	16	576	384	192	96	216	36	72	13	1	1	7	11	13	17			
75	135	45	20	366	330	210	120	165	45	70	15	1	1	7	11	13	14			
80	300	70	40	160	96	160	80	200	45	120	20	1	1	7	9	11	13			
16	24	6	0	96	48	32	8	48	12	16	4	1	1	7	11	13				
3	18	3	4	76	60	36	24	35	9	20	6	1	1	7	11	13				
12	0	9	2	36	72	36	36	18	18	6	6	1	1	7	11	11				
12	21	9	4	32	32	32	24	34	15	18	7	1	1	7	9	11				
0	30	15	10	10	5	30	20	20	15	30	10	1	1	7	8	9				
16	72	18	16	0	0	0	0	48	12	48	12	1	1	5	7	11				
1	3	0	0	18	6	6	0	9	3	6	1	1	1	7	11					
0	0	0	1	12	12	6	6	6	0	4	3	1	1	7	11					
0	2	1	0	8	8	8	4	4	3	4	2	1	1	7	9					
2	0	1	0	4	12	4	8	5	6	0	2	1	1	7	9					
0	3	0	1	4	3	6	6	6	3	6	3	1	1	7	8					
0	0	3	1	0	0	6	6	0	6	3	4	1	1	5	7					
4	6	3	0	0	0	0	0	12	6	4	3	1	1	5	7					
0	6	3	4	0	0	0	0	4	3	12	6	1	1	5	7					
0	0	0	0	4	0	0	0	2	0	2	0	1	1	7						
0	0	0	0	2	2	2	0	1	0	0	1	1	1	7						
0	0	0	0	2	1	2	0	0	1	2	0	1	1	7						
0	0	0	0	0	4	0	2	0	2	0	0	1	1	7						
0	0	0	0	1	2	0	2	2	0	0	1	1	1	7						
0	0	0	0	0	0	2	2	0	0	1	1	1	1	5						
0	0	0	0	0	1	2	0	2	0	1	1	1	1	5						
1	0	0	0	0	0	0	3	3	0	0	1	1	1	5						
	1	0	0	0	0	0	2	1	2	1	1	1	1	5						
		1	0	0	0	0	0	2	0	2	1	1	1	3						
			1	0	0	0	0	0	3	3	1	1	1	5						
				1	0	0	0	0	0	0	0	1	1							
					1	0	0	0	0	0	0	1	1							
						1	0	0	0	0	0	1	1							
							1	0	0	0	0	1	1							
								1	0	0	0	1	1							
									1	0	0	1	1							
										1	1	1	1							
											1	1	1							

D. Three–Dimensional Restrictions

In this appendix we provide additional data for restrictions of irreducible reflection arrangements to subspaces of dimension 3. Two sets of numbers may be associated with a central 3–arrangement. Let n_k be the number of planes which contain k lines of the arrangement. Then $\sum_k n_k = n$ is the cardinality of the arrangement. Let t_p be the number of lines contained in p planes. Then $\sum_p t_p = t$ is the number of lines in the arrangement. Note also that $\sum_k k n_k = \sum_p p t_p$. The tables in Appendix C may be used to determine the n_k, but calculation of the t_p requires additional work. In the tables below we list both sets of numbers. Type A_3 of the symmetric group is isomorphic to type $D_3 = G(2,2,3)$. It follows from Propositions 6.77 and 6.84 that for the infinite families it suffices to consider $\mathcal{A}_3^i(r)$ for $0 \le i \le 3$; see Table D.1. We showed in Proposition 6.85 that these arrangements are free.

In the exceptional groups we identify the restriction by a pair G, T. Here G denotes the reflection arrangement $\mathcal{A} = \mathcal{A}(G)$ in which we are restricting and T is the orbit type of the subspace $X \in L(\mathcal{A})$ whose restriction $\mathcal{A}^X$ is in consideration. We have checked that all these arrangements are free. Their exponents are the corresponding b_i^X in our tables.

In Table D.2 we list restrictions in Coxeter groups. These are real simplicial arrangements, thus their complexifications are $K(\pi, 1)$. In the second column we give Grünbaum's symbol [100] for the arrangement. This symbol contains n in parantheses.

In Table D.3 we list restrictions in the remaining groups. Here (G_{25}, A_0) and (G_{26}, A_0) are $K(\pi, 1)$ by Theorem 6.128. The $K(\pi, 1)$ problem is undecided for the other arrangements.

There are several lattice isomorphisms. In each case, the arrangement on the left appears in one of our tables.

$$\begin{aligned}
\mathcal{A}_3^2(2) &\simeq (E_6, A_3), \\
\mathcal{A}_3^3(2) &\simeq (E_7, D_4), \\
(E_6, A_1 \times A_2) &\simeq (E_7, A_4), \\
(F_4, A_1) &\simeq (E_7, A_1^4) \simeq (E_7, (A_1 \times A_3)') \simeq (E_8, A_1 \times D_4) \simeq (E_8, D_5), \\
(E_7, A_2^2) &\simeq (E_8, A_5), \\
(E_8, A_1^2 \times A_3) &\simeq (E_8, A_2 \times A_3), \\
(G_{26}, A_0) &\simeq (G_{32}, C(3)) \simeq (G_{34}, G(3,3,3)).
\end{aligned}$$

D. Three–Dimensional Restrictions

Table D.1. The infinite families

	n	n_{r+1}	n_{r+2}	t	t_2	t_3	t_r	t_{r+1}	t_{r+2}
$\mathcal{A}_3^0(r)$	$3r$	$3r$		r^2+3		r^2	3		
$\mathcal{A}_3^1(r)$	$3r+1$	$2r$	$r+1$	r^2+r+3	r	r^2	1	2	
$\mathcal{A}_3^2(r)$	$3r+2$	r	$2r+2$	r^2+2r+3	$2r$	r^2		2	1
$\mathcal{A}_3^3(r)$	$3r+3$		$3r+3$	r^2+3r+3	$3r$	r^2			3

Table D.2. Coxeter groups

			t	t_2	t_3	t_4	t_5	t_6
E_6, A_1^3	$A_3(10)$	$n_3=1, n_4=3, n_5=6$	16	6	7	3		
$E_6, A_1 \times A_2$	$A_2(10)$	$n_4=6, n_5=3, n_6=1$	16	6	7	3		
$E_7, (A_1 \times A_3)''$	$A_1(11)$	$n_4=4, n_5=4, n_6=3$	19	7	8	4		
F_4, A_1	$A_2(13)$	$n_4=3, n_6=10$	25	12	4	9		
E_7, A_2^2	$A_1(13)$	$n_4=3, n_6=10$	25	9	12	3		1
$E_7, A_1^2 \times A_2$	$A_3(13)$	$n_4=1, n_5=4, n_6=8$	25	10	10	3	2	
H_3, A_0	$A_1(15)$	$n_6=15$	31	15	10		6	
$E_8, A_1 \times A_4$	$A_5(16)$	$n_6=10, n_8=6$	36	14	16	3	4	
$E_8, A_1^2 \times A_3$	$A_4(17)$	$n_4=1, n_6=6, n_8=10$	41	16	16	7		2
$E_8, A_1^3 \times A_2$	$A_3(19)$	$n_6=4, n_8=15$	49	24	12	6	6	1
$E_8, A_1 \times A_2^2$	$A_1(19)$	$n_6=4, n_8=15$	49	21	18	6		4
H_4, A_1	$A_1(31)$	$n_{10}=6, n_{12}=25$	121	60	40		6	15

Table D.3. The remaining groups

	n		t	t_2	t_3	t_4	t_5	t_6	t_7	t_8
G_{25}, A_0	12	$n_5=12$	21	12	9					
G_{33}, A_2	14	$n_4=2, n_6=9, n_7=3$	28	9	12	6	1			
G_{33}, A_1^2	17	$n_6=6, n_7=8, n_8=3$	37	6	24	3	4			
G_{24}, A_0	21	$n_8=21$	49		28	21				
G_{29}, A_1	21	$n_6=3, n_8=12, n_{10}=6$	55	12	28	9	6			
G_{26}, A_0	21	$n_8=21$	57	36		9	12			
G_{34}, A_3	25	$n_8=7, n_{10}=12, n_{12}=6$	81	24	40	11		6		
$G_{34}, A_1 \times A_2$	30	$n_8=3, n_{10}=9, n_{12}=9, n_{14}=9$	111	28	60	9	11		3	
G_{31}, A_1	31	$n_6=3, n_{12}=16, n_{14}=12$	127	48	64			15		
G_{34}, A_1^2	33	$n_8=3, n_{12}=18, n_{14}=12$	129	36	60	18	12			3
G_{27}, A_0	45	$n_{16}=45$	201		120	45	36			

References

[1] Aigner, M.: Combinatorial Theory. Grundlehren der Math. Wiss. **234**, Springer Verlag, 1979

[2] Aomoto, K.: Un théorème du type de Matsushima–Murakami concernant l'intégrale des fonctions multiformes. J. Math. Pures Appl. **52** (1973) 1–11

[3] Aomoto, K.: Configurations and invariant theory of Gauss–Manin systems. In: Group representations and systems of differential equations. Adv. Studies in Pure Math. **4**, North Holland, 1984, pp. 165–179

[4] Aomoto, K.: Gauss–Manin connection of integral of difference products. J. Math. Soc. Japan **39** (1987) 191–208

[5] Arnold, V. I.: Braids of algebraic functions and cohomologies of swallowtails. Uspehi Mat. Nauk **23**(4) (1968) 247–248

[6] Arnold, V. I.: The cohomology ring of the colored braid group, Mat. Zametki **5** (1969) 227–231 : Math. Notes **5** (1969) 138–140

[7] Arnold, V. I.: Critical points of smooth functions. In: Proc. Intern. Congress Math. **1** (1974) pp. 19–39

[8] Arnold, V. I.: Wave front evolution and equivariant Morse lemma. Comm. Pure Appl. Math. **29** (1976) 557–582

[9] Arnold, V. I.: Index of a singular point of a vector field, Petrovski-Oleinik inequality and mixed Hodge structures. Funct. Anal. Appl. **12** (1978) 1–14

[10] Arnold, V. I.: Indices of singular points of 1-forms on a manifold with boundary, convolution of invariants of reflection groups, and singular projections of smooth surfaces. Russian Math. Surveys **34** (1979) 1-42

[11] Artal–Bartolo, E.: Sur le premier nombre de Betti de la fibre de Milnor du cône sur une courbe projective plane et son rapport avec la position des points singuliers. Preprint

[12] Artin, E.: Theorie der Zöpfe. Hamb. Abh. **4** (1925) 47–72

[13] Arvola, W.: The fundamental group of the complement of an arrangement of complex hyperplanes. Ph.D. Thesis, University of Wisconsin–Madison, 1990

[14] Arvola, W.: The fundamental group of the complement of an arrangement of complex hyperplanes. Topology. To appear

[15] Arvola, W.: Complexified real arrangements of hyperplanes. Manuscripta math. **71** (1991) 295–306

[16] Baclawski, K.: Whitney numbers of geometric lattices. Advances in Math. **16** (1975) 125–138

[17] Bannai, E.: Fundamental groups of the spaces of regular orbits of the finite unitary reflection groups of dimension 2. J. Math. Soc. Japan **28** (1976) 447–454

[18] Barcelo, H.: On the action of the symmetric group on the free Lie algebra and the partition lattice. J. Comb. Theory (A) **55** (1990) 93–129
[19] Barcelo, H., Bergeron, N.: The Orlik–Solomon algebra on the partition lattice and the free Lie algebra. J. Comb. Theory (A) **55** (1990) 80–92
[20] Barnabei, M., Brini, A., Rota, G.-C.: The theory of Möbius functions. Russian Math. Surveys **41**(3) (1986) 135–188
[21] Barthel, G., Hirzebruch, F., Höfer, T.: Geradenkonfigurationen und Algebraische Flächen. Vieweg Publishing, Wiesbaden, 1987
[22] Bayer, M., Sturmfels, B.: Lawrence polytopes. Canad. J. Math. **42** (1990) 62–79
[23] Beilinson, A. A., Goncharov, A. B., Schechtman, V. V., Varchenko, A. N.: Aomoto dilogarithms, mixed Hodge structures and motivic cohomology of pairs of triangles on the plane. In: The Grothendieck Festschrift. Vol. I. Birkhäuser, 1990, pp. 135–172
[24] Beynon, W. M., Spaltenstein, N.: Green functions of finite Chevalley groups of type E_n ($n = 6, 7, 8$). J. Algebra **88** (1984) 584–614
[25] Birkhoff, G.: Lattice Theory. 3rd ed. Colloq. Publ. **25**, Amer. Math. Soc., 1967
[26] Birkhoff, G.: A determinant formula for the number of ways of coloring a map. Annals of Math. **14** (1913) 42–46
[27] Birman, J.: Braids. links and mapping classe groups. Annals of Math. Studies **82**, Princeton Univ. Press, 1975
[28] Björner, A.: Homotopy type of posets and lattice complementation. J. Comb. Theory (A) **30** (1981) 90–100
[29] Björner, A.: On the homology of geometric lattices. Algebra Universalis **14** (1982) 107–128
[30] Björner, A., Las Vergnas, M., Sturmfels, B., White, N., Ziegler, G.M.: Oriented Matroids. To appear
[31] Björner, A., Walker, J. W.: A homotopy complementation formula for partially ordered sets. Europ. J. Combinatorics **4** (1983) 11–19
[32] Björner, A., Edelman, P., Ziegler, G.: Hyperplane arrangements with a lattice of regions. Disc. and Comp. Geometry **5** (1990) 263–288
[33] Björner, A., Ziegler, G. M.: Broken circuit complexes: factorizations and generalizations. J. Comb. Theory (B) **51** (1991) 96–126
[34] Björner, A., Ziegler, G. M.: Combinatorial stratification of complex arrangements. J. Amer. Math. Soc. To appear
[35] de Boor, C., Höllig, K.: B–splines from parallelepipeds. J. D'Anal. Math. **42** (1982) 99–115
[36] de Boor, C., Höllig, K., Riemenschneider, S.: Box splines. Preprint
[37] Bott, R., Tu, L. W.: Differential forms in algebraic topology. GTM **82**, Springer Verlag, 1982
[38] Bourbaki, N.: Groupes et Algèbres de Lie. Chapitres 4,5 et 6, Hermann, Paris 1968
[39] Brieskorn, E.: Die Fundamentalgruppe des Raumes der regulären Orbits einer endlichen komplexen Spiegelungsgruppe. Invent. math. **12** (1971) 57–61
[40] Brieskorn, E.: Singular elements of semi-simple algebraic groups. In: Actes Congrés Intern. Math. **2** (1970) pp. 279–284
[41] Brieskorn, E.: Sur les groupes de tresses. In: Séminaire Bourbaki 1971/72. Lecture Notes in Math. **317**, Springer Verlag, 1973, pp. 21–44

[42] Brieskorn, E., Knörrer, H.: Plane Algebraic Curves. Birkhäuser, Boston 1986
[43] Brylawski, T.: A decomposition for combinatorial geometries. Trans. Amer. Math. Soc. **171** (1972) 235–282
[44] Brylawski, T.: The broken circuit complex. Trans. Amer. Math. Soc. **234** (1977) 417–433
[45] Buck, R. C.: Partition of space. Amer. Math. Monthly **50** (1943) 541–544
[46] Burkhardt, H.: Untersuchungen auf dem Gebiete der Hyperelliptischen Modulfunctionen (Zweiter Teil). Math. Ann. **38** (1891) 161–224
[47] Burnside, W.: Theory of groups of finite order. Cambridge, 1911
[48] Cartier, P.: Les arrangements d'hyperplans: un chapitre de géometrie combinatoire. In: Séminaire Bourbaki 1980/81. Lecture Notes in Math. **901**, Springer Verlag, 1981, pp. 1–22
[49] Chevalley, C.: Invariants of finite groups generated by reflections. Amer. J. Math. **77** (1955) 778–782
[50] Cohen, M.: Simplicial structures and transverse cellularity. Annals of Math. **85** (1967) 218–245
[51] Conway, J. H. A., Sloane, N. J.: The Coxeter–Todd lattice, the Mitchell group, and related sphere packings. Proc. Camb. Phil. Soc. **93** (1983) 421–440
[52] Conway, J. H. A., Sloane, N. J.: Sphere Packings, Lattices and Groups. Grundlehren der Math. Wiss. **290**, Springer Verlag, 1988
[53] Cordovil, R., Guedes de Oliveira, A.: A note on the fundamental group of the Salvetti complex determined by an oriented matroid. Preprint
[54] Coxeter, H. S. M.: Discrete groups generated by reflections. Annals of Math. **35** (1934) 588–621
[55] Coxeter, H. S. M.: The product of the generators of a finite group generated by reflections. Duke Math. J. **18** (1951) 765–782
[56] Coxeter, H. S. M.: Regular Polytopes. 3rd ed., Dover, New York 1973
[57] Coxeter, H. S. M.: Regular Complex Polytopes. Second ed. Cambridge Univ. Press, 1991
[58] Crapo, H.: The Möbius function of a lattice. J. Comb. Theory **1** (1966) 120–131
[59] Crapo, H., Rota, G.-C.: Combinatorial Geometries. MIT Press, Cambridge, MA. 1971
[60] Deheuvels, R.: Homologie des ensembles ordonnés et des espaces topologiques. Bull. Soc. Math. France **90** (1962) 261–321
[61] Deligne, P.: Les immeubles des groupes de tresses généralisés. Invent. math. **17** (1972) 273–302
[62] Deligne, P.: Théorie de Hodge. Publ. Math. IHES **40** (1972) 5–57
[63] Deligne, P., Mostow, G. D.: Monodromy of hypergeometric functions and non-lattice integral monodromy. Publ. Math. IHES **63** (1986) 5–89
[64] tom Dieck, T., Petrie, T.: Contractible affine surfaces of Kodaira dimension one. Japan J. Math. **16** (1990) 147–169
[65] tom Dieck, T., Petrie, T.: Arrangements of lines with tree resolution. Arch. Math. **56** (1991) 189–196
[66] tom Dieck, T., Petrie, T.: Homology planes and algebraic curves I. Preprint
[67] Dold, A.: Lectures on Algebraic Topology. Springer Verlag, 1972.
[68] Dowling, T. A.: A class of geometric lattices based on finite groups. J. Comb. Theory (B) **14** (1973) 61–86. Erratum. ibid **15** (1973) 211

References

[69] Edelman, P.: A partial order on the regions of $\mathbb{R}^n$ dissected by hyperplanes. Trans. Amer. Math. Soc. **283** (1984) 617–631

[70] Edelman, P., Reiner, V.: A counterexample to Orlik's conjecture. Proc. Amer. Math. Soc. To appear

[71] Esnault, H.: Fibre de Milnor d'un cône sur une courbe plane singulière. Invent math. **68** (1982) 477–496

[72] Esnault, H., Schechtman, V., Viehweg, E.: Cohomology of local systems on the complement of hyperplanes. Preprint

[73] Fadell, E., Neuwirth, L.: Configuration spaces. Math. Scand. **10** (1962) 111–118

[74] Falk, M.: Geometry and topology of hyperplane arrangements. Ph.D. Thesis, University of Wisconsin–Madison, 1983

[75] Falk, M.: The minimal model of the complement of an arrangement of hyperplanes. Trans. Amer. Math. Soc. **309** (1988) 543–556

[76] Falk, M.: The cohomology and fundamental group of a hyperplane complement. In: Singularities. Contemporary Math. **90** Amer. Math. Soc., 1989, pp. 55–72

[77] Falk, M.: On the algebra associated with a geometric lattice. Advances in Math. **80** (1990) 152–163

[78] Falk, M.: A geometric duality for order complexes and hyperplane arrangements. Europ. J. Combinatorics. To appear

[79] Falk, M.: Homotopy types of line arrangements. Preprint

[80] Falk, M., Randell, R.: The lower central series of a fiber-type arrangement. Invent. math. **82** (1985) 77–88

[81] Falk, M., Randell, R.: On the homotopy theory of arrangements. In: Complex Analytic Singularities. Adv. Studies in Pure Math. **8**, North Holland, 1987, pp. 101–124

[82] Falk, M., Randell, R.: The lower central series of generalized pure braid groups. In: Geometry and Topology. Lect. Notes in Pure and Appl. Math. **105**, Marcel Decker, New York 1986, pp. 103–108

[83] Falk, M., Randell, R.: Pure braid groups and products of free groups. In: Braids. Contemporary Math. **78**, Amer. Math. Soc., 1988, pp. 217–228

[84] Floyd, E. E.: On periodic maps and the Euler characteristic of associated spaces. Trans. Amer. Math. Soc. **72** (1952) 138–147

[85] Folkman, J.: The homology groups of a lattice. J. Math. and Mech. **15** (1966) 631–636

[86] Fox, R. H., Neuwirth, L.: The braid groups. Math. Scand. **10** (1962) 119–126

[87] Gelfand, I. M.: General theory of hypergeometric functions. Soviet Math. Doklady **33** (1986) 573–577

[88] Gelfand, I. M., Rybnikov, G. L.: Algebraic and topological invariants of oriented matroids. Soviet Math. Doklady **40** (1990) 148–152

[89] Gelfand, I. M., Serganova, V. V.: Combinatorial geometries and torus strata on homogeneous compact manifolds. Russian Math. Surveys **42**(2) (1987) 133–168

[90] Gelfand, I. M., Zelevinsky, A. V.: Algebraic and combinatorial aspects of the general theory of hypergeometric functions. Funct. Anal. and Appl. **20** (1986) 183–197

[91] Goresky, M., MacPherson, R.: Stratified Morse Theory. Springer Verlag, 1988

[92] Greene, C.: On the Möbius algebra of a partially ordered set. Advances in Math. **10** (1973) 177–187
[93] Greene, C.: An inequality for the Möbius function of geometric lattices. Studies in Appl. Math. **54** (1975) 71–74
[94] Greene, C.: Acyclic orientations. In: Higher combinatorics. D. Reidel, 1977, pp. 65–68
[95] Greene, C.: The Möbius function of a partially ordered set. In: Ordered Sets. D. Reidel, 1982
[96] Greene, C., Zaslavsky, T.: On the interpretation of Whitney numbers through arrangements of hyperplanes, zonotopes, non-Radon partitions, and orientations of graphs. Trans. Amer. Math. Soc. **280** (1983) 97–126
[97] Griffiths, P., Morgan, J.: Rational homotopy theory and differential forms. Birkhäuser, Boston 1981
[98] Grove, L. C., Benson, C. T.: Finite reflection groups. Second ed. GTM **99** Springer Verlag, 1985
[99] Grünbaum, B.: Convex polytopes. Interscience, New York 1967
[100] Grünbaum, B.: Arrangements of hyperplanes. In: Proc. Second Lousiana Conf. on Combinatorics and Graph Theory. Baton Rouge 1971, pp. 41–106
[101] Grünbaum, B.: Arrangements and spreads. CBMS Lecture Notes **10**, Amer. Math. Soc., 1972
[102] Grünbaum, B., Shephard, G. C.: Simplicial arrangements in projective 3-space. Mitt. Math. Sem. Univ. Giessen **166** (1984) 49–101
[103] Gudkov, D. A.: Topology of real projective algebraic manifolds. Russian Math. Surveys **29** (1974) 1–79
[104] Hall, P.: A contribution to the theory of groups of prime power order. Proc. London Math. Soc. II. Ser. **36** (1932) 39–95
[105] Hall, P.: The Eulerian functions of a group. Quart. J. Math. Oxford **7** (1936) 134–151
[106] Hamm, H., Lê, D. T.: Un théorème de Zariski du type de Lefschetz. Ann. sci. Éc. Norm. Sup. **6** (1973) 317–366
[107] Harnack, A.: Ueber die Vieltheiligkeit der ebenen algebraischen Curven. Math. Ann. **10** (1876) 189–198
[108] Hattori, A.: Topology of $\mathbb{C}^n$ minus a finite number of affine hyperplanes in general position. J. Fac. Sci. Univ. Tokyo **22** (1975) 205–219
[109] Hattori, A., Kimura, T.: On the Euler integral representations of hypergeometric functions in several variables. J. Math. Soc. Japan **26** (1974) 1–16
[110] Hendriks, H.: Hyperplane arrangements of large type. Invent. math. **79** (1985) 375–381
[111] Hilbert, D.: Ueber die reellen Züge algebraischer Curven. Math. Ann. **38** (1891) 115–138
[112] Hirsch, M.: Differential Topology. Springer Verlag, 1976
[113] Hirzebruch, F.: Arrangements of lines and algebraic surfaces. In: Arithmetic and Geometry. Vol. II. Progress in Math. **36**, Birkhäuser, Boston 1983, pp. 113–140
[114] Hunt, B.: Coverings and ball quotients with special emphasis on the 3-dimensional case. Bonner Math. Schriften **174**, 1986
[115] Jambu, M.: Algèbre d'holonomie de Lie et certaines fibrations topologiques. C. R. Acad. Sci. Paris **306** (1988) 479–482

[116] Jambu, M.: Sur l'algèbre d'Orlik–Solomon. C.R. Acad. Sci. Paris **307** (1988) 125–128
[117] Jambu, M.: Fiber–type arrangements and factorization properties. Advances in Math. **80** (1990) 1–21
[118] Jambu, M., Leborgne, L.: Fonction de Möbius et arrangements d'hyperplans. C. R. Acad. Sci. Paris **303** (1986) 311–314
[119] Jambu, M., Terao, H.: Arrangements libres d'hyperplans et treillis hyper-résolubles. C. R. Acad. Sci. Paris **296** (1983) 623–624
[120] Jambu, M., Terao, H.: Free arrangements of hyperplanes and supersolvable lattices. Advances in Math. **52** (1984) 248–258
[121] Jambu, M., Terao, H.: Arrangements of hyperplanes and broken circuits. In: Singularities. Contemporary Math. **90**, Amer. Math. Soc., 1989. pp. 147–162
[122] Jozsa, R., Rice, J.: On the cohomology of hyperplane complements. Proc. Amer. Math. Soc. To appear
[123] van Kampen, E. R.: On the fundamental group of an algebraic curve. Amer. J. Math. **55** (1933) 255–260
[124] Kelly, L. M.: A resolution of the Sylvester–Gallai problem of J. -P. Serre. Discrete Comp. Geom. **1** (1986) 101–104
[125] Klein, F.: Ueber die transformationen siebenter Ordnung der elliptischen Funktionen. Math. Ann. **14** (1879) 428–471
[126] Kohno, T.: On the minimal algebra and $K(\pi.1)$–property of affine algebraic varieties. Preprint
[127] Kohno, T.: Differential forms and the fundamental group of the complement of hypersurfaces. In: Singularities. Proc. Symp. Pure Math. **40** Part 1. Amer. Math. Soc., 1983. pp. 655–662
[128] Kohno, T.: On the holonomy Lie algebra and the nilpotent completion of the fundamental group of the complement of hypersurfaces. Nagoya Math. J. **92** (1983) 21–37
[129] Kohno, T.: Série de Poincaré-Koszul associée aux groupes de tresses pures. Invent. math. **82** (1985) 57–75
[130] Kohno, T.: Homology of a local system on the complement of hyperplanes. Proc. Japan Acad. Ser. A **62** (1986) 144–147
[131] Kohno, T.: Hecke algebra representations of braid groups and classical Yang–Baxter equations. In: Conformal field theory and solvable lattice models. Adv. Studies in Pure Math. **16**, 1988, pp. 255–269
[132] Kohno, T.: Poincaré series of the Malcev completion of generalized pure braid groups. Preprint
[133] Kohno, T.: Rational $K(\pi.1)$ arrangements satisfy the LCS formula. Preprint
[134] Kohno, T.: Holonomy Lie algebras, logarithmic connections, and the lower central series of fundamental groups. In: Singularities. Contemporary Math. **90** Amer. Math. Soc., 1989, pp. 171–182
[135] Kulikov, V. S.: On the fundamental group of the complement of a hypersurface in $\mathbb{C}^n$. Preprint
[136] Las Vergnas, M.: Matroides orientables. C.R. Acad. Sci. Paris **280** (1975) 61–64
[137] Las Vergnas, M.: Convexity in oriented matroids. J. Comb. Theory (B) **29** (1980) 231–243
[138] Lefschetz, S.: Algebraic Geometry. Princeton Univ. Press, Princeton 1953

[139] Lehrer, G. I.: On the Poincaré series associated with Coxeter group actions on complements of hyperplanes. Preprint
[140] Lehrer, G. I.: On hyperoctahedral hyperplane complements. Preprint
[141] Lehrer, G. I.: Rational tori, semisimple orbits and the topology of hyperplane complements. Preprint
[142] Lehrer, G. I.: The ℓ–adic cohomology of hyperplane complements. Preprint
[143] Lehrer, G. I., Shoji, T.: On flag varieties, hyperplane complements and Springer representations of Weyl groups. J. Austral. Math. Soc. (A) **49** (1990) 449–485
[144] Lehrer, G. I., Solomon, L.: On the action of the symmetric group on the cohomology of the complement of its reflecting hyperplanes. J. Algebra. **104**(2) (1986) 410–424
[145] Libgober, A.: On the homotopy type of the complement to plane algebraic curves. J. Reine Angew. Math. **367** (1986) 103–114
[146] Loeser, F.: Arrangements d'hyperplans et sommes de Gauss. Preprint
[147] Maclane, S.: Homology. Springer Verlag, 1963
[148] Manin, Yu. I., Schechtman, V. V.: Higher Bruhat orders related to the symmetric group. Funct. Anal. and Appl. **20** (1986) 148–150
[149] Manin, Yu. I., Schechtman, V. V.: Arrangements of hyperplanes, higher braid groups and higher Bruhat orders. In: Algebraic number theory–in honor of K. Iwasawa. Adv. Studies in Pure Math. **17**, Academic Press, 1989, pp. 289–308
[150] Maschke, H.: Ueber die quaternäre, endliche, lineare Substitutionsgruppe der Borchadt'schen Moduln. Math. Ann. **30** (1887) 496–515
[151] Maschke, H.: Aufstellung des vollen Formensystems einer quaternären Gruppe von 51840 linearen substitutionen. Math. Ann. **33** (1888) 317–344
[152] Matsumura, H.: Commutative Algebra. Second ed. Benjamin/Cummings, 1980
[153] Matsumura, H.: Commutative Ring Theory. Cambridge Univ. Press, 1986
[154] Melchior, E.: Uber Vielseite der projektiven Ebene. Deutsche Mathematik **5** (1940) 461–475
[155] Milnor, J.: Singular points of complex hypersurfaces. Annals of Math. Studies **61**, Princeton University Press, 1968
[156] Milnor, J., Orlik, P.: Isolated singularities defined by weighted homogeneous polynomials. Topology **9** (1970) 385–393
[157] Mnëv, N. E.: On manifolds of combinatorial types of configurations and convex polyhedra. Soviet Math. Doklady **32** (1985) 335–337
[158] Möbius, A. F.: Ueber eine besondere Art von Umkehrung der Reihen. J. Reine Angew. Math. **9** (1832) 105–123
[159] Moishezon, B., Teicher, M.: Braid group technique in complex geometry I: Line arrangements in $\mathbb{C}P^2$. In: Braids. Contemporary Math. **78**, Amer. Math. Soc., 1988, pp. 425–555
[160] Morgan, J.: The algebraic topology of smooth algebraic varieties. Publ. Math. IHES **48** (1978) 137–204
[161] Munkres, J.: Elements of Algebraic Topology. Addison–Wesley, 1984
[162] Nakamura, T.: A note on the $K(\pi, 1)$–property of the orbit space of the unitary reflection group $G(m, \ell, n)$. Sci. Papers College of Arts and Sciences. Univ. Tokyo **33** (1983) 1–6
[163] Naruki, I.: The fundamental group of the complement of Klein's arrangement of twenty-one lines. Topology and Appl. **34** (1990) 167–181

[164] Nguyễn Viêt Dũng: The fundamental group of the space of regular orbits of the affine Weyl groups. Topology **22** (1983) 425–435
[165] Northcott, G. D.: Lessons on Rings, Modules and Multiplicities. Cambridge Univ. Press, 1968
[166] Orlik, P.: Basic derivations for unitary reflection groups. In: Singularities. Contemporary Math. **90**, Amer. Math. Soc., 1989, pp. 211–228
[167] Orlik, P.: Stratification of the discriminant in reflection groups. Manuscripta math. **64** (1989) 377–388
[168] Orlik, P.: Introduction to arrangements. CBMS Lecture Notes **72**, Amer. Math. Soc. 1989
[169] Orlik, P.: Complements of subspace arrangements. J. Alg. Geom. **1** (1992) 147–156
[170] Orlik, P., Randell, R.: The Milnor fiber of a generic arrangement. Preprint
[171] Orlik, P., Solomon, L.: Combinatorics and topology of complements of hyperplanes. Invent. math. **56** (1980) 167–189
[172] Orlik, P., Solomon, L.: Unitary reflection groups and cohomology. Invent. math. **59** (1980) 77–94
[173] Orlik, P., Solomon, L.: Complexes for reflection groups. In: Algebraic Geometry. Lecture Notes in Math. **862**, Springer Verlag, 1981, pp. 193–207
[174] Orlik, P., Solomon, L.: Arrangements defined by unitary reflection groups. Math. Ann. **261** (1982) 339–357
[175] Orlik, P., Solomon, L.: Coxeter arrangements. In: Singularities. Proc. Symp. Pure Math. **40** Part 2. Amer. Math. Soc., 1983. pp. 269–292
[176] Orlik, P., Solomon, L.: Arrangements in unitary and orthogonal geometry over finite fields. J. Comb. Theory (A) **38**(2) (1985) 217–229
[177] Orlik, P., Solomon, L.: The Hessian map in the invariant theory of reflection groups. Nagoya Math. J. **109** (1988) 1–21
[178] Orlik, P., Solomon, L.: Discriminants in the invariant theory of reflection groups. Nagoya Math. J. **109** (1988) 23–45
[179] Orlik, P., Solomon, L., Terao, H.: Arrangements of hyperplanes and differential forms. In: Combinatorics and Algebra. Contemporary Math. **34**, Amer. Math. Soc., 1984, pp. 29–65
[180] Orlik, P., Solomon, L., Terao, H.: On Coxeter arrangements and the Coxeter number. In: Complex Analytic Singularities. Adv. Studies in Pure Math. **8**, North Holland, 1987, pp. 461–477
[181] Paris, L.: The counting polynomial of a supersolvable arrangement. Preprint
[182] Paris, L.: The covers of a complexified real arrangement of hyperplanes and their fundamental groups. Preprint
[183] Paris, L.: Universal cover of Salvetti's complex and topology of simplicial arrangements of hyperplanes. Preprint
[184] Petrowsky, I.: On the topology of real plane algebraic curves. Annals of Math. **39** (1938) 189–209
[185] Pham, F.: Introduction a l'Étude Topologique des Singularités de Landau. Mémorial des Sci. Math. **164**, Gauthier-Villars, Paris 1967
[186] Quillen, D.: Homotopy properties of the poset of non-trivial p-subgroups of a group. Advances in Math. **28** (1978) 101–128
[187] Randell, R.: On the topology of non-isolated singularities. In: Geometric Topology. Academic Press, New York 1979, pp. 445–473

[188] Randell, R.: On the fundamental group of the complement of a singular plane curve. Quart. J. Math. Oxford (2) **31** (1980) 71–79
[189] Randell, R.: The fundamental group of the complement of a union of complex hyperplanes. Invent. math. **69** (1982) 103–108. Correction: Invent. math. **80** (1985) 467–468
[190] Randell, R.: Lattice-isotopic arrangements are topologically isomorphic. Proc. Amer. Math. Soc. **107** (1989) 555–559
[191] Roberts, S.: On the figures formed by the intercepts of a system of straight lines in a plane, and on analogous relations in space of three dimensions. Proc. London Math. Soc. **19** (1889) 405–422
[192] Rokhlin, V. A.: Congruence modulo 16 in Hilbert's sixteenth problem. Funct. Anal. Appl. **6** (1972) 301–306
[193] Rose, L., Terao, H.: Hilbert polynomials and geometric lattices. Advances in Math. **84** (1990) 209–225
[194] Rose, L., Terao, H.: A free resolution of the module of logarithmic forms of a generic arrangement. J. Algebra **136** (1991) 376–400
[195] Rota, G.-C.: On the foundations of combinatorial theory I. Theory of Möbius functions. Z. Wahrscheinlichkeitsrechnung **2** (1964) 340–368
[196] Rota, G.-C.: On the combinatorics of the Euler characteristic. In: Studies in Pure Math., presented to Richard Rado (L. Mirsky. ed.). Academic Press, London 1971, pp. 221–233
[197] Saito, K.: Regularity of Gauss-Manin connection of flat family of isolated singularities. In: Conference Notes. Centre de Mathématiques de l'Ecole Polytechnique, 1973
[198] Saito, K.: Einfach elliptische Singularitäten. Invent. math. **23** (1974) 289–325
[199] Saito, K.: On the uniformization of complements of discriminant loci. In: Conference Notes. Amer. Math. Soc. Summer Institute, Williamstown, 1975
[200] Saito, K.: On a linear structure of a quotient variety by a finite reflexion group. RIMS Kyoto preprint **288**, 1979
[201] Saito, K.: Theory of logarithmic differential forms and logarithmic vector fields. J. Fac. Sci. Univ. Tokyo Sect.IA Math. **27** (1981) 265–291
[202] Saito, K., Yano, T., Sekiguchi, J.: On a certain generator system of the ring of invariants of a finite reflection group. Communications in Algebra **8**(4) (1980) 373–408
[203] Salvetti, M.: Topology of the complement of real hyperplanes in $\mathbb{C}^N$. Invent. math. **88** (1987) 603–618
[204] Salvetti, M.: Arrangements of lines and monodromy of plane curves. In: Algebraic Geometry. Lecture Notes in Math. **862**, Springer Verlag, 1981, pp. 107–192
[205] Salvetti, M.: Arrangements of lines and monodromy of plane curves. Compositio Math. **68** (1988) 103–122
[206] Salvetti, M.: Generalized braid groups and self-energy Feynmann integrals. In: Braids. Contemporary Math. **78**, Amer. Math. Soc., 1988, pp. 675–686
[207] Schechtman, V. V., Varchenko, A. N.: Arrangements of hyperplanes and Lie algebra homology. Preprint
[208] Scheja, G., Storch, U.: Uber Spurfunktionen bei vollständigen Durchschnitten. J. Reine Angew. Math. **278/279** (1975) 174–190

[209] Shephard, G. C.: Regular complex polytopes. Proc. London Math. Soc.(3) **2** (1952) 82–97
[210] Shephard, G. C., Todd, J. A.: Finite unitary reflection groups. Canad. J. Math. **6** (1954) 274–304
[211] Solomon, L.: Invariants of Euclidean reflection groups. Trans. Amer. Math. Soc. **113** (1964) 274–286
[212] Solomon, L., Terao, H.: A formula for the characteristic polynomial of an arrangement. Advances in Math. **64** (1987) 305–325
[213] Sommese, A.: On the density of ratios of Chern numbers of algebraic surfaces. Math. Ann. **268** (1984) 207–221
[214] Spaltenstein, N.: Open problem. In: Open problems in algebraic groups. Proceedings of the twelfth international symposium, Division of mathematics, The Taniguchi Foundation, R. Hotta and N. Kawanaka organizing committee, 1983
[215] Spanier, E.: Algebraic Topology. McGraw-Hill, New York 1966
[216] Springer, T. A.: Regular elements of finite reflection groups. Invent. math. **25** (1974) 159–198
[217] Stanley, R. P.: Modular elements of geometric lattices. Algebra Universalis **1** (1971) 214–217
[218] Stanley, R. P.: Supersolvable lattices. Algebra Universalis **2** (1972) 214–217
[219] Stanley, R. P.: Acyclic orientations of graphs. Discrete Math. **5** (1973) 171–178
[220] Stanley, R. P.: Relative invariants of finite groups generated by pseudoreflections. J. of Algebra **49** (1977) 134–148
[221] Stanley, R. P.: T–free arrangements of hyperplanes. In: Progress in Graph Theory. Academic Press, 1984, p. 539
[222] Stanley, R. P.: Enumerative Combinatorics I. Wadsworth and Brooks/Cole, Monterey, CA 1986
[223] Steinberg, R.: Invariants of finite reflection groups. Canad. J. Math. **12** (1960) 616–618
[224] Steinberg, R.: Differential equations invariant under finite reflection groups. Trans. Amer. Math. Soc. **112** (1964) 392–400
[225] Sullivan, D.: Infinitesimal computations in topology. Publ. Math. IHES **47** (1977) 269–331
[226] Terao, H.: Arrangements of hyperplanes and their freeness I, II J. Fac. Sci. Univ. Tokyo **27** (1980) 293–320
[227] Terao, H.: Free arrangements of hyperplanes and unitary reflection groups. Proc. Japan Acad. Ser. A **56** (1980) 389–392
[228] Terao, H.: Generalized exponents of a free arrangement of hyperplanes and Shephard–Todd–Brieskorn formula. Invent. math. **63** (1981) 159–179
[229] Terao, H.: On Betti numbers of complements of hyperplanes. Publ. RIMS Kyoto Univ. **17** (1981) 567–663
[230] Terao, H.: The exponents of a free hypersurface. In: Singularities. Proc. Symp. Pure Math. **40** Part 2, Amer. Math. Soc., 1983, pp. 561–566
[231] Terao, H.: Discriminant of a holomorphic map and logarithmic vector fields. J. Fac. Sci. Univ. Tokyo Sect. IA Math. **30** (1983) 379–391
[232] Terao, H.: Free arrangements of hyperplanes over an arbitrary field. Proc. Japan Acad. Ser.A **59** (1983) 301–303
[233] Terao, H.: The bifurcation set and logarithmic vector fields. Math. Ann. **263** (1983) 313–321

[234] Terao, H.: Modular elements of lattices and topological fibration. Advances in Math. **62** (1986) 135–154
[235] Terao, H.: The Jacobians and the discriminants of finite reflection groups. Tohoku Math. J. **41** (1989) 237–247
[236] Terao, H.: Factorizations of Orlik–Solomon algebras. Advances in Math. To appear
[237] Terao, H., Enta, Y.: Basic derivatives for G_{34}. In: Singularities. Contemporary Math. **90**, Amer. Math. Soc., 1989, pp. 225–228
[238] Terao, H., Yano, T.: The duality of the exponents of free deformations associated with unitary reflection groups. In: Algebraic Groups and Related Topics. Adv. Studies in Pure Math. **6**, North Holland, 1985, pp. 339–348
[239] Varchenko, A. N.: Combinatorics and topology of the disposition of affine hyperplanes in real space. Funct. Anal. and Appl. **21** (1987) 9–19
[240] Varchenko, A. N.: On the number of faces of a configuration of hyperplanes. Soviet Math. Doklady **38** (1989) 291–295
[241] Varchenko, A. N.: The Euler beta function, the Vandermonde determinant, Legendre's equation, and critical values of line functions on a configuration of hyperplanes; I Math. USSR Izvestija **35** (1990) 543–572; II ibid **36** (1991) 155–168
[242] Varchenko, A. N.: Multidimensional hypergeometric functions in conformal field theory, algebraic K-theory, algebraic geometry. Proc. Int. Congress Math., Kyoto, Japan, 1990. To appear
[243] Varchenko, A. N.: Bilinear form of real configuration of hyperplanes. Preprint
[244] Varchenko, A. N. and Gelfand, I. M.: Heaviside functions of configurations of hyperplanes. Funct. Anal. Appl. **21** (1987) 255–270.
[245] Viro, O. Ja.: Progress in the topology of real algebraic varieties over the last six years. Russian Math. Surveys **41** (1986) 55–82
[246] van der Waerden, B. L.: Modern Algebra. Ungar, New York 1950
[247] Ward, M.: The algebra of lattice functions. Duke Math. J. **5** (1939) 357–371
[248] Weisner, L.: Abstract theory of inversion of finite series. Trans. Amer. Math. Soc. **38** (1935) 474–484
[249] Wetzel, J. E.: On the division of the plane by lines. Amer. Math. Monthly **85** (1978) 648–656
[250] Whitney, H.: A logical expansion in mathematics. Bull. Amer. Math. Soc. **38** (1932) 572–579
[251] Wiman, A.: Ueber eine einfache Gruppe von 360 ebenen Collineationen. Math. Ann. **47** (1896) 531–556
[252] Yano, T., Sekiguchi, J.: The microlocal structure of weighted homogeneous polynomials associated with Coxeter systems; I Tokyo J. Math. **2**(2) (1979) 193–219; II Tokyo J. Math. **4**(1) (1981) 1–34
[253] Yuzvinsky, S.: Cohen–Macaulay seminormalizations of unions of linear subspaces. Preprint
[254] Yuzvinsky, S.: Cohomology of local sheaves on arrangement lattices. Proc. Amer. Math. Soc. To appear
[255] Yuzvinsky, S.: First two obstructions to the freeness of arrangements. Preprint
[256] Zaslavsky, T.: Facing up to arrangements: Face-count formulas for partitions of space by hyperplanes. Memoirs Amer. Math. Soc. **154**, 1975

[257] Zaslavsky, T.: Counting the faces of cut-up spaces. Bull. Amer. Math. Soc. **81** (1975) 916–918
[258] Zaslavsky, T.: Maximal dissections of a simplex. J. Comb. Theory (A) **20** (1976) 244–257
[259] Zaslavsky, T.: A combinatorial analysis of topological dissections. Advances in Math. **25** (1977) 267–285
[260] Zaslavsky, T.: Arrangements of hyperplanes; matroids and graphs. In: Proc. Tenth Southeastern Conf. on Combinatorics, Graph Theory and Computing (Boca Raton. 1979). Vol. II. pp. 895–911
[261] Zaslavsky, T.: The slimmest arrangements of hyperplanes: I. Geometric lattices and projective arrangements. Geometriae Dedicata **14** (1983) 243–259; II. Basepointed geometric lattices and Euclidean arrangements. Mathematika **28** (1981) 169–190
[262] Zaslavsky, T.: Orientation of signed graphs. Europ. J. Combinatorics. To appear
[263] Zelevinsky, A. V.: Geometry and combinatorics related to vector partition functions. Preprint
[264] Ziegler, G. M.: Algebraic combinatorics of hyperplane arrangements. Ph.D. Thesis. MIT. 1987
[265] Ziegler, G. M.: The face lattice of hyperplane arrangements. Discrete Math. **73** (1989) 233–238
[266] Ziegler, G. M.: Multiarrangements of hyperplanes and their freeness. In: Singularities. Contemporary Math. **90**, Amer. Math. Soc., 1989, pp. 345–359
[267] Ziegler, G. M.: Combinatorial construction of logarithmic differential forms. Advances in Math. **76** (1989) 116–154
[268] Ziegler, G. M.: Matroid representations and free arrangements. Trans. Amer. Math. Soc. **320** (1990) 525–541
[269] Ziegler, G. M.: Some almost exceptional arrangements. Preprint

Index

Actual vertex 180, 187
Acyclic chain complex 62, 63, 68, 83, 90, 133, 149
Acyclic orientation 57
Addition theorem 118
Addition–deletion theorem 18, 118
Additional hyperplane 14
Admissible graph 184
Affine arrangement 11
Affine independent points 168
Aigner, M. 23
Alexander duality 207
Algebra equivalence 78
Algebra(s)
– DG 203
– factorization of 18, 195, 202
– incidence 41
Algebraic variety 196
Annihilator 272
Aomoto, K. 10
Arnold, V. I. 2, 3, 19, 92, 157, 158, 162, 167, 190, 215, 239
Arrangement(s) 10
– affine 11
– B_3– 12, 36, 226
– Boolean 13, 26, 36, 42, 101, 105, 111, 124, 138, 139, 160
– braid 13, 16, 26, 44, 53, 106, 111, 120, 160, 205, 226, 244
– cardinality of 13
– center of 11
– centered 11
– centerless 11
– central 11, 173
– complement of 15, 19
– complexified 16, 19, 168, 173, 178, 184
– Coxeter 225, 257

– diffeomorphic 157, 166
– discriminantal 20, 205
– embedded 171
– empty 11, 105
– essential 24
– fiber type 162, 196, 201, 205
– free 7, 15, 18, 104, 151, 155, 207, 238
    exponents of 108
– general position 5, 164, 205, 209
– generic 20, 165, 209, 210
– graphic 17, 52, 53
– homeomorphic 157
– homotopy equivalent 157
– inductively free 19, 119, 254
– irreducible 27
– $K(\pi,1)$ 160
– ℓ– 11
– $\mathbb{L}$–extended 16
– one–parameter family of 166
– partition of 50
– product 27, 35, 43, 104, 109, 132
– projective 11
– rank of 24
– rational $K(\pi,1)$– 205
– real 168
– recursively free 122
– reducible 27
– reflection 16, 20, 205, 225, 226
– simplicial 3, 163
– singular set of 7, 15
– sub– 112
– subspace 11, 20, 211
– supersolvable 17, 32, 48, 81, 121, 196, 201
Artin, E. 160
Artin group 266

Index

Arvola, W. 8, 19, 157, 168, 174, 175, 176, 178, 184, 190
Associated basic derivations 265
Associated Coxeter group 266
Associated prime 276
Atom 24

Baclawski, K. 99, 142
Bad set 197
Ball quotient 9
Barcelo, H. 10
Basic degrees 224
Basic derivations 234
– associated 265
Basic invariants 224, 257
– flat 281
Basis of free module 271
Birkhoff, G. 54
Birman, J. 160
Björner, A. 8, 10, 136, 142
Boolean arrangement 13, 26, 36, 42, 101, 105, 111, 124, 138, 139, 160
de Boor, C. 8
Braid arrangement 13, 16, 26, 44, 53, 106, 111, 120, 160, 205, 226, 244
Braid group 160
Braid space 259
Brandt, K. 122, 253
Brieskorn, E. 3, 6, 19, 20, 92, 157, 158, 162, 163, 167, 190, 227, 239, 259
Brieskorn's Lemma 3, 78, 191, 195
Broken circuit 67, 72, 75
– basis 10
– module 67, 73
Brylawski, T. 17, 46
Bundle 192
Bundle map 197

Cardinality of arrangement 13
Cartier, P. 7
Category 146
Cell complex 159, 171
Cellular homotopy equivalence 177
Center of arrangement 11
Centered arrangement 11
Centerless arrangement 11
Central arrangement 11, 173

Chain 33, 150, 175
Chamber 4, 28, 51, 57, 174, 175, 196
Characteristic polynomial 43, 47, 51, 56, 145, 153
Chevalley's theorem 224, 237
Chromatic function 54
Chromatic polynomial 17, 56
Circuit 67, 72
Coefficient matrix 103, 218, 240
Coexponents 234
Cohen, M. 176
Cohen-Macaulay ring 277
Cohomology of complement 20, 162, 167, 191, 195, 202
Coloring of graph 53
Complement of arrangement 15, 19
– cohomology of 20, 162, 167, 191, 195, 202
– fundamental group of 8, 19, 177, 190
– homotopy type of 8, 19, 159, 165, 168, 172, 174, 176
– Poincaré polynomial of 20, 191, 195, 205, 206
Complete graph 52
Complex manifold 191
Complex reflection group 7, 259
Complexified arrangement 16, 19, 168, 173, 178, 184
– embedding of 173
Configuration space 16
Coning construction 14, 27, 35, 43, 70, 158
Contractible fiber 176
Contractible space 137
Contraction 55
Contragradient action 216, 218
Contravariant functor 146
Convex hull 168
Covariant functor 146
Covering 210
Coxeter, H.S.M. 266
Coxeter arrangement 225, 257
Coxeter group 3, 12, 162, 163, 224, 226, 286
– associated 266
Coxeter number 257
Cubic 226

Deconing construction 14
Dedekind R. 40
Defining polynomial 11
Degree(s)
– of a homogeneous basis 276
– of derivation 100
– polynomial 100, 124
Deheuvels, R. 142
Deletion 4, 14, 54
Deletion and restriction 17, 56, 74, 95, 138
Deletion theorem 117
Deletion–Restriction Theorem 46
Deligne, P. 3, 19, 123, 157, 158, 162
Dependent hyperplanes 61
Derivation 10, 15, 100
– basic 234
– degree of 100
– Euler 102, 108
– homogeneous 100
– tangent to $\mathcal{A}$ 101
DG algebra 203
– minimal 203
tom Dieck, T. 10
Diffeomorphic arrangements 157, 166
Differential form 18, 92, 93
– regular 124
Differential graded algebra 203
Dimension of module 273
Dirichlet, P. G. L. 40
Dirichlet convolution 40
Discriminant 231, 237, 239
Discriminant locus 260
Discriminant matrix 240
Discriminantal arrangement 20, 205
Distinguished hyperplane 14
Dold, A. 136
Dowling lattice 226
Dual cell complex 171
Duality 128, 257
Duality groups 286
Duality of exponents 224

Edelman, P. 8, 10, 156
Eigenspace 209
Embedded arrangement 171
Embedding of complexified arrangement 173
Empty arrangement 11, 105
Essential arrangement 24
Euler, L. 38, 40
Euler derivation 102, 108
Euler integral representation 5
Exact functor 148
Exact sequence
– for $A(\mathcal{A})$ 76
– for $D(\mathcal{A})$ 115
– for localization 273
– for $R(\mathcal{A})$ 97
– in cohomology 191, 212
Excision 192
Exponents
– duality of 224
– of free arrangement 108
– of reflection group 224
Exterior algebra 60, 233

Face 28
– support of 28
Face poset 17, 19, 28, 168
Factorization of algebra 18, 195, 202
Factorization Theorem 19, 155
Fadell, E. 2, 161
Falk, M. 8, 20, 78, 79, 155, 158, 162, 195, 202, 205, 208
Family of 2–planes 179
Fiber bundle 158
Fiber type arrangement 162, 196, 201, 205
Fibration
– strictly linear 196
Fibration Theorem 201
Finite fields 228
Fixer 216
Flat basic invariants 281
Flat graphing map 182
Folkman, J. 19, 99, 136, 141, 168
Folkman complex 19, 137, 144, 145, 147, 208, 219
– homotopy type of 142
Fox, R. 2, 161
Free arrangement 7, 15, 18, 104, 151, 155, 207, 238
– exponents of 108
Free group 185
– generators of 186

Free module 271
- basis of 271
- rank of 272
Full subcomplex 170, 172
Functor
- contravariant 146
- covariant 146
- exact 148
- local 147
Fundamental group 161, 185
- of complement 8, 19, 177, 190

Gallai, T. 8
Galois group 237
Gelfand, I. M. 10
General position 1, 61
General position arrangement 5, 164, 205, 209
Generators
- adapted to v 186
- minimal set of 274
- of free group 186
- of group 178
Generic arrangement 20, 165, 209, 210
Generic property 134
Geometric lattice 24
Geometric linking 20, 208
Geometric poset 24
Glauberman, G. 66
Good subspace 198
Goresky, M. 8, 20, 158, 168, 195, 211
Graded module 274
Grading 219, 239
- of $\Omega^p(\mathcal{A})$ 126
Graph 52, 180
- admissible 184
- coloring of 53
- complete 52
- node of 2- 182
- orientation of 57
- real 178
- regular 2- 181
Graphic arrangement 17, 52, 53
Graphing map 179
- flat 182
- proper 180
- regular 181
Greene, C. 38

Griffiths, P. 203
Group
- action 159, 217
  on $L(\mathcal{A})$ 243
- Artin 266
- braid 160
- Coxeter 3, 12, 162, 163, 224, 226, 286
- free 185
- generators of 178
- irreducible 224
- monomial 226, 245
- nilpotent 204
- presentation of 178
- pure braid 160
- reducible 224
- reflection 16, 224
- relators of 178
- representation 10
- Shephard 163, 265, 286
- symmetric 3, 86, 161, 221, 226, 245, 259
Grünbaum, B. 2, 3, 164, 301

Hall, P. 41
Harnack, A. 6
Hasse diagram 26
Hattori, A. 5, 19, 157, 158, 164, 165, 166, 209
Hecke algebra 10
Height of prime ideal 273
Hereditarily free 156, 253
Hereditarily inductively free 253
Hermitian form 216, 225
Hessian configuration 227
Hessian determinant 226
Hessian map 218
Hessian matrix 235
Hilbert, D. 2, 6
Hilbert's Nullstellensatz 135
Hirsch extension 203
Hirzebruch, F. 9, 227
Holonomy Lie algebra 10
Homeomorphic arrangements 157
Homogeneous basis 105, 275
- degrees of 276
Homogeneous derivation 100
Homogeneous polynomial 100
Homotopy equivalence 176, 185

Homotopy equivalent arrangements 157
Homotopy type 210
– of complement 8, 19, 159, 165, 168, 172, 174, 176
– of Folkman complex 142
– of order complex 136
Hopf bundle 158, 210
Hopf trace formula 220
Horizontal space 197
Hunt, B. 9
Hurewicz isomorphism 165
Hypergeometric function 5, 9, 60
Hyperplane(s) 10
– additional 14
– dependent 61
– distinguished 14
– independent 61
– reflecting 16, 224
– trace of 181

Ideal
– prime 147, 272
  height of 273
  localization at 272
– variety of 272
Incidence algebra 41
Indecomposable elements 203
Independent hyperplanes 61
Independent partition 50, 83
Induced partition 50
Induced representation 219
Induction table 119
Inductively k–free 253
Inductively free arrangement 19, 119, 254
Inflection point 226
Interior product 127
Intersection poset 4, 17, 24
Invariant derivations 217
Invariant differentials 217
Invariant polynomials 217
Invariants
– basic 224, 257
  flat 281
– relative 228
Irreducible arrangement 27
Irreducible group 224

Jacobian 230
Jacobian matrix 218, 240
Jambu, M. 10
Jamison, F. 1
Janssen, W. A. M. 66
Join 24
Jordan, C. 227
Jozsa, R. 191, 195

van Kampen, E. R. 178
van Kampen's theorem 140, 185
Keaty, J. 122, 167
Kelly, L. M. 9
Kervaire, M. 94
Klein, F. 228, 241, 264
Klein's simple group 228
Knot theory 186
Knörrer, H. 227
Kohno, T. 8, 10, 20, 158, 202, 205
Krull dimension 273

Las Vergnas, M. 4
Lattice
– Dowling 226
– equivalence 24
– geometric 24
– homology 10
– isotopy 166
– partition 26, 45, 221, 244, 245
– same 166
Lawrence, J. 168
LCS formula 205
Lefschetz number 211
Lehrer, G. I. 10, 253
Lie algebra 204
– holonomy 10
– nilpotent 204
Linear character 228, 237
Linear order in $\mathcal{A}$ 67
Linking
– geometric 20, 208
– of loops around hyperplanes 189
Liouville J. 40
Local functor 147
Localization at prime ideal 272
Logarithmic form 19, 124
– polynomial degree of 126

Loop passing 186
Lower central series 204

MacPherson, R. 8, 20, 158, 168, 195, 211
Malcev, A. 204
Manifold 159
Manin, Yu. 20, 158, 205, 206
Maschke, H. 228, 268
Matroid 8
– oriented 17, 29
Maximal element 24, 168, 206
Maximal ideal 135
Meet 24
Melchior, E. 3, 9
Milnor, J. 210
Milnor fiber 20, 158, 168, 210
Milnor fibration 158, 209
Minimal DG algebra 203
Minimal model 20, 203
Minimal set of generators 274
Miyaoka, Y. 9
Miyaoka–Yau inequality 9
Mnëv, N. E. 8
Möbius, A. F. 39
Möbius function 17, 32, 38, 221
Möbius inversion 34, 154
Modular element 31, 80, 199
Modular pair 30
Module
– dimension of 273
– free 271
    basis of 271
    rank of 272
– graded 274
– support of 273
Module of $\mathcal{A}$-derivations 15, 101, 216
Monodromy 159, 209, 210
Monomial group 226, 245
Morgan, J. 203
Multiple point 178, 179

Neuwirth, L. 2, 161
Nguyễn V. D. 66
Nice partition 17, 50, 84, 86
Nilpotent group 204
Nilpotent Lie algebra 204
Node of 2–graph 182

Noetherian ring 271
Noncoplanar faces 174
Nonsingular cubic 226

One–parameter family of arrangements 166
Orbit 221
Orbit map 260
Orbit space 223
Orbit stratification 260
Order complex 19, 136, 168, 175, 211
– homotopy type of 136
Order complex with functor coefficients 146
Order of pole 150
Orientation
– acyclic 57
– of graph 57
Oriented matroid 17, 29
Orlik, P. 6, 7, 8, 10, 17, 60, 156
Orlik–Solomon algebra 17, 60
Overcrossing 183

Parallel subspaces 168
Partial order 23
Partition
– independent 50, 83
– induced 50
– nice 17, 50, 84, 86
– of arrangement 50
– section of 82
Partition lattice 26, 45, 221, 244, 245
Path connected 139
Pencil of lines 188
Petrie, T. 10
Petrowsky, I. 6
Pham, F. 95
Poincaré polynomial 223
– of $\mathcal{A}$ 17, 42, 51
– of $A(\mathcal{A})$ 74
– of algebra 17
– of $B(\mathcal{A})$ 91
– of complement 2, 20, 167, 191, 195, 205, 206
– of $\mathcal{H}(\mathcal{A}; \mathcal{K})$ 144
– of partition 82
– of $R(\mathcal{A})$ 98

Index   321

– of restriction 253
Poincaré series 145, 242, 275
– pole of 275
Pole 243
Pole of Poincaré series 275
Polynomial algebra 11
Polynomial degree(s) 100, 124
– of $\mathcal{A}$ 18
– of logarithmic form 126
Poset
– face 17, 19, 28, 168
– geometric 24
– intersection 4, 17, 24
Presentation of group 178
Prime ideal 147, 272
– height of 273
– localization at 272
Product arrangement 27, 35, 43, 104, 109, 132
Projective arrangement 11
Proper graphing map 180
Pure braid group 160
Pure braid space 2, 16, 167

Quadratic form 216
Quillen, D. 136, 142

Radical of ideal 135
Randell, R. 8, 19, 155, 157, 162, 166, 178, 205
Rank
– of arrangement 24
– of free module 272
Rank function 24
Rational $K(\pi, 1)$–arrangement 205
Rational homotopy type 8
Real arrangement 168
Real graph 178
Recursively free arrangement 122
Reducible arrangement 27
Reducible group 224
Reflecting hyperplane 16, 224
Reflection 16, 224
Reflection arrangement 16, 20, 205, 225, 226
Reflection group 16, 224
– complex 7, 259

– exponents of 224
Regular 2–graph 181
Regular cell complex 172, 173, 175
Regular complex polytope 265
Regular differential form 124
Regular graphing map 181
Regular representation 224, 232
Regular sequence 277
Reiner, V. 10, 156
Relative invariants 228
Relators of group 178
Representation
– induced 219
– regular 224, 232
– ring 219
Residue 97
Restriction 4, 14
– deletion and 17, 56, 74, 95, 138
– Poincaré polynomial of 253
de Rham cohomology 191
de Rham complex 92
Rice, J. 191, 195
Riemann, G. 41
Riemann ζ function 41
Roberts, S. 2
Rose, L. 78
Rota, G. -C. 38, 136

Saito, K. 7, 15, 123, 164, 215, 230, 281
Saito's criterion 105, 130
Salvetti, M. 8, 19, 157, 168, 175, 178
Salvetti complex 175
Same lattice 166
Schechtman, V. 10, 20, 158, 205, 206
Scheja, G. 230
Schläfli, L. 1
Section of partition 82
Segment 26
Separable extension 236
Separator 47
Serre, J.P. 9
Shephard, G. C. 8, 20, 215, 226, 232, 268
Shephard group 163, 265, 286
Shoji, T. 10, 253
Shuffle product 87
Simple pole 96
Simplicial arrangement 3, 163
Simplicial complex 168

Simplicial map 176
Simplicial triangulation 170
Simply connected 140
Singular set of arrangement 7, 15
Smith, P. A. 6
Solomon, L. 6, 7, 8, 10, 17, 60, 234
Sommese, A. 9
Spaltenstein, N. 253
Spanier, E. 136
Stabilizer 219
Standard tuple 67, 72
Stanley, R. 17, 23, 38, 48, 57, 58, 155, 228, 229
Star of a vertex 139
Steinberg, R. 230
Storch, U. 230
Stratification of V 163, 260
Stratified Morse theory 8, 211
Strictly linear fibration 20, 162, 196
Strictly linearly fibered 201
Strong deformation retract 207
Strong deformation retraction 170, 172, 175
Subarrangement 14, 112
Subspace arrangement 11, 20, 211
Sullivan, D. 203, 204
Supersolvable arrangement 17, 32, 48, 81, 121, 196, 201
Support of face 28
Support of module 273
Sylvester, J. J. 8
Sylvester–Gallai configuration 8
Symmetric group 3, 86, 161, 221, 226, 245, 259

Tangent space 163
Terao, H. 7, 10, 78, 155
Thom class 193
Thom isomorphism 192, 212
Todd, J. A. 20, 215, 226, 232, 268
Total degree 101, 124
Trace 221

Trace of hyperplane 181
Transversality 181
Triangulated cell 170
Triple 14
Tubular neighborhood 191
Two–planes
– family of 179

Undercrossing 183
Universal cover 165, 210

Varchenko, A. 10
Variety of $\mathcal{A}$ 15
Variety of ideal 272
Vector product 29
Vertex
– actual 180
– virtual 181, 187
Virtual vertex 181, 187

Walker, J. 142
Ward, M. 41
Wedge of spheres 141
Weighted homogeneous 240
Weil, A. 210
Weisner, L. 34, 41
Wetzel, J. 2
Whitney complex 142, 144, 145, 147
Woodbridge, J. L. 1
Wreath product 226

Yau, Sh.T. 9
Yuzvinsky, S. 10, 19, 99

Zariski, O. 177
Zariski topology 134, 200
Zaslavsky, T. 4, 8, 46, 51, 168
Zelevinsky, A. V. 8, 10
Zero section 192
Ziegler, G. 8, 10, 113

Index of Symbols

$\mathcal{A}$ 10, 13
A 181
$\mathcal{A}_1 \times \mathcal{A}_2$ 27
$A(\mathcal{A})$ 61, 70, 204
$(\mathcal{A}, \mathcal{A}', \mathcal{A}'')$ 4, 14, 46, 74, 76, 97, 113, 138, 191
$\mathcal{A}$–equivalence 78
$\mathcal{A}(G)$ 16, 225
$AO(G)$ 57
a_S 61
A_v 197
$\mathcal{A}^X$ 14
$\mathcal{A}_X$ 14
A_X 65
$A_X(\mathcal{A})$ 77
$\mathcal{A}/Y$ 197

$b(\mathcal{A})$ 114
$B(\mathcal{A})$ 88, 142
b_S 88

$\mathbb{C}$ 10
$\mathbf{c}\mathcal{A}$ 14, 18, 19, 35, 43, 70, 74, 80, 158, 195
$C(\mathcal{A})$ 67, 73
$\mathcal{C}(\mathcal{A})$ 4
$(\mathcal{C}.(\mathcal{A}, F), \partial)$ 148
$(\tilde{\mathcal{C}}.(\mathcal{A}, F), \tilde{\partial})$ 148
$ch(L)$ 33
$ch[X, Y]$ 33
C_k 200
$\mathbf{c}S$ 70

$\mathbf{d}\mathcal{A}$ 14
$D(\mathcal{A})$ 101, 163
$D(\mathcal{A})_p$ 103
$D(\mathcal{A}_X)$ 148

$D(\mathcal{A}')\alpha_0$ 114
$\mathrm{Der}_{\mathbb{K}}(S)$ 15, 100
Der_S^G 217
D_{f_i} 236
D_i 100
$D_R(\delta)$ 237
$D_S(\mathcal{A})$ 15
$D_S(f)$ 237
d_χ 228

$E(\mathcal{A})$ 60
e_H 228
e_S 61
E_X 64
$\exp \mathcal{A}$ 108

$\mathsf{F}(\mathcal{A})$ 137
$\mathrm{Fix}(g)$ 216
$\mathbb{F}_q$ 10

$[g]$ 218
G–action 217
G_H 228
G–module 217
G_U 216
G_v 216

$\mathrm{Hess}(f)$ 218
$\mathsf{H}(f)$ 218
Hor 197

$I(\mathcal{A})$ 61, 70

J 231
$\mathsf{J}(f_1, \ldots, f_\ell)$ 218

$J(\mathcal{F})$ 230

$\mathbb{K}$ 10
k-free 156
$\mathsf{K}(\mathcal{L}(\mathcal{H}))$ 168
$K(\pi,1)$-arrangement 160
$K(\pi,1)$ space 2, 8, 159, 161, 162, 163, 164, 165, 207, 259, 301

$\mathbb{L}$ 10
$L(\mathcal{A})$ 4, 14, 24
$\mathcal{L}(\mathcal{A})$ 28
L-equivalence 24
L^g 221
$L_p(\mathcal{A})$ 26
L_v 197
L^X 26
L_X 26

$M(\mathcal{A})$ 15
max$S$ 67
$\mathcal{M}_{\mathcal{H}}(\mathcal{A})$ 171
$\mathsf{M}_{\mathcal{H}}(\mathcal{A})$ 172
$M_\ell(S)$ 218
M-property 6, 8, 20, 196
M^X 163, 260
M_Δ 240
$\mathsf{M}(\theta_1,\ldots,\theta_\ell)$ 103

$\mathbb{N}$ 10
$N(\mathcal{A})$ 15
$\mathcal{N}_{\mathcal{H}}(\mathcal{A})$ 171

pdegω 124
pdegθ 100
Poin$(A(\mathcal{A}),t)$ 74
Poin(M,x) 145
Poin$(M(\mathcal{A}),t)$ 2, 195

$\mathbb{Q}$ 10
$Q(\mathcal{A})$ 11
$\mathbb{Q}$-polynomial forms 203

$\mathbb{R}$ 10
$r(\mathcal{A})$ 24

$R(\mathcal{A})$ 93
res(ϕ) 97
$(R$-Mod$)$ 146
$r(X)$ 24
$R = S^G$ 217, 224

S 61
S 61
s_H 228
S_p 61
Spec$R$ 147, 272
S_X 64
S_X 219
$Sym(p)$ 86
$S = (H_1,\ldots,H_p)$ 61
$S = \mathbb{K}[x_1,\ldots,x_\ell]$ 11
$\cap S$ 61
$\vee_m S^k$ 141

$\mathcal{T}$ 86
$T(\mathcal{A})$ 24
tdegθ 101
tdegω 124

$\mathcal{U}$ 87

$[w_1,\ldots,w_k]$ 178
$\mathsf{W}(\mathcal{A})$ 175

$X(\wp)$ 147
$[X,Y]$ 26
$[X,Y]$ 26

$\mathbb{Z}$ 10

$\doteq$ 11
$\prec$ 67

$(\alpha_0, b(\mathcal{A}))$ 114
α_H 11
∂_A 63
∂_E 60
∂_η 134
δ 231
Δ 231

Index of Symbols

ζ function 210
η-complex 134, 152
η_H 192
$[\eta_H]$ 192
$\langle\eta_H\rangle$ 192
$\bar{\theta}$ 115, 236
θ_E 102
$\lambda : \mathcal{A}' \to \mathcal{A}''$ 75
$\mu(\mathcal{A})$ 36
$\mu(X)$ 35
$\mu(X,Y)$ 32
$\pi(\mathcal{A},t)$ 42
$\pi(\mathcal{A},t)$ factors 155
π–equivalence 47

Φ_ℓ 11
$\chi(\mathcal{A},t)$ 43
χ–independent 67, 72
$\Psi(\mathcal{A};x,t)$ 151
ω_H 93
$\Omega^{\cdot}(\mathcal{A})$ 19, 130
$(\Omega^{\cdot}(\mathcal{A}),\partial)$ 133, 152
$(\Omega^{\cdot}(\mathcal{A}),\partial_\eta)$ 134
$\Omega^p(\mathcal{A})$ 124, 147
$\Omega_q^p(\mathcal{A})$ 126
$\Omega_q^p[V]$ 124
Ω_S^G 217
$\Omega(V)$ 92